Propagation Engineering in Radio Links Design

Abdollah Ghasemi • Ali Abedi • Farshid Ghasemi

Propagation Engineering in Radio Links Design

Springer

Abdollah Ghasemi
TEC Telecom Engineering Co.
Briarhill Lane NE 1334
Atlanta, GA, USA

Ali Abedi
University of Maine
Barrows Hall 5708
Orono, ME, USA

Farshid Ghasemi
Briarhill Lane NE 1334
Atlanta, GA, USA

ISBN 978-1-4614-5313-0 ISBN 978-1-4614-5314-7 (eBook)
DOI 10.1007/978-1-4614-5314-7
Springer New York Heidelberg Dordrecht London

Library of Congress Control Number: 2012951128

Printed on acid-free paper

Springer is part of Springer Science+Business Media (www.springer.com)

Foreword

To:
All scientists and experts
who work hard to promote
the living standards of the
world community.

To meet telecommunications demands of the world community, it is crucial to employ radio services. Among vast and fast expansions during recent decades, the satellite services, navigational aids, remote sensing, telemetering, audio and video broadcasting, high-speed data communications, mobile radio systems, and some other special radio services may be addressed.

Radiowaves, propagating between the transmitter and receiver antennas, are subject to a number of phenomena which should be studied and differentiated carefully for designing a reliable radio link. Engineering of radiowave propagation as an outstanding and highly specialized issue is required for all types of radio communications. The pressure to provide data for more effective use of the frequency spectrum, as a natural and limited source of radio systems, requires wider improved prediction methods specially for new bands.

The subject of radiowave propagation is now too large to be treated in a single volume book, encompassing all theoretical and practical aspects. The purpose of this book is to deal in brief with the basic principles needed for understanding of radiowave propagation for common frequency bands used in radio communications. It includes descriptions of new achievements and newly developed propagation models. The provided materials are intended to bridge the gap between theoretical calculations and approaches and applied procedures needed for radio link design in a proper manner.

The intention of authors is to emphasize the propagation engineering, giving sufficient fundamental information, but going on to explain the use of basic principles together with technical achievements in this field and formulation of prediction models and planning tools for radio network design. To do this, study and analysis of main propagation phenomena and mechanisms in a professional way is a fundamental requirement for which the recommendations prepared by international telecommunication union (ITU) are powerful technical procedures.

This book has been prepared as a complementary volume to the book *Propagation Engineering in Wireless Communications, A. Ghasemi et al., Springer, 2011.* It is strongly recommended to consider both volumes for a full course on radiowave propagation.

To use the book in an efficient manner, the following points should be taken into account:

1. The primary objective of the book is to introduce the propagation phenomena and mechanisms likely to be encountered practically and to present fundamental principles. It serves to introduce and orient the reader to those aspects of propagation that must be considered in the design and evaluation of a radio link of a given type and operating frequency.
2. The content of the book along with those presented in the book *Propagation Engineering in Wireless Communications, A. Ghasemi et al., Springer, 2011* covers most topics required for academic or applied courses regarding radiowave propagation. For a better understanding the reader is required to have a good background of advanced and applied mathematics, electromagnetic theory, and principles of radio fields and waves.
3. Some parts of the book are based on the studies conducted by radio sector of ITU, presented through series-P of ITU-R recommendations, which have been revised in accordance with new achievements in the existing fields or newly developed radio services.
4. A great effort has been made to clarify obscure points and improve existing gaps between pure theoretical approaches and practical procedures. The reader must look elsewhere for more details of the theory of radiowave propagation and analysis of phenomena such as reflection, refraction, diffraction, absorption, attenuation, precipitation, focusing/defocusing, fading, scintillation, scattering, dispersion, and depolarization.
5. The aim throughout the book is to give the simplest and most direct account of the applied procedures. There are over 110 solved examples distributed in this volume to encourage students and experts to use the relevant procedures and basic principles by themselves.
6. Full content of the *Radiowave Propagation Engineering* has been organized in two volumes containing totally 17 chapters. The first one includes basic principles, tropospheric and ionospheric propagation, MF/HF ionospheric links, and mobile and fixed line of sight tropospheric propagation. The second volume deals with propagation mechanisms related to radar, satellite, short distance, broadcasting, and trans-horizon radio links complete with two chapters dedicated to radio noises and main parameters of radio link design.
7. The structure of each chapter consists typically of an introduction, definitions, basic formulas and expressions, applied relations, calculation procedures, tables, figures, examples, questions, and problems related to the chapter topic.

8. This volume includes over 270 illustrations, 50 tables, 110 solved examples, 200 questions, 170 problems, acronyms, and appendices.
9. In case of any particular requirement for a dedicated radio network, specific combination of chapters may be selected. The following selections are some examples:

   - For terrestrial fixed radio relay links select Chaps. 1, 2, 3, and 7 from volume 1, and Chaps. 2, 8, and 9 from volume 2.
   - For terrestrial mobile radio networks select Chaps. 1, 2, 3, and 6 from volume 1 and Chaps. 2, 7, 8, and 9 from volume 2.
   - For satellite links select Chaps. 1, 2, 3, and 4 from volume 1 and Chaps. 2, 3, 8, and 9 from volume 2.

10. There are five appendices providing additional information regarding logarithmic system of units, ITU-based terms and definitions, series of ITU-R recommendations, list of ITU-R recommendations related to propagation (P-series), references, software and web sites, acronyms, and book index.

I am most grateful to my main partners in this project, Dr. A. Abedi from University of Maine and F. Ghasemi from Georgia Institute of Technology, for their valuable efforts and providing necessary motivation to compile this book. I would like to appreciate also the efforts of all involved parties in the project including experts and typing, drafting, and publishing groups.

Atlanta, GA A. Ghasemi
April 2012

Contents

Chapter 1
Introduction to Radiowaves Propagation

1.1 Introduction

In 1865, James Clerk Maxwell introduced the notion of electromagnetic (EM) waves propagating with constant speed in homogenous media, based on relations between varying electric and magnetic fields. The speed of EM waves in free space corresponds to the speed of light and is equal to 3×10^8 m/s. Several years later, a German scientist named Hertz found out that radiowaves have a nature similar to EM waves but they are invisible.

Radiowaves radiating from transmitter antenna are collected by receiver antenna after propagating between antennas. They are affected by several phenomena during propagation. The imposed effects are due to transmission media as a function of radiowave characteristics such as frequency bandwidth, polarization, and type of signal.

In general, radiowaves are some type of electromagnetic waves in a specific frequency band. Although a distinctive spectrum is not determined for radiowaves, in this book, propagational phenomena are limited to frequency bands devised by International Telecommunications Union (ITU), that is, 3 KHz to 275 GHz.

Due to the key role of radio communications in a wide range of applications, a lot of studies, researches, and efforts have been spared by competent experts, institutes, organizations, and administrations all over the world for radiowave propagation and using higher-frequency bands.

Major results and achievements in radiowave propagation are collected and summarized in series P of ITU-R recommendations which are revised and updated periodically. In this book, the last version of these recommendations is employed. As a fast reference, the lists of series P of ITU-R recommendations are given in the Appendix B.

This chapter includes general issues related to the radiowaves propagation, transmission media, troposphere, and ionosphere layers. Frequency bands, main kinds of propagations modes, radio services, equivalent radiated power (ERP),

A. Ghasemi et al., *Propagation Engineering in Radio Links Design*,
DOI 10.1007/978-1-4614-5314-7_1,

and radio noises are listed and discussed in brief. Also, tropospheric and ionospheric impacts on the radiowaves are outlined.

It is noted that for details, the reader can refer to the first 4 chapters of the book "Propagation Engineering in Wireless Communications" written by A. Ghasemi et al., published in 2011 by Springer.

1.2 Earth Atmosphere Layers

Earth atmosphere consists of different gases, vapors, meteors, hydrometeors, and dust particles. Some items are permanent and fixed, but some others are temporary and varying. The permanent components of the atmosphere under influence of sun and other stars make different layers with particular features. Major Earth atmosphere layers as depicted in Fig. 1.1 are troposphere, stratosphere, ionosphere, and magnetosphere.

- *Troposphere Layer*

  This layer extends from the Earth surface up to approximately 20 km above it and includes climatic phenomena such as rain, snow, cloud, fog, wind, and storm.

  One of the main features of wet particles like rain and snow is their frequency dependant attenuation. The index of relative permittivity can be introduced as follows:

$$\varepsilon_r = \varepsilon_r' - j\varepsilon_r'' \tag{1.1}$$

  ε_r' is real part of relative permittivity index and causes reflection and scattering of radiowaves, while ε_r'' is its imaginary part which introduces absorption of radiowave power and finally imposes attenuation on it.
- *Stratosphere Layer*

  This layer is above troposphere and extends up to approximately 50 km above mean sea level (AMSL). Main features of this layer are:

  - Contains a big portion of atmospheric gases
  - Low temperature variations per height

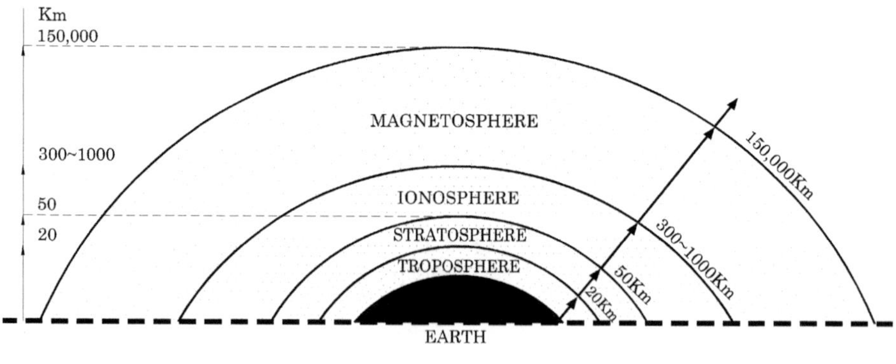

Fig. 1.1 Earth atmosphere layers

- *Ionosphere Layer*

  This layer is located above stratosphere and extends up to approximately 300 km AMSL (this value is specified up to 1,000 km in some references). Ionosphere has great impact on radiowave propagation and consists of three sub-layers named D, E, and F with the following main characteristics:

  Three distinct sub-layers called D, E, and F, which the last one is divided to F_1 and F_2 sub-layer in daytime.

  Height and depth of each main layer is shown in Fig. 1.1; however, it should be noted that different values are given in references.
- *Magnetosphere Layer*

  This layer is above ionosphere and extends up to 150,000 km above the Earth surface acting as an antimagnetic protective shield for it.

1.3 Radio Frequency Bands

1.3.1 Electromagnetic Spectrum and Radiowaves

J.C. Maxwell using mathematical theories proved that EM waves consist of time varying electric and magnetic fields propagating in the media. In their simple form, radiowaves are time harmonic fields in sinusoidal form with frequency f. When the medium is not dispersive, its velocity is not dependent on the frequency but is only related to the medium parameters.

EM waves spectrum includes a wide range of frequencies and as shown on Fig. 1.2 covers all infra radio, ultra radio, infrared, laser, visible light, ultraviolet, X-ray, and γ-ray bands.

1.3.2 Role of Frequency in Radio Communications

Frequency, as a natural resource, has a key role in radio communications. In this regard, some of the major facts to clarify its crucial role in radio networks are:

- Most radiowave propagation phenomena depend on the frequency in linear or nonlinear forms.
- Dependence of technical characteristics of radio equipment and their applications and service quality on operating frequency.
- Applications of some frequency bands are exclusively allocated to the specific services.
- Limitation of frequency resources.
- Ever-increasing radio communications requirements.

Emerging advanced technologies and creation of new services significantly increase the demand for more spectrums. In order to meet the new requirements,

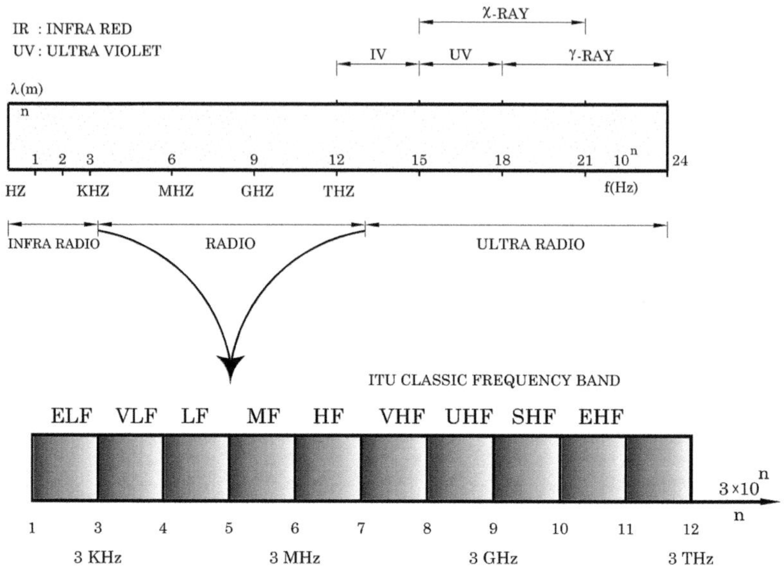

Fig. 1.2 Electromagnetic spectrum and radio frequency bands

R & D centers and professional institutes at national and international levels have conducted extensive studies and investigation which among them the following items can be addressed:

- Employing new technologies to improve frequency utilization efficiency such as higher digital modulation levels, TDMA, CDMA, and compression techniques.
- Using other transmission media such as cable TV, SDH over fiber optics cable, WDM, and DWDM.
- Manufacturing RF components in higher-frequency bands.
- Using network and bandwidth management systems.
- Using automatic transmitter power control (ATPC) and frequency reuse techniques.

1.3.3 Classification of Frequency Bands

Major frequency bands based on ITU regulations are listed in Table 1.1. Lower and upper extremes of the ITU classic frequency bands are defined by the relation given below:

$$F/B = 3 \times 10^{n-1} \sim 3 \times 10^{n} \qquad 4 \leq n \leq 11 \tag{1.2}$$

F/B is frequency band in Hertz and n is relevant band number as per Table 1.1.

Table 1.1 ITU classic frequency bands

Wavelength, metric equivalent	Frequency range	Designation	Band number
Myriametric waves	3–30 KHz	VLF	4
Kilometric waves	30–300 KHz	LF	5
Hectometric waves	300–3,000 KHz	MF	6
Decametric waves	3–30 MHz	HF	7
Metric waves	30–300 MHz	VHF	8
Decimetric waves	300–3,000 MHz	UHF	9
Centimetric waves	3–30 GHz	SHF	10
Millimetric waves	30–300 GHz	EHF	11
Decimillimetric waves	300–3,000 GHz	–	–
Micrometric waves	>300 THz	–	–

Table 1.2 Initial applied frequency bands

Frequency range (GHz)	Band designation
0.225–0.390	P-band
0.390–1.550	L-band
1.550–3.900	S-band
3.900–6.200	C-band
6.200–10.900	X-band
10.900–36.00	K-band
36.000–46.000	Q-band
46.000–56.000	V-band
56.000–100.000	W-band

Table 1.3 New applied frequency bands

Frequency range (GHz)	Band designation
0.100–0.250	A-band
0.250–0.500	B-band
0.500–1.000	C-band
1.000–2.000	D-band
2.000–3.000	E-band
3.000–4.000	F-band
4.000–6.000	G-band
6.000–8.000	H-band
8.000–10.000	I-band
10.000–20.000	J-band
20.000–40.000	K-band
60.000–80.000	L-band
80.000–100.000	M-band

As indicated in Tables 1.2 and 1.3, there are other classifications known as initial and new models, respectively. These frequency bands are usually employed by radio manufacturers and users. For example, L-band in maritime satellite system, C-band in satellite networks, and C- and S-bands in radar technology.

1.4 Frequency Allocation

1.4.1 Introduction

Radio frequency band, as a limited natural radio resource, shall be allocated to a variety of radio services stated in Sect. 1.2. Frequency band allocation shall be studied carefully and coordinated on worldwide basis. Major significant considerations are:

- Geographical distribution
- Types of radio services
- Volume of communications traffic
- Technical limitations
- Radiowaves propagation characteristics
- Edge of radio technologies

ITU has spent a lot of efforts to prepare the Article 5 of radio regulation for frequency allocation ranging from 9 KHz to 275 GHz bands. The table and related footnotes have been revised several times and also will be modified in the future, taking into account technical achievements and new arising operational requirements.

1.4.2 Frequency Registration

Frequency registration shall be performed by every competent authorized administration for all assigned frequencies within each country. This shall be done for protection of the assigned frequencies from harmful interference.

Also, on worldwide basis, main aspects of assigned frequencies in the selected services shall be registered in the International Frequency Registration Board (IFRB) for global coordination. For more details of procedures and regulations, reference is made to the ITU website and publications.

1.4.3 ITU Regions for Frequency Allocation

A chart has been prepared by ITU for the Earth regional classification. As shown in Fig. 1.3, the whole Earth is divided by three lines A, B, and C into three regions named region 1, 2, and 3. Each country is located in one of the regions. For more details, see the Article 5 and Appendix 24 of the radio regulations.

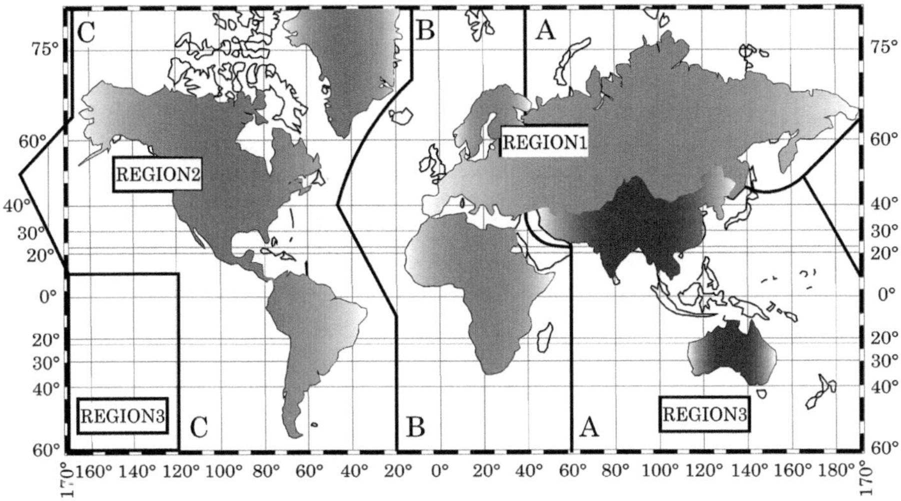

Fig. 1.3 ITU regions chart for frequency allocation

1.4.4 Frequency Assignment

In addition to the Article 5, for frequency bands allocation, ITU and ITU-R have prepared several tables and arrangements through a number of appendices and recommendations as follows:

- Appendices 16, 17, 18, 25, 26, and 27 of radio regulations including tables of RF carrier frequencies for maritime and aeronautical mobile radio services in MF, HF, and VHF bands.
- RF channel arrangements for satellite services included in Appendices 30, 30A, and 30B of the radio regulations.

Example 1.1. Two types of RF channel arrangement for fixed microwave relay systems extracted from ITU-R, F.386 in 8 GHz frequency band are given below:

1. 6-RF channels with 14 MHz separation and following criteria (Fig. 1.4):

$$f_n = f_0 - 108.5 + 14\,n, \quad f_0 = 8,387.5\,\text{MHz}$$
$$f'_n = f_0 + 10.5 + 14\,n, \quad n = 1,2,\ldots,6$$

Using the above formula, carrier frequencies of transmit (TX) and receive (RX) channels are:

$f_1 = 8293$	$f_2 = 8307$	$f_3 = 8321$
$f'_1 = 8412$	$f'_2 = 8426$	$f'_3 = 8440$

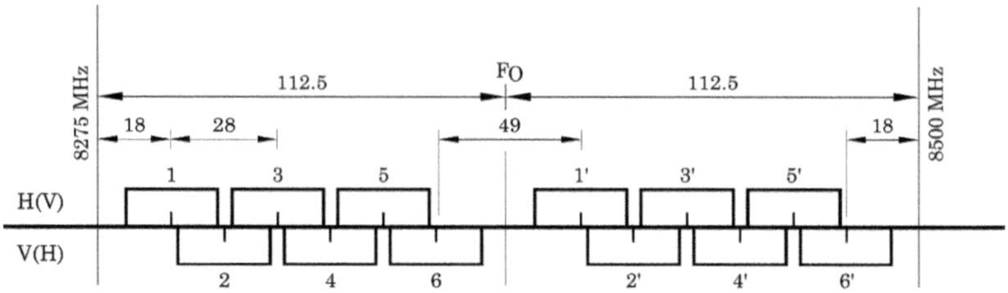

Fig. 1.4 RF channel arrangement for 8 GHz band (6-channel, 14 MHz separation)

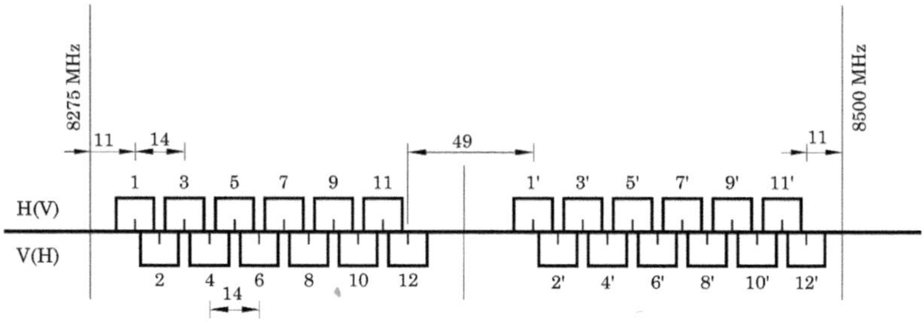

Fig. 1.5 RF channel arrangement for 8 GHz band (12-channel, 7 MHz separation)

$$\left| \begin{matrix} f_4 = 8335 \\ f_4' = 8454 \end{matrix} \right. \quad \left| \begin{matrix} f_5 = 8349 \\ f_5' = 8468 \end{matrix} \right. \quad \left| \begin{matrix} f_6 = 8363 \\ f_6' = 8482 \end{matrix} \right.$$

2. 12-RF channels with 7 MHz separation and the following criteria (Fig. 1.5):

$$f_n = f_0 - 108.5 + 7n, \quad f_0 = 8,387.5\,\text{MHz}$$
$$f_n' = f_0 + 17.5 + 7n, \quad n = 1,2,\ldots,12$$

Using the above formula, carrier frequencies of TX and RX channels are:

$$\left| \begin{matrix} f_1 = 8286 \\ f_1' = 8412 \end{matrix} \right. \quad \left| \begin{matrix} f_2 = 8293 \\ f_2' = 8419 \end{matrix} \right. \quad \cdots \quad \left| \begin{matrix} f_{12} = 8363 \\ f_{12}' = 8489 \end{matrix} \right.$$

■

Example 1.2. A GEO satellite is equipped with totally nine transponders including three units with 72 MHz bandwidth and six units with 36 MHz bandwidth. In case of using satellite C-band, that is, 3,700 ~ 4,200 MHz for downlinks and 5,925 ~ 6,425 MHz for uplinks, answer the following questions:

1. How much is the available bandwidth and is it possible to design a suitable RF channel arrangement by one type polarization?
2. Is it possible to increase frequency utilization efficiency?

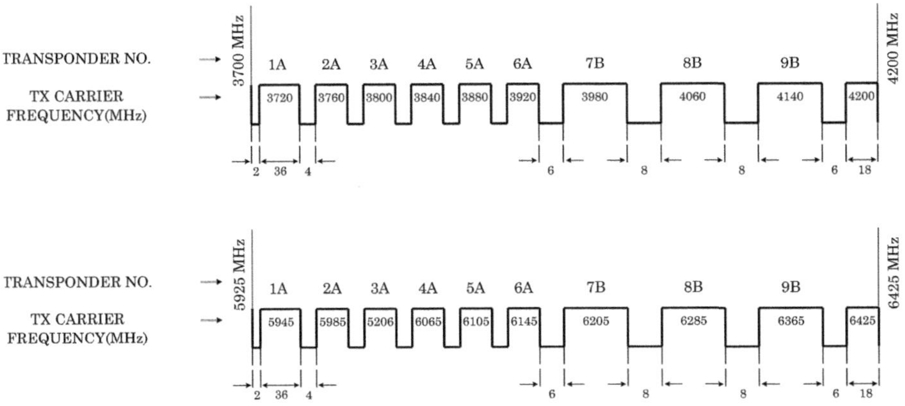

Fig. 1.6 Satellite C-band RF channel arrangement

Solution. Total available bandwidth is 500 MHz which is sufficient for the required RF channel arrangement with one type polarization like right-hand circular (RHC) as shown in Fig. 1.6.

Also, frequency utilization efficiency may be increased by using the same RF channel arrangement but with another polarization such as left-hand circular (LHC) polarization. ■

1.5 International Telecommunication Union

1.5.1 Objectives

In radio communications field, codes and standards have an outstanding position due to a number of reasons including, but not limited to, the following:

- Ever-increasing demand for a variety of services.
- Growing requirements for more traffic and higher capacities.
- Desired signal when radiated from transmitting antenna is beyond control and may act as an interfering source for a number of receivers except desired one(s).
- There are two parties for each radio link, that is, transmitter and receiver. For a perfect and proper operation, their specifications should be standard and compatible; otherwise, they cannot work satisfactorily.
- In fact, standards are minimum technical requirements which shall be observed by telecommunications industries and service providers.
- Better standards result in better service quality.

1.5.2 Radio Regulations

ITU radio regulations are recognized internationally and applied by all member states/countries. The radio regulations are arranged by ITU in nine chapters as follows:

Chapter I: Terminology and technical characteristics
Chapter II: Frequencies
Chapter III: Coordination, notification, and recording of frequency assignments and plan modifications
Chapter IV: Interferences
Chapter V: Administrative provisions
Chapter VI: Provisions for services and stations
Chapter VII: Distress and safety communications
Chapter VIII: Aeronautical services
Chapter IX: Maritime services

It is followed by 42 appendices and a number of resolutions and recommendations as well. The ITU radio regulations usually are updated and modified accordingly through World Administrative Radio Conferences (WARC) held by ITU secretariat.

1.5.3 ITU-R Recommendations

In radio sector of ITU (known as ITU-R), a radio communications study group has been for many years involved in introducing some technical guidelines under several recommendations on a worldwide basis. The products of ITU-R are prepared in different series as listed in the next page.

Series P is dedicated to the radiowaves propagation issues for which all relevant recommendations are listed in Appendix B. These recommendations which will be referred frequently in this book are rich from technical points of view and supported by many experts in the international level.

1.6 Types of Radiowave Propagation

Different types of radiowaves may be propagating between transmitter and receiver radio units. As shown in Fig. 1.7, the main types of radio links are:

- Ground waves commonly used for AM broadcasting, radio-navigational aids, and short wave radio systems.

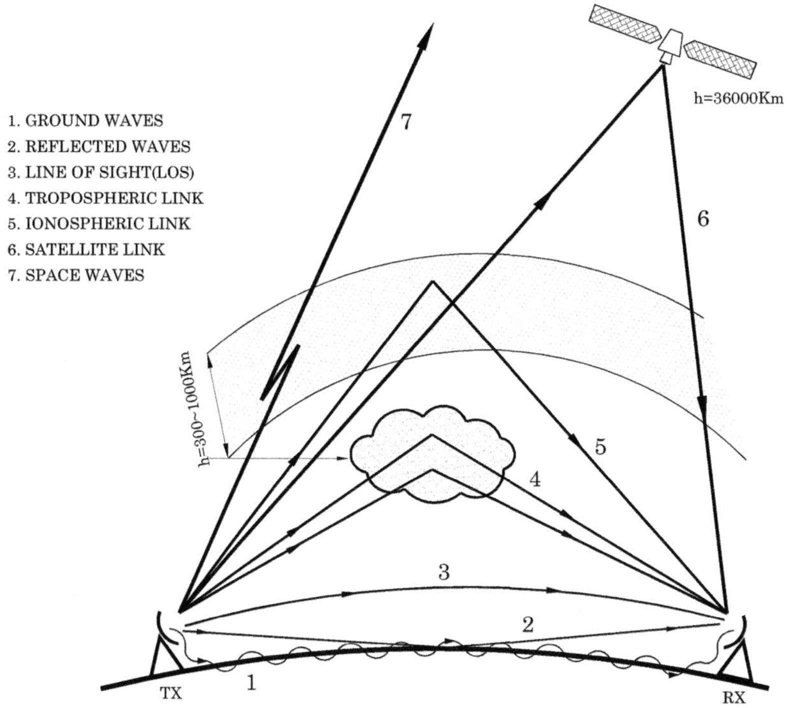

Fig. 1.7 Main types of radio links

Series	Subject
BO	Broadcasting satellite service (sound and television)
BR	Sound and television recording
BS	Broadcasting service (sound)
BT	Broadcasting service (television)
F	Fixed service
IS	Interservice sharing and compatibility
M	Mobile, radio determination, amateur, and related satellite services
P	Radiowave propagation
RA	Radio astronomy
S	Fixed-satellite service
SA	Space applications and meteorology
SF	Frequency sharing between the fixed-satellite service and the fixed service
SM	Spectrum management
SNG	Satellite news gathering
TF	Time signals and frequency standards emissions
V	Vocabulary and related subjects

- Reflective waves producing multipath links along with the main route which are common in UHF and microwave radio links including FM broadcasting in TV networks as well.

- Line-of-sight (LOS) links used in terrestrial microwave, UHF, and radar networks.
- Tropospheric links for point-to-point telecommunications by refracting/reflecting waves through troposphere layer and using over-horizon troposcatters.
- Ionospheric links employed for long-distance telecommunications using reflection from ionosphere D, E, and F layers in MF, HF, and public audio broadcasting networks.
- Satellite links for communications between satellites and ground stations with a distance ranging from several hundred up to around 40,000 km.
- Radio links for space telecommunications between ground and spacecraft stations.

1.7 Radio Services

Radiowaves are used for transmitting various kinds of audio, video, data, control, and navigational signals. Among numerous applications are the following groups:

- Aeronautical, land, and maritime mobile services
- Search and rescue and navigational aids
- Different types of satellite services
- Fixed services of low, medium, and high capacity
- Audio and video broadcasting
- Telemetry, SCADA, and remote sensing
- Aeronautical, land, and maritime traffic control systems
- Radio special services for industrial, scientific, research, medical, and social applications

Radio services based on the ITU classification for frequency allocation included in the Article 5 are as follows:

1. Aeronautical mobile service
2. Aeronautical radio navigation
3. Aeronautical radio navigation satellite
4. Amateur satellite service
5. Amateur service
6. Broadcasting satellite service
7. Broadcasting service
8. Down link service
9. Earth exploration satellite service
10. Fixed satellite service
11. Fixed service
12. Inter satellite service
13. Land mobile service
14. Maritime mobile service
15. Maritime radio navigation
16. Meteorological aids service
17. Meteorological satellite service
18. Mobile excerpt aeronautical service
19. Mobile satellite service
20. Mobile service
21. Radio astronomy
22. Radio determination satellite
23. Radio location
24. Radio navigation
25. Radio navigation satellite
26. Secondary (non-satellite) service
27. Secondary (satellite) service
28. Space operation service
29. Space research service
30. Standard frequency & time signal
31. Standard frequency & time signal satellite
32. Up link service

1.8 Tropospheric Propagation

1.8.1 Troposphere Layer

Most of the radio transmissions are carried out in the lower layers of the Earth atmosphere and in fact in the troposphere. Some of radio systems in point-to-point, point-to-multipoint, and point-to-area types include line-of-sight radio links in UHF, SHF, and EHF bands; mobile radio networks in VHF and UHF bands; and TV and FM audio broadcasting. Due to vast usage of these communication systems, the study of troposphere layer and its impacts on wave propagation is extremely important to radio experts and scientists.

In addition to these systems whereby both sides of transmission links are situated in troposphere layer, there are some other communication systems that only one side is situated in this layer, so they are somehow, but not as much as the above-mentioned systems, affected by this layer. Some examples of these systems are satellite communications for voice, video, and data; navigational aids; SAR; positioning and telemetry; ionospheric communications in HF and MF bands; as well as space communication systems.

There are a number of natural phenomena that occur in the path of radiowaves and affect the signal level, quality, and performance. Considering these facts, recognition, study, and survey of these phenomena are very important and have been thoroughly investigated in Chap. 3 of "Propagation Engineering in Wireless Communications" complete with some useful tables and graphs related to this topic.

Basically, the wave propagation phenomena are affected by one or a set of the following main factors:

- The Earth and its natural and physical properties such as bulge, roughness, structure, material, and flatness.
- Environmental conditions such as temperature, humidity, stable atmosphere composition, and air molecules such as oxygen, nitrogen, and water vapor.
- Atmospheric climate phenomena such as wind, storm, dust, thunderstorm, cloud, rain, snow, and hail.
- Celestial phenomena such as magnetic storms, sunspots, diurnal, and seasonal effects.
- Artificial items and artifacts such as skyscrapers, chemical particles, and gases generated by large factories and urban facilities.

The effects of the above-mentioned phenomena mostly depend on the frequency of concerning wave. Therefore, recognition of these phenomena and their impacts on wave propagation is important to combat their adverse effects and find suitable counter measures.

Recently, several study topics and research projects have been defined by ITU in order to recognize the wave propagation behavior in the Earth atmosphere. The results of these studies are collected and provided in a set of ITU recommendations, called P series. A complete list of these recommendations and their subjects is presented in appendix B at the end of this book. It is notable that some of the reports and recommendations are currently under revision by ITU-R study groups.

1.8.2 Major Effects of Troposphere on Radiowaves

Radiowaves passing through the Earth atmosphere are subject to the following major effects:

- Absorption and attenuation
- Reflection
- Refraction
- Changes of polarization
- Scattering and diffusion
- Atmospheric ducting

1.8.3 Standard Earth Atmosphere

The standard Earth atmosphere is defined in order to make it possible to calculate the radiowave loss due to gas compounds in the Earth atmosphere and also to define the temperature, atmospheric pressure, and water vapor pressure versus altitude. After a long survey and thorough study based on ITU-R, P.835-8 recommendation, finally, the reference atmosphere condition is defined according to the reference location of the United States standards atmosphere. In this standard, the temperature and Earth surface pressure is equal to

$$P_0 = 1{,}013.25\text{ hpa}, \quad T_0 = 288.15\,^{\circ}\text{K}.$$

Based on this definition, the Earth atmosphere is divided into seven consecutive layers. The temperature variation rates of these layers versus altitude are depicted in Fig. 1.8. According to this model, the relationship between temperature rate $T(h)$ in terms of Kelvin and altitude h in terms of kilometer is equal to

$$T(h) = T_i + L_i(h - H_i) \tag{1.3}$$

$$T_i = T(H_i), \tag{1.4}$$

where L_i is the temperature gradient in altitude H_i and its values for different layers and heights are presented in Table 1.4. Also, atmospheric pressure depends on the location (latitude) and the date (the season of the year) in addition to the height.

Thermodynamic stability of atmosphere breaks down in the heights above 85 km, and the hydrostatic equations that are the basis of these equations are not valid

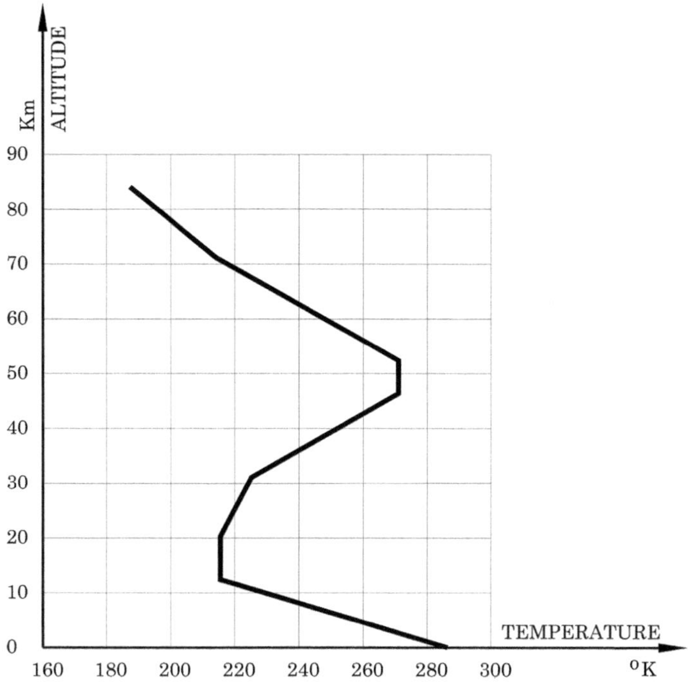

Fig. 1.8 Atmosphere temperature variations of standard Earth (Ref.: ITU-R, P.835–4)

Table 1.4 Temperature gradient in atmosphere sub-layers (Ref.: ITU-R, P.835–4)

i	H_i (km)	Termperature gradient, L_i (°K/km)
0	0	−6.5
1	11	0
2	20	+1.0
3	32	+2.8
4	47	0
5	51	−2.8
6	71	−2.0
7	85	–

anymore. Changes in water vapor distribution of atmosphere are approximated by the following equation:

$$\rho(h) = \rho_0 e^{(-h/h_0)}, \tag{1.5}$$

where $h_0 = 2\,\text{km}$ and $\rho_0 = 7.5\ \text{g/m}^3$ are similar to the atmospheric pressure; the water vapor pressure also depends on the location, height, and the season of the year. It should be noted that the water vapor pressure decreases exponentially as height increases up to a particular value of the height.

To calculate attenuation caused by other atmospheric gases, similar procedure may be followed with $h_0 = 6\,\mathrm{km}$.

1.8.4 Refraction of Radiowaves in Troposphere

The radiowaves propagating in the Earth atmosphere always experience the wave refraction phenomenon. As the height increases, the air density and consequently its refractive index decreases. This nonhomogeneous characteristic of air in the atmosphere causes deviation in wave propagation path, so they do not travel further on a straight direction. When the rate of refraction index changes linearly, the ray path would be an arc of a circle with a constant radius.

1.8.4.1 Refractive Index of Air

The refractive index of every environment is related to its relative permittivity according to the following equation:

$$n = \sqrt{\varepsilon_\mathrm{r}}. \tag{1.6}$$

Therefore by applying appropriate relation for ε_r, we found:

$$n = \sqrt{\varepsilon_\mathrm{r}} = \left\{1 + \frac{155.1}{T}\left[P + \frac{4810e}{T}\right] \times 10^{-6}\right\}^{1/2}. \tag{1.7}$$

Using the binomial expansion for the above expression and selecting the first two terms, we have

$$n = 1 + \frac{77.6}{T}\left[P + \frac{4810e}{T}\right] \times 10^{-6}. \tag{1.8}$$

Since it is difficult to use the value of this term in the calculations, another parameter called refractivity number and denoted by N is defined as follows:

$$N = (n-1) \times 10^6. \tag{1.9}$$

According to this definition, the refractivity number of the air at sea level and altitude equal to 1 km is

$$h = 0 \quad \Rightarrow \quad N_0 = 289$$
$$h = 1 \quad \Rightarrow \quad N_1 = 251.$$

Also, a new parameter called modified refractive index and denoted by n_M is defined according to the following equation:

$$n_\mathrm{M} = n + h/R_\mathrm{e}. \tag{1.10}$$

In the last equation, n is refractive index of the air, h is the height of the location, and R_e is the actual Earth radius where the last two parameters have the same unit. There is another parameter called refractive modulus and is defined according to the following equations:

$$M = (n_M - 1) \times 10^6 \tag{1.11}$$

$$M = N + 10^6 h/R_e. \tag{1.12}$$

Example 1.3. Calculate the value of n and N for air at the altitude of 2 km based on the formula (1.7) at the standard atmosphere conditions.

Solution.

$$h = 2\,\text{km} \quad \Rightarrow \quad T = 277\,^\circ\text{K},\ \rho = 716\,\text{mb}\ ,\ e = 2\,\text{mb}$$

$$\varepsilon_r = 1 + \frac{151.1}{277}\left(716 + \frac{4,810 \times 2}{277}\right) \times 10^{-6} = 1.00042$$

$$n = \sqrt{\varepsilon_r} = 1.00021$$

$$N = (n-1) \times 10^6 \quad \Rightarrow \quad N = 210$$

■

1.8.4.2 Wave Path and Effective Earth Radius

Radiowave refraction in the Earth atmosphere has several effects on its propagation and can result in transition of the wave over the horizon. It can be assigned to a straight line for propagation path of waves provided that a new equivalent radius is assumed for the Earth which is denoted by R'_e instead of the actual Earth radius.

Formula for the equivalent radius of the Earth was derived in Chap. 2. For standard atmosphere, calculations resulted in $R'_e = 8,500\,\text{km}$, while actual radius of the Earth is 6,370 km.

1.8.5 K-Factor

1.8.5.1 Definition

To analyze radiowave path clearance in line-of-sight links, it is preferred to assume straight path for radiowave propagation rather than its actual curved trajectory. This assumption can be satisfied by considering a factor named K-factor which is defined as the ratio of the equivalent Earth radius (R'_e) to its actual value (R_e).

K-factor is a dynamic parameter which is dependent on the refractive index variations of the ray path in the atmosphere. Some relations and formulas for K-factor have been stated in the said book. K-factor values shall be taken into account where radiowaves path calculations are necessary.

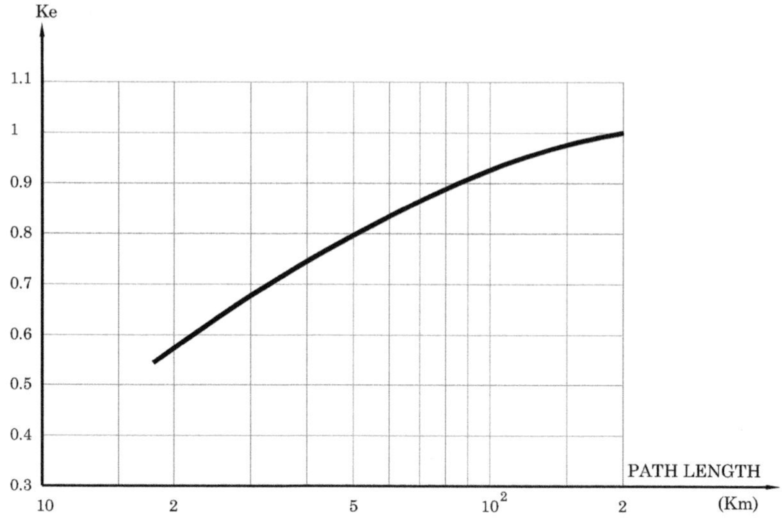

Fig. 1.9 Variation of K_e (Ref.: ITU-R, P.530–12)

1.8.5.2 Variation Range of *K*-Factor

K-factor values have a key role in radio link design. Its actual values and range of variations are related to the site-specific conditions. The following different values of *K*-factor are used to analyze propagation phenomena in the terrestrial line-of-sight radio links:

- The standard value of *K*, equals to $4/3$, is generally used to set path clearance criteria of LOS propagation.
- The lowest value of *K*, for instance $K = 2/3$, causes the effective Earth radius to decrease, and consequently, the ray path between two particular points with a specific distance has more bulges on it. Therefore, higher transmitter and receiver antennas are required for LOS transmission resulting in higher implementation cost. For the minimum value of *K*, the least path clearance of LOS transmission is obtained, so suitable predictions and calculations should be performed to avoid obstruction of the paths.
- The high values of *K*, for instance $K = 2$ or more, cause an abnormal long range of the LOS path that results in undesired formation of reflective paths.
- The effective value of *K* which is called K_e and is based on an approximate graph for conventional environmental conditions and widely used in fixed radio system design at frequency range of 2–10 GHz. Figure 1.9 shows K_e variation for terrestrial LOS links based on ITU-R recommendations.

As shown in Table 1.5, the value of *K* depends on some factors such as the variation of the rate of *N*, curvature radius, and relative wave path length.

Table 1.5 Ray path condition versus N'

REFRACTION TYPE	N'=dN/dh	K FACTOR	RADIUS OF RAY R(Km)	EQUIVALENT EARTH RADIUS R'e(Km)	RAY PROFILE VS EARTH
SUB-REFRACTION	N'>0	—	—	—	
LESS THAN STANDARD	N'=0	K=1	R= ∞	$R'_e > R_e$	
STANDARD	N'=-39	$K=\frac{4}{3}$	R=2500	$R'_e = 8500$	
CRITICAL	N'=-157	K= ∞	$R=R_e$	$R'_e = \infty$	
SUPER REFRACTION	N'<-157	K<0	$R<R_e$	$R'_e < R_e$	

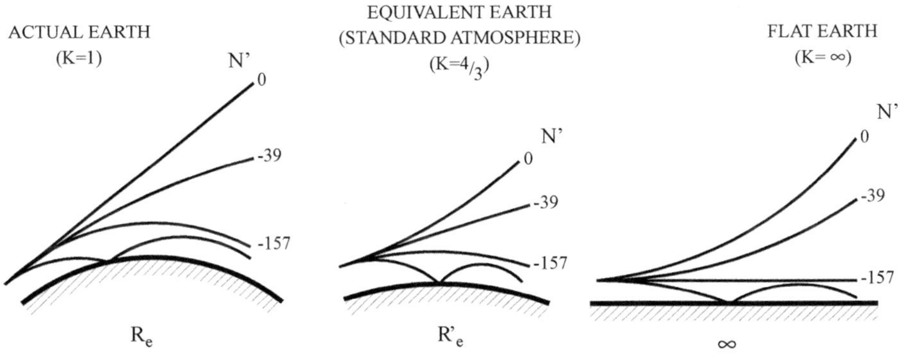

Fig. 1.10 Ray trajectory relative to Earth

If the vertical gradient of refractivity number is presented by $N' = dN/dh$, different conditions based on this value can be classified as follows:

- $N' > 0$: Subrefraction condition
- $N' = -39$: Standard refraction condition
- $N' = -157$: Critical condition ($K = \infty$) which means that wave propagates parallel to the Earth surface
- $N' < -157$: Superrefraction condition which causes duct effect

To provide a good perception of this subject, the relative status of propagating wave and Earth surface is depicted in Fig. 1.10 for three cases of the Earth including actual, equivalent, and flat conditions.

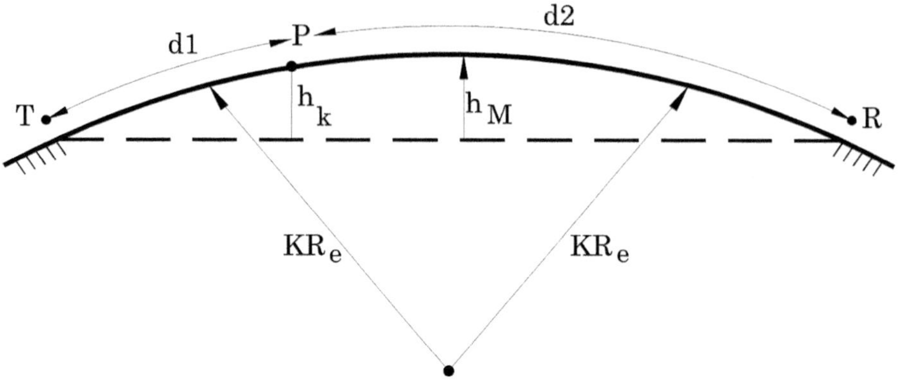

Fig. 1.11 Earth bulge geometry

1.8.5.3 The Earth Bulge

The bulge between transmitter and receiver is defined as the height of every point of equivalent Earth regarding the straight line of TR according to Fig. 1.11. This parameter can be calculated as

$$h_K = \frac{500\, d_1\, d_2}{K\, R_e}. \tag{1.13}$$

The maximum value of bulge occurs in the middle of the path which equals to

$$h_M = \frac{125\, d^2}{K\, R_e}. \tag{1.14}$$

In the two latter equations, h is the bulge of Earth in meter, R_e is the actual Earth radius, d_1 and d_2 are distance between point P from the transmitter and the receiver respectively, and d is the sum of d_1 and d_2, all in km.

According to the mentioned equations, the value of bulge is inversely proportional to the value of K. Therefore, as K decreases, the bulge of Earth and the required height of transmitter and receiver antennas increase.

Example 1.4. If the relative permittivity of a medium is 1.001, the vertical gradient of refractive index is -35×10^{-6}, and the distance between transmitter and receiver is 20 km:

1. Find the Earth bulge in the middle of the path.
2. Compare the Earth bulge at a point 5 km off transmitter toward receiver with the same point in standard atmosphere condition.

Solution.

$$\varepsilon_r = 1.001 \quad n = \sqrt{\varepsilon_r} = 1.0005$$

$$R = \frac{n}{|\frac{dn}{dh}|} = \frac{1.0005}{35 \times 10^{-6}} = 28,586\,\text{km}$$

$$K = \frac{R}{R - R_e} = 1.287.$$

Earth bulge value in the middle of path is:

$$h_M = \frac{125 \times 20^2}{1.287 \times 6370} = 6.1\,\text{m}.$$

2. The bulge value for the given point is

$$d_1 = 5\,\text{km}, \quad d_2 = 15\,\text{km}$$

$$h_1 = \frac{500 \times 5 \times 15}{1.287 \times 6370} = 4.58\,\text{m}.$$

The bulge value for the standard atmosphere condition ($K = 1.3333$) at the same point is

$$h_1' = \frac{500 \times 5 \times 15}{1.333 \times 6370} = 4.42\,\text{m}.$$

Therefore, the bulge value for this medium in the given point is 16 cm greater than the corresponding value at standard atmosphere. ■

1.9 Ionospheric Propagation

1.9.1 Introduction

The fundamental phenomena of troposphere were investigated and their effects on radiowaves propagation explained briefly in 1.8. This section is dedicated to the study of main phenomena of radiowave propagation in ionosphere. As shown in Fig. 1.12, this layer basically starts from the height of 50 km above the Earth and extends up to the height of 600 km. Although, in some references it is addressed up to the height of 1,000 km, but empirically the main effects of these phenomena appear at the heights of up to 600 km.

In case of the radiowaves passing through this layer or somehow penetrate into it, they will be affected by the specific phenomena of this layer. Some of the most common communication systems that experience these effects are those working in MF and HF radio transmission bands as well as satellite and space communication systems where one or both end point terminals are located on the ground.

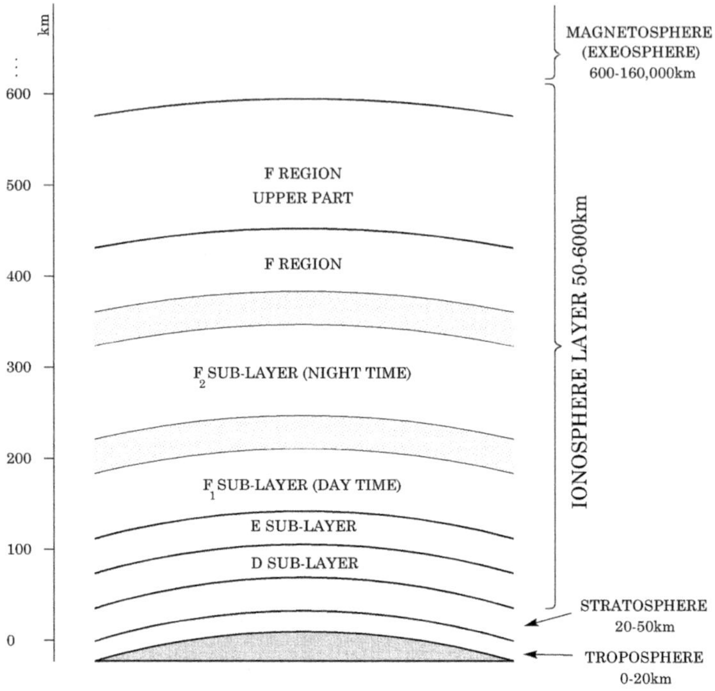

Fig. 1.12 Earth ionosphere structure

The main ionospheric phenomena, similar to other Earth atmosphere layers, are highly dependent on radio frequency by a nonlinear relation. Some of the main considerations regarding the ionosphere layer are outlined in this section. Also, more details are given in Chap. 4 of the book "Propagation Engineering in Wireless Communications."

1.9.2 Sub-layers of Ionosphere

Ionosphere includes three distinct sub-layers called D, E, and F with the following main characteristics:

- *D Sub-layer*

    1. Its height reaches to 70 km AMSL during daytime.
    2. Electron content and density is directly related to the sun's activities.
    3. Its impact on radiowave propagation during daytime is more effective than night period with a maximum at noon and minimum in sunset.
    4. Its impact during summer is more significant than winter.
    5. Power absorption of D sub-layer is most significant in HF low frequencies.

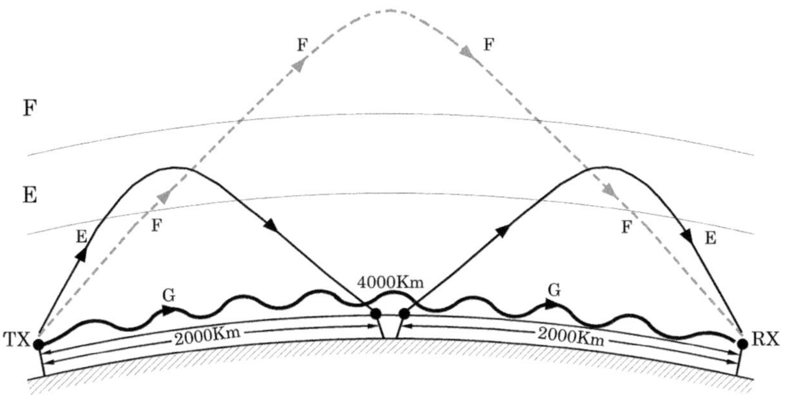

Fig. 1.13 Single hops of E and F layers

- *E Sub-layer*

  1. E sub-layer is located above D sub-layer with altitude up to 100 km/AMSL.
  2. It is the atmosphere's lowest layer that is able to refract/reflect radiowaves.
  3. Its impact on radiowaves propagation during daytime is more than night period with a maximum at noon and a minimum in sunset.
  4. Its impact during summer is greater than winter.
  5. Considering the refractivity nature of radiowaves, it can cover communications some 2,000 km in MF/HF band (see Fig. 1.13).

- *F Sub-layer*

  1. F sub-layer is located above E sub-layer and extends up to $300 \sim 1,000$ km AMSL.
  2. It provides long-distance radio communications up to 4,000 km in HF band at nighttime by single radio hop (see Fig. 1.13).
  3. During daytime, it is divided into F_1 and F_2 layers and merging into one layer during nighttime.
  4. F_1 acts similar to E sub-layer.
  5. F_2 has the greatest density of atmospheric ionization.
  6. F_2 is able to provide communications up to 4,000 km with single radio hop.

1.9.3 Ionization and Plasma State

The most important property of ionosphere is its ionized gases. The amount of ionized gases in this layer is much higher than the inert gases. This layer can change effectively the electrical properties which are dominant factors and have a great influence on radiowaves propagation. The ionosphere contains a specific state of material known as the forth or plasma state.

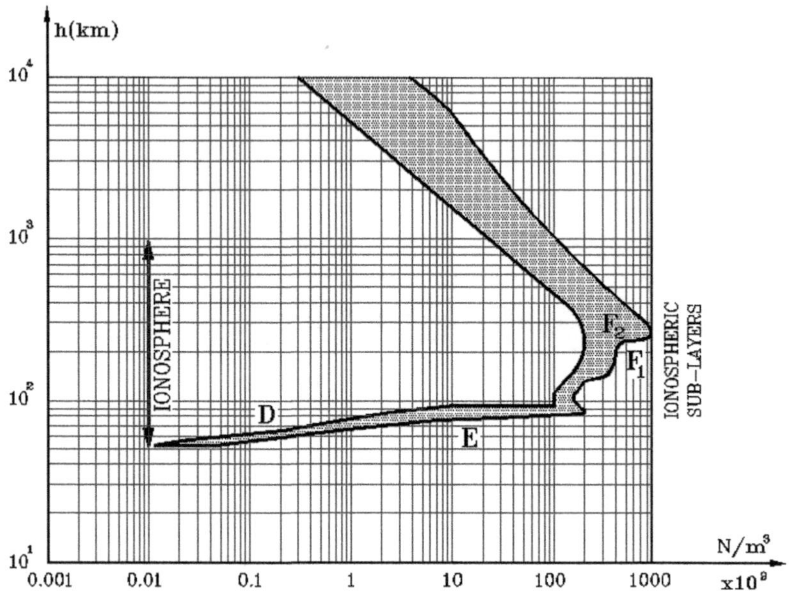

Fig. 1.14 Ionospheric electron density

Since the ionosphere layer has the most of plasma properties, therefore, ionosphere researchers and plasma experts are very close to each other. Due to the fact that major part of the visible world is in the plasma state, the study of the plasma is valuable not only for radiowave propagation but also for perception of plasma world.

The plasma environment of Earth and in better expression the ionosphere is not a simple and static layer, and it should be studied by periodic monitoring and testing. The Earth is affected by radiation and bombardment of different cosmic rays. Some of these rays have large amplitude and convey considerable amount of energy enabling electrons to be separated from molecules and to form positive and negative ions. The number of free electrons in unit volume is called the plasma density. This parameter is plotted as a function of height in Fig. 1.14.

1.9.4 Ionosphere layer Classification

Various types of cosmic rays ionize different molecules in the air atmosphere. This stratifies the ionosphere at different heights into different sub-layers. The plasma density has usually a considerable value in the range of $10^9\,\mathrm{el/m^3}$ at heights from 50 to 70 km above the Earth surface which is called D-layer. At the altitudes between 70 and 100 km which is called E-layer, the plasma density reaches its relative maximum value, and after that, it starts to decrease, and again, it reaches its absolute maximum value at heights up to 1,000 km which is called F-layer and beyond it is a layer called magnetosphere.

During daytime, F-layer is divided into two sub-layers called F1 and F2 due to sunlight. These two sub-layers merge again at nighttime to form a single layer.

1.9.5 Ionospheric Phenomena

When the radiowaves pass through or penetrate into the ionosphere, they are affected by different phenomena. Some of the most important ones are listed below:

- Faraday rotation
- Propagation delay and group delay
- Refraction/Reflection
- Dispersion
- Absorption
- Scintillation

The first four items depend on the plasma state of ionosphere. The total number of electrons in a cylinder with a cross section of one square meter in solar zenith direction is called TEC and measured in terms of el/m^2. TEC is used as an essential parameter to evaluate the quality of ionosphere. The nominal value for different states of ionosphere layer is in the range of 10^{16}–$10^{18}\,\mathrm{el}/\mathrm{m}^2$. Most of the ionospheric phenomena have statistical properties and depend on different factors such as:

- Geographical latitude and longitude and geomagnetic location
- Earth orbital motion or seasonal effects
- Earth rotation motion or diurnal effects
- Solar activities, especially the number of sunspots
- Magnetic storms
- Geomagnetic field effects

Considering the above factors and their statistical relationship to each other, the ionospheric telecommunications have wide range of changes which shall be taken into account ensuring a reliable and efficient communications.

Example 1.5. 1. Determine the approximate value of plasma density at an altitude of 400 km above the ground level.
2. If the height of D, E, and F layers are 70, 100, and 300 km, respectively, find the value of TEC for each of these layers.
3. Find the effective value of TEC for a satellite transmission using radiowaves with elevation angle equal to 40°.

Solution. 1. Considering the graph in Fig. 1.14, it results in

$$h = 400\,\text{km} \Rightarrow \quad 1.1 \times 10^{11} \le P_\text{D} \le 5 \times 10^{11}\,\text{el/m}^3$$

2. The thickness of each layer is

$$W_\text{D} = 20\,\text{km}, \quad W_\text{E} = 40\,\text{km}, \quad W_\text{F} = 190\,\text{km}$$

Considering the graph in Fig. 1.14, the average density of each of the above-mentioned layers is approximately

$$P_{\text{D/D}} \approx 1 \times 10^8\ \text{N/m}^3,\ P_{\text{D/E}} \approx 1\ \times 10^{10}\ \text{N/m}^3,\ P_{\text{D/F}} \approx 3 \times 10^{11}\ \text{N/m}^3$$

Therefore,

$$\text{TEC/D} = 20 \times 10^3 \times 1 \times 10^8 = 2 \times 10^{12}\ \text{el/m}^2$$

$$\text{TEC/E} = 40 \times 10^3 \times 1 \times 10^{10} = 4 \times 10^{14}\ \text{el/m}^2$$

$$\text{TEC/F} = 190 \times 10^3 \times 3 \times 10^{11} = 5.7 \times 10^{16}\ \text{el/m}^2$$

3. If the number of electrons is ignored in the space above 300 km, then the value of TEC for vertical radiation is calculated as

$$\text{TEC} = \text{TEC/D} + \text{TEC/E} + \text{TEC/F} \approx 5.75 \times 10^{16}\ \text{el/m}^2$$

In the given satellite communication, since the radiation path is inclined, the effective value of TEC is equal to

$$(\text{TEC})_\text{e} = (\text{TEC})\ \times \text{Sec}\ 45^\circ = 8.1 \times 10^{16}\ \text{el/m}^2$$ ■

1.10 Free Space

Free space possesses an ideal condition, that is, without any energy absorption or adverse propagation effects. When radiowaves are radiated in the space by an isotropic antenna, they will propagate identically in all directions.

1.10.1 Free-Space Loss

Radiation of radio power P_t by an isotropic antenna in free space results in power flux density P_0 at a distance d:

$$P_0 = \frac{P_\text{t}}{4\pi d^2} = \frac{E_0^2}{2\eta_0} \tag{1.15}$$

In the above formula, P_t is the transmitter power in Watts, d is the distance from antenna in m, E_0 is the electric field magnitude in V/m, and η_0 is free-space intrinsic impedance equal to 120π ohms. Applying G_t as TX antenna gain, power flux density P will be

$$P = \frac{P_t \cdot G_t}{4\pi d^2}. \tag{1.16}$$

Using a receiving antenna with effective aperture area A_e, the received signal power would be

$$P_r = P \cdot A_e. \tag{1.17}$$

According to the EM waves theory, A_e is

$$A_e = \frac{G_r \cdot \lambda^2}{4\pi}. \tag{1.18}$$

By manipulating the last three relations, the following formula is derived:

$$P_r = \frac{P_t \cdot G_t}{4\pi d^2} \times \frac{G_r \cdot \lambda^2}{4\pi} = \frac{P_t \cdot G_t \cdot G_r \cdot \lambda^2}{(4\pi d)^2}. \tag{1.19}$$

To calculate free-space loss (FSL) by using the above relation and assuming $G_t = G_r = 1$,

$$L_{fb} = \text{FSL} = 10 \log \frac{P_t}{P_r} = -10 \log \frac{\lambda^2}{(4\pi d)^2} \tag{1.20}$$

$$\Rightarrow \quad \text{FSL} = 20 \log \frac{4\pi d}{\lambda}. \tag{1.21}$$

Considering $\lambda = c/f$ we have

$$\text{FSL} = 20 \log \frac{4\pi f \cdot d}{c}. \tag{1.22}$$

The above formula is a generic form of FSL in metric system of units. Since in actual links, the frequency is in MHz or GHz and distance in km, by putting $c = 3 \times 10^8$ m/s, then FSL is specified by one of the following formulas:

$$\text{FSL[dB]} = 32.4 + 20 \log f[\text{MHz}] + 20 \log d[\text{km}] \tag{1.23}$$

$$\text{FSL[dB]} = 92.4 + 20 \log f[\text{GHz}] + 20 \log d[\text{km}]. \tag{1.24}$$

1.10.2 Power Flux Density

Instantaneous power flux density W_i of a plane wave at any location in the space based on electromagnetic theory is

$$W_i = |\bar{E} \times \bar{H}| = c\varepsilon_0 |\bar{E}|^2 = c\mu_0 |\bar{H}|^2. \tag{1.25}$$

E and H denote electric and magnetic fields in V/m and A/m, respectively; W_i is maximum power flux density in W/m^2. Applying relevant equation yields

$$\frac{1}{c\varepsilon_0} = c\mu_0 = \eta_0 = 120\pi\Omega. \tag{1.26}$$

Mean value of W_i based on sinusoidal nature of radiowaves, denoted by W, is expressed by

$$W = \frac{1}{2}W_i = \frac{1}{2\eta_0}|E|^2 = \frac{\eta_0}{2}|H|^2. \tag{1.27}$$

Example 1.6. At a given location, the mean value of power flux density is 100 pW/m^2, calculate effective values of E and H.

Solution.

$$W = 100\,\text{pW/m}^2 \Rightarrow E_m = \sqrt{2\eta_0 \cdot W}$$

$$E_m = 275\,\mu\text{V/m} \Rightarrow E_e = \frac{E_m}{\sqrt{2}} = 196\,\mu\text{V/m}$$

$$H_m = \sqrt{\frac{2\times 10^{-10}}{120\pi}} = 0.728\,\mu\text{A/m} \Rightarrow H_e = 0.52\mu\text{A/m}$$ ■

1.11 ITU-R Formulas

Considering that free-space propagation is a fundamental reference for engineering of radio links, ITU-R assembly recommends Rec. P.525 that the following methods be used for the calculation of attenuation in free space.

1.11.1 Point-to-Area Links

If there is a transmitter serving several randomly distributed receivers (broadcasting, mobile service), the field strength is calculated at a point located at some appropriate distance from the transmitter by the expression

$$e = \frac{\sqrt{30p}}{d}, \tag{1.28}$$

where:

e: r.m.s. field strength (V/m)
p: Equivalent Isotropically Radiated Power (EIRP) of the transmitter in the direction of the point in question
d: Distance from the transmitter to the point in question (m)

Equation (1.28) is often replaced by (1.29) which uses practical units:

$$e_m[\mathrm{V/m}] = 173\frac{\sqrt{p\ [\mathrm{kW}]}}{d\ [\mathrm{km}]}. \tag{1.29}$$

For antennas operating in free-space conditions, the cymomotive force as defined by ITU-R may be obtained by multiplying together e and d in (1.28). Its dimension is volts, and to apply the above formulas, the following points shall be taken into account:

- If the wave is elliptically polarized and not linear, and if the electric field components along two orthogonal axes are expressed by e_x and e_y, the left-hand term of (1.28) should be replaced by $\sqrt{e_x^2 + e_y^2}$ or it may be simplified only if the axial ratio is known. The term e should be replaced by $e\sqrt{2}$ in the case of circular polarization.
- In the case of antennas located at ground level and operating on relatively low frequencies with vertical polarization, radiation is generally considered only in the upper half-space. This should be taken into account in determining the EIRP (see recommendation ITU-R PN.368).

1.11.2 Point-to-Point Links

With a point-to-point link, it is preferable to calculate the free-space attenuation between isotropic antennas, also known as the free-space basic transmission loss (symbols: L_{bf} or A_0), as follows:

$$L_{\mathrm{bf}} = 20\log\left(\frac{4\pi}{\lambda}\right), \tag{1.30}$$

where:

L_{bf}: Free-space basic transmission loss (dB)
d: Distance
λ: Wavelength

d and λ are expressed with the same unit.

Equation (1.30) can also be written using the frequency instead of the wavelength.

$$L_{\mathrm{bf}} = 32.4 + 20\log f + 20\log d, \tag{1.31}$$

where:

f: Frequency (MHz)
d: Distance (km)

1.11.3 Radar Links

Radar systems represent a special case because the signal is subjected to a loss while propagating both from the transmitter to the target and from the target to the receiver. For radars using a common antenna for both transmitter and receiver, a radar-free space basic transmission loss, L_{br}, can be written as follows:

$$L_{\mathrm{br}}[\mathrm{dB}] = 103.4 + 20\log f + 40\log d - 10\log \sigma, \tag{1.32}$$

where:

σ: Radar target cross section (m^2)
d: Distance from the radar to the target(km)
f: Frequency of the system (MHz)

The radar target cross section of an object is the ratio of the total isotropically equivalent scattered power to the incident power density.

1.11.4 Power Flux Density

There are also relations between the characteristics of a plane wave (or a wave which can be treated as a plane wave) at a point:

$$S = \frac{e^2}{120\,\pi} = \frac{4\pi p_r}{\lambda^2}, \tag{1.33}$$

where:

S: Power flux density ($\mathrm{W/m^2}$)
e: r.m.s. field strength (V/m)
p_r: Power (W) available from an isotropic antenna located at this point
λ: Wavelength (m)

1.11.5 Conversion Relations

On the basis of free-space propagation, the following conversion formulas may be used. Field strength for a given isotropically transmitted power

$$E = P_{\mathrm{t}} - 20\log d + 74.8. \tag{1.34}$$

Isotropically received power for a given field strength

$$P_{\mathrm{r}} = E - 20\log f - 167.2. \tag{1.35}$$

Free-space basic transmission loss for a given isotropically transmitted power and field strength

$$L_{bf} = P_t - E + 20\log f + 167.2. \tag{1.36}$$

Power flux density for a given field strength

$$S = E - 145.8, \tag{1.37}$$

where:

P_t: Isotropically transmitted power (dB(W))
P_r: Isotropically received power (dB(W))
E: Electric field strength (dB(μV/m))
f: Frequency (GHz)
d: Radio path length (km)
L_{bf}: Free-space basic transmission loss (dB)
S: Power flux density (dB(W/m^2))

Note that (1.34) and (1.36) can be used to derive equation (1.31).

1.12 Equivalent Radiated Power

1.12.1 ERP and EIRP

ERP, includes all gain and loss factors on the transmitting side and usually expressed in dB$_m$ or dB$_w$. In fact, ERP is the product of TX output power and antenna gain in the desired direction by taking into account all losses regarding RF feeder, connectors, etc., simply expressed by

$$\text{ERP} = \frac{G_t \cdot P_t}{L_t}. \tag{1.38}$$

Reference antenna in the above expression is half-wave dipole, and its logarithmic form is

$$\text{ERP}[\text{dB}_m] = P_t[\text{dB}_m] + G_t[\text{dB}_d] - L_t[\text{dB}]. \tag{1.39}$$

In the case of selecting isotropic antenna as reference, then it is called Equivalent Isotropic Radiated Power denoted by EIRP and expressed by

$$\text{EIRP}[\text{dB}_m] = P_t[\text{dB}_m] + G_t[\text{dB}_i] - L_t[\text{dB}], \quad \text{or} \tag{1.40}$$

$$\text{EIRP}[\text{dB}_w] = P_t[\text{dB}_w] + G_t[\text{dB}_d] - L_t[\text{dB}]. \tag{1.41}$$

1.12.2 Electric Field Intensity

Electric field intensity can be expressed versus EIRP. To find its relation, first it should be noted that

$$W = R_e\{\bar{E} \times \bar{H}\} = \frac{|E_d^2|}{\eta_0}. \tag{1.42}$$

In line-of-sight radio propagation in free space, power flux density at a distance d is

$$W = \frac{P_t \cdot G_t}{L_t \cdot 4\pi d^2} = \frac{\text{EIRP}}{4\pi d^2}. \tag{1.43}$$

Considering $\eta_0 = 120\pi$ in free space and using the last two formulas,

$$\frac{|E_d|^2}{\eta_0} = \frac{P_t \cdot G_t}{L_t \cdot 4\pi d^2} = \frac{\text{EIRP}}{4\pi d^2} \tag{1.44}$$

$$\Rightarrow \quad |E_d| = \frac{\sqrt{30\ \text{EIRP}}}{d}. \tag{1.45}$$

It is noted that in (1.45), numerical (not logarithmic) value of EIRP shall be employed in proper system of units.

Example 1.7. Radiowaves are radiated by a 10 W transmitter connected to a 5 dB_i antenna

1. Find EIRP if feeder loss is 2 dB.
2. Find $|E_d|$ and W at a location of 8 km from TX.

Solution. 1.

$$\text{EIRP}[\text{dB}_w] = P_t[\text{dB}_w] + G_t[\text{dB}_i] - L_t[\text{dB}] = 13\ \text{dB}_w$$

$$\Rightarrow \quad \text{EIRP} = \text{Antilog}1.3 = 20\,\text{W}$$

2. Applying (1.45) and $\eta_0 = 377\ \Omega$

$$|E_d| = \frac{\sqrt{30 \times 20}}{8000} = 3.06 \times 10^{-3}\,\text{V/m} = 3.06\,\text{mV/m}$$

$$W = \frac{1}{2\eta_0}|E_d|^2 \quad \Rightarrow \quad W = 12.4\,\text{nW/m}^2$$

■

1.13 Transmission Loss

1.13.1 Loss Terms in Radio Links

Radiowaves propagating between transmitting and receiving antennas, in addition to free-space loss, are subject to excess attenuations including, but not limited to, the following parameters:

- RF feeder loss
- Antenna gain/loss
- Propagation mechanisms losses
- Depolarization loss

For actual calculations and radio design, all factors shall be taken into account. To describe and standardize the terminology and notations employed to characterize transmission loss and its component, ITU-R under Recommendation No. P.341 has defined the concept of transmission loss for radio links.

As shown in Fig. 1.15, the following types of losses are indicated:

- Free-space loss, of L_{bf}
- Basic transmission loss, L_b
- Transmission loss, L

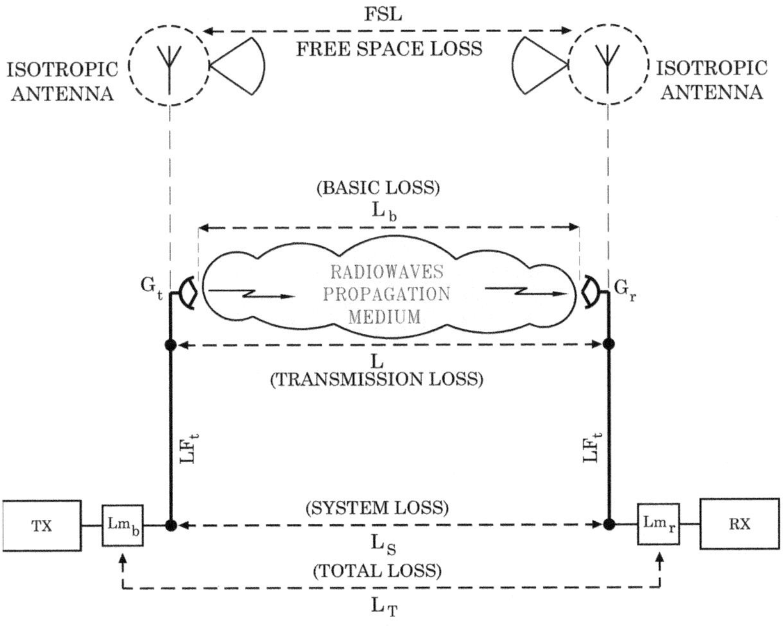

Fig. 1.15 ITU-based concept of transmission loss (Ref.: ITU-R, P.341–5)

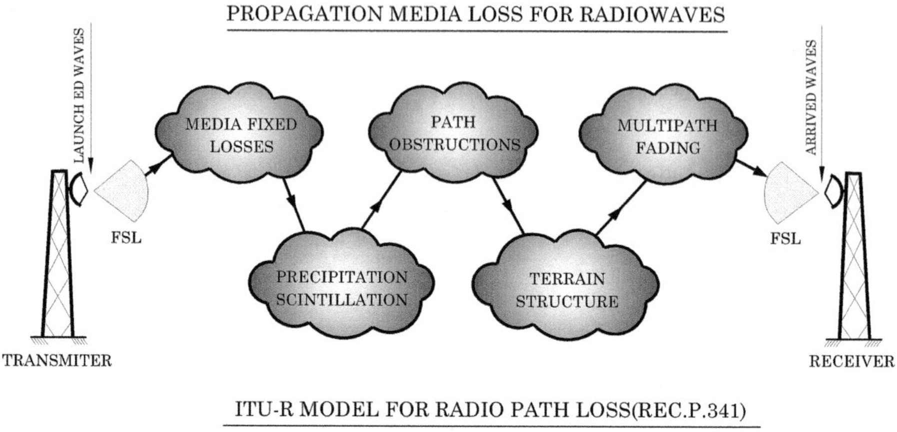

Fig. 1.16 Loss of propagation media for radiowaves

- System loss, L_b
- Total loss, L_t

Full definitions are given in ITU-R, P.341 for each of the above parameters.

1.13.2 Basic Transmission Loss

Free-space loss as permanent radiowaves loss is calculated by equations stated in (1.30) and (1.31). In actual transmission condition, some other attenuation factors are imposed on radiowaves due to medium effects. The sum of FSL and medium loss L_m is defined as basic transmission loss L_b:

$$L_b = \mathrm{FSL} \times L_m. \tag{1.46}$$

The above formula in logarithmic form is

$$L_b[\mathrm{dB}] = \mathrm{FSL}[\mathrm{dB}] + L_m[\mathrm{dB}]. \tag{1.47}$$

A number of medium losses are (Fig. 1.16):

- Atmospheric absorption loss due to gases, vapor, and aerosols
- Reflection loss, including focusing or defocusing due to curvature of reflecting layer
- Scattering of radiowaves due to irregularities in the atmospheric refractive index or by hydrometeors
- Diffraction loss due to obstructions
- Radio precipitation due to rain and snow
- Temporal climatic effects such as fog and cloud
- Antenna to medium coupling loss

- Polarization coupling loss
- Multipath adverse effects

By considering TX and RX antennas, then transmission loss L is calculated:

$$L[\mathrm{dB}] = L_{\mathrm{b}}[\mathrm{dB}] - G_{\mathrm{t}}[\mathrm{dB}_i] - G_{\mathrm{r}}[\mathrm{dB}_i]. \tag{1.48}$$

1.13.3 System and Total Losses

In accordance with ITU transmission loss concept, system loss Ls is defined as the following ratio:

$$L_{\mathrm{b}}[\mathrm{dB}] = 10\ \log\left(\frac{P_{\mathrm{t}}}{P_{\mathrm{a}}}\right), \tag{1.49}$$

where:

P_{t}: Transmitter power delivered to the input of TX antenna
P_{a}: Received signal level (RSL) at the output of RX antenna

Combining the mentioned relations yields

$$L_{\mathrm{b}}[\mathrm{dB}] = L[\mathrm{dB}] + L_{\mathrm{tc}}[\mathrm{dB}] + L_{\mathrm{rc}}[d] = P_{\mathrm{t}}[\mathrm{dB}_m] - P_{\mathrm{a}}[\mathrm{dB}_m]. \tag{1.50}$$

Total loss L_{t} is defined as the ratio of signal levels at selected points within transmitter and receiver systems. Exact points shall be indicated to avoid misunderstanding.

Example 1.8. A radio link is characterized by

$$\mathrm{FSL} = 128\ \mathrm{dB},\ \ L_{\mathrm{b}} = 135\ \mathrm{dB},\ \ L_c = L_{\mathrm{tc}} + L_{\mathrm{rc}} = 5\ \mathrm{dB},\ \ G_{\mathrm{t}} = G_{\mathrm{r}} = 30\ \mathrm{dB}_i$$

1. Find medium loss
2. Find P_{t} for the RSL of $-60\,\mathrm{dB}_m$

Solution. 1.

$$L_{\mathrm{m}}[\mathrm{dB}] = L_{\mathrm{b}} - \mathrm{FSL} = 7\ \mathrm{dB}$$

$$L_{\mathrm{m}} = \mathrm{Antilog}\ 7 = 5$$

Thus, the related medium attenuates radiowave five times more.

2.

$$P_{\mathrm{t}}[\mathrm{dB}_m] - P_{\mathrm{r}}[\mathrm{dB}_m] = L_c[\mathrm{dB}] + L_{\mathrm{b}}[\mathrm{dB}] - G_{\mathrm{t}}[\mathrm{dB}_i] - G_{\mathrm{r}}[\mathrm{dB}_i]$$

$$P_{\mathrm{t}} + 60 = 5 + 135 - 30 - 30 \quad \Rightarrow \quad P_{\mathrm{t}}[\mathrm{dB}_m] = 20\ \mathrm{dB}_m$$

$$P_{\mathrm{t}} = \mathrm{Antilog}\ 2.0 \quad \Rightarrow \quad P_{\mathrm{t}} = 100\,\mathrm{mW}$$

■

1.14 Radio Noises

All kinds of unwanted emissions added to the desired signals or radiowaves which impair either the proper detection or quality of the main signal are considered as noise. Among a variety of noises, some types with electromagnetic nature are called radio noises. The original signal may be affected by noise during different stages of generation, processing, transmission, propagation, amplification, modulation/demodulation, and multiplexing/demultiplexing.

Major sources of noise are given in Fig. 1.17 which can be listed in the following categories:

- Natural noises produced by ground surface, atmosphere, sky, sun, radio stars, moon, and other cosmic (galaxy) origin.
- Man-made noise due to the operations of electrical or electro mechanical devices.
- Urban, sub urban, and industrial noises.
- Equipment noises including transmitter and receiver units.
- Temporary noises due to lightning and rainfall.
- Spurious and radio frequency interference (RFI) because of poor design of radio links or radio equipment malfunctioning.

Major effects of noise on the radiowaves can be summarized as follows:

- Limitation in bandwidth allocation to different services.
- Limitation in frequency assignment to radio channels.
- Reduction in signal to noise ratio (SNR).
- Reduction in quality of the received signal.
- Increasing of the equivalent system noise temperature resulting from the use of higher-performance equipment and increased cost.

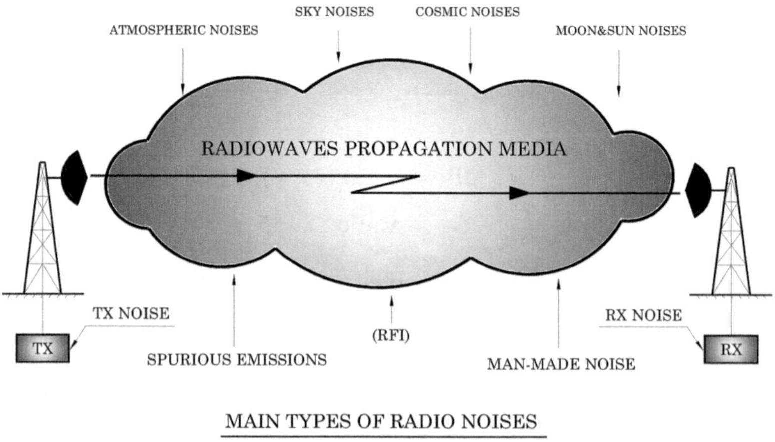

Fig. 1.17 Main types of radio noises

1.15 Exercises

Questions

1. Explain main applications of radiowaves in the present time and its future prospects.
2. Specify different kinds of satellite services for which there is no substitute other than radio systems.
3. Describe atmosphere layers and indicate their heights and applications in radiowaves propagation.
4. Investigate appendices to the radio regulations and ITU-F recommendations and prepare a brief report containing applications and limitations of RF channel arrangements in 1 to 10 GHz frequency band.
5. Explain the key role of frequency spectrum in radio communications and reuse techniques in frequency planning.
6. Investigate ITU-based chart of regions for frequency allocation. In what region is your country located?
7. Indicate main factors affecting the frequency band selection in the satellite communications and broadcasting systems as well.
8. Refer to P-series of ITU-R recommendations list in the appendix B of this book and arrange them based on the tropospheric and ionospheric propagation issues.
9. What is the variation rate of K-factor for different radio communications systems?
10. List the factors affecting Earth bulge and state their relations.
11. Investigate the effect of frequency on different loss sources for the radiowaves propagating in the troposphere layer.
12. Specify different regions of the ionosphere and determine their approximate heights from the Earth surface.
13. List the main ionospheric phenomena affecting the radiowave propagation.
14. List the main effects of ionosphere layer in satellite radio networks especially mobile satellite communications.
15. Referring to the ITU-based concept of transmission loss, determine their normal rough values for LOS microwave links working at 8 GHz and satellite links working in Ku-band.
16. Define EIRP by specifying its formula in logarithmic form. Determine maximum value of EIRP in dB_m for the following cases:

    - $G_a = 54\ dB_i, \quad P_t = 2\ kW$
    - $G_a = 42\ dB_i, \quad P_t = 500\,mW$
    - $G_a = 10\ dB_d, \quad P_t = 25\,W$

Problems

1. A P-MP radio network topology is depicted in Fig. 1.18. The system works at 3.5 GHz band, and a maximum of ten pairs of frequencies may be assigned.

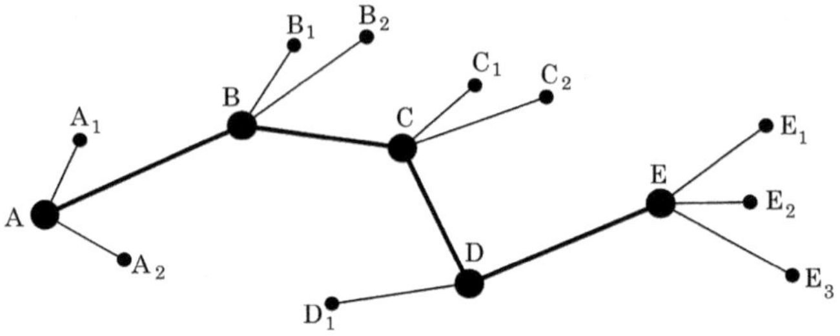

Fig. 1.18 P-MP radio network topology

   (a) Specify an optimized frequency plan using only one type polarization (say vertical).
   (b) Find the optimized plan by using frequency reuse techniques. Assume that only adjacent stations are in line-of-sight condition and other routes are obstructed.

2. A GEO satellite covering Atlantic Ocean Region (AOR) includes 12 transponders working in Ku-band. Bandwidth of each transponder is 36 MHz, and H-type polarization is employed.

   (a) Design a suitable frequency plan for the satellite.
   (b) Propose the procedure to duplicate the system capacity by similar transponders.

3. (a) Calculate values of ε_r, n, and N in the Earth atmosphere at an altitude of 2 km.
   (b) If the refractivity number of air at sea level and at altitude of 1.5 km are $N_0 = 289$ and $N_1 = 241$, respectively, find R (radius of radio path curvature) at the height of 200 m above the ground level.
4. Assuming that radius of radio path curvature is 26,333 km, find:

   (a) K-factor
   (b) Equivalent Earth radius

5. The K value range is between 0.67 and 1.3 for 40 km path length.

   (a) Find minimum and maximum values of equivalent Earth radius.
   (b) Find effective value of K for the path at C-band radiowaves.

6. The distance between transmitter and receiver is 30 km, find the Earth bulge for points with 5 and 15 km distance from transmitter for the following conditions:

   (a) Standard atmosphere.
   (b) $K = \frac{1}{2}$ and compare the results with standard atmosphere.

7. Specify the height from the ground level in which the electron density is maximum, find:

   (a) The variation range of this parameter at the height of 300 km aboveground surface.
   (b) If D, E, and F sub-layers are assumed to be 20, 40, 180 km respectively, calculate the value of TEC in a cylinder perpendicular to the ionosphere layer.

8. Effective amplitude of electric field at radiation source is 100 V/m and its value at a location 8 km far from the radiating source is $200\,\mu\text{V/m}$. At frequency $f = 300\,\text{MHz}$, find:

   (a) FSL
   (b) Transmission basic loss
   (c) Transmitting power of the source and receiving power at the distant location

9. Calculate free-space and basic transmission losses for:

   (a) $d = 30\,\text{km}, \quad f = 450\,\text{MHz}, \quad L_p = 12$ dB
   (b) $d = 20\,\text{km}, \quad f = 8.4\,\text{MHz}, \quad L_p = 5$ dB

10. Find ERP for 2 W transmitter connected by an RF feeder with 6 dB loss to an antenna with $G_i = 1{,}000$. How much is the amplitude of E-field at locations of 1 and 6 km away from the transmitter?

Chapter 2
Antennas and Passive Reflectors

2.1 Introduction

Radio signals produced in the transmitter are amplified through the last stage of RF part to increase its level to the required output power. As shown in the Fig. 2.1, the high-power RF signal will be fed to the antenna system for coupling to the medium and propagating toward the specified receiving station(s).

Converting of the RF signal to radiowaves and radiation in the open media will be performed by antennas. In the receiving station, the required radiowaves will be coupled to the receiver for further processing.

Sometimes the required radiowaves cannot be detected directly from radiowaves radiated by the transmitter, and depending on the radio path topography, one or more relay stations may be essential. A relay station called repeater can be "active" or "passive" type. Main component in passive repeaters is called passive reflector.

The main purpose of this chapter is to provide a comprehensive and short investigation of what are important for radiowaves propagation engineering, while theoretical basics, expressions, and formulas are left to other references and antenna books. A considerable portion of the chapter is devoted to introduce different types of antennas used practically for various applications in common frequency bands. This chapter is then followed by presenting passive repeaters and reflectors. In addition, some examples are given to provide the reader with more details.

2.2 RF Transmission Devices

In addition to antennas, some passive devices should be employed for proper operations based on system design considerations. Among a variety of RF passive devices, the following items are more popular:

- RF feeder such as coaxial cable, waveguide, and heliax cable.
- Connectors, flanges, and terminations.

A. Ghasemi et al., *Propagation Engineering in Radio Links Design*,
DOI 10.1007/978-1-4614-5314-7_2,

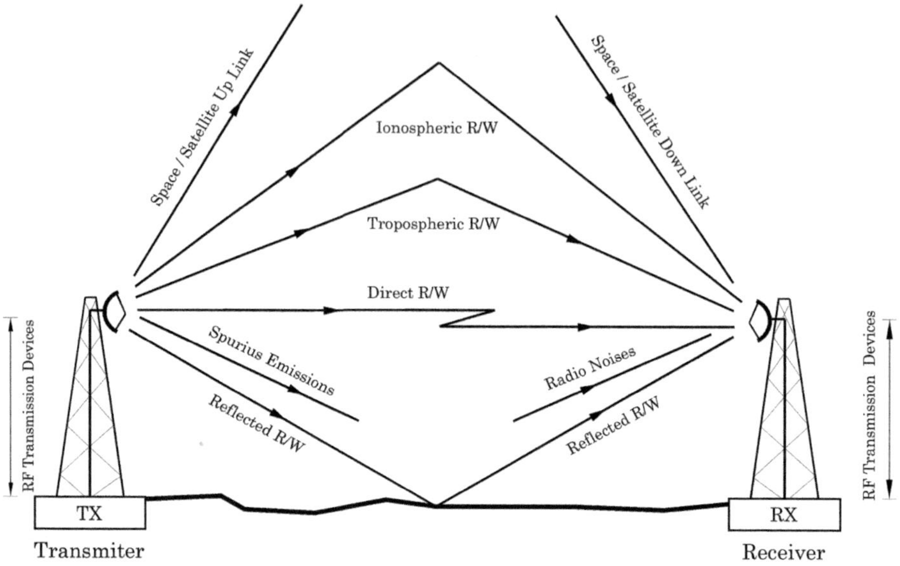

Fig. 2.1 Role of antenna in radio communications

- Couplers or multicouplers.
- Hybrid junctions.
- Power dividers and splitters.
- Attenuators.
- Circulators or branching circuits.
- Antenna tuning unit.

For more details of the above components and their applications, reference is made to the related handbooks and manufacturer manuals and data sheets. It should be noted that each passive device imposes some attenuation and impact on radiowaves which shall be taken into account for technical calculations including link power budget calculation.

2.3 Antenna

2.3.1 Definition

Antenna is a metallic structure specially designed to radiate radiowaves of electromagnetic nature into the medium in a proper and efficient way. It is an interface between the transmitting/receiving system and the propagation medium (normally atmosphere). In other words, an antenna couples radio units, either transmitter or receiver, to the propagation medium. Some of the main aspects of radiowaves, such as polarization, radiated power, and direction, are shaped and controlled by the related antenna.

Antenna is an outdoor-type device which shall be able to withstand against ambient harsh conditions such as temperature variations in the specified range, humidity, wind loads, wind gust, sandstorm, rain, snow, hail, fog, and earthquakes. Some of antenna main characteristics having technical impacts on the radio system design are stated below:

- Frequency band and polarization
- Directivity and gain
- Radiation pattern and half-power beamwidth (HBW)
- Efficiency and power handling capability
- Effective area of the receiving antenna and noise temperature

Today numerous types of antennas are manufactured to meet various applications. These antennas can be categorized generally as wire, aperture, array, reflector, lens, smart, and microstrip antennas.

2.3.2 Classification

Antennas may be classified in different ways based on some antenna specific aspects such as:

- *Frequency band*: including MF, HF, VHF, UHF, SHF, microwave, satellite Ku and Ka bands, and radar S and X bands.
- *Directivity*: including directional, omnidirectional, and sectoral.
- *Performance*: including standard (S), high performance (HP), superhigh performance (SHP), and ultrahigh performance (UHP).
- *Application*: including broadcasting, radar, satellite, maritime, and aeronautical.
- *Structure*: including dipole, horn, Yagi, parabolic, log-periodic, rhombic, L-type, and T-type.

A number of different types of antennas are given in Sect. 2.5.11.

2.3.3 Antenna Radome

Antenna is normally an outdoor unit, and to protect it against ambient harsh conditions such as temperature variation, humidity, atmosphere precipitations, wind gusts, and sandstorms, its main structure is enclosed by a radome. Main applications of the radomes are summarized as follows:

- Wind load reduction
- Prevention of ice or freezing rain formation

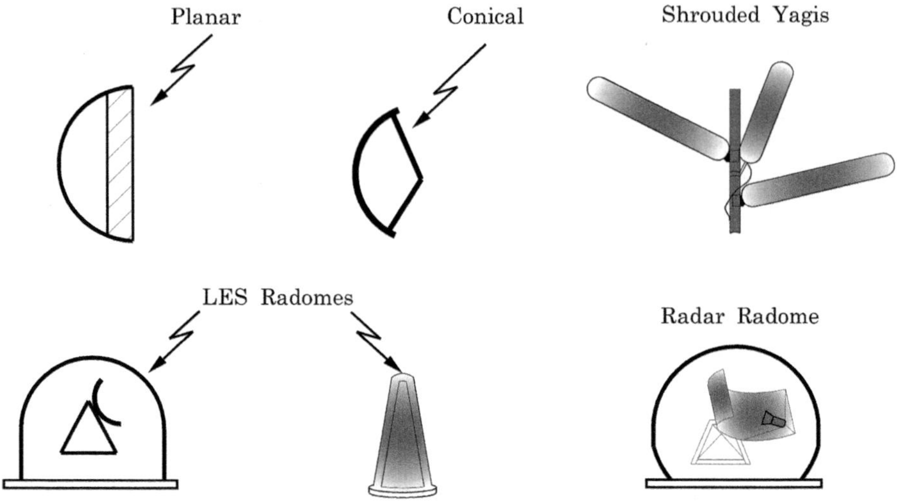

Fig. 2.2 Typical antenna radomes

- Protecting antenna structure and related auxiliary devices against harsh climatic conditions
- Protecting the antenna from debris and rotational irregularities due to wind
- Providing safety for nearby personnel of being accidentally struck by fast rotating antennas
- Supporting the antenna against birds and animal impairments

Usually a radome is a weather-proof enclosure that protects some types such as microwave or radar antennas. Also they are widely used to protect marine satellite antennas installed on the ships when they experience pitch, roll, and yaw movements. A variety of antenna radomes are available which some types are shown in Fig. 2.2. General requirements for antenna radomes are listed below:

- Using suitable anticorrosion material with small attenuation and low adverse impacts on the radiowaves propagation. It should be transparent to radio signal with attenuation in the order of 0.5–1 dB.
- Good mechanical and chemical properties to support the antenna and auxiliary devices for safe and efficient operations during its lifetime.
- Good aerodynamic shape with low wind loads. Some more popular shapes include spherical, geodesies, and planar, depending on the particular application.

2.3.4 Auxiliary Assemblies

Some mechanical or electrical subsystems are employed for antenna operations in professional applications. These subsystems include:

- Tracking system
- Tuning/alignment unit
- Antenna supports on the tower structure
- Pedestal and rotating mechanism
- Grounding material
- Deicing device

2.4 Radio Power Radiation

Antenna main parameters affecting radiowaves propagation and link power budget calculations are briefly explained in this section. For more theoretical details, reference is made to antenna textbooks.

2.4.1 Radiation Pattern

One of the basic characteristics of each antenna is its radiation pattern which is defined as relative distribution of radiated power (or field strength) from the antenna in terms of direction in surrounding space. For example, as depicted in Fig. 2.3, the radiation pattern of an ideal isotropic antenna, which radiates equally in all directions, would be a sphere, while radiation pattern of an elementary dipole or a long wire would be different.

Although radiation patterns of all types of antennas are three-dimensional, it is a common practice to show planar sections of it. As shown in Fig. 2.4, the direction containing maximum radiation power is principal axis called boresight axis (Fig. 2.5).

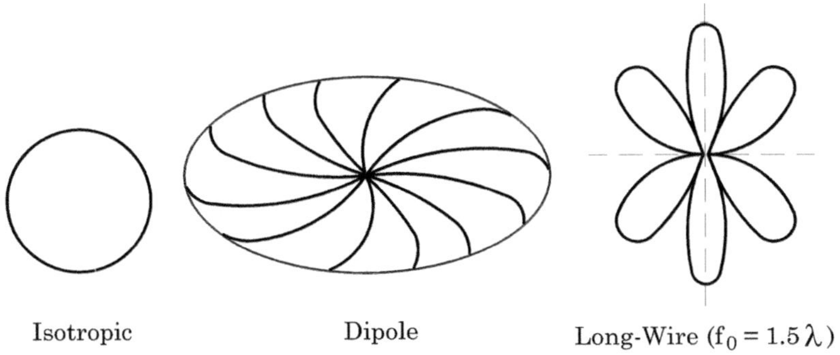

Fig. 2.3 Typical antenna three-dimensional radiation patterns

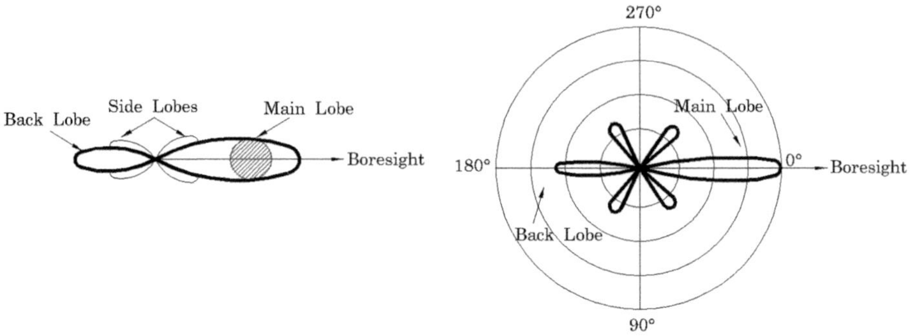

Fig. 2.4 Sample of radiation pattern and principal axis

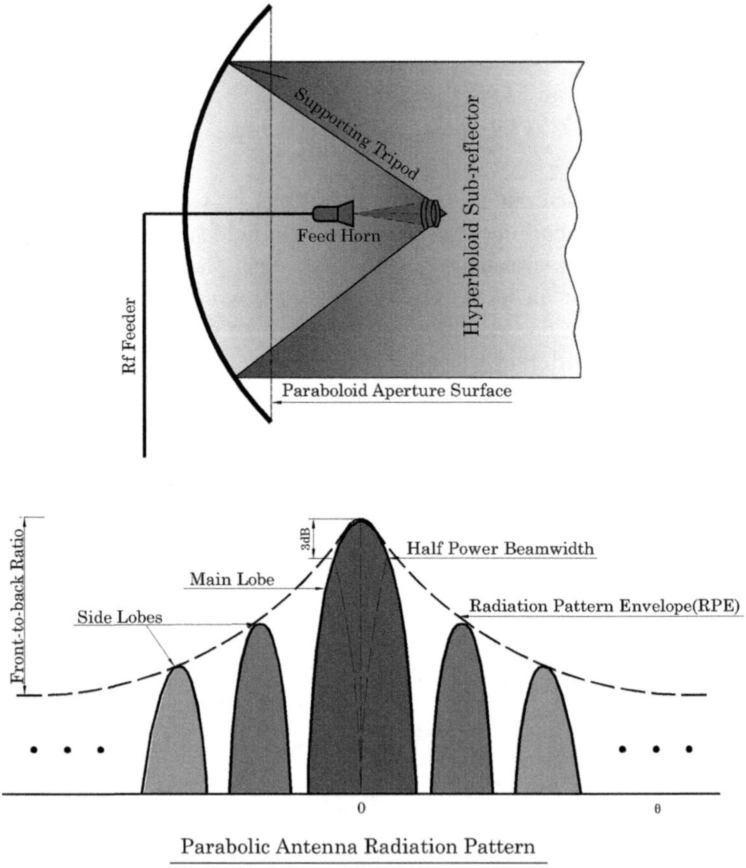

Fig. 2.5 Radiation pattern of a typical parabolic antenna

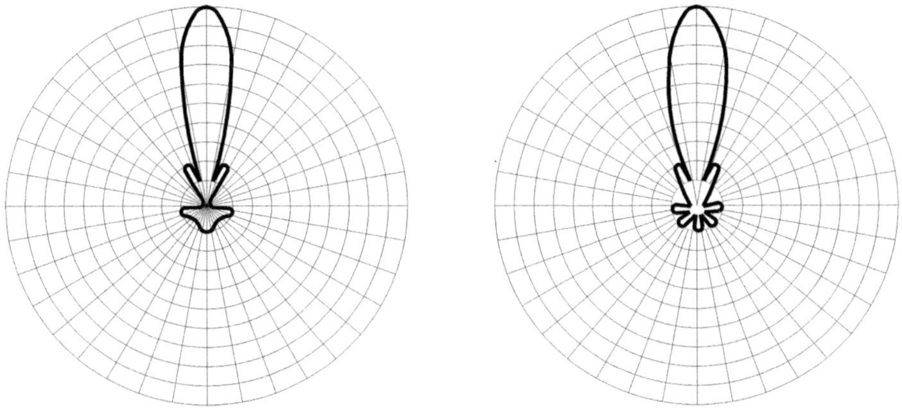

Fig. 2.6 Sample of radiation patterns in E and H planes (UHF Yagi antenna)

The two most popular presentations are the principal E-plane and H-plane patterns. The E-plane pattern is a view of the radiation pattern obtained from a section containing the principal axis of the radiated field and in which the electric field lies in the selected plane. Similarly, the H-plane pattern is a sectional view in which the magnetic field lies in the plane of the section when it is so selected to contain the principal axis (Fig. 2.6).

Example 2.1. Normalized radiation pattern of an antenna is expressed by $D_n = \sin^2\theta$; assuming $D = 10$ at $\theta = 45°$, find:

1. The maximum directivity
2. Direction of the principal axis

Solution. 1. $D = A_0 \cdot D_n = A \cdot \sin^2\theta$

$$(\theta = 45°,\ D = 10) \Rightarrow A_0 = 20\ ,\ D_{max} = 20$$

2. The principal axis is in direction of the maximum radiation; thus,

$$D = D_{max} \Rightarrow \theta = 90°$$

■

2.4.2 *Radiation Power Density and Intensity*

To elaborate further on the antenna power radiation, some commonly used terms such as "radiation power density," "total radiated power," and "radiation intensity" should be defined clearly.

Radiation power density shown by W (watt per m^2) is based on the Poynting vector and is written as

$$W = \frac{1}{2} R_e [E \times H^*] \tag{2.1}$$

Total radiated power, P_r, is defined as the total electromagnetic power radiated by an antenna in all directions. This parameter can be expressed in the following integral forms:

$$P_r = \oint \oint_s W \cdot d_s = \int_0^{2\pi} \int_0^{\pi} W \cdot r^2 \sin\theta \cdot d\theta \cdot d\varphi \tag{2.2}$$

$$P_r = \frac{1}{2} \int_0^{2\pi} \int_0^{\pi} R_e[E \times H^*] \cdot r^2 \sin\theta \cdot d\theta \cdot d\varphi \tag{2.3}$$

Radiation intensity is an antenna far field parameter defined as the power radiated from it per unit solid angle, in a given direction. Radiation intensity denoted by U in watt per unit solid angle is expressed by the following formula:

$$U = r^2 \cdot W \quad \Rightarrow \quad W = \frac{U}{r^2} \tag{2.4}$$

$$P_r = \int_0^{2\pi} \int_0^{\pi} U \cdot \sin\theta \cdot d\theta \cdot d\varphi \tag{2.5}$$

Example 2.2. Radiated power density of an antenna is expressed by

$$W = 10 \frac{\sin\,\theta}{r^2} \quad \text{W/m}^2\,, \quad 0 \le \theta \le \pi\,, \quad 0 \le \varphi \le 2\pi$$

1. Find radiation power density and intensity in the $\theta = 90°$ direction at $r = 2\,\text{m}$
2. Calculate the total radiated power

Solution. 1.

$$(\theta = 90°\,,\ r = 2\,\text{m}) \Rightarrow W = 2.5\ \text{W/m}^2$$

$$U = r^2 W \Rightarrow U = 10\ \sin\ \theta = 10\,\text{W/unit solid angle}$$

2.

$$P_r = \int_0^{2\pi} \int_0^{\pi} U \cdot \sin\ \theta \cdot d\theta \cdot d\varphi$$

$$= 2\pi \int_0^{\pi} 10\ \sin^2\theta \cdot d\theta = 98.7\,\text{W}$$

■

2.4.3 Effective Radiated Power

Equivalent radiated power (ERP) includes all gain and loss factors on the transmitting side and usually expressed in dB_m or dB_w. In fact, ERP is the product of TX output power and antenna gain in the desired direction taking into account all losses regarding the RF feeder, connectors, etc., simply expressed by

$$\mathrm{ERP} = \frac{G_t \cdot P_t}{L_t} \tag{2.6}$$

Reference antenna in the above expression is half-wave dipole, and its logarithmic form is

$$\mathrm{ERP}[\mathrm{dB}_m] = P_t[\mathrm{dB}_m] + G_t[\mathrm{dB}_d] - L_t[\mathrm{dB}] \tag{2.7}$$

In the case of selecting isotropic antenna as a reference, then it is called equivalent isotropic radiated power denoted by EIRP and expressed by

$$\mathrm{EIRP}[\mathrm{dB}_m] = P_t[\mathrm{dB}_m] + G_t[\mathrm{dB}_i] - L_t[\mathrm{dB}] \quad , \quad \text{or} \tag{2.8}$$

$$\mathrm{EIRP}[\mathrm{dB}_w] = P_t[\mathrm{dB}_w] + G_t[\mathrm{dB}_d] - L_t[\mathrm{dB}] \tag{2.9}$$

2.4.4 Electric Field Intensity

Electric field intensity can be expressed versus EIRP. To find its relation, first it should be noted that

$$W = \frac{1}{2} R_e \{\bar{E} \times \bar{H}\} = \frac{|E_d^2|}{\eta_0} \tag{2.10}$$

In line-of-sight (LOS) radio propagation in free space, power flux density at a distance d is

$$W = \frac{P_t \cdot G_t}{L_t \cdot 4\pi d^2} = \frac{\mathrm{EIRP}}{4\pi d^2} \tag{2.11}$$

Considering $\eta_0 = 120\pi$ in the free space and using the last two formulas yields

$$\frac{|E_d|^2}{\eta_0} = \frac{P_t \cdot G_t}{L_t \cdot 4\pi d^2} = \frac{\mathrm{EIRP}}{4\pi d^2} \tag{2.12}$$

$$\Rightarrow \quad |E_d| = \frac{\sqrt{30\,\mathrm{EIRP}}}{d} \tag{2.13}$$

It is noted that in (2.13), numerical value of EIRP (not logarithmic) shall be employed in proper system of units.

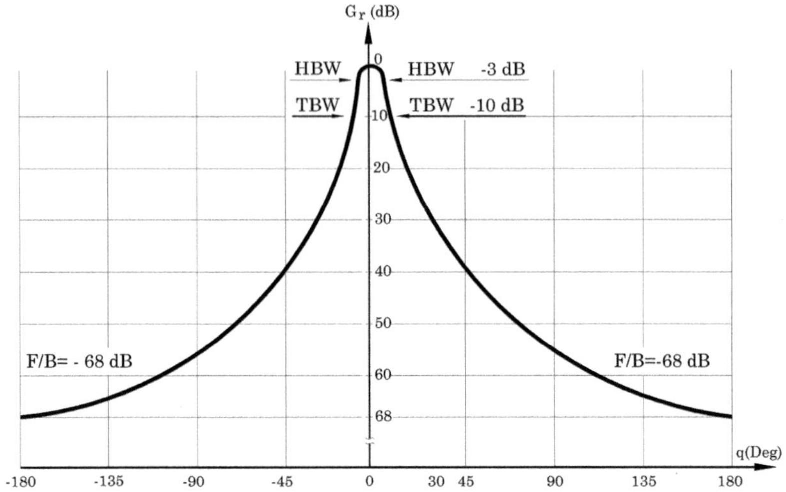

Fig. 2.7 Typical RPE for parabolic antenna

Example 2.3. Radiowaves are radiated by a 10-W transmitter connected to a 5-dB$_i$ antenna; find:

1. EIRP if feeder loss is 2 dB
2. $|E_d|$ and W at a location of 8 km from TX

Solution. 1.

$$\mathrm{EIRP}[\mathrm{dB}_w] = P_t[\mathrm{dB}_w] + G_t[\mathrm{dB}_i] - L_t[\mathrm{dB}] = 13\ \mathrm{dB}_w$$
$$\Rightarrow \quad \mathrm{EIRP} = \mathrm{Antilog}13 = 20\,\mathrm{W}$$

2. Applying (2.13) and $\eta_0 = 377\ \Omega$,

$$|E_d| = \frac{\sqrt{30 \times 20}}{8000} = 3.06 \times 10^{-3}\ \mathrm{V/m} = 3.06\ \mathrm{mV/m}$$
$$W = \frac{1}{\eta_0}|E_d|^2 \quad \Rightarrow \quad W = 24.8\ \mathrm{nW/m^2}$$ ■

2.4.5 *Radiated Peak Envelope*

Radiated peak envelope (RPE) of a directional antenna is typically shown in Fig. 2.7 which is a curve containing all maxima of its main and side lobes. The RPE is used instead of actual radiation pattern for radio-frequency interference (RFI) calculation in the design and engineering of radio networks.

Example 2.4. RPE presented in Fig. 2.7 is typical for a 44-dB$_i$ parabolic antenna working at frequency $f = 7.5\,\text{GHz}$; find:

1. How much is the maximum EIRP at a direction of 30° off-axis when a transmitter with 500-mW output power is connected to the antenna input. Assume 5-dB loss for RF feeder and connectors.
2. How much is F/B ratio as defined by (2.14).

Solution. 1. The maximum gain at 30° off-axis orientation is

$$G_\text{m}(30^\circ) = 44 - 30 = 14\ \text{dB}_i$$

$$P_\text{t} = 500\,\text{mW} \ \Rightarrow\ P_\text{t}[\text{dB}_m] = 27\ \text{dB}_m$$

$$\text{EIRP} = P_\text{t}[\text{dB}_m] + G_\text{m}(30^\circ) - L_f = 36\ \text{dB}_m$$

2. $F/B = G/(\theta = 0) - G/(\theta = 180^\circ) = 0 - (-68) = 68\ \text{dB}$ ■

2.5 Antenna Basic Characteristics

Characteristics of antenna basic parameters are outlined in this section. These parameters have crucial effects on the radio link design and must be taken into account properly.

2.5.1 *Beamwidth*

The radiation of a number of antennas specially high-gain directive ones such as parabolic, Cassegrain, radar, and Yagi. contains repetitive maxima and minima as illustrated in Fig. 2.8. This is due to the interfering pattern of emitted waves by different parts of the antenna. Every part of the pattern located within two successive minima is called a "lobe" among which the biggest one is called "main lobe," while others with smaller magnitude are "side lobes" except a lobe located in opposite side of the main lobe named "back lobe."

The front-to-back ratio of a directional antenna is important in RFI calculations. This ratio is defined as follows:

$$F/B = \frac{\text{maximum magnitude of main lobe}}{\text{maximum magnitude of back lobe}} = \frac{P_\text{t}}{P_b} \tag{2.14}$$

Beamwidth of an antenna is related to its main lobe and is defined as the angular width between symmetrical points at which the radiated power (per unit area) reduces to a certain level of its maximum value. Widely used beamwidth as shown

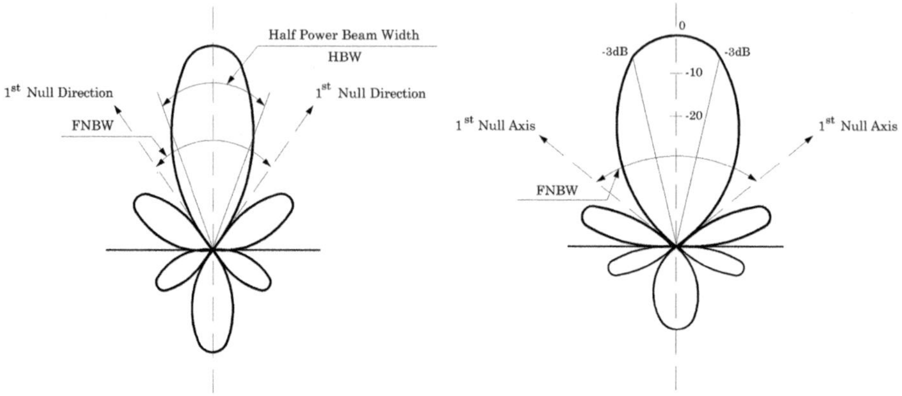

Fig. 2.8 Concept of antenna beamwidth

in Fig. 2.8 is half-power beamwidth denoted by HBW and defined by the angular width between symmetrical points on the antenna main lobe where the radiated power reduces to the half value of its maximum.

The HBW is usually given for both the principal E- and H-plane patterns. Although they are generally of different values, but in some cases, they can be equal. For a dipole antenna, the HBW in E plane is 90°, while in H plane it does not include HBW, since the pattern is constantly circular (normally called omnidirectional).

Another important beamwidth is the first-null beamwidth denoted by FNBW and defined as the angular separation between the first nulls of its radiation pattern. Other beamwidth is the tenth-power beamwidth denoted by TBW and defined as angular width between symmetrical points on the antenna main lobe where they are 10 dB less than its maximum.

In parabolic antennas used in terrestrial microwave networks and satellite Cassegrain antennas, the HBW is in the order of few degrees and its lower value corresponds to higher gain. In this type of antennas, the HBW and TBW are given roughly by

$$\text{parabolic reflector HBW} = 70\frac{\lambda}{D} \tag{2.15}$$

$$\text{parabolic reflector TBW} = \frac{60\lambda}{D} \tag{2.16}$$

where λ is the wavelength and D is the diameter of the antenna reflector and both are of the same unit.

Example 2.5. Radiation intensity of the main lobe of an antenna is approximated by $U = 10 \cos \theta$, and the FNBW is $\pi/10$. In the case of total radiated power $P_r = 4$ W, find:

1. The percentage of the P_r concentrated in the main lobe
2. Antenna directivity

Solution. 1.

$$P_{r0} = \int\int U d\Omega = \int_0^{2\pi}\int_0^{\pi/10} 10 \cos\theta \cdot \sin\theta \cdot d\theta \cdot d\varphi$$
$$= 10\pi \sin^2\theta \,|_0^{\pi/10} = 10\pi \sin^2\frac{\pi}{10} \approx 3W$$

Main lobe radiation % $= \frac{P_{r0}}{P_r} \times 100 = 75\%$

2.

$$P_r = 4\,\mathrm{W}\ ,\ U_{max} = 10\,\mathrm{W/unit\ solid\ angle}$$
$$D_{max} = \frac{4\pi U_{max}}{P_r} = 31.4$$
$$D_{max}(\mathrm{dB}_i) = 10 \log D_{max} = 14.7\ \mathrm{dB}_i$$

■

2.5.2 *Directivity*

Radiation of an antenna is not uniform in all directions except for the isotropic one which is a hypothetical radiation source. For an antenna, the variation of power intensity with direction (in the space) is expressed by the directivity function $D(\theta,\varphi)$:

$$D(\theta,\varphi) = \frac{\text{radiated power per unit solid angle}}{\text{average power radiated per unit solid angle}} \tag{2.17}$$

$$D(\theta,\varphi) = \frac{U}{P_r/4\pi} = 4\pi\frac{U}{P_r} \tag{2.18}$$

In other words, an antenna directivity is the ratio of power radiation intensity in a given direction to power radiation of an isotropic antenna when fed with the same source. Also directivity is defined in the ITU-R, F.162 recommendation as "In a given direction, 4π times the ratio of the intensity of radiation (power per unit solid angle (steradian)), in that direction, to the total power radiated by the antenna."

If the direction is not specified, the direction of the maximum radiation intensity will be used. Thus, directivity which often implies the maximum directivity is a measure of an antenna ability to concentrate the radiated power in a given direction.

Example 2.6. Radiation power density of an antenna is expressed by

$$W = 10\,\frac{\sin^2\theta}{r^2}\ \ (\mathrm{W/m^2})$$

Find:

1. Total radiated power
2. Maximum directivity and its function
3. Radiation patterns in φ and θ fixed planes

Solution. 1.

$$P_r = \oint \oint W \cdot r^2 \sin\theta \cdot d\theta \cdot d\varphi$$
$$= \int_0^{2\pi} d\varphi \int_0^{\pi} 10 \sin^2\theta \cdot d\theta$$
$$P_r = \frac{80\pi}{3} \quad W$$

2.

$$D_{max} = 4\pi \frac{U_{max}}{P_r} \quad \Rightarrow \quad D_{max} = 3/2$$
$$U = r^2 W = 10 \sin^3\theta$$
$$D = 4\pi \frac{U}{P_r} = 4\pi \frac{\sin^2\theta}{P_r} = 1.5 \sin^2\theta$$

3. In the fixed φ plane, $D_\theta = 1.5 \sin^2\theta$, while in the θ fixed plane, say θ_0, $D_\varphi = 1.5 \sin^2\theta_0$; thus, the pattern is a circle. ■

2.5.3 *Efficiency*

Some portion of radio signal energy will dissipate during radiation process through antenna. The capability of an antenna for power radiation is measured by its efficiency denoted by η and defined as follows:

$$\eta = \frac{P_r}{P_{in}} \quad \text{or} \quad \eta\% = \frac{P_r}{P_{in}} \times 100 \tag{2.19}$$

Efficiency of every antenna is a function of several parameters which shall be considered case by case. In some cases, it is close to unity, but drops to less than 0.5 in some others. The amount of efficiency for low RF power radiation is not too critical, but when it is used for radiation of high RF power, it will affect power handling capability of the antenna.

As an example, efficiency for parabolic antennas can be expressed by the following equation:

$$\eta = \eta_\mathrm{f} \cdot \eta_\mathrm{s} \cdot \eta_\mathrm{A}\,, \tag{2.20}$$

where

η_f: RF feed efficiency (feed losses)
η_s: radiowave efficiency (≈ 1 for large reflectors)
η_A: aperture efficiency

The η_A may be expressed as a function of the following factors:

$$\eta_\mathrm{A} = \eta_\mathrm{i} \cdot \eta_\mathrm{p} \cdot \eta_\mathrm{x}\,, \tag{2.21}$$

where

η_i: illumination loss
η_p: phase-error loss
η_x: cross polarization loss

For aperture antenna with parabolic reflector, the overall efficiency is in the range of 0.5–0.7 with average value of 0.55.

2.5.4 Gain

The gain of an antenna is a significant performance characteristic which is used widely in radio link calculations such as power budget calculation. As defined in the ITU radio regulation RR.166, "A gain of an antenna: The ratio, usually expressed in decibels, of the power required at the input of a loss-free reference antenna to the power supplied to the input of the given antenna to produce, in a given direction, the same field strength or the same power flux density at the same distance. When not specified otherwise, the gain refers to the direction of maximum radiation. The gain may be for a specified polarization."

The gain of an antenna is defined in a manner similar to the directivity, except for the reference power. For the "gain" reference, the antenna input power (P_in) must be considered, while for the "directivity," the radiated power (P_r) must be taken into account. Simply, it should be noted that for the "gain" estimation, antenna efficiency is incorporated, while the "directivity" must be regarded as loss of the efficiency; thus,

$$\begin{aligned} G(\theta,\varphi) &= 4\pi \frac{\text{radiated power per unit solid angle}}{\text{input power}} \\ &= 4\pi \frac{U}{P_\mathrm{in}} = 4\pi \frac{U}{P_\mathrm{r}}\eta \\ &= \eta \cdot D(\theta,\varphi) \end{aligned} \tag{2.22}$$

Although the gain of an antenna can be applied to every direction of its radiation pattern, it is normally referred to the maximum directivity, and when direction is not mentioned, it implies the direction of maximum radiation intensity (maximum directivity).

Antenna gain is defined as "the ratio of the power required at the input of a loss-free reference antenna to the power supplied to the input of the given antenna to produce the same field strength or the same power flux density at the same distance and in the desired direction."

Usually, antenna gain is expressed in decibels and refers to the direction of maximum radiation. The amount of gain is proportional to its dimension compared to the RF wavelength (λ) and design efficiency. In general, antennas being used in higher-frequency bands such as SHF and EHF are high gain, while antennas in LF/MF/HF bands are normally medium-/low-gain types.

A variety of antennas are employed in radio networks among which the following types are more common:

- T-type, inverted L, conical, biconical, rhombic, and log-periodic used in LF, MF, and HF radio communications and AM broadcasting.
- Whip, collinear, Yagi, and corner reflector used in VHF/UHF radio systems and FM and TV broadcasting.
- Horn and panel antennas in UHF/SHF bands used in LOS radio links.
- Directional antennas of high gain such as parabolic or Cassegrain used in microwave and satellite telecommunications.
- Special antennas for specific applications such as radio navigations, GPS, NDB, position locating systems.

Antenna gain, depending on the selected reference antenna, may be expressed in different ways as follows:

1. When reference antenna is an isotropic antenna isolated in free space, then gain is denoted by G_i and known as isotropic or absolute gain.
2. When reference antenna is a half-wave dipole isolated in free space whose equatorial plane contains the given direction, then the gain is denoted by G_d and known as dipole-related gain. Between dipole and isotropic gains for a given antenna, the following relation exists:

$$G[\mathrm{dB}_i] = G[\mathrm{dB}_d] + 2.15 \tag{2.23}$$

3. When the reference antenna is a Hertzian dipole or short vertical conductor (monopole) much shorter than one-quarter of the RF wavelength, and normal to the surface of perfectly conducting plane containing the given direction, then the gain is denoted by G_v and known as short vertical/hertzian dipole-related gain. Between G_v and G_i of a given antenna, the following relation exists:

$$G[\mathrm{dB}_i] = G[\mathrm{dB}_v] + 4.8 \tag{2.24}$$

General applications of the mentioned gains are:

1. G_{i} at UHF, SHF, and EHF bands
2. G_{d} at VHF and UHF bands
3. G_{v} at LF and MF bands

Example 2.7. In a radio link of 40-km length and working at 7.5 GHz, 60% of free-space loss is compensated by using high-gain TX and RX antennas.

1. How much is the received signal level (RSL) at the output of RX antenna with 1-W TX output power and considering 15-dB additional losses?
2. Find fade margin of the link if RX threshold is to be $P_{\mathrm{th}} = -78\ \mathrm{dB}_m$.

Solution. 1.

$$\mathrm{FSL} = 92.4 + 20\ \log f \cdot d = 141\ \mathrm{dB}$$
$$P_{\mathrm{t}} = 1\,W \quad \Rightarrow \quad P_{\mathrm{t}}[\mathrm{dB}_m] = 30\ \mathrm{dB}_m$$
$$P_{\mathrm{r}} = \mathrm{RSL} = P_{\mathrm{t}}[\mathrm{dB}_m] - 0.4\ \mathrm{FSL} - L_{\mathrm{a}}$$
$$= 30 - 56.8 - 15 = -41.4\ \mathrm{dB}_m$$

2.

$$\mathrm{FM} = P_{\mathrm{r}}[\mathrm{dB}_m] - P_{\mathrm{th}}[\mathrm{db}_m] = -41.4 + 78 = 36.6\ \mathrm{dB}$$ ■

Example 2.8. The maximum directivity of a parabolic antenna is $D_{\max} = (4\pi^2R^2)/\lambda^2$, where λ is the wavelength of radiated radiowaves and R is reflector radius (both with the same unit); assuming $R = 1.2\,\mathrm{m}$, $f = 7.5\,\mathrm{GHz}$, find:

1. Antenna gain if $\eta = 60\%$.
2. EIRP when the antenna is connected to a 500-mW transmitter through RF feeder with 3-dB loss.
3. Maximum radiated power from a side lobe where its peak is 15 dB less than the peak of the main lobe.

Solution. 1.

$$f = 7.5\ \mathrm{GHz} \quad \Rightarrow \quad \lambda = 0.04\,\mathrm{m}$$
$$D_{\max} = \frac{\pi^2 \cdot (2.4)^2}{(0.04)^2} \quad \Rightarrow \quad D_{\max} = (60\pi)^2$$
$$g = \eta \cdot D_{\max} \quad \Rightarrow \quad g = 0.6(60\pi)^2$$
$$G[\mathrm{dB}_i] = 10\ \log\ g \approx 43.3\ \mathrm{dB}_i$$

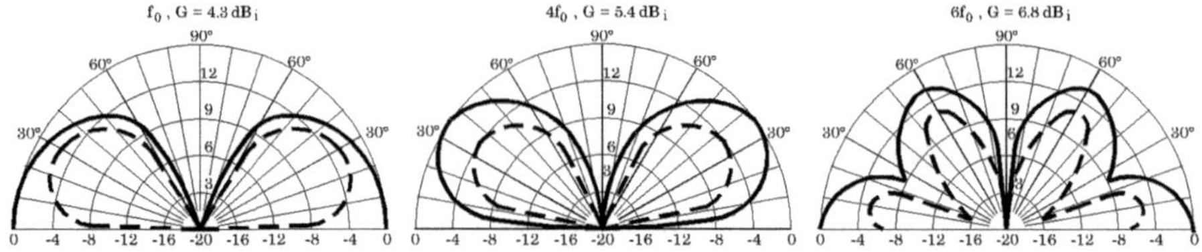

Fig. 2.9 Radiation pattern of horizontal log-periodic antenna versus frequency

2.

$$500\,\text{mW} \quad \Rightarrow \quad P_t = 100 \log 500 = 27 \text{ dB}_m$$

$$\text{EIRP} = 27 + 43.3 - 3 = 67.3 \text{ dB}_m$$

3.

$$\text{EIRP}_s = P_t + (G_t - 15) - L = 52.3 \text{ dB}_m$$

■

2.5.5 Bandwidth

In practical applications, every antenna shall maintain its performance within operating frequency range. The bandwidth of an antenna is defined as the frequency range over which its performance conforms to a specific criterion.

The bandwidth of antenna usually refers to some characteristic (such as pattern, gain, beamwidth, polarization, impedance) within the range of frequencies on either side of the center frequency. Due to variations of antenna parameters with respect to frequency, in most cases, a unique characterization of the bandwidth is not possible.

Usually there is a distinction between pattern and impedance variations with frequency reflecting in the useful bandwidth. The pattern bandwidth deals with gain, side lobe level, beamwidth, polarization, and beam direction, while the impedance bandwidth and radiation efficiency are related to the antenna input impedance.

As an example, radiation patterns of an MF/HF horizontal log-periodic antenna with center frequency adjusted at f_0 are given in Fig. 2.9, for $4f_0$ and $6f_0$. This antenna is considered as broadband with respect to its gain, while it is a narrowband in terms of radiation pattern or directivity.

2.5.6 Polarization

2.5.6.1 Definition of Polarization

Polarization is a characteristic of radiowaves which describes time-dependent variations of electric field vector $\bar{E}$ at a point in the space. To specify the polarization of a radiowave, it is not necessary to define the magnetic field vector $\bar{H}$ separately because for a plane wave in homogeneous isotropic medium, $\bar{E}$ and $\bar{H}$ are perpendicular to each other and the ratio of their magnitudes is equal to intrinsic impedance of the medium. The main types of radiowaves polarization are:

1. If $\bar{E}$ is continually horizontal, its polarization is called horizontal.
2. If $\bar{E}$ is continually perpendicular to the horizontal plane, its polarization is called vertical.
3. If the projection of the head of $\bar{E}$ on a plane perpendicular to radiowave propagation direction at any point in the space follows a circle, its polarization is circular. In the case of rotation by time (for an observer looking in the direction of propagation) in clockwise (CW) direction, it is known as right-hand circular polarization (RHCP). Similarly if the specified rotation is counterclockwise (CCW), it is known as left-hand circular polarization (LHCP).
4. If the projection of the head of $\bar{E}$ on a plane perpendicular to radiowave propagation direction at any point in the space follows an ellipse, its polarization is called elliptical. In the case of rotation by time (for an observer looking in the direction of propagation) in clockwise (CW) direction, it is known as right-hand elliptical polarization (RHEP). Similarly if the specified rotation is CCW, it is known as left-hand elliptical polarization (LHEP).

The polarization of an antenna is the orientation of the electric field (E plane) of the radiated waves with respect to the Earth's surface and is determined by the physical structure of the antenna and its orientation. A simple straight wire antenna will have a specific polarization when mounted vertically and a different polarization when mounted horizontally. "Electromagnetic wave polarization filters" are structures which can be employed to act directly on the electromagnetic wave to filter out wave energy of an undesired polarization and to pass through wave energy of a desired polarization.

Reflections affect polarization generally. For radiowaves the most important reflector is the ionosphere which the reflected waves from it will have their polarization changed unpredictably. For signals which are reflected by the ionosphere, polarization cannot be relied upon. For line-of-sight communications for which polarization can be relied upon, it can make a large difference in signal quality to have the transmitter and receiver using the same polarization.

Polarization is largely predictable from antenna structure, but especially in directional antennas, the polarization of side lobes can be quite different from that of the main propagation lobe. For radio antennas, the polarization corresponds to the orientation of the radiating element in an antenna. For example, a vertical omnidirectional WiFi antenna will have vertical polarization.

2.5.6.2 Main Types of Radiowaves Polarization

In an overall classification, radiowave polarizations are divided into linear and nonlinear types. As shown in Fig. 2.10, linear types include the following three groups:

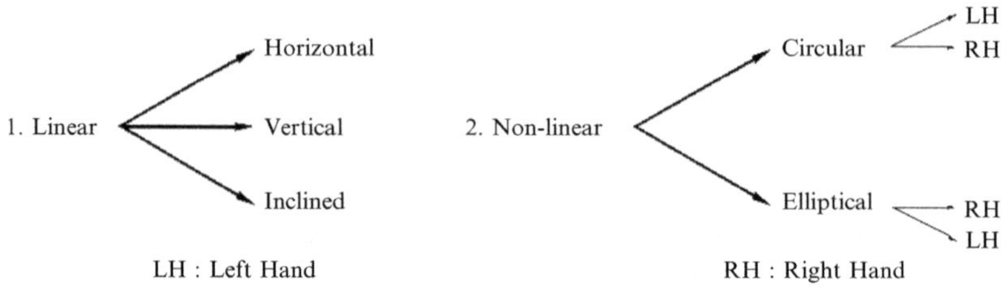

Fig. 2.10 Basic types of polarization

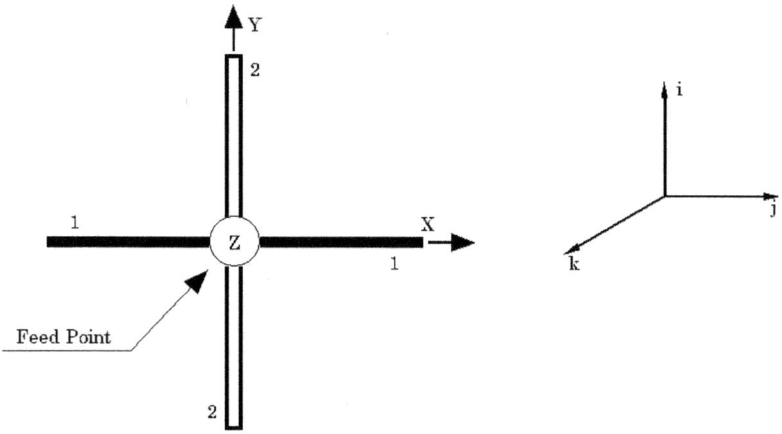

Fig. 2.11 The basic concept of generating polarized waves

- Horizontal
- Vertical
- Inclined

Also, nonlinear polarizations mainly consist of the following groups:

- RHCP
- LHCP
- RHEP
- LHEP

2.5.6.3 Basic Polarized Radiowaves

To produce basic polarized radiowaves, a simple practical way is to employ two orthogonal dipole antennas. As shown in Fig. 2.11, antenna 1 is a horizontal dipole, and antenna 2 is a vertical dipole.

By proper feeding of the antennas, the following radiowaves will be generated:

$$\bar{E}_1 = E_x \times \hat{\alpha}_x = E_A \sin(\omega t - \beta z) \cdot \hat{\alpha}_x \tag{2.25}$$

$$\bar{E}_2 = E_y \times \hat{\alpha}_y = E_B \sin(\omega t - \beta z + \bar{\varphi}) \cdot \hat{\alpha}_y \tag{2.26}$$

In the plane $Z = 0$, the above relations result in

$$Z = 0 \Rightarrow \left| \begin{array}{l} \bar{E}_1 = E_A \sin(\omega t) \cdot \hat{\alpha}_x \\ \bar{E}_2 = E_B \sin(\omega t + \phi) \cdot \hat{\alpha}_y \end{array} \right. \Rightarrow \left| \begin{array}{l} E_X = E_A \sin(\omega t) \\ E_Y = E_B \sin(\omega t + \phi) \end{array} \right. \tag{2.27}$$

$$\Rightarrow \sin \omega t = \frac{E_X}{E_A}, \quad \cos \omega t = \sqrt{1 - (\frac{E_X}{E_A})^2} \tag{2.28}$$

$$E_Y = E_B \sin \omega t \ \cos \phi + E_B \cos \omega t \ \sin \phi \tag{2.29}$$

$$\frac{E_Y}{E_B} = \frac{E_X}{E_A} \cos \phi + \sqrt{1 - (\frac{E_X}{E_A})^2} \cdot \sin \phi \tag{2.30}$$

By squaring the relation (2.30) and simple manipulations, we have

$$\frac{E_Y^2}{E_B^2} + \frac{E_X^2}{E_A^2} - 2\frac{E_X}{E_A} \times \frac{E_Y}{E_B} \cos \phi = \sin^2 \phi, \quad \text{or} \tag{2.31}$$

$$E_Y^2 \left(\frac{1}{E_B^2 \sin^2 \phi} \right) + E_X^2 \left(\frac{1}{E_A^2 \sin^2 \phi} \right) - 2E_X E_Y \frac{\cos \phi}{E_A \cdot E_B \sin^2 \phi} = 1 \tag{2.32}$$

The last expression is in the general form of $aE_X^2 + bE_Y^2 - cE_X E_Y = 1$ from which basic polarized radiowaves may be generated for various values of a, b, and c, as follows:

1. For horizontal polarization, only dipole 1 shall be fed, that is, $E_A \neq 0, E_B = 0$.
2. For vertical polarization, only dipole 2 shall be fed, that is, $E_A = 0, E_B \neq 0$.
3. For inclined linear polarization, both dipoles shall be fed in phase with a fixed amplitude ratio, that is, $\Phi = 0$ and $\frac{E_A}{E_B} = cte.$
4. For circular polarization, both dipoles shall be fed by equal amplitudes and $\pm 90°$ phase difference.
5. If none of the above conditions are met, the polarization would be elliptical.

Example 2.9. Find polarization of a uniform plane radiowave with magnetic field expressed by $\bar{H} = 10^{-6}(\hat{\alpha}_Y + j\hat{\alpha}_Z)\cos \beta x$ A/m

Solution. Propagation of the specified radiowave is in the x-axis direction, and due to its nature, only E_y and E_z components exist. To determine the type of polarization, E-field variations versus time should be studied in a plane perpendicular to the propagation direction, say $x = 0$:

$$\bar{E} = -\eta \hat{\alpha}_x \times \bar{H} = 10^{-6} \times 120\pi[\hat{\alpha}_z - j\hat{\alpha}_y] \cos \beta x \quad \text{V/m}$$

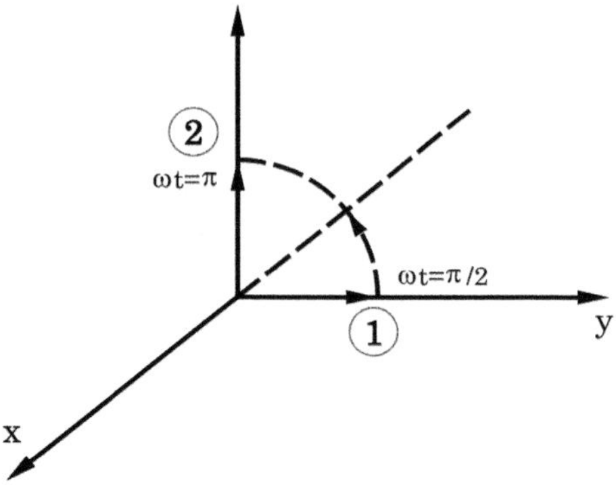

Fig. 2.12 Rotation of $\bar{E}$ with time ($\omega t = \pi/2$ and $\omega t = \pi$)

$\bar{E}(x,t)$ is

$$\bar{E}(x,t) = -10^{-6} \times 120\pi[\hat{\alpha}_z \cos(\omega t - \beta x) - \hat{\alpha}_y \sin(\omega t - \beta x)]$$

$$\text{At plane } x = 0 \ \Rightarrow \ \left| \begin{array}{l} E_Z = -10^{-6} \times 120\pi \cos \omega t \\ E_Y = 10^{-6} \times 120\pi \sin \omega t \end{array} \right.$$

$$\Rightarrow \quad (E_Y)^2 + (E_Z)^2 = (10^{-6} \times 120\pi)^2 = R^2$$

It is clear that the polarization is circular. To fix rotation of $\bar{E}$ with time, its position shall be investigated by a reference observer. By using Fig. 2.12, and fixing the $\bar{E}$ positions at say $\omega t = \pi/2$ and $\omega t = \pi$, respectively, direction of radiowave rotation can be determined.

In Fig. 2.12, $\bar{E}$ are shown by **1** for $\omega t = \pi/2$ and by **2** for $\omega t = \pi$; thus, the rotation of $\bar{E}$ for reference observer is clockwise, and it is concluded that the radiowave polarization is RHCP. ■

Example 2.10. Determine polarization of a radiowave expressed by $\bar{E} = (A\hat{\alpha}_X + jB\hat{\alpha}_Y)e^{-j\beta z}$.

Solution. Propagation is in the Z-axis direction, and at plane $Z = 0$, the $\bar{E}$ field components are

$$\left| \begin{array}{l} E_X = A \cos \omega t \\ E_Y = B \sin \omega t \end{array} \right. \Rightarrow \left(\frac{E_X}{A}\right)^2 + \left(\frac{E_Y}{B}\right)^2 = 1$$

Polarization of the radiowave will change based on A and B conditions as follows:

$A = 0, B \neq 0$ vertical (linear)
$A \neq 0, B = 0$ horizontal (linear)
$A = B \neq 0$ circular
$A \neq B \neq 0$ elliptical

To determine rotation of $\bar{E}$, its position at two different times, say $t = 0$ and $t = \pi/2$, shall be studied:

$$\left| \begin{array}{lll} \omega t = 0 & \Rightarrow & E_X = A,\ E_Y = 0 \\ \omega t = \pi/2 & \Rightarrow & E_X = 0,\ E_Y = B \end{array} \right.$$

$A > 0,\ B > 0 \Rightarrow$ CW, right hand
$A > 0,\ B < 0 \Rightarrow$ CCW, left hand
$A < 0,\ B > 0 \Rightarrow$ CCW, left hand
$A < 0,\ B < 0 \Rightarrow$ CW, right hand ∎

2.5.6.4 Polarization Loss

Polarization mismatch between incident radiowaves and the related receiving antenna generates additional loss which is referred as polarization loss. This means that the amount of power extracted by the antenna from the incoming signal will not be maximum because of the polarization loss.

The effect of the polarization mismatch is expressed by the polarization loss factor (PLF). Assuming the electric field of incident wave and receiving antenna by

$$\bar{E}_{\mathrm{i}} = \hat{\rho}_{\mathrm{i}} \cdot E_{\mathrm{i}} \tag{2.33}$$

$$\bar{E}_{\mathrm{a}} = \hat{\rho}_{\mathrm{a}} \cdot E_{\mathrm{a}} \tag{2.34}$$

$\hat{\rho}_{\mathrm{i}}$ and $\hat{\rho}_{\mathrm{a}}$ are unit vectors of the incident wave and receiving antenna, then

$$\mathrm{PLF} = |\hat{\rho}_{\mathrm{i}} \cdot \hat{\rho}_{\mathrm{a}}|^2 = |\cos\ \psi_p|^2 \tag{2.35}$$

where ψ_p is the angle between the two unit vectors. The PLF in the logarithmic scale can be expressed as follows:

$$\mathrm{PLF} = 20\ \log\ |\cos\ \psi_p| \tag{2.36}$$

Example 2.11. The electric field of a linearly polarized electromagnetic wave is expressed by $\bar{E}_{\mathrm{i}} = \hat{a}_x\ E_0\ \mathrm{e}^{-\mathrm{j}kz}$. In the case of receiving the waves by a linearly polarized antenna specified by $\bar{E}_{\mathrm{a}} = (\hat{a}_x + \hat{a}_y)\ E_1$, find PLF:

Solution.

$$\hat{\rho}_{\mathrm{i}} = \hat{a}_x \quad , \quad \hat{\rho}_{\mathrm{a}} = \frac{\hat{a}_x + \hat{a}_y}{\sqrt{2}}$$

$$\cos\ \psi_p = \hat{\rho}_{\mathrm{i}} \cdot \hat{\rho}_{\mathrm{a}} = \frac{1}{\sqrt{2}}$$

$$\Rightarrow \mathrm{PLF} = 20\ \log\ |\cos\ \psi_p| = -3\ \mathrm{dB}$$ ∎

2.5.7 *Input Impedance*

Electromagnetic waves travel through the different parts of the transmission system including radio, RF feeder, antenna, and free space. It may encounter differences in impedance (E/H, V/I, etc.). At each interface, depending on the impedance mismatch, some fraction of the wave energy will be reflected back to the source, forming a standing wave in the feed line.

The ratio of maximum power to minimum power in the wave is called the standing wave ratio (SWR). An SWR of 1:1 is ideal, and the ratio of 1.5:1 is considered to be marginally acceptable in low-power applications where power loss is more critical, although an SWR as high as 6:1 may still be usable with the right equipment. Minimizing impedance differences at each interface (impedance matching) reduces the SWR and maximizes power transfer through each part of the antenna system.

Complex impedance of an antenna is related to the electrical length of the antenna at the wavelength in use. The impedance of an antenna can be matched to the feed line and the radio by adjusting the impedance of the feed line (using the feed line as an impedance transformer). More commonly, the impedance is adjusted at the load with an antenna tuner, a balun, a matching transformer, matching networks composed of inductors and capacitors, or matching sections such as the gamma match.

2.5.8 *Power Handling Capability*

One of the antenna characteristic parameters is its power handling capability which is specific to the transmitting antennas. Although this parameter is not critical for low-power radiation of radiowaves, it is a crucial factor in the radiation of high-power radio units. For example, when a high-power radar with 250-kW output is connected to a rotatable directional antenna with 80% efficiency, it means that 50 kW of radio power will be lost in the antenna system which may damage its sensitive or weak parts.

With respect to the above fact, in selecting of the transmitter antenna, the power handling requirements shall be specified clearly. This is an outstanding issue for antennas in the broadcasting, radar, MF/HF radio units, and high-power satellite ground stations.

2.5.9 *Effective Area*

The effective area or effective aperture of a receiving antenna indicates the portion of the power of an electromagnetic wave which can be delivered to its terminals, expressed in terms of an equivalent area. For instance, if a radiowave passing through a given location has a flux of 1 pW/m^2 (10^{-12} W/m^2) and an antenna has an effective area of 12 m^2, then the antenna would deliver 12 pW of RF power to the receiver input (30 μV rms at 75 ohms). Since the receiving antenna is not equally sensitive to signals received from all directions, the effective area is a function of the direction to the source.

Due to the reciprocity theorem, the gain of an antenna used for transmitting must be proportional to its effective area when used for receiving. Consider an antenna with no loss, that is, 100% efficiency. It can be shown that its effective area, averaged over all directions, must have an efficiency equal to $\lambda^2/4\pi$. Gain is defined in such a way that the average gain over all directions for an antenna with 100% electrical efficiency is equal to 1. Therefore, the effective area denoted by A_e in terms of the gain G in a given direction is stated as

$$A_e = \frac{\lambda^2}{4\pi} G \tag{2.37}$$

For an actual antenna with an efficiency of less than 100%, both the effective area and gain are reduced by the same amount. Therefore, the above relationship between gain and effective area is still valid; that is, there are two different ways of expressing the same quantity. A_e is especially convenient when computing the power that would be received by an antenna of a specified gain, as illustrated by the above example.

2.5.10 Antenna Noise Temperature and G/T

The receiving antennas intercept some radio noises along with the desired signal. To differentiate the desired signal, it is required that the received signal be greater than the noise level to some extent, say 10 dB; otherwise, the desired signal cannot be detected properly. This means that in some types of radio communications, specially satellite communications, quality of reception is a key factor, where intercepted radio noises are considerable and may affect the performance of the received signals.

2.5.10.1 Receiving G/T

The G/T is a measure to express the merit of receiving system for detecting the main signal. This parameter is defined in the logarithmic scale as

$$G/T = G[\mathrm{dB}_i] - T_s[\mathrm{dB}_K] \tag{2.38}$$

where

G: gain of receiving antenna (dB$_i$)
T_s: receiving system noise temperature (°K)

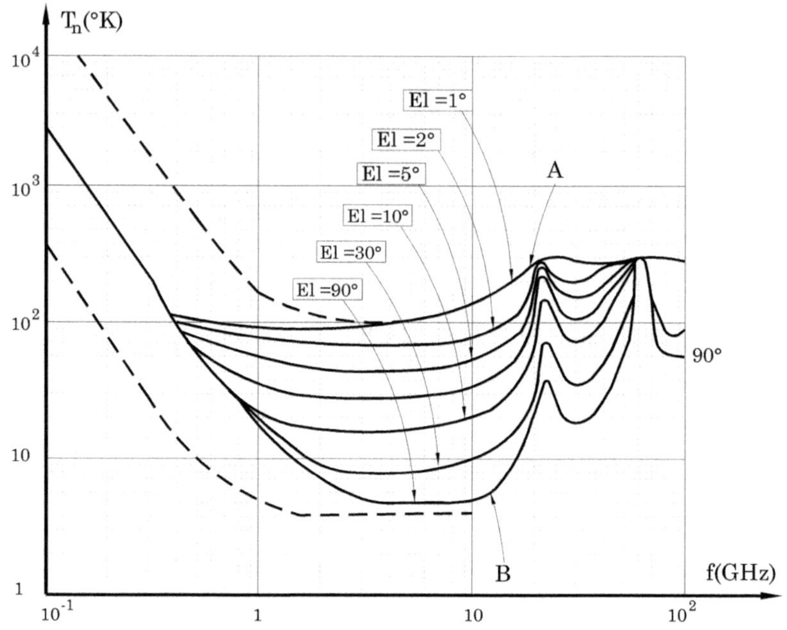

Fig. 2.13 Sky noise vs. frequency and antenna elevation angle

Example 2.12. A ground station antenna with 56-dB$_i$ gain is directed toward GEO satellite with elevation angle of $\theta = 10^\circ$; find:

1. Sky-noise temperature at C band using the Fig. 2.13.
2. The G/T of the ground station receiving system in case of 150°K total system noise temperature

Solution. 1.

$$\text{C} - \text{band} \implies 4\text{ GHz} < f < 6\text{ GHz}$$

$$E_l = \theta = 10^\circ \overset{\text{Fig. (2.13)}}{\Longrightarrow} T_\text{A} \approx 20\ ^\circ\text{K}$$

2.

$$G/T = G[\text{dB}_i] - T_\text{s}[\text{dB}_K] = 56 - 10 \log 150 = 34.24 \text{ dB}/K$$ ■

2.5.10.2 Antenna Noise Temperature

An object with a physical temperature radiates energy which is usually expressed by an equivalent noise temperature named brightness temperature and denoted by T_B. The brightness temperature is defined as

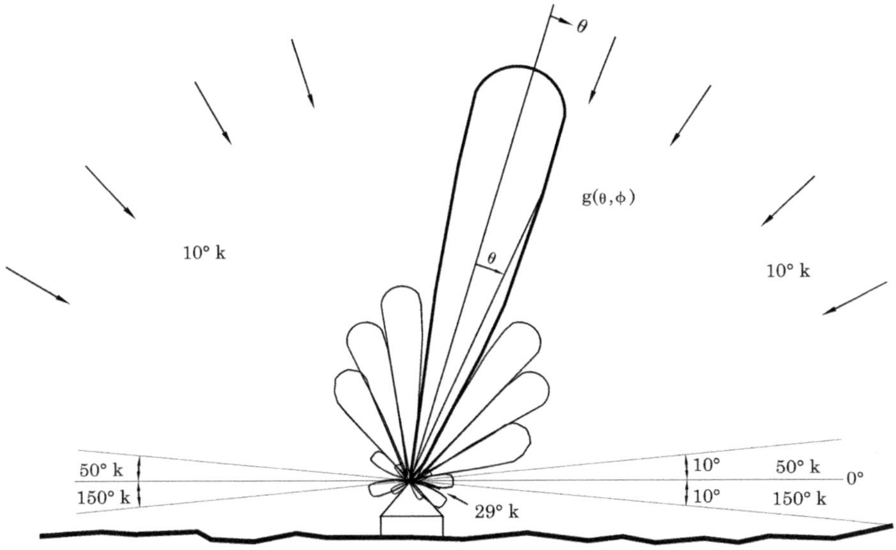

Fig. 2.14 Sky and ground equivalent noise temperature vs. satellite antenna elevation angle

$$T_B(\theta,\varphi) = \varepsilon(\theta,\varphi)T_m = (1-|R|^2)T_m \tag{2.39}$$

where

T_B: brightness temperature in °K as a function of θ and φ
ε: emissivity coefficient ($0 \sim 1$)
T_m: molecular temperature in °K
R: reflection coefficient of the surface for radiowave polarization as a function of θ and φ

Emissivity coefficient of each body like an antenna is in the range of zero to one. This means that thc maximum value of T_B is equal to T_m. Usually the emissivity is a function of frequency and polarization of the radiowaves in interest and the molecular structure of the body as well.

Sky noise as indicated in Fig. 2.13 is composed of components emitted by different sources. This noise is intercepted by antenna and appears at its terminal as antenna noise temperature. The amount of sky noise depends on frequency and antenna elevation angle.

In addition to sky noise, another natural source of radio noise is the ground surface. As depicted in Fig. 2.14, some of natural noise emitters in the satellite C, Ku, and K_a frequency bands are:

- Ground surface with equivalent noise temperature of about 290°K.
- Sky with equivalent noise temperature of about 5°K when looking toward zenith ($\theta \approx 90°$).

- Sky with equivalent noise temperature of about 100 ~ 150°K when looking toward horizon ($\theta < 10°$).

The noise temperature appearing at a receiving antenna output after being weighted by the antenna gain pattern is given by the following equation:

$$T_{\mathrm{A}} = \frac{\int_0^{2\pi} \int_0^{\pi} T_{\mathrm{B}}(\theta,\varphi)\, G(\theta,\varphi)\, d\Omega}{\int_0^{2\pi} \int_0^{\pi} G(\theta,\varphi)\, d\Omega}, \text{ or} \tag{2.40}$$

$$T_{\mathrm{A}} = \frac{\int_0^{2\pi} \int_0^{\pi} T_{\mathrm{B}}(\theta,\varphi)\, G(\theta,\varphi)\, \sin\theta \cdot d\theta \cdot d\varphi}{\int_0^{2\pi} \int_0^{\pi} G(\theta,\varphi)\, \sin\theta \cdot d\theta \cdot d\varphi} \tag{2.41}$$

where

T_{A}: antenna equivalent noise temperature in °K
G: antenna gain pattern

2.5.10.3 Antenna Noise Power

When the transmission line is considered lossless, antenna noise power is expressed by

$$P_{\mathrm{n}} = k \cdot T_{\mathrm{A}} \cdot \Delta f \tag{2.42}$$

where

P_{n}: antenna noise power (W)
k: Boltzmann constant $(1.38 \times 10^{-23}\,\mathrm{J/°K})$
T_{A}: antenna noise temperature °K
Δf: bandwidth (Hz)

When an antenna and its related transmission line are at a certain physical temperature including some losses, the antenna noise temperature, expressed by (2.42), cannot be used and a modified formula should be taken into account.

2.5.11 Samples of Practical Antennas

Antennas play a significant role in the radio networks, and numerous types of antennas have been designed and fabricated to meet a wide range of applications.

A number of antennas commonly used in the radio stations are introduced in this section. The advanced antenna books are addressed for theoretical discussions, while practical and professional information can be found in the manufacturers' data sheets and manuals.

In the following pages, some pictorial samples of professional antennas are presented. These types of antennas are employed by operating agencies to provide telecommunications services in different fields of radio links. The samples include the following types:

- MF/HF antennas
- VHF/UHF antennas
- SHF/EHF reflective antennas
- Special antennas
- Microstrip antennas
- Basic antennas

Samples of Antenna

MF/HF Band

- *Frequency range*: 300 KHz to 30 MHz
- *Wavelength*: 1,000 m to 10 m
- *Main applications*:
    - Maritime radio services
    - Ground waves in short distance
    - Radio broadcasting
    - Radio navigation
    - Ionospheric hops
- *Types*:
    - Log-periodic, dipole
    - Inverted L, T-type, rhombic
    - Conical, inverted conical, biconical
- *Gain range*: up to $\approx 10 \sim 12$ dB_i (Figs. 2.15 and 2.16)

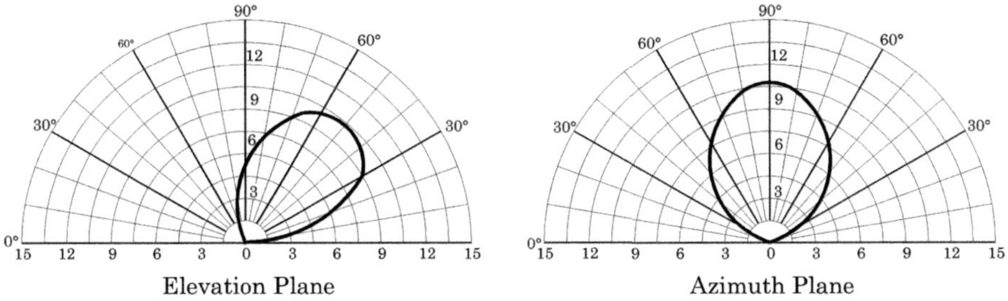

Fig. 2.15 Typical radiation pattern for horizontal log-periodic antenna

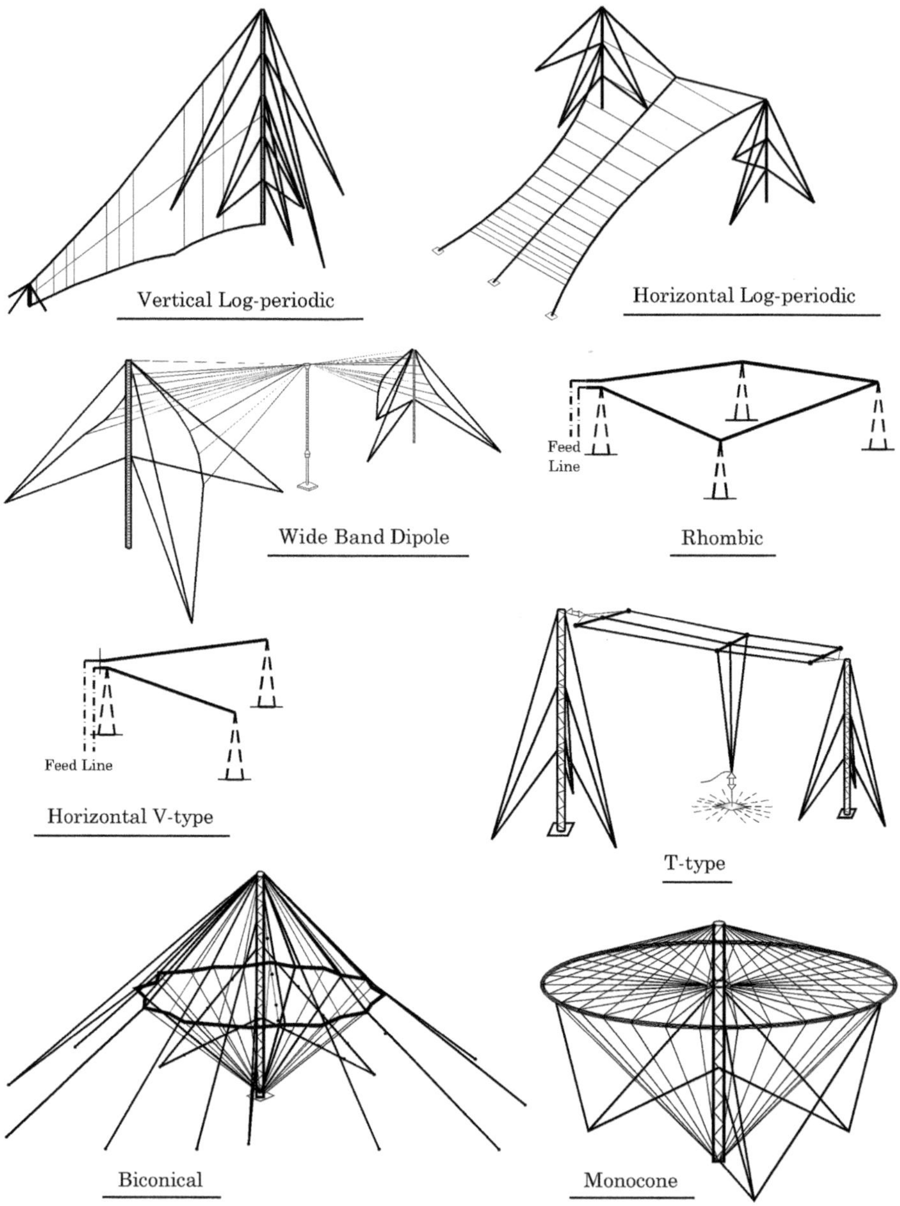

Fig. 2.16 Samples of MF/HF antennas

Samples of Antenna

VHF/UHF Band

- *Frequency band*: 30 MHz $\sim$ 3 GHz
- *Wavelength*: 10 m $\sim$ 10 cm
- *Main applications*:
  - Short/medium-distance communications
  - Audio and video broadcasting
  - Aeronautical and maritime radio services
  - Troposcatters
  - Cellular mobile radio services
  - Radar and radio navigation
  - Private mobile radio services
  - P-P and P-MP LOS radio links
  - LEO and maritime GEO satellite communications
- *Types*:
  - Center/end-fed dipole
  - Collinear
  - Stacked folded dipoles
  - Wideband discones
  - Corner reflectors
  - Yagi antennas
  - UHF sectoral antennas
  - UHF horn and panel directional antennas
  - UHF solid/grid directional parabolic antennas
- *Gain range*:
  - Up to 10 dB_i for VHF
  - Up to 35 dB_i for UHF (Figs. 2.17 and 2.18)

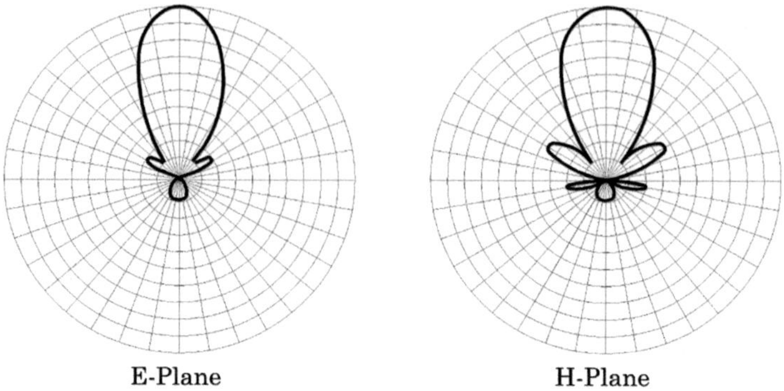

Fig. 2.17 Typical radiation pattern for UHF directional Yagi antenna

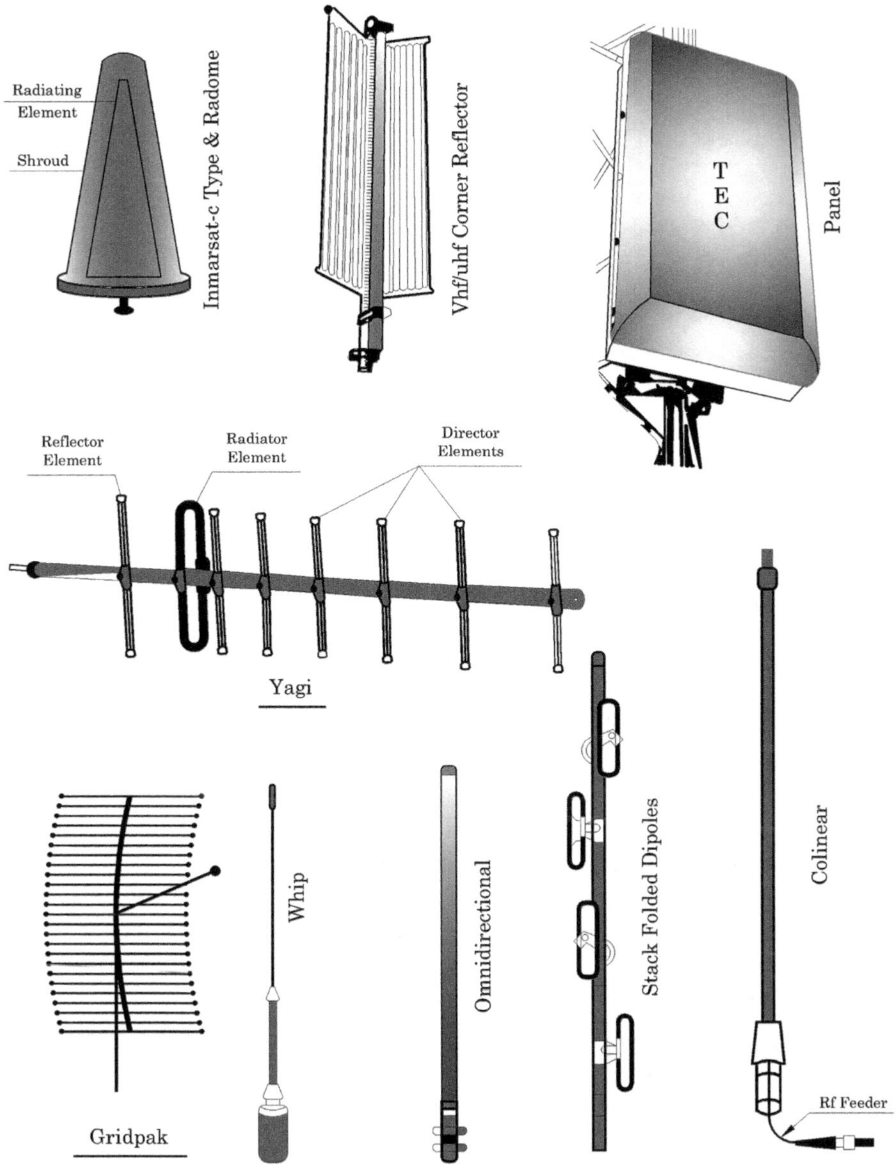

Fig. 2.18 Samples of VHF/UHF antennas

Samples of Antenna

SHF/EHF Reflective Antennas

- *Frequency band*: 3 $\sim$ 300 GHz
- *Wavelength*: 10 cm $\sim$ 100 mm
- *Main applications*:
  - LOS microwave links
  - Fixed and mobile satellite links
  - Radar and military services
  - Over horizon by troposcatter
  - P-P, P-MP, and fixed radio access
  - TV satellite broadcasting
  - Satellite remote sensing
  - Wireless access links
  - WiMAX
- *Types*:
  - Standard, HP, SHP, and UHP parabolic antennas
  - Radar rotating antenna
  - Satellite space/ground antennas
  - Satellite
  - Cassegrain
- *Gain range*: few dB_i (Figs. 2.19–2.24)

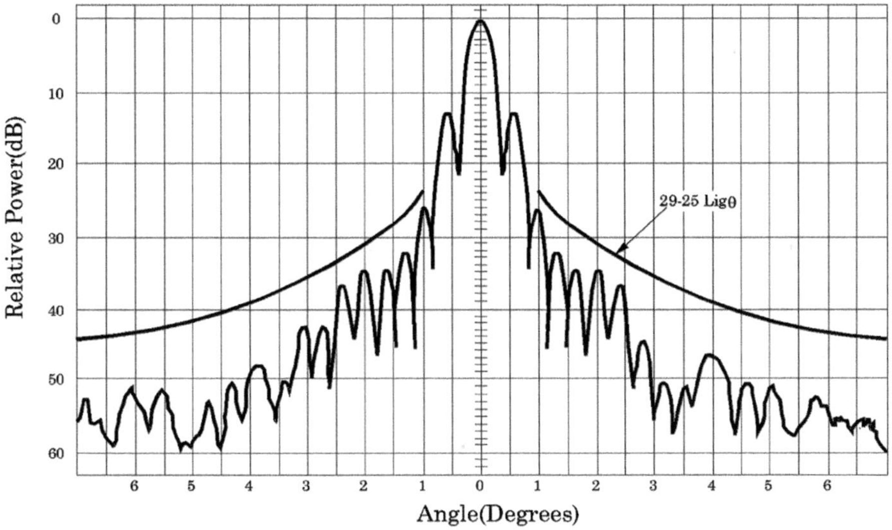

Fig. 2.19 Typical radiation pattern for & satellite Cassegrain antenna

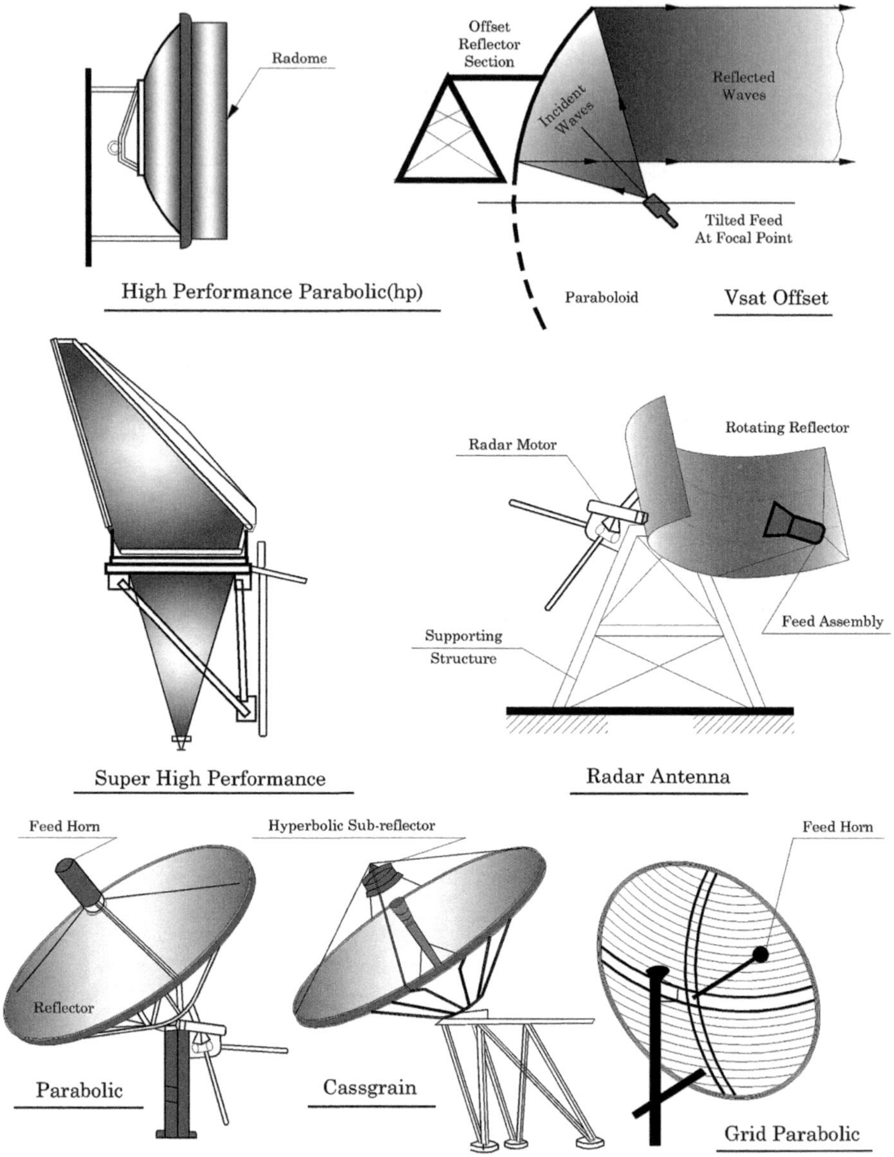

Fig. 2.20 Samples of UHF/SHF/EHF antennas

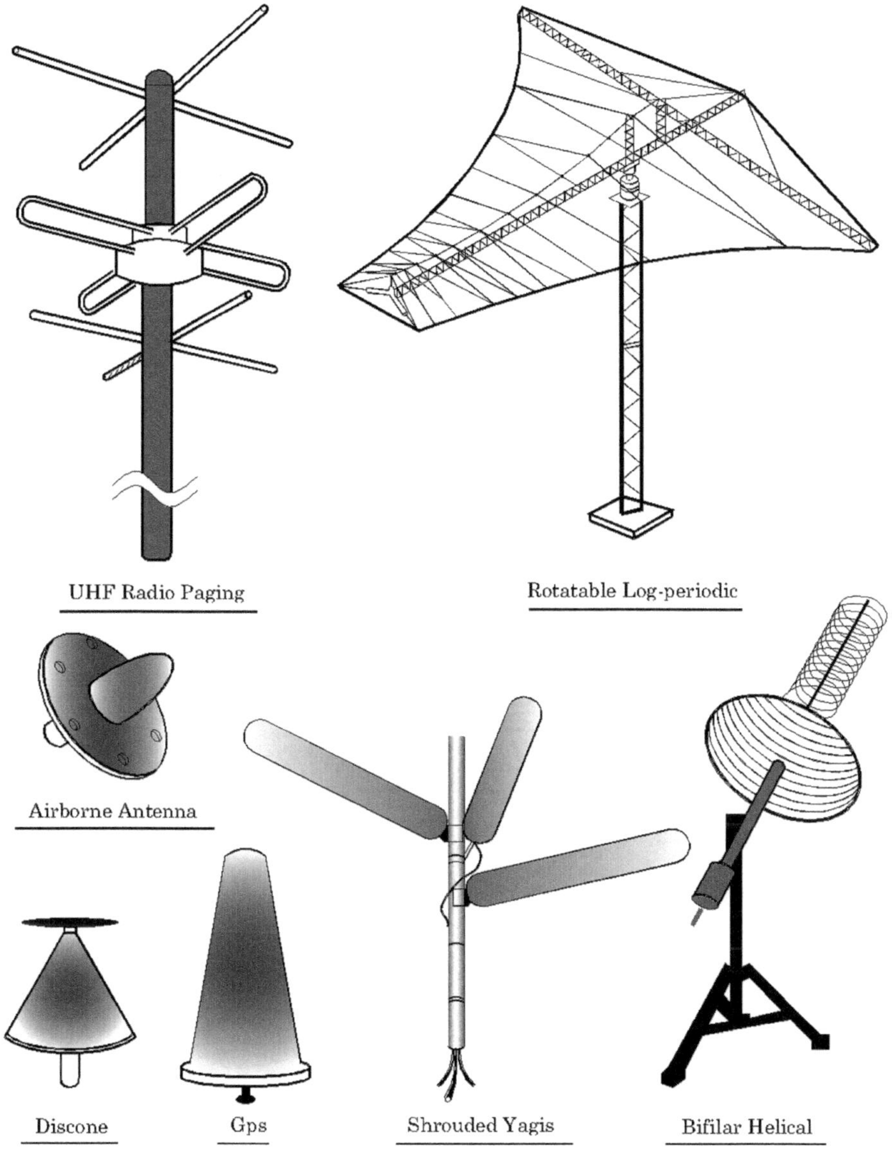

Fig. 2.21 Samples of special antennas/1

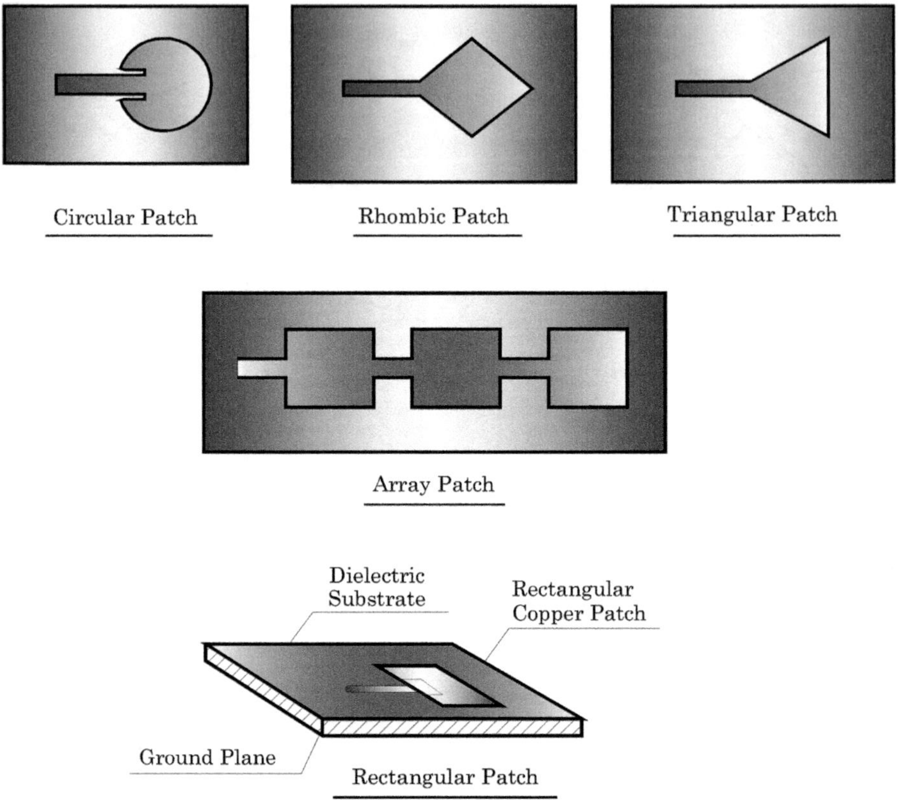

Fig. 2.22 Samples of microstrip antennas

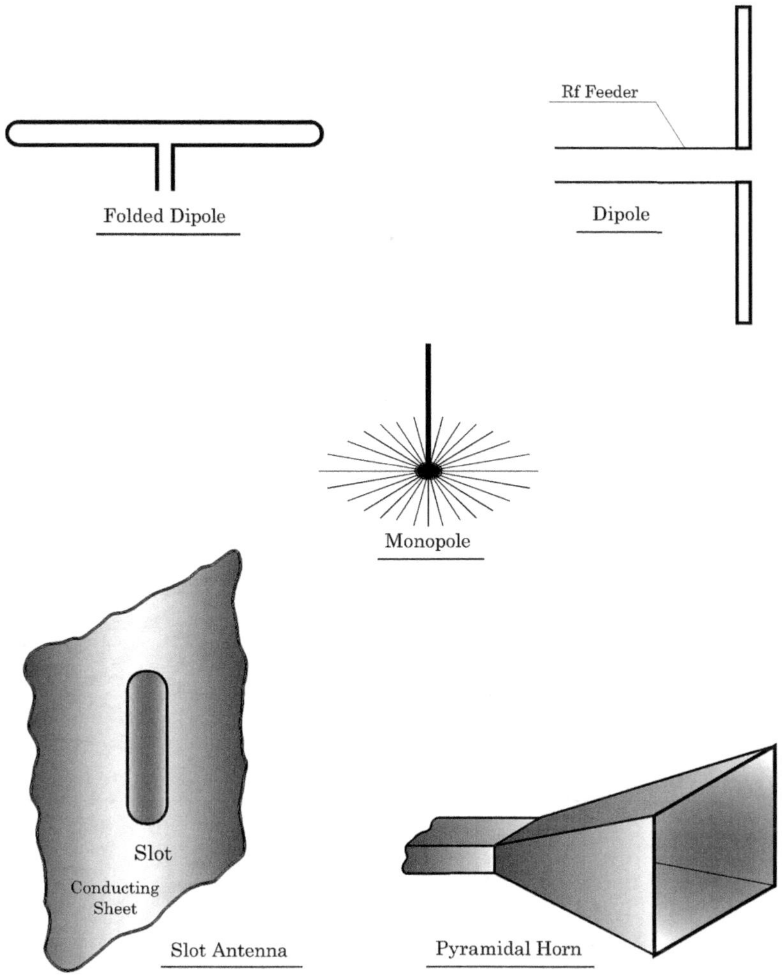

Fig. 2.23 Samples of basic antennas

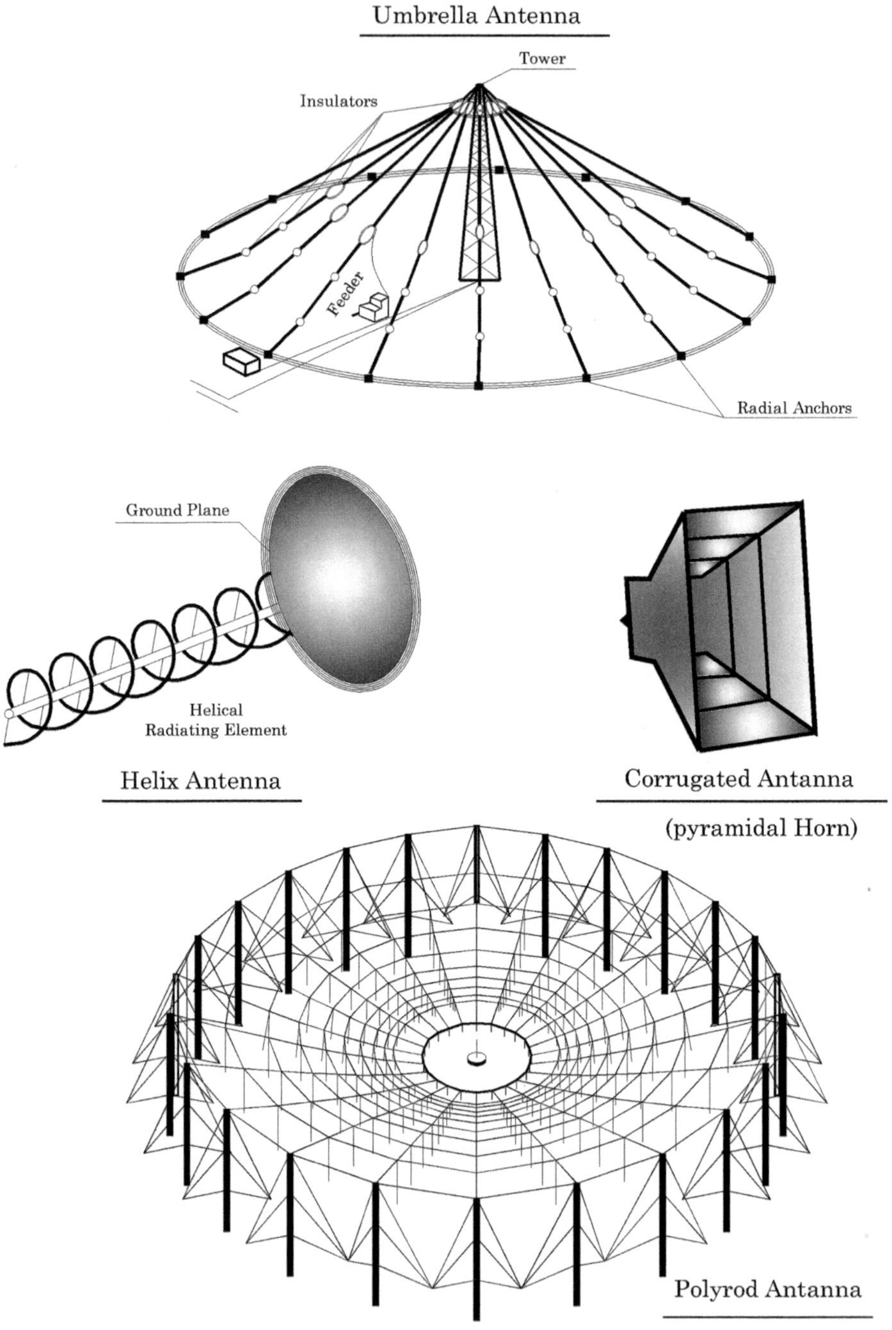

Fig. 2.24 Samples of special antennas/2

2.6 Passive Repeaters

2.6.1 Radio-Relay Network

To establish communications between various stations where the LOS condition is not met, the use of one or more repeaters is a crucial requirement. As shown in Fig. 2.25, the main intention of selecting a repeater is to collect radiowaves from the transmitting station and then redirecting them toward the receiving station.

To meet the requirements for passing the terrain obstructions, it is a normal practice to employ repeaters. In an overall classification, the repeaters are divided into "active" and "passive" types.

This section is devoted to introduce passive repeaters and discuss their applications, configurations, and calculation procedures. Some curves and data provided here are based on the "Passive Repeater Engineering document prepared by Microfleet Co., 1989."

2.6.2 Objectives

The main objectives of using passive repeaters in the radio-relay networks are as follows:

- Passing through natural or artificial obstructions on the radio path
- Redirecting radiowaves toward receiving station
- Increasing the network availability and reliability
- Reducing RF feeder or waveguide length resulting in lower attenuation

2.6.3 Types of Passive Repeaters

The main kinds of passive repeaters as depicted in Fig. 2.26 are:

- Passive reflectors

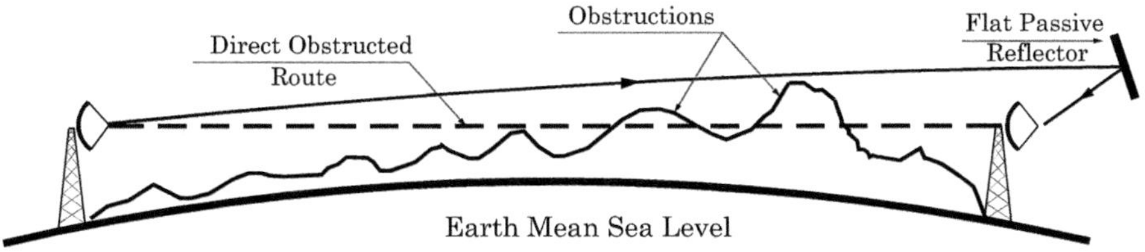

Fig. 2.25 Role of passive repeater in radio links

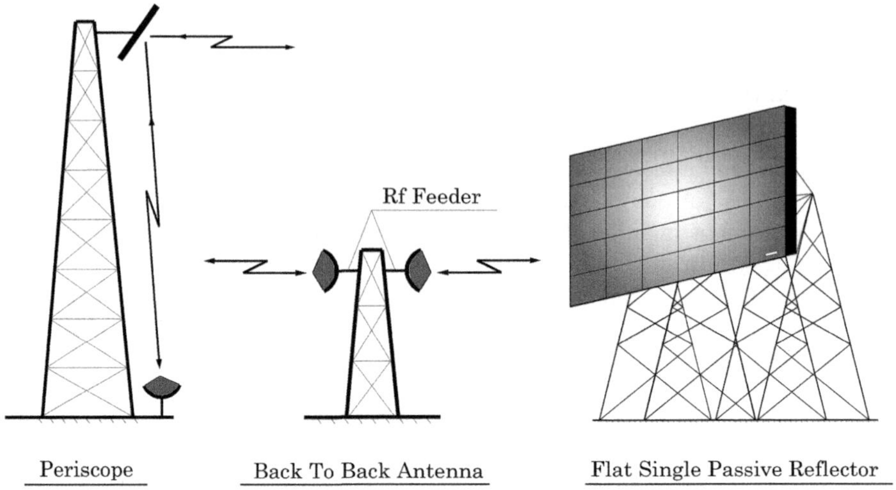

Fig. 2.26 Types of passive repeaters

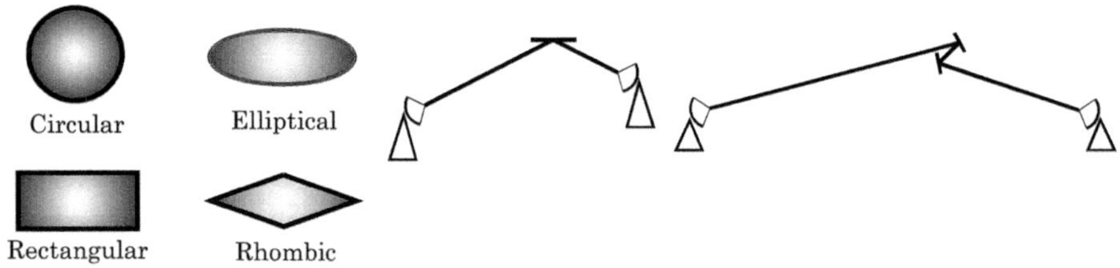

Fig. 2.27 Types of passive reflectors

- Periscopes
- Back-to-back antennas

2.6.4 Types of Passive Reflectors

Usually a passive reflector is referred to its reflector shape such as rectangular, circular, elliptical, and rhombic. Also, as shown in Fig. 2.27, they are further divided into single or double types.

It should be noted that the relative position of a passive reflector to the antenna of an active station is important and may be considered near field (NF) if it meets the relevant criterion; otherwise, it is regarded as far field (FF), for which different methods of calculation are applied.

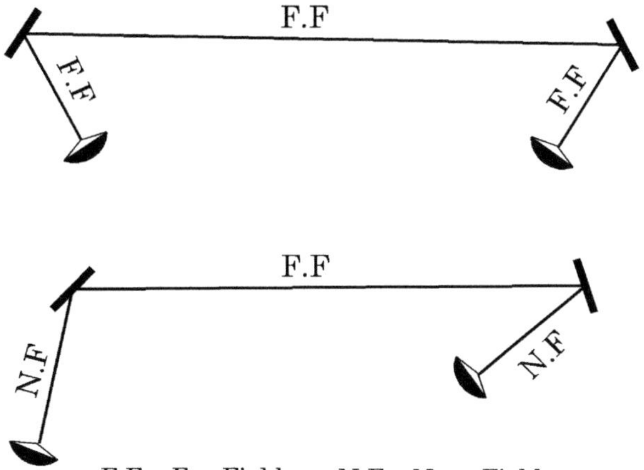

Fig. 2.28 Sample of radio links with single passive repeaters

2.6.5 Features

The outstanding features of the passive repeaters can be listed as follows:

- More lifetime because of using nonactive elements
- Low maintenance resulting in low running cost and more profit
- No need for electric power supply
- No need for access road to reach the passive repeater site
- No need for building and guardroom
- No environmental pollution
- Low initial investment in comparison with the active repeaters

2.6.6 Link Configuration with Passive Repeaters

The passive repeaters can be used in a variety of configurations which mostly depend on the network structure and terrain topology along radio routes. A number of real configurations are illustrated for radio links with single passives in Fig. 2.28 and also for radio links with double passives in Fig. 2.29.

2.7 Gain and Radiation Pattern of Passive Reflectors

The gain and radiation pattern of the passive reflectors may be analyzed by applying different calculation procedures based on distinguished conceptual approaches. The readers are addressed to the professional references for detailed theoretical discussions.

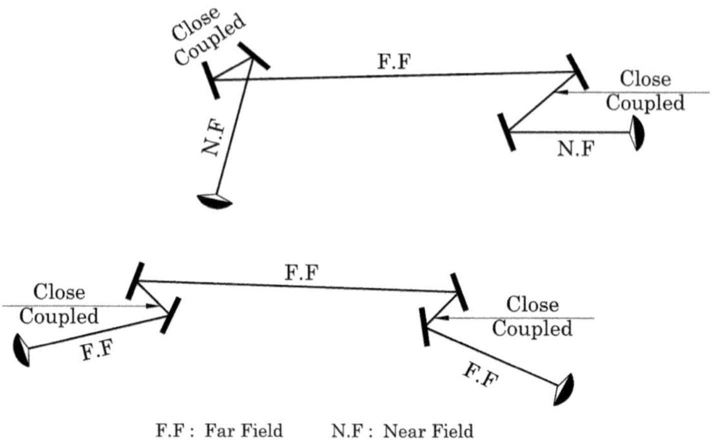

Fig. 2.29 Sample of radio links with double passive repeaters

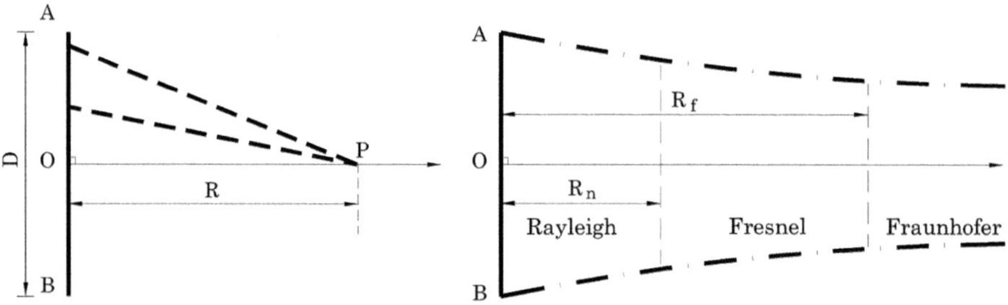

Fig. 2.30 Radiation regions of flat reflector

2.7.1 Near and Far Fields

To study the radiowaves radiation by a parabolic or flat passive reflector, the following three regions with distinct properties are known:

- Far field or Fraunhofer region
- Near field or Fresnel region
- Reflector close zone or Rayleigh region

Figure 2.30 is selected to present the above regions.

The far field is defined as all locations at which the difference between each of them from two distinct points on the reflector surface is not more than $\lambda/16$ (λ is the wavelength of reflected waves). In the worst case and observing Fig. 2.30, we can write

$$AP - OP < \lambda/16 \Rightarrow \sqrt{R^2 + \frac{D^2}{4}} - R < \lambda/16 \tag{2.43}$$

$$D \ll R \Rightarrow \sqrt{R^2 + \frac{D^2}{4}} \approx \sqrt{R^2 + \frac{D^2}{4} + \frac{D^4}{64R^2}} = R + \frac{D^2}{8R}$$

$$\Rightarrow AP - OP = R + \frac{D^2}{8R} - R < \frac{\lambda}{16} \tag{2.44}$$

$$\Rightarrow R_f > \frac{2D^2}{\lambda} \tag{2.45}$$

Thus, the Fraunhofer region called far fields (FF) is defined as locations which meet the criterion specified in (2.45).

When the distance of a point is less than R_f, the received signals are considered out of phase, and an interfering pattern including repeating local maxima and minima will occur. This condition is acceptable up to points with distance equal to R_n, where the distance difference between this point from two distinct points on the reflector surface is less than $\lambda/4$ for which the following results can be concluded:

$$AP - OP > \lambda/4 \Rightarrow \sqrt{R^2 + \frac{D^2}{4}} - R > \frac{\lambda}{4} \tag{2.46}$$

$$D \ll R \Rightarrow \frac{D^2}{8R} > \frac{\lambda}{4} \Rightarrow R < \frac{D^2}{2\lambda} \tag{2.47}$$

Thus, the Fresnel region called near fields (NF) is defined as locations which meet the following criterion:

$$\frac{D^2}{2\lambda} < R_n < \frac{2D^2}{\lambda} = 4 \times \frac{D^2}{2\lambda} \tag{2.48}$$

Finally, the Rayleigh region which is of low interest for the actual links when passive reflectors are used can be defined as very close locations to the reflector surface and meeting the following criterion:

$$R_\mathrm{r} < \frac{D^2}{2\lambda} \tag{2.49}$$

2.7.2 *Parabolic Reflector Antenna*

Any antenna with metallic reflector using a portion of paraboloid surface is commonly called parabolic or dish antenna. It is highly directional in UHF, SHF, and EHF bands and used widely in the microwave band for LOS point-to-point or point-to-multipoint radio links. A simple form of parabolic antenna is shown in Fig. 2.31.

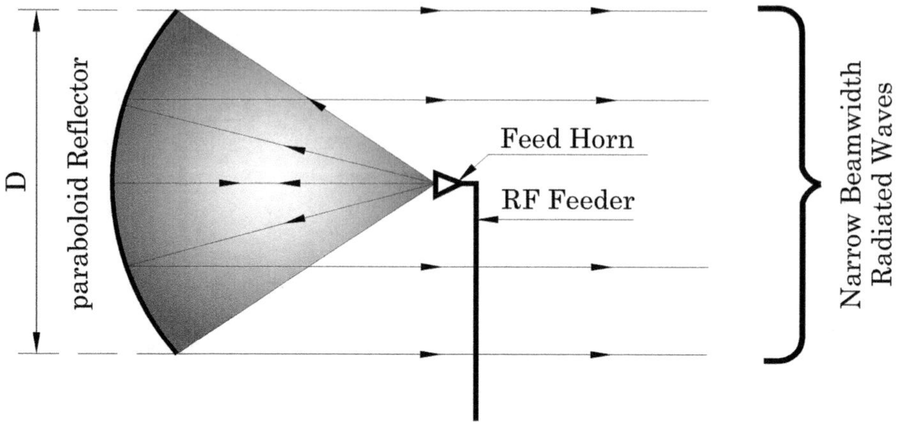

Fig. 2.31 Simple structure of parabolic antenna

The feed horn is connected to the radio unit (transmitter or receiver) and acts as a radiator of electromagnetic waves. The radiated waves are collected by the parabolic reflector and are concentrated in a narrow beamwidth providing a high-gain antenna in the direction of its principal axis (boresight).

The gain of a parabolic antenna is expressed by the following formula:

$$g = \varphi \left(\frac{\pi^2 \cdot D^2}{\lambda^2} \right) \tag{2.50}$$

where D and λ are the reflector diameter and wavelength, respectively, both with the same unit. Also, φ is the antenna efficiency in the range $0 \leq \varphi \leq 1$.

Equation (2.50) can be written in the logarithmic system as follows:

$$G[\mathrm{dB}_i] = 10 \log (g) = 20 \log \frac{\pi D}{\lambda} + 10 \log \varphi, \text{ or} \tag{2.51}$$

$$G[\mathrm{dB}_i] = 20 \log (f \cdot D) + 10 \log \left(\frac{100\pi^2}{9} \varphi \right), \quad (f \text{ in GHz}) \tag{2.52}$$

The far field radiation pattern of a parabolic dish antenna as specified in Fig. 2.32 includes a main lobe and a number of side lobes.

2.7.3 Rectangular Flat Passive Reflector

Rectangular flat passive reflector can be considered as an extreme state of the parabolic reflector when its focal point is displaced to the infinity. This means that the reflector radiates the collected radiowaves in a direction based on the Snell's

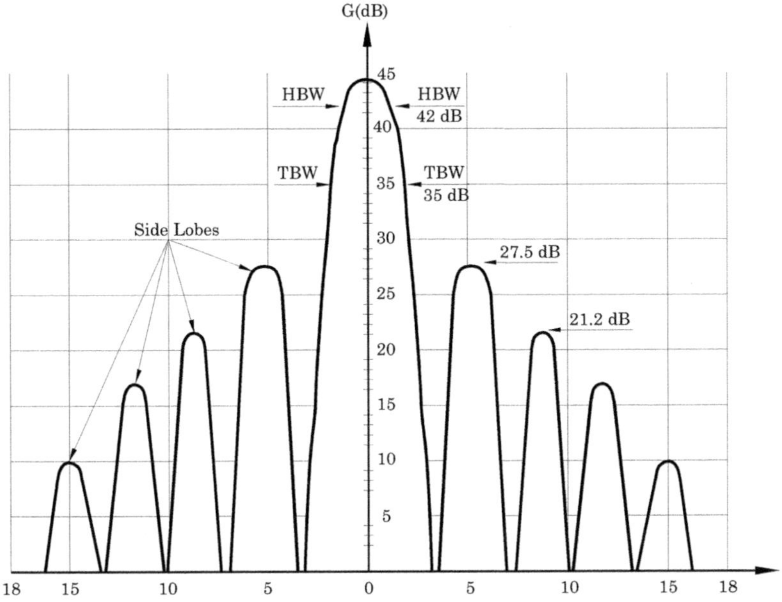

Fig. 2.32 Typical radiation pattern of parabolic antenna

reflection rule. In addition to the reflected waves, there are some scattered waves in other directions resulting in the radiation pattern specified in Fig. 2.33.

The pattern is similar to the radiation pattern of the parabolic antenna with some minor differences. The following formulas, based on parameters defined in Fig. 2.33, are used to calculate some of the main parameters pertaining to the rectangular flat passive repeater:

$$a = w \cdot \cos \alpha \tag{2.53}$$

$$A = w \cdot b \tag{2.54}$$

$$A_e = A \cdot \cos \alpha = w \cdot b \cdot \cos \alpha \tag{2.55}$$

$$P(U) = \left(\frac{\sin U}{U}\right)^2, \quad U = \frac{\pi a}{\lambda} \sin \theta \tag{2.56}$$

$$E(U) = 20 \log u, \quad U = \frac{\pi a}{\lambda} \sin \theta \tag{2.57}$$

In the above equations, w and b are the actual width and length of the reflector, A and A_e are the actual and effective areas of the reflector, and $P(U)$ and $E(U)$ are the normalized radiation power and the envelope of peak points (related to the side lobes), respectively.

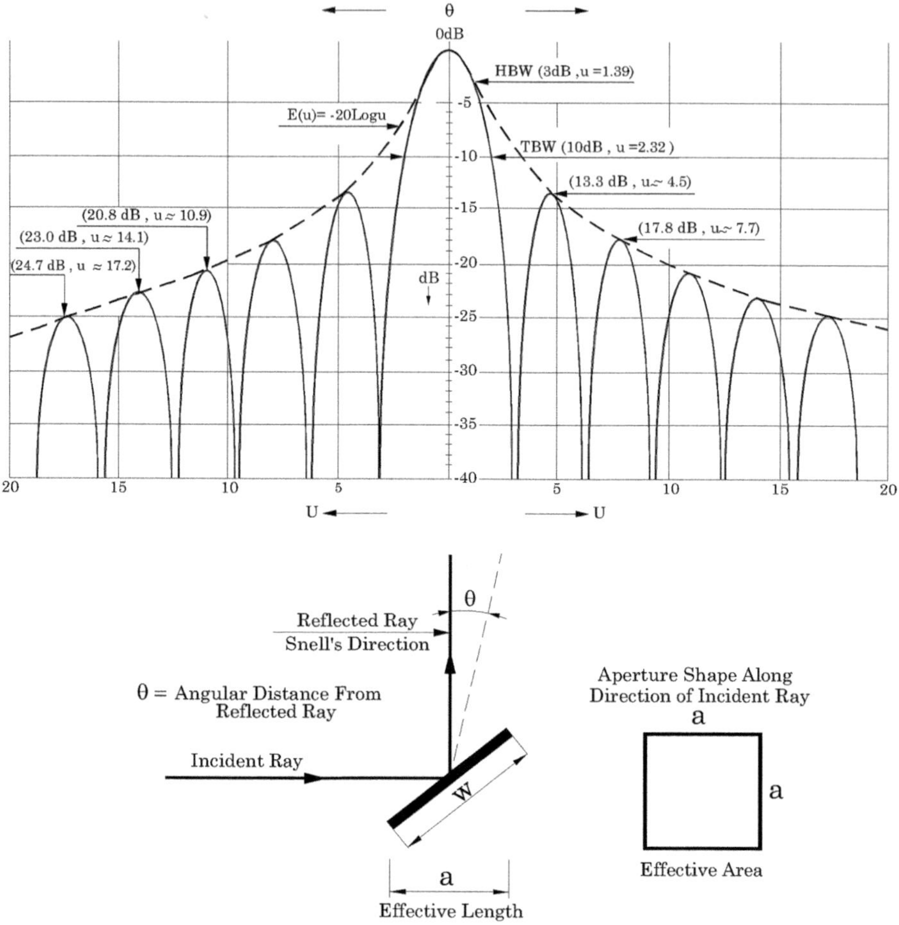

Fig. 2.33 Radiation pattern of rectangular flat passive reflector

Example 2.13. For a $4 \times 6\,\mathrm{m}^2$ rectangular flat reflector used in a radio link working at $f = 6\,\mathrm{GHz}$ and where the incident angle is 60°, find:

1. The reflector ideal and actual gains ($\varphi = 90\%$)
2. The local nulls and maxima points
3. The half-power (3-dB) and tenth-power (10-dB) points and related beamwidths

Solution. 1. $f = 6\ \mathrm{GHz} \quad \Rightarrow \quad \lambda = 0.05\,\mathrm{m}$

$$G_p = 20\ \log\left(\frac{4\pi \times 4 \times 6}{(0.05)^2}\right) = 101.6\ \mathrm{dB}_i \quad \text{(ideal gain)}$$

$$G'_p = 20\ \log\left(\frac{4\pi \times A_\mathrm{e}}{\lambda^2}\right) + 20\ \log\ \varphi \approx 95.1\ \mathrm{dB}_i \quad \text{(actual gain)}$$

2. For null and peak points,

$$\sin U = 0 \quad \Rightarrow \quad U = K\pi \ , \ K = 1,2,3,\ldots \quad \text{null points}$$

$$\sin^2 U = 1 \quad \Rightarrow \quad U = K\pi + \frac{\pi}{2} \ , \ K = 1,2,3 \quad \text{peak points}$$

3. As indicated in Fig. 2.33 for HBW,

$$U = 1.39, \quad a = w \cos \alpha = 3\,\text{m}, \quad \lambda = 0.05\,\text{m}$$

$$U = \frac{\pi a}{\lambda} \sin \theta \quad \Rightarrow \quad \theta = 0.42^\circ$$

$$\text{HBW} = 2\theta = 0.84^\circ$$

Also for TBW,

$$U = 2.32 \quad \Rightarrow \quad \theta' = 0.7^\circ$$

$$\text{TBW} = 2\theta' \quad \Rightarrow \quad \text{TBW} = 1.4^\circ$$

■

2.7.4 *Elliptical Flat Passive Reflector*

Elliptical flat passive reflector as specified in Fig. 2.34 also acts in a similar way like the rectangular reflector. Considering the following formulas for calculation of its related parameters when the effective aperture along the incident wave is circular with a diameter equal to D,

$$A = \pi a \cdot b \tag{2.58}$$

$$A_{\text{e}} = A \cdot \cos \alpha \tag{2.59}$$

$$D^2 = 4\, a \cdot b \cos \alpha \tag{2.60}$$

$$P(U) = \left[\frac{2}{U} J_1(U)\right]^2 \ , \ U = \frac{\pi D}{\lambda} \sin \theta \tag{2.61}$$

$$J_1(U) = \frac{U}{2 \cdot 0! \cdot 1!} - \frac{U^3}{2^3 \cdot 1! \cdot 2!} + \frac{U^5}{2^5 \cdot 2! \cdot 3!} + \cdots \tag{2.62}$$

$$E(U) = 30 \log \left(\frac{e^2 \cdot U}{10}\right) \ , \ U = \frac{\pi D}{\lambda} \sin \theta \tag{2.63}$$

In the above equations, $2a$ and $2b$ are the major axis and minor axis of the ellipse, D is diameter of equivalent circle, α and λ are incident angle and wavelength of the radiowaves, and $P(U)$ and $E(U)$ are normalized radiation power and the envelope

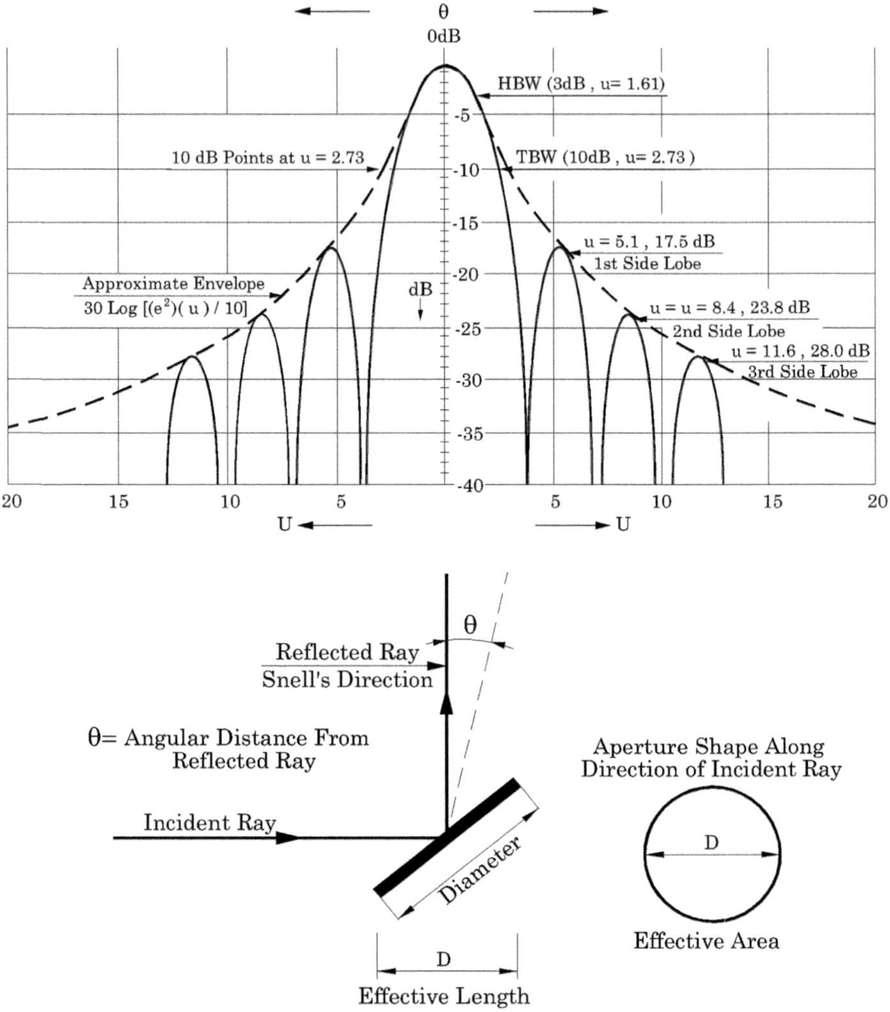

Fig. 2.34 Radiation pattern of elliptical flat passive reflector

of peak points (related to the side lobes), respectively. Also $J_1(U)$ is the Bessel function of the first kind and order one.

Example 2.14. For an elliptical flat reflector of $2a = 4\,\text{m}$ and $2b = 6\,\text{m}$ used in a radio link at $f = 10\,\text{GHz}$ where the incident angle is $\alpha = 36.86°$, find:

1. The ideal and actual gains ($\varphi = 95\%$) of the reflector
2. The local nulls and the peaks of side lobes
3. The half-power (3-dB) and tenth-power (10-dB) points and related beamwidths

Solution. 1. $f = 10\,\text{GHz} \Rightarrow \lambda = 0.03\,\text{m}$

$$G_p = 20\ \log \frac{4\pi A}{\lambda^2} = 111.9\ \text{dB}_i$$

$$G'_p = G_p + 20\ \log(\cos\ \alpha) + 20\ \log\ \varphi = 109.5\ \text{dB}_i$$

2. For null and peak points,

$$(\text{null points}) \Rightarrow J_1(U) = 0 \Rightarrow U = 3.7,\ 7,\ \ldots$$

$$(\text{peak points}) \Rightarrow U = 5.1,\ 8.4,\ \ldots$$

3. As indicated in Fig. 2.34, for HBW,

$$A_e = \pi \cdot a \cdot b\ \cos\ \alpha = 4.8\ \pi$$

$$U = 1.61\ ,\ D = \sqrt{4\ a\ b\ \cos\ \alpha} = 4.38\ ,\ \lambda = 0.03\,\text{m}$$

$$U = \frac{\pi D}{\lambda} \sin\ \theta \Rightarrow \theta \approx 0.201^\circ$$

$$\text{HBW} = 2\theta \Rightarrow \text{HBW} = 0.402^\circ$$

Also for TBW,

$$U = 2.73 \Rightarrow \theta' = 0.341^\circ$$

$$\text{TBW} = 2\theta' = 0.682^\circ$$

■

2.8 Design Calculations

Passive repeaters are used to change the direction of LOS radio links to overcome obstructions and reduce the number of active repeaters enabling the radio engineer to select more convenient locations for them close to the existing public utilities such as roads and power lines. The passive repeater requires no access road or electric power and virtually no maintenance resulting in reduced cost compared to similar systems not using passive repeaters.

To employ passive repeaters in the radio link design, some technical calculations are required to ensure that the network will operate properly with a minimum guaranteed fade margin. The basic calculation procedures are outlined in this section which will be completed with a number of practical examples presented at the end of this section.

2.8.1 Gain of Reflector

Back-to-back parabolic dish antennas or flat passive reflectors in different shapes and configurations may be used for repeating the radiowaves. Two main factors affecting the link design are gain and radiation efficiency of the reflectors explained in this section.

- *Parabolic Dish Antenna*

  The gain of an ideal parabolic dish antenna with 100% efficiency in the metric system of units is expressed by the following formula:

$$G_A = \frac{\pi^2 D^2}{\lambda^2} \tag{2.64}$$

  Actual efficiency of a parabolic antenna is normally 50–70% with an average value of 55% for which the gain in terms of dB_i can be written as the following equation:

$$G_A[dB_i] = 7.5 + 20 \log f(\text{GHz}) + 20 \log D(\text{feet}), \quad \text{or} \tag{2.65}$$

$$G_A[dB_i] = 17.8 + 20 \log f(\text{GHz}) + 20 \log D(\text{m}) \tag{2.66}$$

  For example, when a back-to-back passive repeater employs two dishes with 10-feet diameter and subject to neglecting connector loss, the total gain of the repeater will be 95 dB_i at 10 GHz.

- *Passive Reflector*

  The gain of a flat passive reflector in the best position, that is, when the incident radiowaves are perpendicular to its surface, can be expressed by the following formula:

$$G_P = \frac{4\pi A}{\lambda^2} \tag{2.67}$$

  This formula is valid when 100% efficiency is considered; otherwise, it will be changed to

$$G_P = \frac{4\pi A}{\lambda^2} \times \eta\,, \quad \text{or} \tag{2.68}$$

$$G_P[dB_i] = 20 \log G_P = 20 \log \left(\frac{4\pi A}{\lambda^2}\right) + 20 \log \eta \tag{2.69}$$

  For practical application of the above formula, assuming $\eta \approx 1$ and f in GHz and replacing area (A) with effective area (A_e) yields

$$G'_P[dB_i] = 20 \log \left(\frac{4\pi A_e}{\lambda^2}\right) + 20 \log \eta \tag{2.70}$$

2.8.2 *Passive Reflector Efficiency*

The efficiency of a passive repeater can be written as

$$\eta = \eta_P \cdot \eta_A = \eta_P \cdot \eta_f \cdot \eta_i \quad (2.71)$$

η_P is the polarization efficiency which is usually considered 100% on LOS radio systems where every effort is made to match polarization by adjustment of antenna feed.

η_A is the aperture efficiency which depends on the flatness of passive panel (η_f) and the incident angle of the radiowaves η_i. The panel flatness efficiency can be improved by good workmanship in panel manufacturing, and η_i may be considered in the aperture effective area, A_e, instead of its physical (actual) area.

2.8.3 *Effective Area of Reflector*

Effective area of a passive reflector depends on its position in a radio link. Since the optimum orientation of a passive reflector is achieved when the angle of incident and reflected waves are equal, thus as shown in Fig. 2.35, the bisector line is perpendicular to its surface. In this case, the effective area of the passive reflector is

$$A_e = A \cdot \cos \alpha \quad (2.72)$$

For α greater than zero, the effective area will be less than physical area of the reflector resulting in gain reduction. The actual gain of a flat passive reflector when used in a real link is given by

$$G'_P = \frac{4\pi A_e}{\lambda^2} \cdot \eta \quad (2.73)$$

$$G'_P[\text{dB}_i] = G_P[\text{dB}_i] + 20 \log (\cos \alpha) \quad (2.74)$$

2.8.4 *Double Passive*

As indicated in Fig. 2.35, the following relation exists between the transfer angle ψ and the incident angle α:

$$\psi = 180 - 2\alpha \quad (2.75)$$

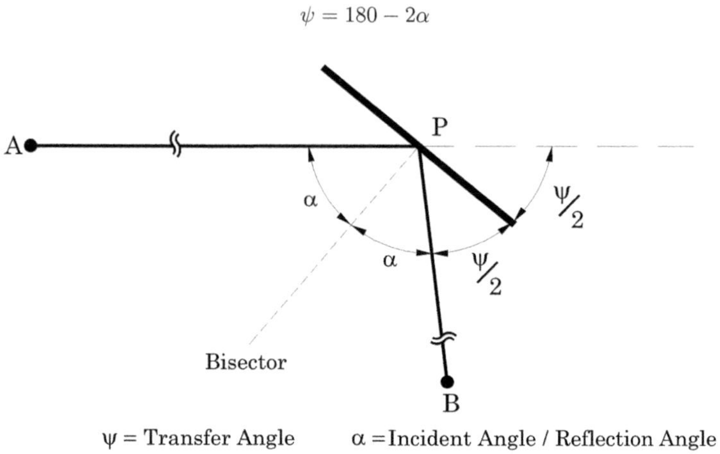

Fig. 2.35 Transfer angle in flat passive reflector

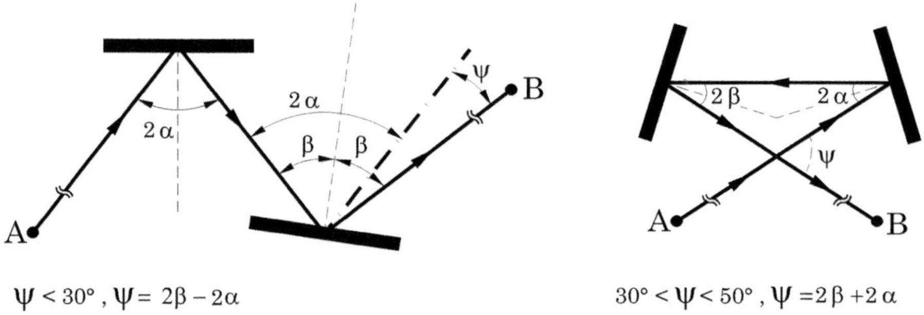

Fig. 2.36 Arrangements of closely coupled passive reflectors

Considering the formula (2.74), the gain of a passive reflector is proportional to $\cos\,\alpha$ resulting in a loss equal to $20\,\log(\cos\,\alpha)$. When the incident angle reaches to $65°$, the gain of passive reflector will decrease significantly by the following value:

$$\text{gain reduction} = 20\,\log(\cos\,\alpha) = -7.8\ \text{dB} \tag{2.76}$$

In the case of increasing α, the passive reflector gain reduction is not acceptable, and using closely coupled double passive reflectors instead of a single passive is a normal practice. Some common arrangements for closely coupled double passive reflectors are given in Fig. 2.36.

When a closely coupled passive reflector is used instead of a single passive, the smaller effective area will be the basis of calculation, and optimization should be made to employ similar passives for a cost-effective solution. The step-by-step procedure for selection and calculation is recommended as follows:

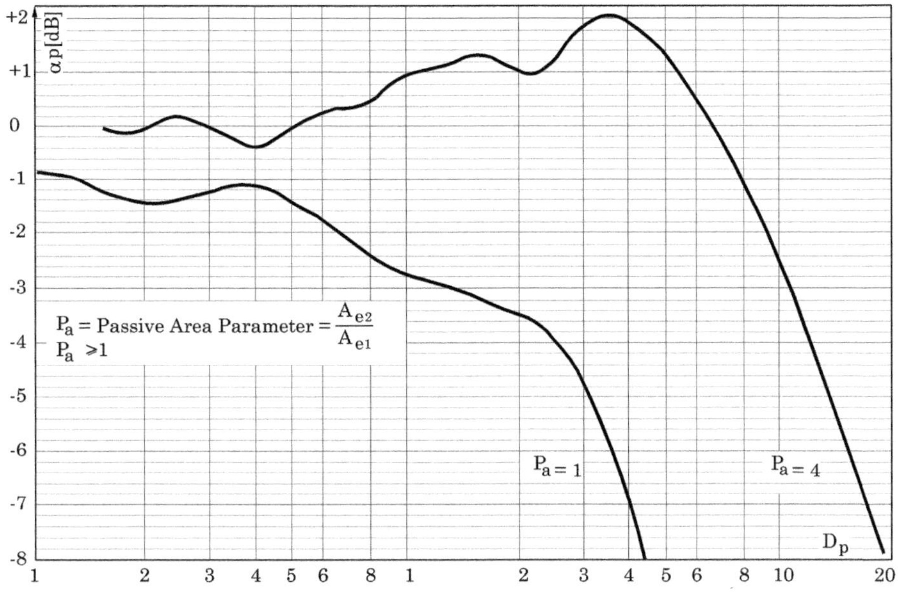

Fig. 2.37 Double passive loss

Step 1: When transfer angle (ψ) is small, select closely coupled double passive reflectors, that is,

$$\psi \leq 50^\circ \quad \Rightarrow \quad \text{select a double passive}$$

Step 2: For double passive arrangement, observe the following criteria:

$$30^\circ \leq \psi \quad \Rightarrow \quad \text{arrangement No.1 (Fig. 2.36)}$$

$$50^\circ \geq \psi > 30^\circ \quad \Rightarrow \quad \text{arrangement No.2 (Fig. 2.36)}$$

Step 3: Calculate the effective areas (A_{e1} and A_{e2}) of each passive considering their actual areas and position parameter.

Step 4: Calculate the closely coupled double passive parameter:

$$\mathrm{DP} = \frac{2\lambda d}{A_{e1}} \tag{2.77}$$

Step 5: Based on the ratio of A_{e2}/A_{e1}, select proper curve from Fig. 2.37 and determine the loss value denoted by α_P:

$$P_a = A_{e2}/A_{e1} \quad , \quad P_a \geq 1 \tag{2.78}$$

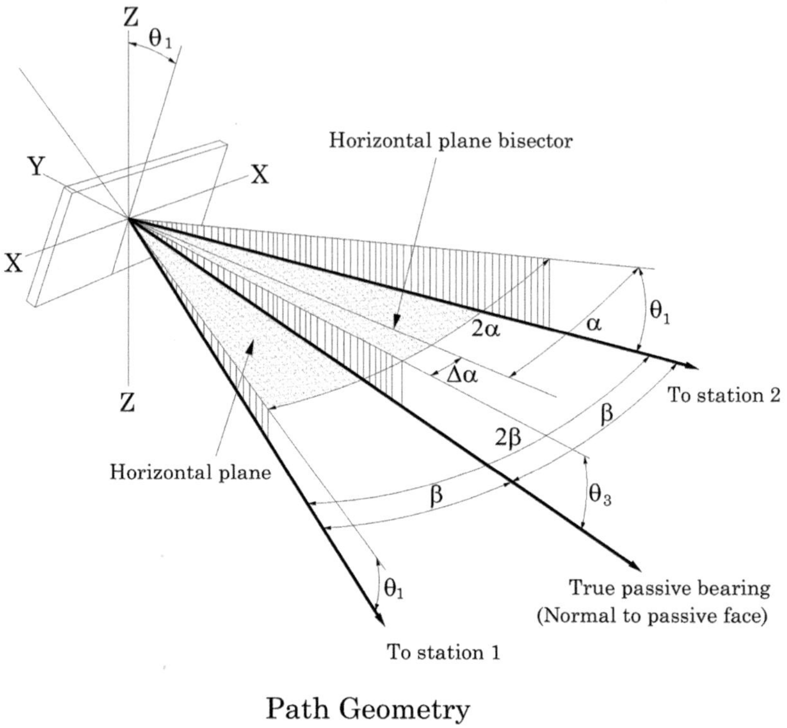

Fig. 2.38 Angle of passive reflector orientation

where A_{e1} is the smaller effective area of the two passives and A_{e2} is the larger one.

λ , d, and A_{e1} are the RF wavelength, distance between two passive reflectors, and the smaller effective area, respectively, and all with the same units. Examples (2.21) and (2.22) have been prepared for more details and clarifications.

2.8.5 *Angles of Reflector Orientation*

Proper operations of a passive reflector, either single or double configuration, are subject to accurate alignment of its panel toward the transmitting and receiving antennas. To do this, the horizontal and vertical bearings should be calculated where only geographical coordinates and altitude of passive and related active stations are known. Referring to Fig. 2.38 and applying geometry rules and trigonometric equations, the following formulas are derived for calculation of the reflector panel elevation angle denoted by θ_3:

$$\tan\ \theta_3 = \frac{(\sin\ \theta_1 + \sin\ \theta_2) \times \cos(\Delta\alpha)}{(\cos\ \theta_1 + \cos\ \theta_2) \times \cos\ \alpha} \tag{2.79}$$

$\Delta\alpha$ can be calculated by

$$\tan(\Delta\alpha) = \frac{(\cos\ \theta_1 - \cos\ \theta_2) \times \tan\ \alpha}{\cos\ \theta_1 + \cos\ \theta_2} \tag{2.80}$$

It should be noted that in the above formulas, $\theta_2 \geq \theta_1$ and sign of downward angles related to the horizontal plane is positive (+), while sign of upward angles is negative (−). Then the actual angle between the incident and reflected waves (2β) is derived as follows:

$$\cos\ \beta = \frac{\sin\ \theta_1 + \sin\ \theta_2}{2\ \sin\ \theta_3} \tag{2.81}$$

Finally, for calculation of passive reflector gain, using the above formulas yields

$$G_{\mathrm{P}} = 20\ \log\left[\frac{4\pi A}{\lambda^2} \cdot \cos\ \beta\right] \tag{2.82}$$

Also, the horizontal orientation of the reflector is normal to the bisector line of the angle between the incident and reflected directions.

Example 2.15. A rectangular flat reflector $a \times b = 10' \times 16'$ is used in a radio link where the incident angle is 20° with $\eta = 90\%$ for $\alpha = 0$ at $f = 10\,\text{GHz}$; find:

1. Maximum gain of the reflector
2. Effective area of the reflector
3. Actual gain of the passive reflector in the specified radio link

Solution. 1. $f = 10\ \text{GHz} \Rightarrow \lambda = 3\,\text{cm} \Rightarrow \lambda \approx 0.1'$

$$\begin{aligned} G_{\mathrm{P}} &= 20\ \log\left(\frac{4\pi A}{\lambda^2}\right) + 20\ \log\ \eta \\ &= 20\ \log\left(\frac{4\pi \times 10 \times 16}{(0.1)^2}\right) + 20\ \log\ 0.9 \\ &= 105.1\ \text{dB}_i \end{aligned}$$

2.

$$A_{\mathrm{e}} = A \cdot \cos\ \alpha \Rightarrow A_{\mathrm{e}} = 10 \times 16 \times \cos\ 20^\circ = 150.35\ \text{square feet}$$

3.

$$G'_{\mathrm{P}} = G_{\mathrm{P}} + 20\ \log\ \cos\ \alpha \Rightarrow G'_{\mathrm{P}} = 104.56\ \text{dB}_i$$

■

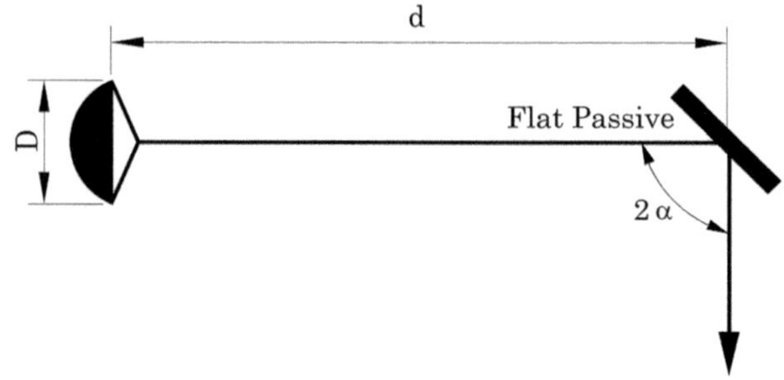

Fig. 2.39 Determination of field type index, F_t

2.8.6 Near Field Reflector

When a passive reflector is placed in the near field of an antenna, the calculation procedure to determine its gain is quite different, when compared to the gain formula given by 2.82. To differentiate the case, as the first step, the field-type index denoted by F_t (1/K in some references) should be calculated using the following equation:

$$F_t = 1/K = \frac{\pi \lambda d}{4\,a^2}\;,\quad a^2 = A \cdot \cos\,\alpha \tag{2.83}$$

where all parameters are defined in Fig. 2.39 and all with the same unit. Practically the value of F_t greater than 2.5 indicates that the passive reflector is in the far field; otherwise, it is located in the near field and the relevant procedure should be followed.

In the case of satisfying the near field criterion, instead of the passive gain given by (2.82), the overall near field passive gain or loss (α_P) is determined using the graphs presented by Fig. 2.40. For selecting the proper graph, the near field parameter l is required which is expressed by the following formula:

$$l = \frac{D}{2\,a}\sqrt{\pi} = \frac{D}{2}\sqrt{\frac{\pi}{A_e}} \tag{2.84}$$

Figure 2.41 consists of a set of graphs indicating the maximum distance of passive reflector from the closer active terminal versus its effective area (A_e) in terms of the applied frequency bands.

Example 2.16. Find the gain or loss of the passive reflector specified in the previous example when it is placed at 1.2 km from a parabolic antenna ($D = 3$ m).

Solution.

$$\lambda = 0.03\,\text{m}$$

$$F_t = \frac{\pi d}{4\,a^2} = \frac{\pi \times 0.03 \times 1200}{4 \times 16\,\cos\,20} \approx 2$$

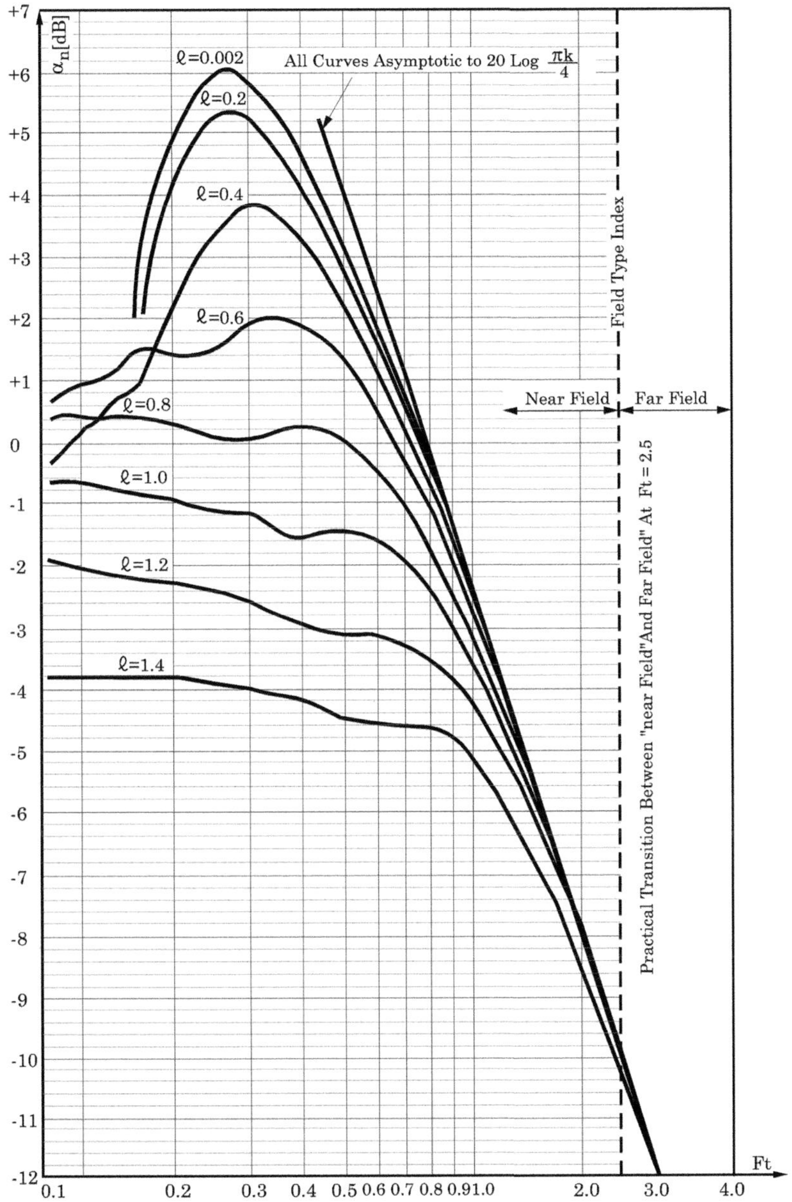

Fig. 2.40 Gain and loss of near field rectangular passive reflector

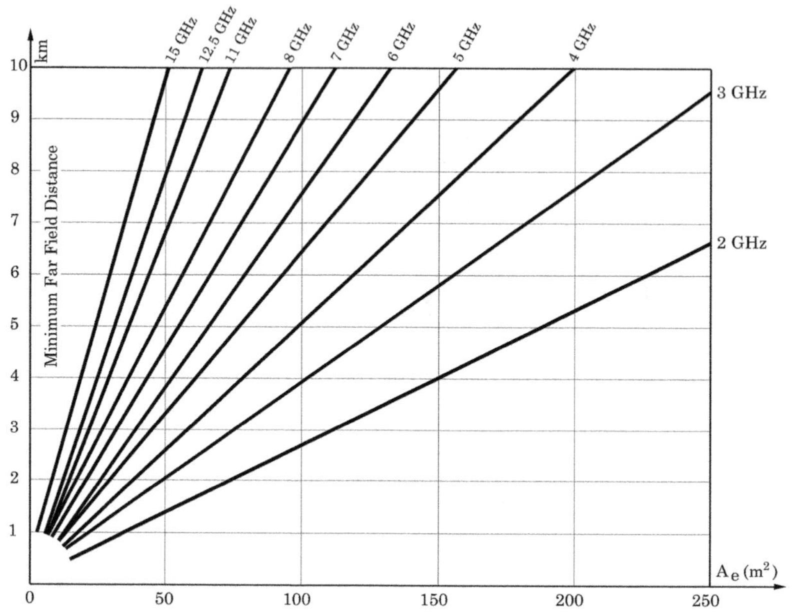

Fig. 2.41 Maximum near field distance of passive reflector

The value of $F_t = 2$ indicates that the passive reflector is placed in the near field region. Thus,

$$l = \frac{D}{2}\sqrt{\frac{\pi}{A_e}} \quad \Rightarrow \quad l \approx 0.7$$

$$(l = 0.7\ ,\ \text{Fig. 2.40}) \quad \Rightarrow \quad \alpha_n = -8.5\ \text{dB}$$

■

2.8.7 Periscope

The periscope is a special near field single passive reflector located in the active station to exchange the radiowaves with the active antenna (either transmitter or receiver) as illustrated in Fig. 2.42.

In addition to the general procedure for calculation of the near field loss or gain explained in Sect. 2.8.6 and the examples (2.20) and (2.22), a simplified method may be applied based on the charts given in Fig. 2.43 as follows:

Step 1: Calculate the value for $l = \frac{D}{W}$.

Step 2: Mark point A on chart "1" for selected d and passive size.

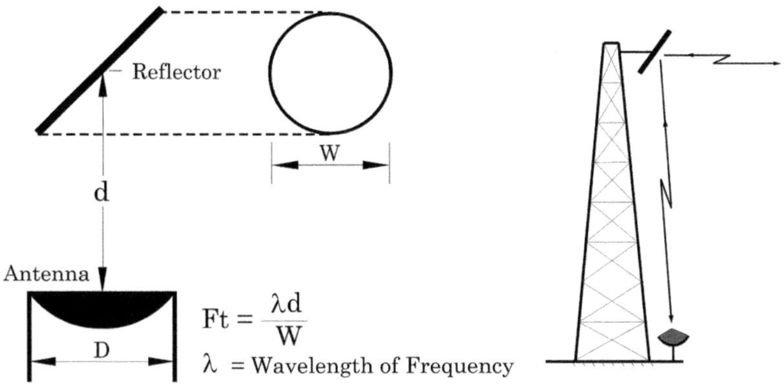

Fig. 2.42 Role of periscope in radio links

Step 3: Reflect the result of step 2 on chart "2" to indicate the related value for d/W^2 at point B.

Step 4: Determine the value of α_n on chart "3" for the calculated l and selected curve at point C.

Example 2.17. For a periscope of $10' \times 15'$ size operating at 11 GHz with $100'$ spacing with an active parabolic antenna with $10'$ diameter, find:

1. The periscope loss
2. The net gain of the antenna and periscope
3. The RF feeder loss in the case of using waveguide instead of the periscope

Solution. 1. • $l = \frac{D}{W} = \frac{10}{10} = 1$

- Mark point A on chart "1," point B on chart "2," and point C on chart 3 of Fig. 2.43
- Finally the periscope loss is found $\alpha_n = -1$ dB

2. The gain of the antenna using formula (2.65) is

$$G_A[\mathrm{dB}_i] = 7.5 + 20\,\log(f \cdot D)$$
$$= 48.3\ \mathrm{dB}_i$$
$$\text{net gain} = G_A + \alpha_n = 47.3\ \mathrm{dB}$$

3. Assuming 0.03 dB/FT loss for the RF feeder, then

$$l_f = 0.03 \times 100 = 3\ \mathrm{dB}$$

It is found that the RF feeder loss is greater than the loss of the selected periscope. ■

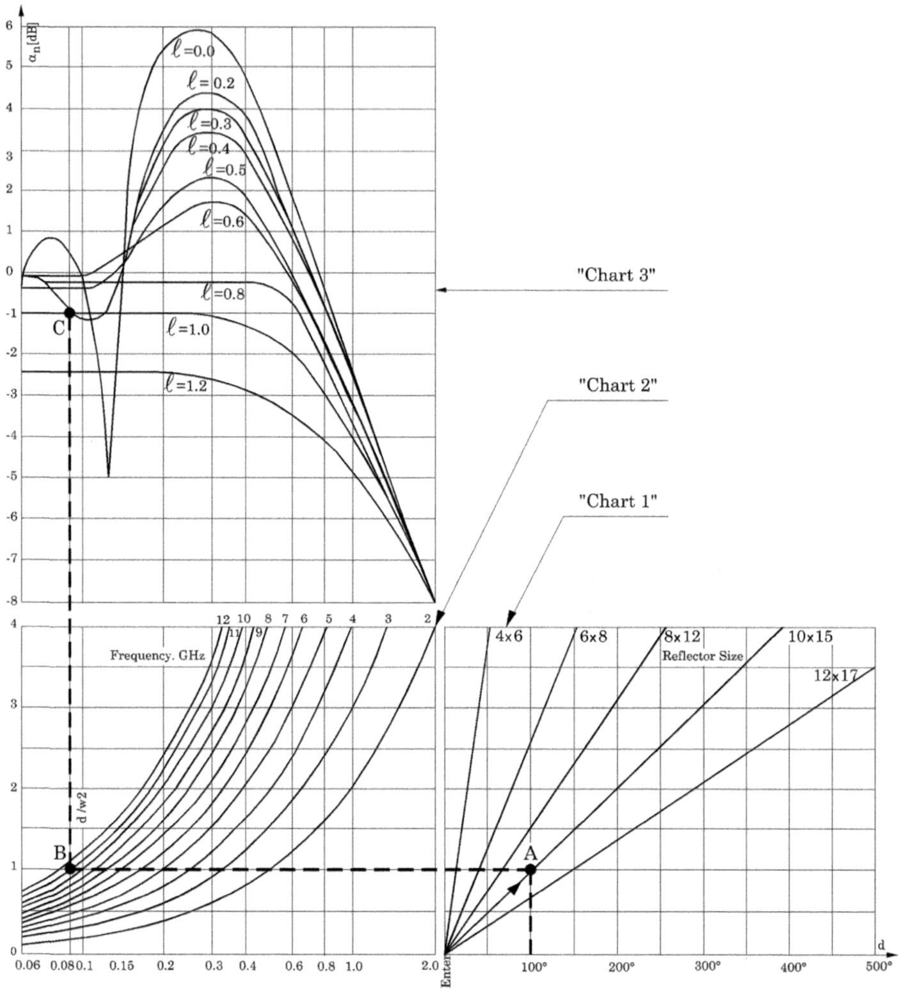

Fig. 2.43 Charts for periscope calculations

2.8.8 *Radio Links with Passive Reflectors*

Five radio links have been presented in this section including different configuration of passive reflectors. All of the examples are actual cases under operations employed by telecom operators all over the world. Each one exhibits the related link indicating the required dimensions and assumptions.

The step-by-step procedure has been followed to calculate the final fade margin for the following examples:

Example 2.18. Typical calculation for a radio link using back-to-back antenna at 7.5 GHz

Example 2.19. Typical calculation for a radio link using a single passive reflector working at 6,725 MHz

Example 2.20. Typical calculation for a radio link using a near field single passive reflector working at 11.2 GHz

Example 2.21. Typical calculation for a radio link using a far field double passive reflector working at 6,175 MHz

Example 2.22. Typical calculation for a radio link using a near field double passive reflectors working at 6,175 MHz

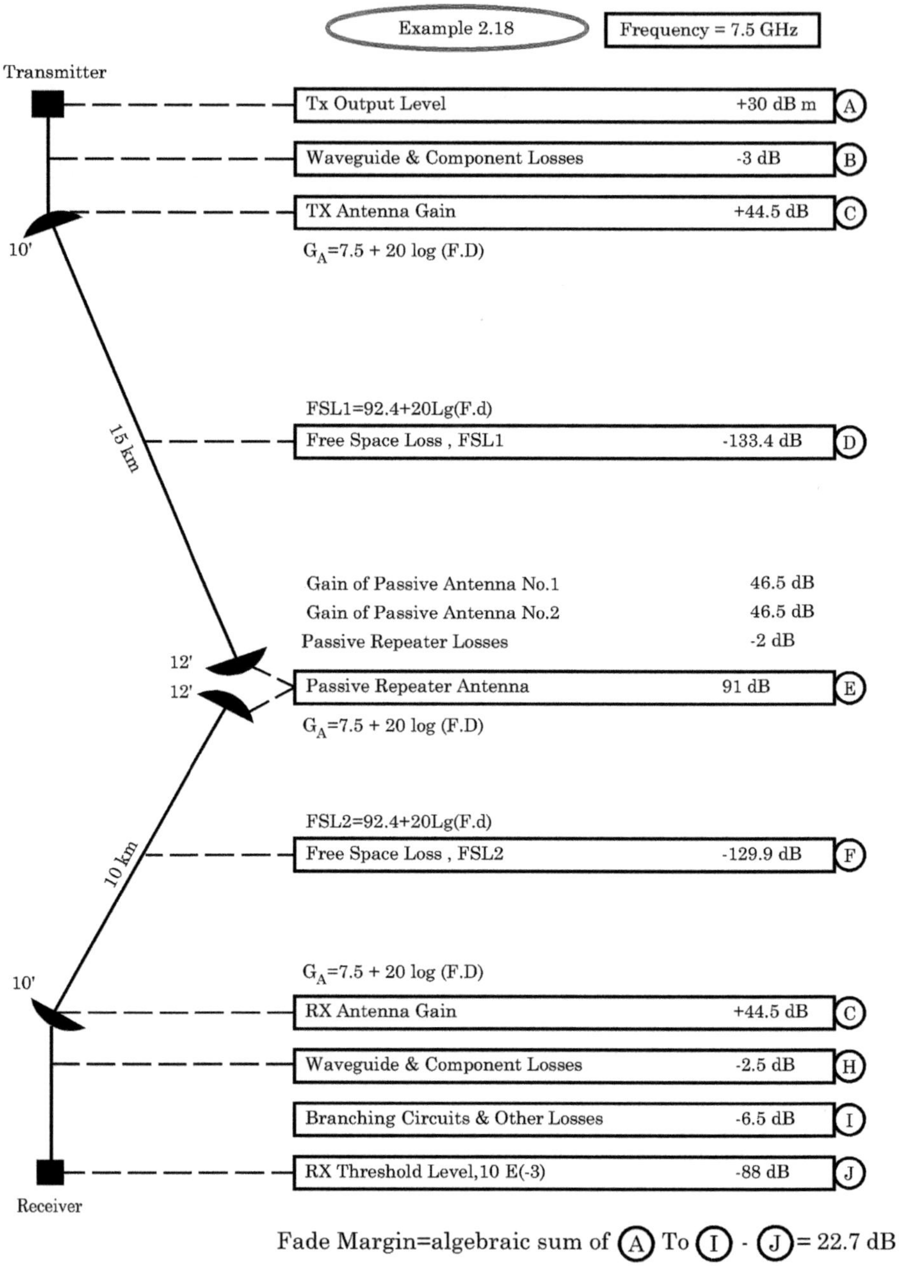

Fig. 2.44 Radio link for example 2.18

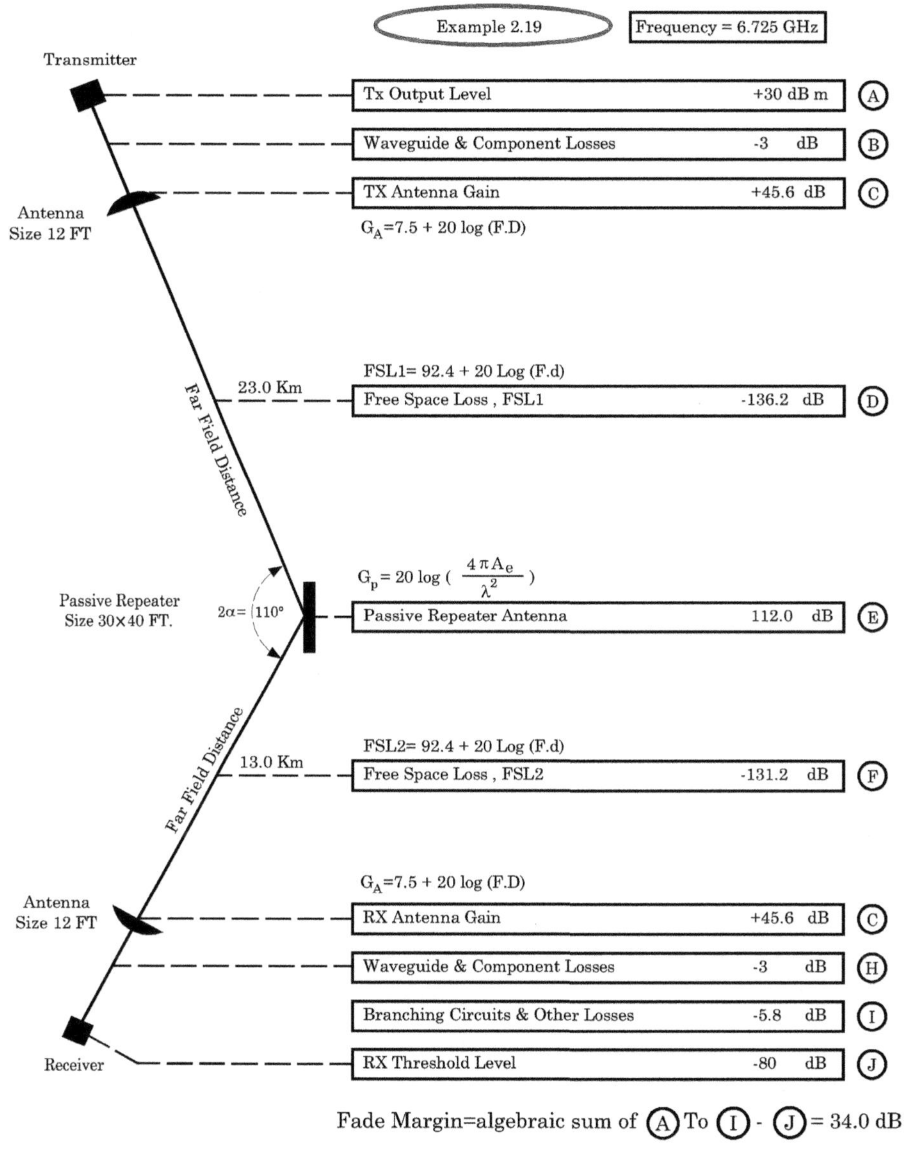

Fig. 2.45 Radio link for example 2.19

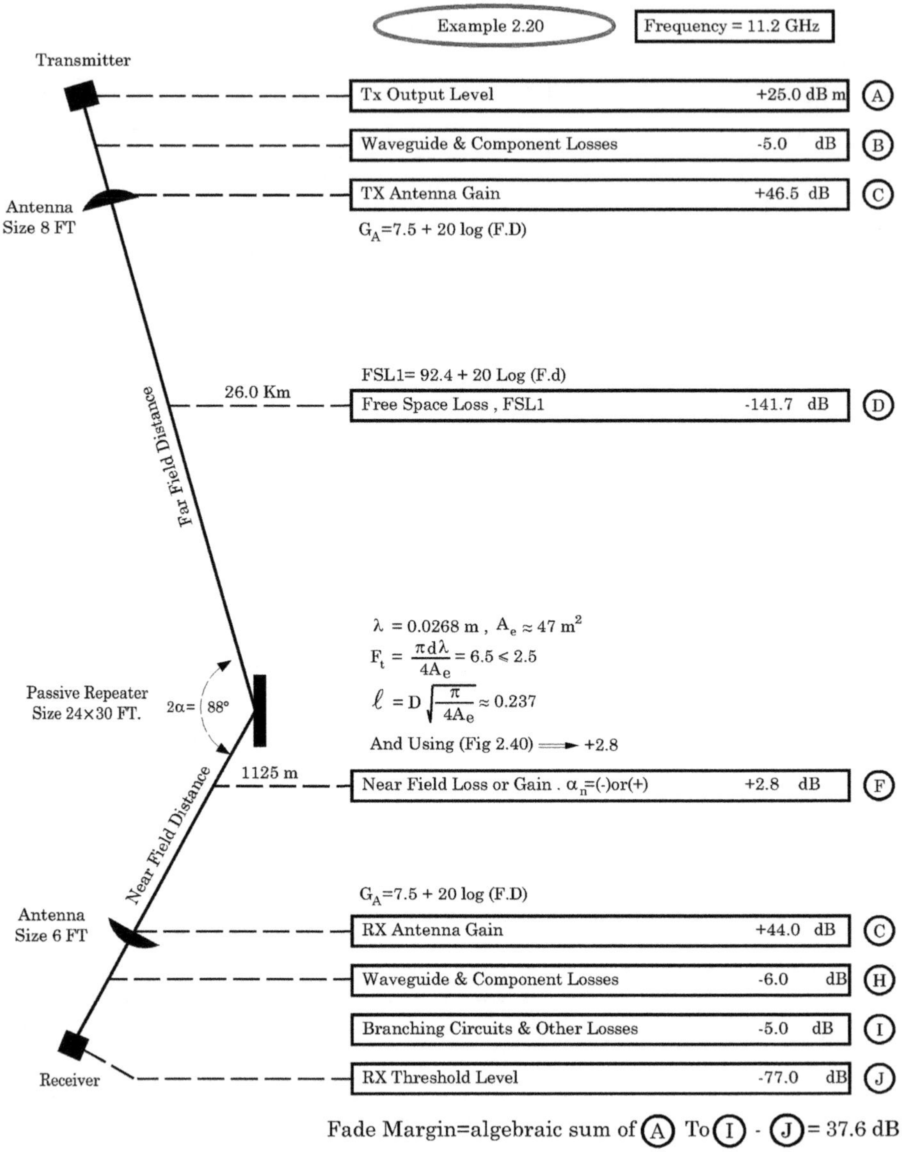

Fig. 2.46 Radio link for example 2.20

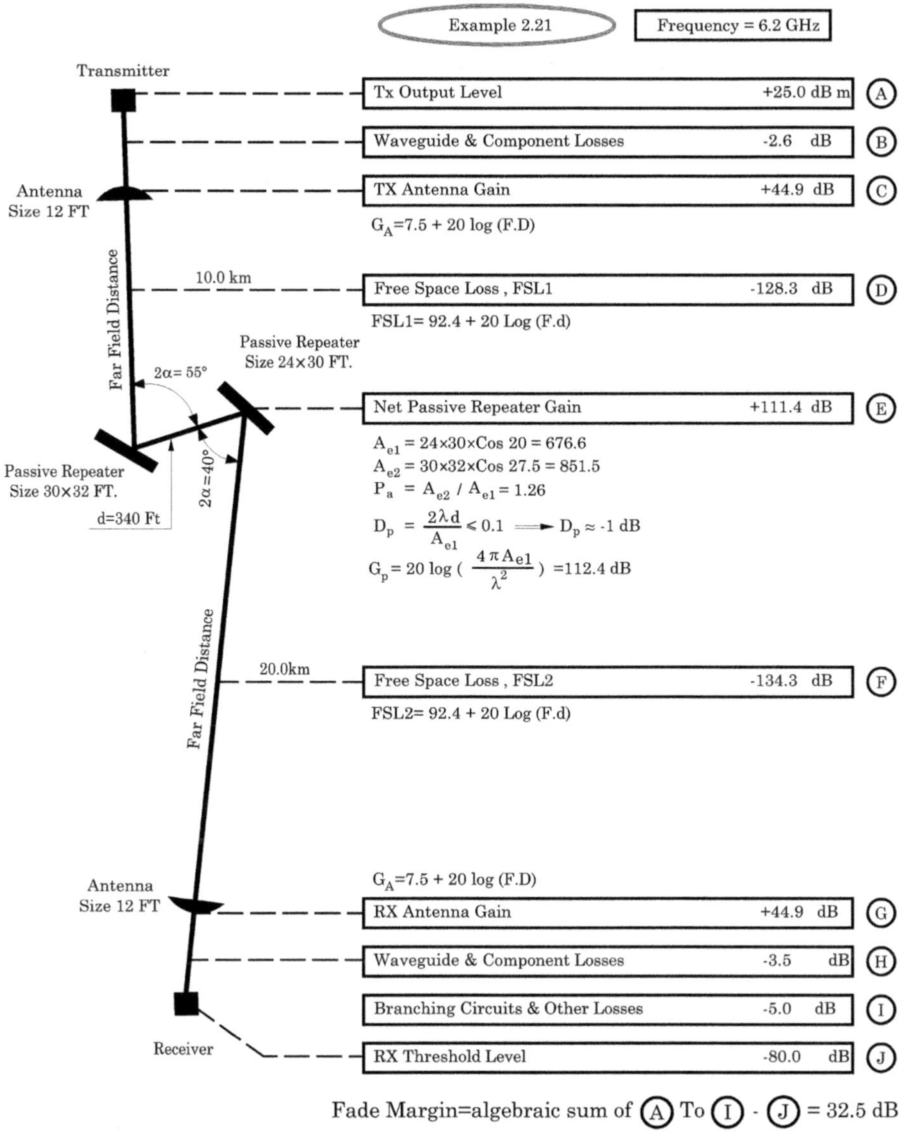

Fig. 2.47 Radio link for example 2.21

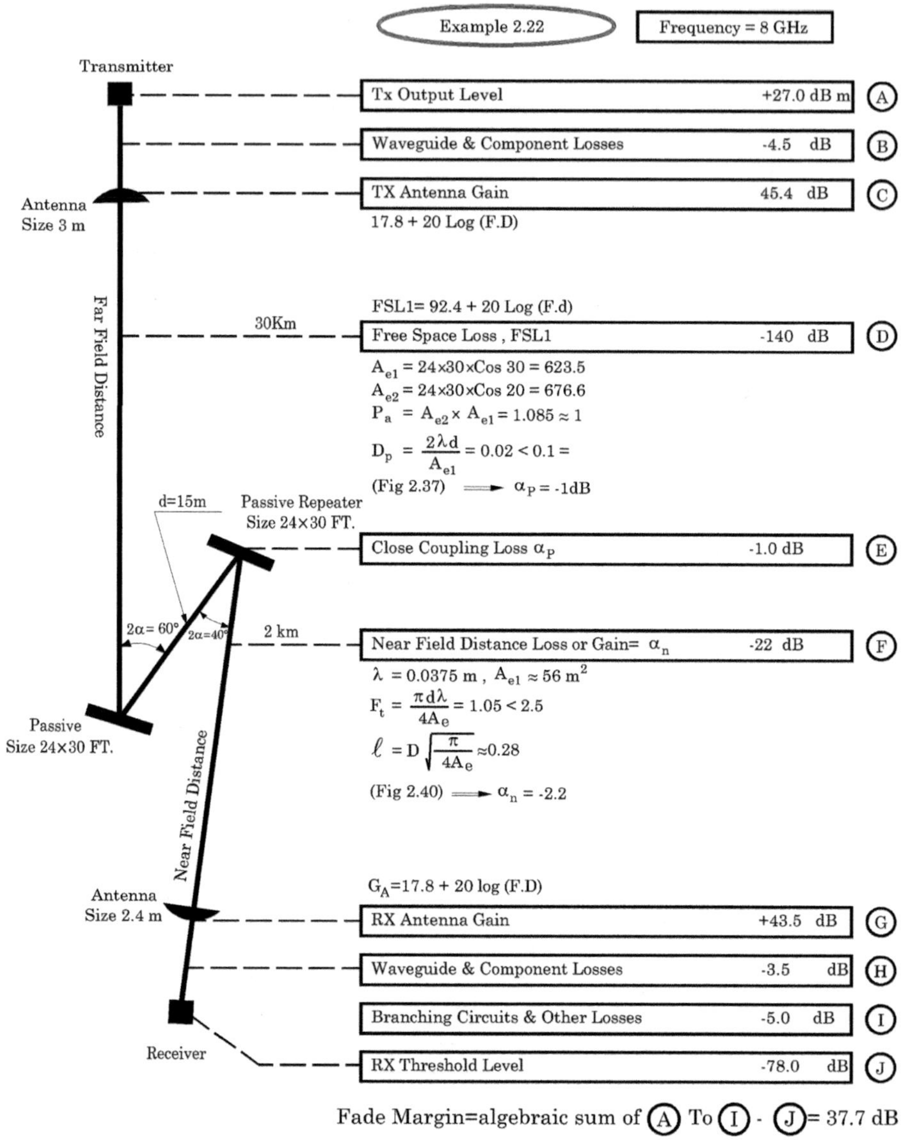

Fig. 2.48 Radio link for example 2.22

2.9 Exercises

Questions

1. Study the advantages of using antenna in the radio communications, and list its outstanding roles to enhance the range and quality of radio links.
2. List the major components or devices required to exchange RF signal between radio units (either transmitter or receiver) and associated antenna in the SHF band used for the terrestrial LOS links, radar systems, and satellite radio links.
3. Explain the main limitations of radio communications in very low frequency (such as ELF and VLF bands) and very high frequency (such as EHF and THz bands).
4. What are the main reasons to use antenna radomes? How do they improve the performance of radio communications in the following fields:
   - Radar station
   - Satellite ground terminal stations
   - Satellite mobile Earth stations
5. Specify main technical aspects of antennas which should be taken into account when selecting a suitable one for specific radio link.
6. Investigate the following technical issues referring to the ITU-R recommendations such as ITU-R Rec. P.310 and V.573 and prepare a report:
   - Cross polarization discrimination, XPD
   - Cross-polarization isolation
   - Directivity and antenna directivity diagram
   - Gain of an antenna
7. Investigate the following technical issues and discuss about them:
   - PLF
   - Antenna to medium coupling loss
   - Arrival and launch radiowave angles to/from an antenna
   - Antenna efficiency and its range
   - RPE and its application
8. Explain different bases to specify the antenna gain and express their relations as defined by the ITU-R Rec. P.341 and V.573.
9. An antenna of 50-dB_i gain generates the required RF power 100,000 times more than isotropic antenna with the same input power. Explain how the power amplification is done in a passive device.
10. Define the following terms for antenna using suitable figures:
    - Antenna boresight (principal axis)
    - HBW
    - FNBW
    - RPE
    - Main lobe and side lobes.

11. Explain the exact meaning of "radiation power density," "total radiated power," and "radiation intensity" and specify equations indicating their relations.
12. The impacts of antenna on the radiation and reception of radiowaves are included in the ERP and Ae, respectively. Express the relation of these items in terms of the antenna gain.
13. Define the directivity of an antenna and express its relation with the gain. What is its application in the omnidirectional antennas?
14. What is the polarization of an antenna and how it will affect the reception of the radiowaves?
15. Explain linear circular and elliptical polarized radiowaves and the conditions which should be provided for maximum reception for different types of polarizations.
16. Discuss about the loss imposed on the radiowaves detection due to the polarization mismatching between intercepted radiowaves and receiver antenna. Define PLF and its formula.
17. What is noise equivalent temperature and how it is related to the merit of reception in the satellite ground stations?
18. Prepare a report regarding the antennas used in the MF/HF bands including basic technical data of their radiation patterns, gain, polarization, applications, power handling capacity, and HBW for directional types.
19. Repeat question no. 18 for VHF and UHF bands focusing on the panel directional antennas.
20. Repeat question no. 18 for SHF and EHF bands focusing on the following antennas:

    - Satellite ground station of the Cassegrain type and the associated auxiliary devices
    - High-gain directional of parabolic type used in the LOS radio links

21. Explain the role of repeater stations in radio-relay networks indicating the main difference between active and passive repeaters.
22. Determine types of passive repeaters and discuss about major objectives for using them.
23. Define the far field (Fraunhofer region) and the near field (Fresnel region) for a passive reflector and derive their criteria in terms of its dimension and operating frequency.
24. Answer the following questions with reference to Fig. 2.33:

    - Is it a normalized radiation pattern or not?
    - Determine the passive gain.
    - In case of selecting the passive size equal to $a \times b = 6 \times 8\,\mathrm{m}^2$, how much is the ideal gain of this passive reflector?
    - How this curve is related to the orientation of the passive when used in a radio link?
    - Indicate the relation between U and θ (if $a = b \cos \alpha$).

25. Define the HBW of a passive reflector and indicate the affecting parameters.

26. Determine the ratio of a rectangular passive reflector of the size $a \times b$ and an elliptical passive reflector of $2a$ and $2b$ as its length of major axis and minor axis, respectively.
27. Repeat question no. 24 for the elliptical passive reflectors.
28. Discuss about the effects of wavelength on the gain, minimum far field distance, gain of side lobes, HBW, and TBW of a passive reflector.
29. Study about common size of the passive reflectors which are used in practical radio links. Why cannot they be used normally for frequencies less than 1 GHz or more than 20 GHz?
30. Define transfer angle and indicate its normal range and effects on the selection of passive arrangement.
31. Define field-type index (F_t) and near field parameter (l) in the flat reflectors. Which one is related to λ?
32. With reference to Fig. 2.37 for closely coupled double passive arrangement, indicate:

   - Major parameters affecting the overall gain/loss
   - Steps which should be followed to calculate α_n

Problems

1. Plot the radiation pattern of a reflector expressed by the following equations and indicate the value of local maxima:

   (a) $D(\theta,\varphi) = (\sin 5\theta)^2/5\theta^2$ $\quad -\frac{\pi}{2} \le \theta \le +\frac{\pi}{2}$
   (b) $D(\theta,\varphi) = K\cos^2 3\theta \cdot \sin^2\varphi$ $\quad 0 \le \theta \le \frac{\pi}{2},\ 0 \le \varphi \le \pi$

2. Repeat problem 1 for the following equations:

   (a) $D(\theta,\varphi) = K\sin^4\theta$ $\quad 0 \le \theta \le \pi$
   (b) $D(\theta,\varphi) = \left[\frac{J_1(\theta)}{\theta}\right]^2$ $\quad 0 \le \theta \le 4\pi$

   $J_1(\theta)$ is the Bessel function of the first kind and order one.
3. Normalized power density of radiowaves radiated by a TX antenna is specified by $W_n = (\sin\theta)/r^2$; assuming $10\,\mathrm{W/m^2}$ for the peak power density, find:

   (a) The radiated power in a region determined by $0 \le \theta \le \pi/3$ and $0 \le \varphi \le 2\pi$
   (b) The total radiated power

4. Power density of an omnidirectional antenna is expressed by

$$W = \frac{4}{r^2}\sin\theta\,,\quad 0 \le \theta \le \pi\,,\quad 0 \le \varphi \le 2\pi$$

   Find:

   (a) The radiation intensity in the direction of ($\varphi = 45°,\ \theta = 90°$)
   (b) Total radiated power
   (c) Plot normalized radiation pattern

5. The normalized radiation intensity of a directional antenna is expressed by $U(\theta) = \cos^2\theta \cdot \cos^2 3\theta$ in the range $0 \leq \theta \leq \pi/2$

   (a) Plot variations of U in terms of θ
   (b) Plot two-dimensional power pattern in the logarithmic scale
   (c) Find equation giving peak values of side lobes

6. For an antenna with the normalized radiation intensity given in problem 5, find:

   (a) Radiation intensity at $\theta = \pi/6$ for $U_{max} = 5$ W/unit solid angle
   (b) Power density at $\theta = \pi/6$ and $r = 2$ m
   (c) Total radiated power

7. The power density function of an antenna is given by $W = (10\cos^2 2\theta)/r^2$; find:

   (a) Radiation intensity function
   (b) Plot the radiation pattern in terms of θ

8. 500-mW output power of a microwave transmitter is connected to 40-dB$_i$ directional antenna via a feeder with 6-dB loss.

   (a) Find receiving power and electric field amplitude at a distance of 40 km in LOS condition.
   (b) Assuming total path loss equal to 140 dB, is it possible to detect the received signal by a receiver with −80-dB$_m$ threshold level connecting to an omnidirectional antenna ($G = 0$ dB$_i$) via RF feeder with 2-dB loss?
   (c) Calculate antenna gain required for receiving a signal with 30 dB more power than threshold level.

9. Radiowaves are radiated by a 20-W transmitter connected to an 8-dB$_i$ antenna; find:

   (a) The EIRP if the feeder loss is 2 dB
   (b) The value of $|E_d|$ and power flux density at a location 8 km from TX

10. The RPE of a 1.2-m parabolic dish antenna is given in Fig. 2.49; find:

   (a) The gain using formula (2.66), for 1,800- and 2,300-MHz center frequencies
   (b) The front-to-back ratio for the specified cases

11. The radiation pattern and related peak envelope are presented in Fig. 2.19 for a typical Cassegrain satellite ground station antenna. Considering G_a = 56 dB$_i$, find:

   (a) Level of the first side lobe
   (b) Maximum level of the antenna gain at $\theta = 2°$ off-axis direction
   (c) Front-to-back ratio

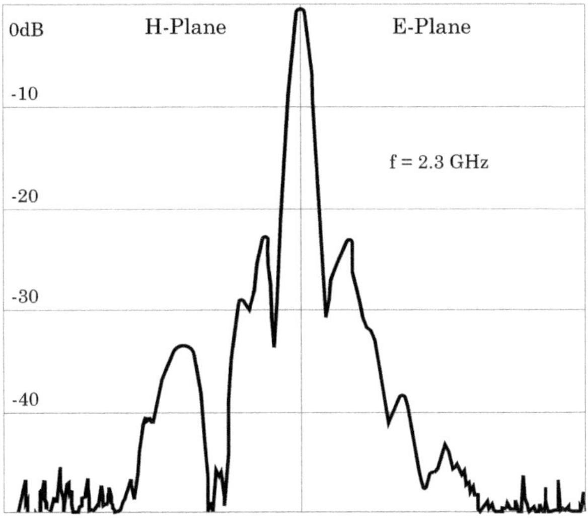

Fig. 2.49 Radiated peak envelope of 1.2-m parabolic antenna

12. Considering the antenna specified in problem 7, find:
    (a) HBW, TBW, and FNBW
    (b) Total radiated power
    (c) Radiated power through the HBW of the main lobe
    (d) Average gain of the main lobe
    (e) Antenna gain in the principal direction

13. The power density of an antenna is expressed by $W = (A_0 \sin\theta)/r^2$; find:
    (a) Maximum value of the radiated intensity
    (b) Maximum directivity and its function
    (c) Directivity in the direction of $(\theta = 30°,\ \varphi = 45°)$

14. The radiated power density of a dipole antenna ($l \ll \lambda$) at a distance r can be estimated by $W = (5\sin^2\theta)/r^2(\mathrm{W/m^2})$; find:
    (a) Total radiated power
    (b) Maximum directivity and its function in terms of θ
    (c) Radiation pattern

15. The radiation intensity of the main lobe of an antenna in the hemisphere is specified by

$$U = 10\ \cos\ 2\theta\ , \quad 0 \le \theta \le 90°\ , \quad 0 \le \varphi \le 360°$$

    (a) Calculate the beam solid angle
    (b) Plot 3-D pattern
    (c) Determine the maximum directivity

16. Find the HBW and the FNBW for the following normalized radiation intensities where $0 \leq \theta \leq 90°$ and $0 \leq \varphi \leq 360°$:

$$a \cdot \ \cos^2\theta, \quad b \cdot \ \cos 2\theta, \quad c \cdot \ \cos 3\theta$$

17. Find the HBW and the FNBW for the following normalized radiation intensities when $0 \leq \theta \leq 90°$ and $0 \leq \varphi \leq 360°$:

$$a \cdot \ U(\theta) = \cos\theta \ \cos 2\theta, \quad b \cdot \ U(\theta) = \cos^2\theta \ \cos^2 2\theta$$
$$c \cdot \ U(\theta) = \cos 2\theta \ \cos 3\theta$$

18. A lossless $\lambda/2$ dipole antenna with input impedance of around 73 ohms is connected to a transmission line whose characteristic impedance is 50 ohms. The power intensity of the antenna is approximated by $U = B_0 \sin^3\theta$; find:

    (a) Maximum absolute gain of the antenna
    (b) Reduction of overall gain due to the impedance mismatch

19. Maximum intensity of an antenna with a 90% efficiency is 200 mW per unit solid angle. Calculate the directivity and the gain when:

    (a) The input power is equal to 120 mW
    (b) The radiated power is equal to 120 mW

20. A directional lossless antenna is characterized by the following radiation intensity function:

$$U(\theta) = K\cos^2\theta, \ (0 \leq \theta \leq 90°, \quad 0 \leq \varphi \leq 360°)$$

    In the case of radiating power equal to 10 W, find:

    (a) Maximum power density (W/m^2) at a distance 1,000 m
    (b) Direction at which the maximum power density exists
    (c) Beam solid angle Ω_A
    (d) Directivity and gain of the antenna

21. Repeat problem 20 when $U(\theta) = K\cos^3\theta$
22. The E-plane and H-plane normalized radiation patterns and the gain variations over $\lambda/2$ dipole for a shrouded unidirectional cellular radio antenna are presented in Figs. 2.50 and 2.51; determine:

    (a) The gain in dB$_i$ at 930 MHz
    (b) The E-plane and H-plane HBWs
    (c) Frequency range for $G \geq 15.25$ dB$_i$

23. The normalized radiation intensity of an antenna is given by:

$$a \cdot \ U(\theta, \varphi) = \sin\theta \ \sin\varphi, \quad b \cdot \ U(\theta, \varphi) = \sin^2\theta \ \sin\varphi$$
$$c \cdot \ U(\theta, \varphi) = \sin^2\theta \ \sin^2\varphi$$

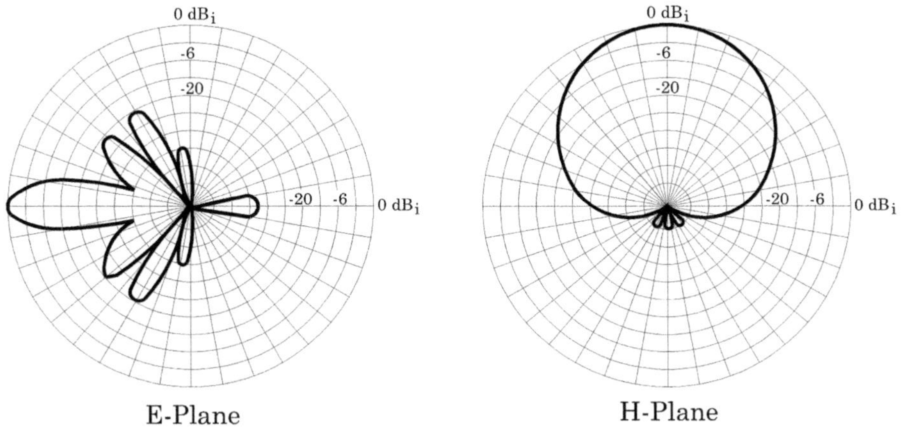

Fig. 2.50 Radiation patterns of unidirectional cellular antenna

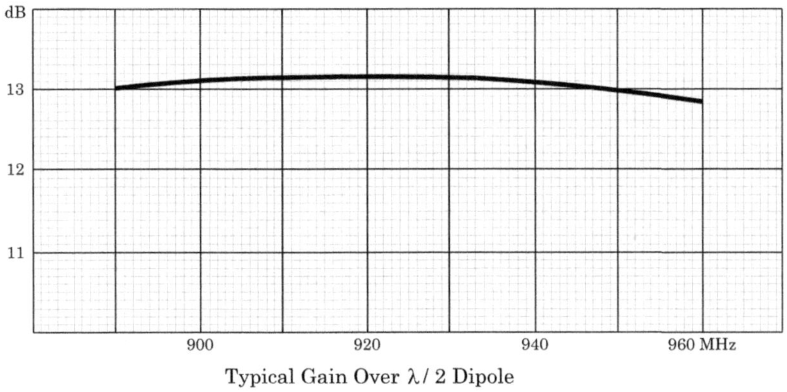

Fig. 2.51 Gain variations of antenna over $\lambda/2$ dipole

(a) Find the directivity
(b) Calculate the HBW in azimuthal and elevation planes

24. Repeat example (2.5) for $U(\theta) = 5\,\cos^2\theta$ and $P_r = 5\,\text{W}$.
25. The polar diagram of an HF spiral antenna radiating circular polarized radiowave is specified in Fig. 2.52; determine:

(a) The antenna gain and its directions
(b) The HBW and TBW

26. A LOS radio-relay network operating at $f = 10\,\text{GHz}$ is used to relay TV and audio signals. Each repeater station consists of TX, RX antenna, and associated equipment. The TX and RX antennas are identical, each having a gain of 15 dB over isotropic antenna. The adjacent repeaters are separated with a distance of 10 km. For acceptable signal-to-noise ratio (SNR), the received power level

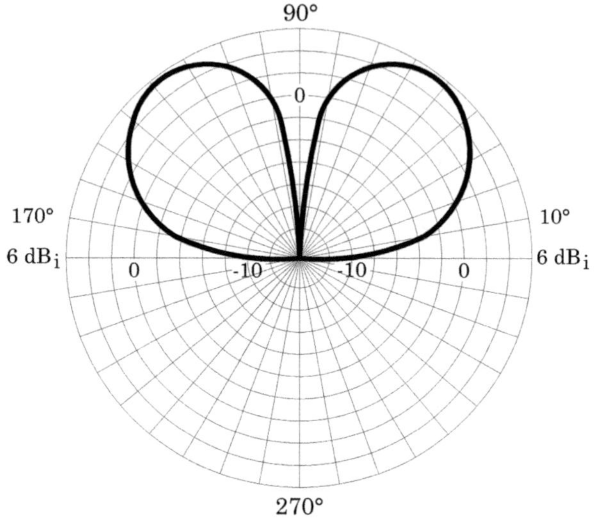

Fig. 2.52 Polar radiation diagram of HF spiral antenna

must be greater than 10 nW. Loss due to polarization mismatch is not expected to exceed 3 dB. Assume matched loads and free-space propagation condition and determine the minimum TX output power.

27. The electric field of a uniform plane radiowave traveling along the negative z-direction is expressed by

$$\bar{E}_a = (\hat{a}_x + j\hat{a}_y)E_0 e^{+jkz}$$

The radiowaves are received by an antenna placed at the origin and characterized by:

$$\bar{E}_a = (2\hat{a}_x + 3\hat{a}_y)E_1 \frac{e^{-jkr}}{r}$$

Determine the following:

(a) Polarization of the incident radiowaves and receiver antenna
(b) Losses due to the polarization mismatch between the incident radiowaves and the receiving antenna

28. An elliptically polarized radiowave traveling in the negative z-direction is detected by a circularly polarized antenna whose main lobe is along the $\theta = 0$ direction. The unit vector indicating the polarization of the incident radiowave is given by

$$\hat{\rho_w} = \frac{2a_x \hat{+} j\hat{a}_y}{\sqrt{5}}$$

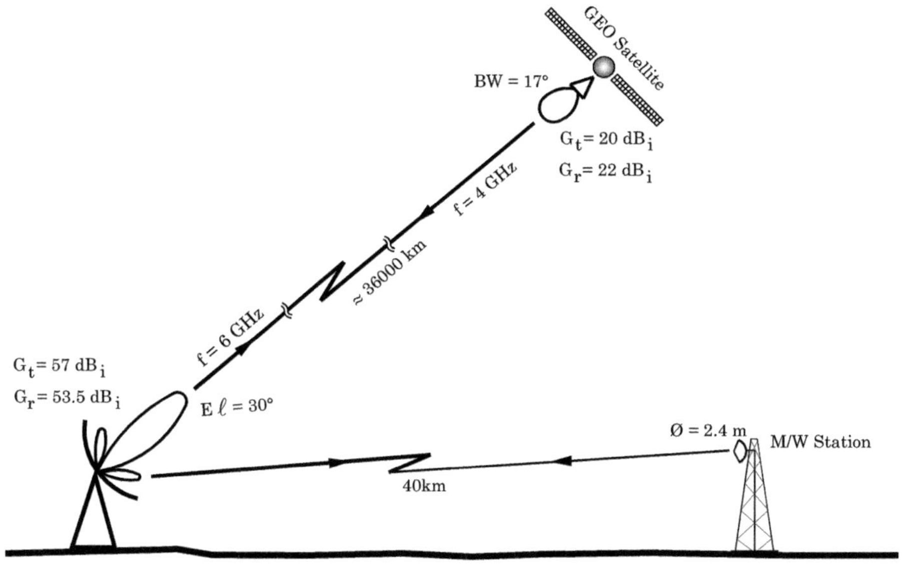

Fig. 2.53 Satellite radio link

Calculate the PLF, when the radiated radiowaves are:

(a) Right-hand circular polarized (RHCP)
(b) Left-hand circular polarized (LHCP)

29. A communications satellite as specified in Fig. 2.53 is located in the Earth stationary orbit about 36,000 km above its surface. The satellite transmitter generates 8-W output power signal connected to an antenna with $G = 20$ dB$_i$. The ground station receiver is a 53.5-dB$_i$ ($D \approx 15$ m) Cassegrain antenna. Assume no resistive losses in either antenna; perfect polarization match and perfect impedance match at both ends working at 4 GHz for down link determine:

    (a) The power density incident on the receiving antenna at the ground station in terms of W/m^2
    (b) The power received by the ground station antenna and its effective area

30. For the satellite link presented in problem 29 and based on the data fixed on Fig. 2.53, find:

    (a) The HBW of the main lobe related to the ground station antenna. Justify the big difference with and the HBW of satellite antenna.
    (b) The spurious emissions generated by the ground station in a direction toward a terrestrial microwave fixed station, when TX output power is 2 kW.

(c) Unwanted signal level detected by a terrestrial microwave antenna (ϕ = 2.4 m working at C band) when it is directed toward the ground station at LOS condition. Assume 45-dB lower antenna gain for off-axis radiation by the ground station.

31. For the satellite link presented in Fig. 2.53, find:

    (a) The sky-noise temperature for the satellite C, Ku, and Ka bands
    (b) The G/T of the ground station assuming the total system noise temperature equal to 60°K
    (c) The G/T in case of using Ku band, while the antenna gain and total system noise temperature increase to 57 dB$_i$ and 80°k, respectively

32. Find antenna system equivalent noise temperature of the following satellite ground stations:

    (a) $G/T = 28.2$ dB/K, $D = 7.5$ m , Ku band
    (b) $G/T = -2$ dB/K, $D = 1$ m , INMARSAT L band

33. Calculate the far field (Fraunhofer) and the near field (Fresnel) distances for the following antenna systems with parabolic reflectors:

    (a) $D = 7.5$ m, $f = 12.5$ GHz
    (b) $D = 3$ m, $f = 7.5$ GHz
    (c) $D = 0.9$ m, $f = 1{,}500$ MHz

34. Figure 2.32 presents the radiation pattern of a typical parabolic dish antenna; determine:

    (a) Antenna gain in dB$_i$
    (b) The first side lobe gain and its HBW
    (c) The HBW and TBW beamwidths
    (d) The ratio of radiated power in the principal axis and maximum first lobe directions

35. A flat rectangular passive reflector is used in a microwave radio link working at 8.4 GHz. The passive reflector size is $6 \times 8\,\mathrm{m}^2$, and the incident angle is 43°; find:

    (a) The reflector ideal and actual gains ($\eta = 90\%$)
    (b) The local nulls and maxima points
    (c) The half-power (3-dB) and tenth-power (10-dB) points and related beamwidths
    (d) The radiation pattern in terms of θ

36. The RPE of the passive reflector characterized in problem 35 is expressed by $E(U) = 20 \log U$ for $U \geq 3$. Calculate the front-to-back ratio.
37. Repeat problem 35 for a flat elliptical passive reflector with the following assumptions:

$$2a \times 2b = 5 \times 6\,\mathrm{m} \times \mathrm{m}\,, \quad f = 7.5\ \mathrm{GHz}\,, \quad \alpha = 53^\circ$$

38. The RPE of the passive reflector can be approximated by $30\ \log(0.1\ e^2 \cdot U)$ for $U \geq 5$; find the gain of first, second, and third side lobes in dB_i.
39. Assume a passive reflector with 30-m^2 physical area; find:

    (a) The reflector gain in dB_i for $f = 7.5$ GHz
    (b) The overall efficiency if it provides 100 dB_i gain in a specific position
    (c) In the case of incident angle $\alpha = 0$, determine the aperture efficiency if actual gain measured is equal to 107 dB_i

40. Determine a suitable and practical arrangement for passive reflector(s) based on its position in a radio link, where:

    (a) Incident angle $\alpha = 50°$
    (b) Incident angle $\alpha = 70°$
    (c) Incident angle $\alpha = 80°$

    Also find the transfer angle for each case including imposed loss in case of using only single passive reflector.
41. For a double passive arrangement including similar units of rectangular shape with physical area of 30 m^2, find:

    (a) Closely coupled parameter for double passive arrangement
    (b) Loss of double passive, α_p in dB, where $\alpha_1 = 10°$ and $\alpha_2 = 40°$
    (c) Net passive repeater gain

42. A rectangular passive reflector is located 1.2 km from the active antenna in a radio link at $f = 8$ GHz. Physical area of the passive is 50 m^2, and incident angle of radiowaves is 37°.

    (a) Calculate the passive-type index, F_t, and verify its position to meet the near field criterion
    (b) Calculate the near field parameter, l
    (c) Determine the passive net gain or loss

43. Repeat example 2.15 for $a \times b = 20' \times 24'$, $\alpha = 30°, \eta = 90\%$, and $f = 6$ GHz
44. The vertical path angles of a passive reflector toward associated active antennas are $\theta_1 = 2°$ and $\theta_2 = 12°$; when $\alpha = 35°$, find:

    (a) Vertical angle of the passive face
    (b) Reduction ratio in the passive gain due to actual incident angle β compared with its value where $\alpha = 0$

45. Follow the step-by-step method explained in Sect. 2.8.7 to calculate a periscope gain/loss with the following assumptions:

$$\text{Passive size}: 8' \times 12' \quad \text{Passive and antenna spacing}: d = 150'$$

$$\text{Operating frequency}: \ f = 8\ \text{GHz}$$

46. A radio link includes a back-to-back antenna as a passive repeater, considering the following assumptions:

    - Repeater distance from active terminals: $d_1 = 8\,\text{km}$, $d_2 = 12\,\text{km}$
    - Operating frequency: $f = 8\,\text{GHz}$
    - Active antenna diameter: $D_1 = 8'$, $D_2 = 10'$
    - Passive antenna diameter: $D_3 = D_4 = 12'$

    find the link fade margin if other parameters are as indicated in Fig. 2.44.
47. A typical radio link using a single passive reflector is shown in Fig. 2.45. Considering the following assumptions:

    - Passive distance from active terminals: $d_1 = 15\,\text{km}$, $d_2 = 20\,\text{km}$
    - TX output power: 500 mW
    - Operating frequency: $f = 7.5\,\text{GHz}$
    - Active antenna diameters: $D_1 = 10'$, $D_2 = 12'$
    - Passive size: $A = 72\,\text{m}^2$
    - Incident radiowave angle: $\alpha = 37^\circ$

    Find the link fade margin if other parameters are as indicated in Fig. 2.45.
48. A typical radio link using a near field single passive reflector is given in Example 2.20 with the following assumptions:

    - Passive distance from active terminals: $d_1 = 25\,\text{km}$, $d_2 = 1{,}000\,\text{m}$
    - RX threshold level: $-80\ \text{dB}_m$
    - Operating frequency: $f = 11\,\text{GHz}$
    - Active antenna diameters: $D_1 = 2.4\,\text{m}$, $D_2 = 3\,\text{m}$
    - Passive size: $A = 6 \times 8\,\text{m}^2$
    - Incident radiowave angle: $\alpha = 40^\circ$

    (a) Calculate the passive-type index (F_t) and verify its position to meet the near field criterion.
    (b) Find the link fade margin if other parameters are as indicated in Fig. 2.46.

49. A typical radio link using far field double passive repeater (as depicted in Fig. 2.47) is characterized by the following parameters:

    - Repeater distance from related active terminals: $d_1 = 15\,\text{km}$, $d_2 = 12\,\text{km}$
    - TX output power: 500 mW
    - Distance between passive reflectors: 20 m
    - Active antenna diameters: $D_1 = 3\,\text{m}$, $D_2 = 3.6\,\text{m}$
    - Passive no. 1, size and angle: $4 \times 8\,\text{m}^2$, $\alpha = 20^\circ$
    - Passive no. 2, size and angle: $6 \times 8\,\text{m}^2$, $\alpha = 30^\circ$

    Find the link fade margin if other parameters are unchanged.
50. A typical radio link using a near field double passive repeater is given in example (2.22) with the following assumptions:

    - Repeater distance from active terminals: $d_1 = 25\,\text{km}$, $d_2 = 1.5\,\text{km}$
    - RX threshold level: $-82\ \text{dB}_m$
    - Active antenna diameters: $D_1 = 8'$, $D_2 = 10'$

- Operating frequency: $f = 7.5\,\text{GHz}$
- Size and incident angle/passive no. 1: $A = 60\,\text{m}^2$, $2\alpha = 75°$
- Size and incident angle/passive no. 2: $A = 60\,\text{m}^2$, $2\alpha = 60°$

(a) Calculate the passive-type index (F_t) and verify its position to meet the near field criterion
(b) Calculate the near field parameter, l
(c) Determine the overall passive repeater gain or loss in dB
(d) Find the link fade margin if other parameters are the same as indicated in Fig. 2.48.

Chapter 3
Radiowave Propagation in Satellite Communication

3.1 Introduction

3.1.1 Overall Objectives

The first space craft was launched in the 1960s and placed successfully in an orbit around the Earth. Further achievements in the space telecommunications technologies led scientists and radio experts to employ space facilities for expansion of the telecommunications services to meet ever-increasing requirements of the world community.

Using space technology in the telecommunications fields created a new branch of radio links called "satellite communications." The main objective of satellite communications is to provide reliable and good-quality radio services with vast coverage for large-scale traffic capacity.

In a satellite communications link, the radiowaves are subject to a number of propagation phenomena in various atmospheric layers of the Earth, especially in the Earth-to-space direction. Although satellite communications is a kind of radio link with line-of-sight propagation mode, it possesses different nature due to its propagation medium.

Satellites in telecommunication networks employ an Earth orbit in equatorial, polar, or inclined planes. Earth orbital paths include low, medium, and geostationary (LEO, MEO, and GEO) orbits among which GEO is of high interest. This is a unique circular orbit located in the equatorial plane for which the satellite and Earth rotations are synchronous, resulting in a fixed satellite location as seen by a ground station.

This chapter is devoted to particular issues related to the radiowaves propagation in satellite communications. For general phenomena of troposphere and ionosphere layers and basic principles of radiowaves propagation, you can refer to other references including "Propagation Engineering in Wireless Communications, by A. Ghasemi et al. 2011 Springer."

A. Ghasemi et al., *Propagation Engineering in Radio Links Design*,
DOI 10.1007/978-1-4614-5314-7_3,

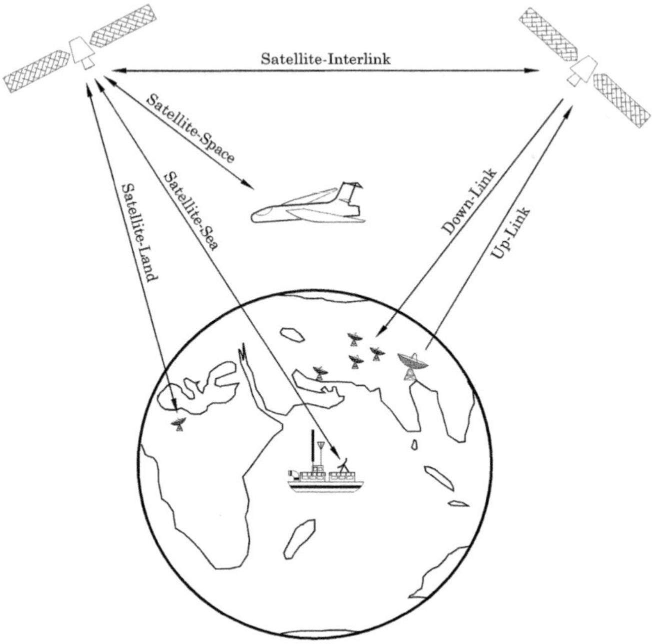

Fig. 3.1 Satellite radio links

3.1.2 Basic Concept

At initial stage, passive satellites and other passive reflectors are considered as a potential solution to fulfill overall objectives of satellite communications to expand worldwide requirements by increasing the distance of a single radio hop, covering vast areas, and improving the service quality as well.

Big reflective balloons and "the Moon" were examined and a number of experimental tests conducted. As depicted in Fig. 3.1, for this type of communications, similar to radar links, the following expression can be used to calculate the received signal power level:

$$P_r = \frac{P_t \cdot G_t \cdot G_r \cdot \sigma}{(4\pi)^2 \cdot d_t^2 \cdot d_r^2} \tag{3.1}$$

P_t, P_r: transmitter and receiver power
G_t, G_r: gain of TX and RX antennas
d_t, d_r: distance of ground stations from satellite
σ: satellite gain/loss factor

Assuming $d_r \approx d_t = d$ for long distance satellite links, then

$$P_r = K \cdot \frac{\sigma}{d^4} \cdot P_t \tag{3.2}$$

The satellite gain/loss factor is a function of its shape, material, size, and frequency of radiowaves. Equation (3.2) reveals that the received power level is inversely proportional to the fourth power of the distance. Finally, based on the following facts, passive satellites are not a practical solution to meet telecommunications requirements:

- Requiring high-power transmitter and very sensitive receivers
- Requiring very large directional antennas with high gains
- Movements of reflectors in the space

Together with research on space passive reflectors, a lot of studies were performed using active satellites. These types of satellites include telecommunication transponders equipped with transmitters, receivers, auxiliary devices, and electronics circuitry to act as an active radio repeater. For active satellite repeaters, we can use the following expression to calculate the received signal level in the ground station:

$$P_r = \frac{P_t \cdot G_t \cdot G_r \cdot g_s \cdot A_s}{(4\pi)^2 \cdot d_t^2 \cdot d_r^2} \tag{3.3}$$

In the above relation, g_s is the transponder amplification coefficient, and A_s is the effective area of satellite antenna. Assuming that g_s is proportional to $4\pi d_r^2$, then (3.3) is converted into the following form:

$$P_r = K' \cdot \frac{A_s}{d^2} \cdot P_t \tag{3.4}$$

In this situation, the received signal power is inversely proportional to the second power of the distance, which is considered a great advantage for active satellites. Based on potential capabilities of active satellites, they grew rapidly, expanding the existing facilities and presenting new services with local, national, regional, and global coverage.

3.1.3 Main Applications

Nowadays, satellites have provided immense, and at some instances unique, applications in different equatorial orbits in the form of LEO (low Earth orbit), MEO (medium Earth orbit), and GEO (geostationary orbit) including inclined and polar orbits on circular or elliptical paths. Main applications of the satellite communications include but not limited to the following items:

- International networks for handling of audio, video, and data traffic
- Global coverage for navigational aids
- Audio and video public broadcasting networks
- Mobile communications networks
- Remote sensing, meteorology, and telemetering networks
- Satellite dedicated and multipurpose national/regional networks

3.1.4 Features of Satellite Communications

Main characteristics of the satellite communications which have created a special and exclusive place for it are very diverse and versatile. The most significant of these features are listed below:

- Extensive coverage without geographic or political restrictions
- Flexibility and capability of providing different services
- Limitations of natural adverse effects and man-made artifacts
- Wide bandwidth utilizing various technologies
- Evolution of different access technologies and optimum use of the available bandwidths
- Possibility of using different and adaptive structures for different projects requirements
- Relative independence with respect to distance for users in the coverage region of the satellite
- Cost-effective solution for the links between users separated with very long distances and low traffic volume
- High reliability and good quality of communications

The above-mentioned features have made satellite communications a distinguished and the best alternative to provide services in many fields of telecommunications.

3.1.5 Satellite Links

As indicated in Fig. 3.2, the satellite links can be divided into two general groups:

- Earth satellite link
- Intersatellite link (ISL)

Each bidirectional satellite link consists of one uplink and one downlink. In uplink, the ground station acts as transmitter (TX) and the satellite station acts as a receiver (RX), while in the downlink, it operates vice versa.

ISLs can be used for direct contact between satellites. In the past time, because of satellite mobility, long distance between them, limitation of available electric power, and saving in the weight of satellite, the links between satellites were established through ground stations. But nowadays, using LEO satellites for continuous coverage makes ISLs necessary. This is feasible due to achievements in telecommunications technology and manufacturing of high-quality RF components and equipments. This process can be done by installing related on-board equipment in the satellite.

ITU-R has suggested several hypothetical reference digital paths for the design of telecommunications satellite networks. The HRDP circuit is used as a basis for

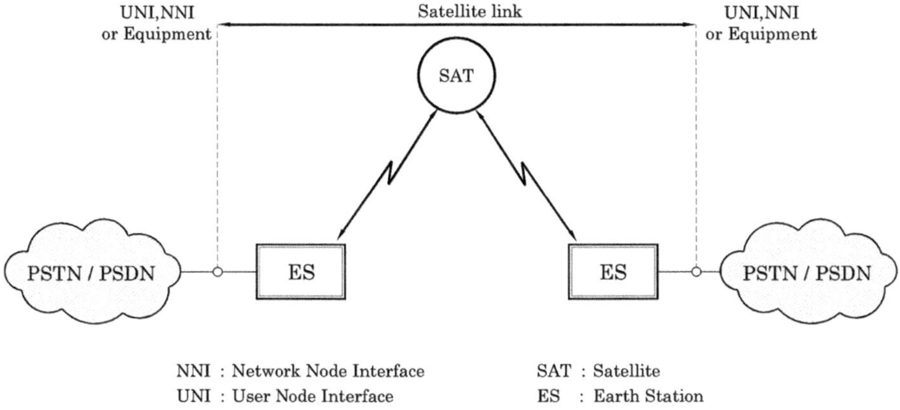

Fig. 3.2 Hypothetical reference path for satellite radio link (Ref.: ITU-R, S.521-4)

Table 3.1 Main services by satellite communications

• Broadcasting satellite service	• Fixed satellite service
• Amateur satellite radio	• Mobile satellite service
• Satellite Earth exploration service	• Space research service
• Radio determination satellite	• Radio navigation satellite
• Meteorology satellite service	• Inter satellite service

calculations and design criteria for different conditions of satellite communications network.

A typical example of the satellite HRDP circuits is given in Fig. 3.2, and for more details, you can refer to the ITU-R recommendation S.521 in P-series.

3.1.6 Frequency Bands

3.1.6.1 Telecommunications Services

Communications services which are provided by satellites are very diverse and expanding constantly. Noting the type of service based on a list prepared by ITU-R and presented in Table 3.1, the suitable frequency bands are dedicated based on the provisions of article 5 of radio regulations. Also, ITU-R recommendations include a number of RF channel arrangements that may be used by radio experts in manufacturing, design of satellite network, and frequency planning.

Numerous bands and sub-bands have been allocated to the main satellite services which are indicated below:

- Fixed services
- Maritime, aeronautical, and land mobile services

- Military services
- Public broadcasting services
- VSAT services
- Intersatellite services
- Special telecommunications services like navigational aids, location positioning, and remote sensing.

3.1.6.2 Satellite Frequency Bands

Some portions of VHF, UHF, SHF, and EHF bands have been allocated to provide services related to the satellite communications. These bands, including their application and priorities, are indicated in the Article 5 of the ITU radio regulations for different regions of the world.

Technical and operational aspects of each sub-band should be observed by radio designers and competent authorities. Among different limitations for each sub-band, there are also total bandwidth, center frequency, number of RF channels, RF channel space, guard bands, polarization, and frequency separation of each pair of frequencies for go-and-return RF channels.

Addition to regulatory restrictions for selecting a suitable frequency band and RF channel arrangement, the following points must be taken into account for a satellite link design:

- Types and capacity of traffic requirements
- Radiowaves propagation and its different phenomena such as attenuation, atmospheric phenomena, composition of atmospheric layers, and polarization of the waves
- Type of satellites and circuits being used including integration of services and determination of required transponders
- Type of service, such as telephony, data, video, and multimedia

Some of frequency bands used for satellite services are:

- *UHF band* for meteorological, special services, and military.
- *L band* in the range of 1.5–1.6 GHz for communications of mobile units (particularly maritime).
- *C band* in the range of 4–6 GHz for intersatellite and Earth stations link in the fixed and mobile services.
- *7 and 8* GHz *bands* for military use.
- *Ku band* in the range of 12–14 GHz for intersatellite and Earth stations in the fixed and VSAT services.
- *Ka band* in the range of 20–30 GHz for fixed and mobile intersatellite and Earth exploration services.

3.2 Geometry of Satellite Orbits

3.2.1 Earth Orbits

3.2.1.1 Types of Orbits

A satellite orbit is defined by the center of it subjected to forces of natural origin, mainly the force of gravity. The path may be adjusted occasionally by low-energy corrective forces exerted by a propulsive device in order to achieve and maintain a desired path. Simply, a satellite orbit is a path which satellite travels around the Earth in its stable condition.

The Earth orbits in satellite communications as shown in Fig. 3.3 are of the following types:

- *Equatorial orbit* of a satellite is an orbit which its plane coincides with the equator of the Earth.
- *Polar orbit* of a satellite is an orbit which its plane contains the polar axis of the Earth (a line passing the reference points of the North and South Poles).
- *Inclined orbit* of a satellite is an orbit which is neither equatorial nor polar.

In accordance with Fig. 3.3, the inclination angle of the orbital plane relative to the equator plane is denoted by i and in this condition:

$i = 0 \implies$ orbit in equatorial plane
$i = 90^\circ \implies$ orbit in polar plane
$0 < i < 90 \implies$ orbit in inclined plane

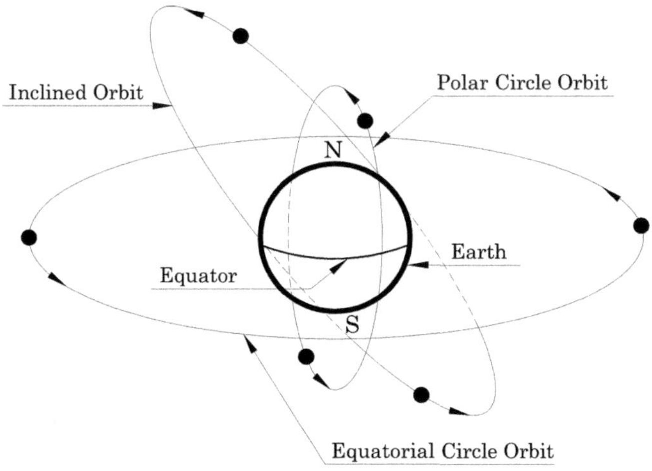

Fig. 3.3 Main satellite orbits

Another classification of communications satellites is based on their revolving orbit altitude from the Earth's surface such as:

- *GEO satellites*
  A satellite revolving in the *geo-synchronous Earth orbit* is defined as a satellite for which the mean sidereal period of revolution is equal to the sidereal period of rotation of the Earth about its own axis.
  The GEO is a circular orbit in the equatorial plane around 36,000 km above the Earth's surface. This is a unique orbit regarded as a natural resource of the satellite communications. The use of the GEO needs to be coordinated internationally, and special permit should be given by the relevant authorities.
- *HEO satellites*
  A satellite revolving in a *highly-inclined Earth orbit* which is defined as an elliptical orbit most typically with a perigee of 500 km or more and an apogee of 50,000-km or less altitude above the Earth's surface with an inclination angle greater than 40° from the equatorial plane.
- *LEO satellites*
  A satellite revolving in *LEO* which is defined as a circular or elliptical orbit of about 700–3,000 km (below about 2,000 km based on ITU-R recommendation S.673) altitude above the Earth's surface.
- *MEO satellites*
  A satellite revolving in *MEO* which is defined as a circular or elliptical orbit about 8,000–20,000 km (about 10,000 km based on ITU-R recommendation S.673) altitude above the Earth surface.

3.2.1.2 Basic Principles of Satellite Motion

For satellite motion in its orbital plane, the following principles must be considered:

1. Center of the Earth is situated in the satellite orbital plane. Otherwise, the satellite will not be stabilized on its orbital plane.
2. In the ideal case, the satellite motion is along an elliptical path with eccentricity of e $(0 \leqslant e \leqslant 1)$ which, in the extreme conditions, will be converted to a circular path $(e = 1)$ or a straight line $(e = 0)$.
3. Earth center is located on one of the focal points of the related ellipse of the satellite orbit.
4. On satellite orbit, the resultant force exerted to it must be always equal to zero; otherwise, satellite will be diverted from its intended trajectory. These forces mainly consist of the satellite weight and centripetal force.
5. Kepler's rules govern the motion of satellites, and based on these rules, it can be concluded that

$$T = 2\pi\sqrt{\frac{a^3}{\mu}} \tag{3.5}$$

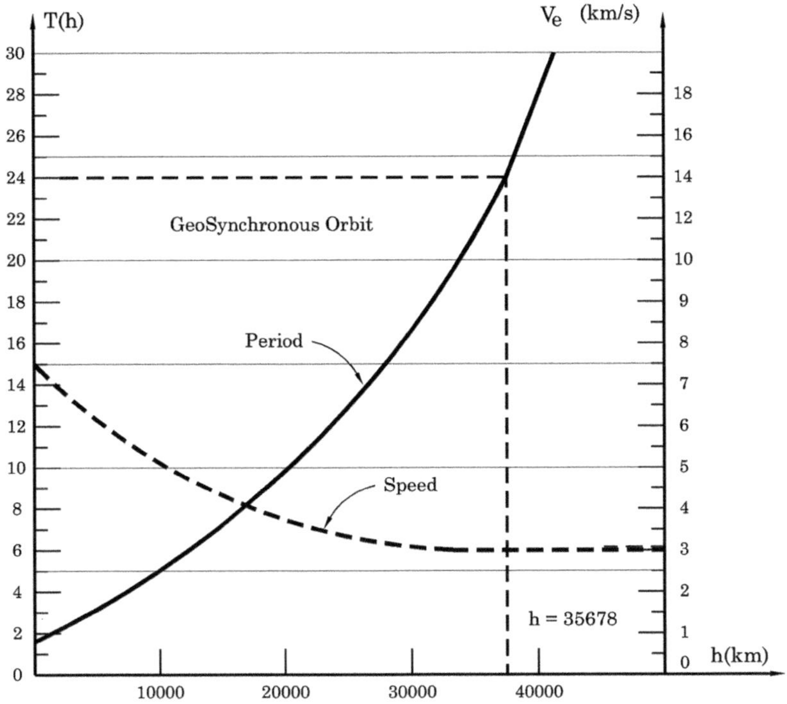

Fig. 3.4 Variation of satellite period and velocity vs. height

T: revolving period of the satellite around the Earth in seconds.
a: half of the major axis of the elliptical path in m.
μ: Keppler's constant and equivalent to the product of the mass and gravity force constant of the Earth and equal to $\mu = 3.99 \times 10^{14} \mathrm{m}^3/\mathrm{s}^2$.

The following equation determines satellite speed at any instant:

$$V_e = \sqrt{\mu \left(\frac{2}{r} - \frac{1}{a} \right)} \tag{3.6}$$

V_e: speed of the satellite in m/s
r: distance of the satellite from the center of Earth in m

The graphs of Fig. 3.4 show variations of T and V_e in terms of the satellite altitude. As it can be noted in order to have the identical period times for satellite the Earth's rotation, that is, equal to 24 h, it is essential that the satellite be positioned

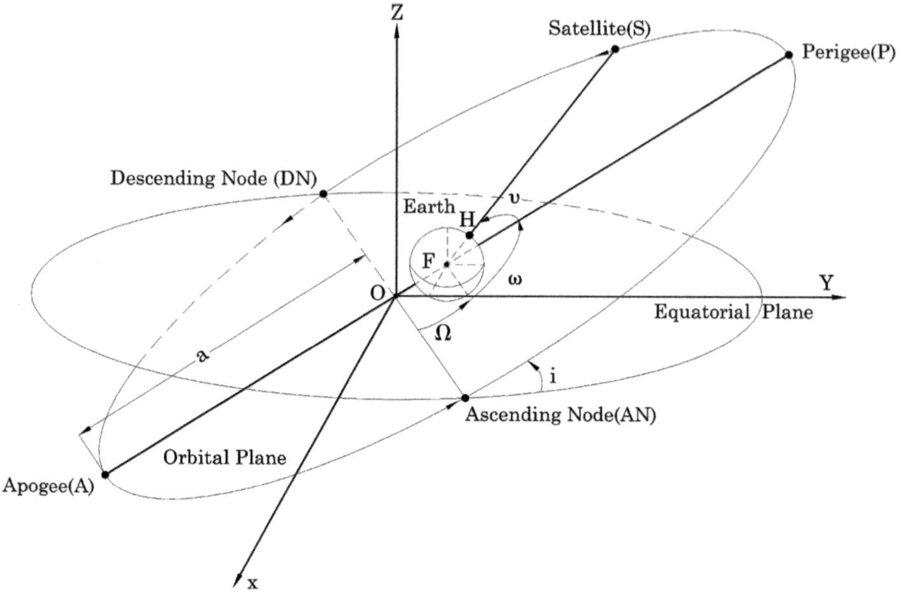

Fig. 3.5 Geometry of satellite orbits

at a height of 35,678 (approximately 36,000) km above the Earth's surface in the equatorial plane.

In other words, to obtain stationary condition of the satellite relative to the ground observer, it is required that the revolving period of the satellite around the Earth be equal to the Earth's rotation period around its axis passing the poles.

3.2.1.3 Orbital Parameters

Generally, to define the exact specifications of the satellite orbital parameters, the following two defaults must be considered:

- Elliptical shape of revolving satellite orbit around the Earth
- Adjusting the Earth's center on one of the focal points of the ellipse

On the basis of principles of solid geometry and noting Fig. 3.5, it is essential to elaborate on the following three cases:

1. Orbital plane, including flat plane containing satellite orbit.
2. Main characteristics of the ellipse on the plane.
3. Specifications of the elliptical position and major axis angle of the ellipse relative to the baseline.

To determine each of the above cases, two factors among the following six factors should be indicated:

- Inclination angle made between orbital plane and equatorial plane which is denoted as the angle i in Fig. 3.5.
- Angle between the ascending direction (interconnecting line of the Earth's center at point O and satellite transversing equatorial plane on a specified direction) denoted as Ω.
- Eccentricity of the elliptical orbit denoted as e and having a value between zero and one.
- Half of major axis of the elliptical orbit denoted as a.
- Argument of meridian point indicated by ω.
- Mean anomaly indicated by M.

3.2.2 Satellite Coverage

3.2.2.1 Geographic Coverage

Geographic coverage of a satellite transponder is defined as a region of the Earth where establishing direct link between terrestrial stations and satellite is possible. The geographic coverage of stationary satellites is nearly fixed, whereas for a moving satellite, it is a function of time. In general, the geographics coverage of satellites can be categorized in the following four groups:

- Global coverage (Fig. 3.6)
- Regional coverage (Fig. 3.7)
- National coverage (Fig. 3.7)
- Domestic/local coverage (Fig. 3.7)

The area of satellite coverage is one of the most significant issues in satellite network design. For the planning of different segments of satellite network either space or ground stations, the following basic data are required:

- Number of satellites
- Satellite orbit
- RF channel arrangement for satellite transponders
- Link budget calculations

Geometrical concept of a satellite coverage and related radius on a smooth Earth are given in Figs. 3.8 and 3.9. Geographic coverage of each satellite is a function of its altitude from the Earth's surface, the least elevation angle of terrestrial antennas, EIRP, and G/T of the satellite footprints.

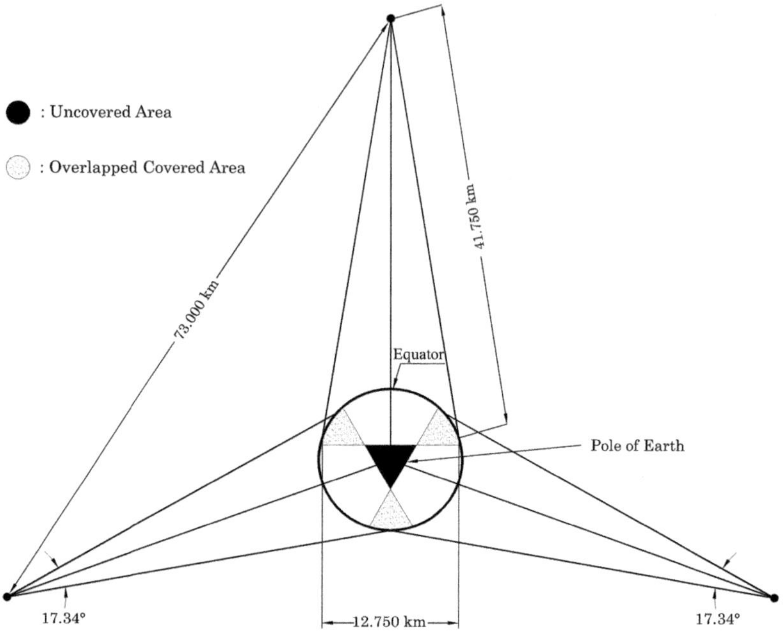

Fig. 3.6 Satellite global coverage

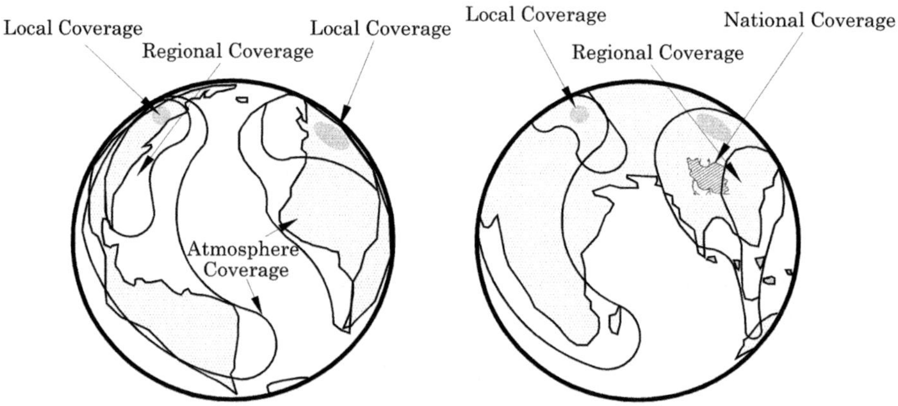

Fig. 3.7 Types of satellite geographical coverage

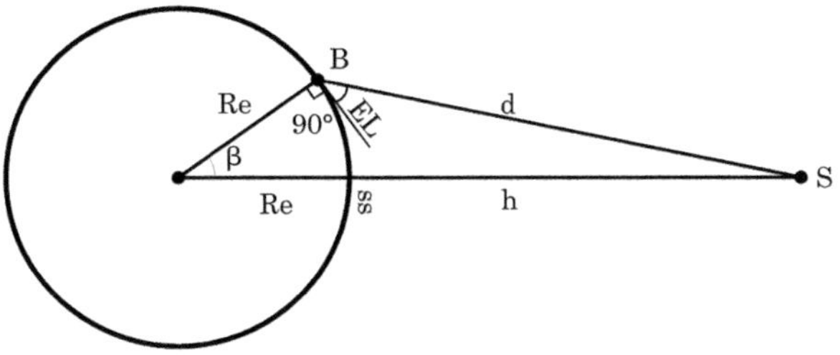

Fig. 3.8 Geometry of satellite coverage

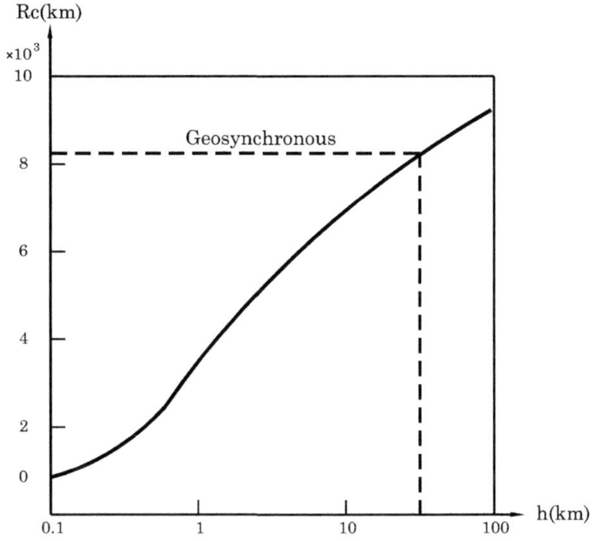

Fig. 3.9 Radius of ideal satellite coverage vs. height

For calculation of the radius of covered area of a satellite noting the quoted figure, its markings, and using trigonometric relations, we can write

$$\frac{R_e}{R_e + h} = \frac{\cos(\beta + E_l)}{\cos E_l} \tag{3.7}$$

In the above relation, E_l is the elevation angle of terrestrial antenna, β is look angle of the satellite, h is the height of satellite from the ground, and R_e is the Earth's radius. For an extreme condition where $E_l = 0$, that is, SB line tangent to the Earth, then the largest coverage area, called the ideal coverage area, is obtained for which the following relation is valid:

$$\beta_0 = \text{Arccos}\frac{R_e}{R_e + h} \tag{3.8}$$

Therefore, surface of the covered area is

$$S_M = 2\pi R_e^2 (1 - \cos \beta_0) \tag{3.9}$$

It may be explained that the radius of ideal satellite coverage in terms of its height from the Earth surface is displayed by the graph of Fig. 3.9.

Example 3.1. In a global positioning system (GPS), satellites are located at a distance of 25,000 km from the Earth's center. In case the least elevation angle of the wave path is 20° and path deviation caused by troposphere and ionosphere layers is neglected, then find:

1. The coverage area of each satellite
2. Effective beamwidth of the satellite antenna

Solution. 1.

$$E_l = 20^\circ, \quad R_e = 6,370\,\text{km}, \quad R_e + h = 25,000\,\text{km}$$

Using (3.7), it can be concluded that

$$(\beta_0 + 20^\circ) = \text{Arccos}[(6,370)/(25,000) \times \cos\,20^\circ] \implies \beta_0 = 56^\circ$$

$$S_M = 2\pi R_e^2(1 - \cos\,\beta_0) = 1.1238 \times 10^8\,(\text{km})^2$$

2. Noting Fig. 3.9 results in

$$\angle S = 180 - \beta - E_l - 90^\circ \implies \angle S = 14^\circ$$

$$\text{BW} = 2 \times \angle S = 28^\circ$$

■

3.2.2.2 Global Coverage

In international telecommunications services, it is required to provide a satellite network to cover possibly the entire Earth surface or at least a big portion of it. This type of service is called "global coverage" and includes normally many countries spread over different continents and oceans. At present time, there are several satellite networks operating globally among which the following are more popular ones:

- International telecommunications satellite (INTELSAT) network
- International maritime satellite (INMARSAT) network
- GPS

- *Global Coverage by GEO Satellites*

  Global coverage by GEO satellites is available through three satellites positioned in geostationary orbit (GSO). These satellites, as shown in Fig. 3.6, should be spaced with 120° spacing at an altitude around 36,000 km above mean sea level (AMSL) equal to 41,750 km from the Earth's center. A typical coverage of global network such as INTELSAT is given in Fig. 3.10.

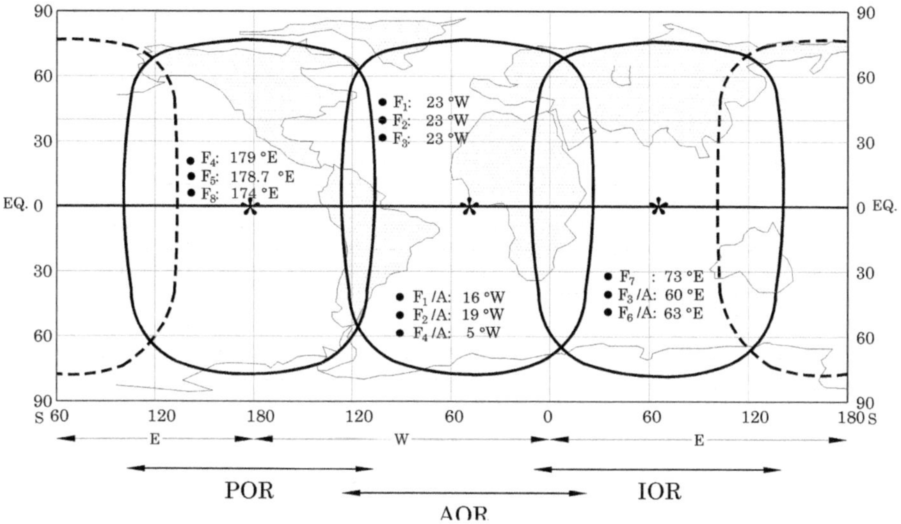

Fig. 3.10 INTELSAT-5 global coverage

Normally such coverage of the Earth is divided into three regions as described below, and each group is served by one satellite positioned on an equatorial synchronous orbit:

- Atlantic Ocean Region (AOR)
- Indian Ocean Region (IOR)
- Pacific Ocean Region (POR)

On the basis of calculations, the coverage area of each GEO satellite with an antenna having 17.3° beamwidth is equivalent to 216 million km^2. Considering the entire surface of the Earth around 510 million km^2, then three satellites arranged in proper positions are able to meet the global coverage requirements with suitable percentage of overlapping at the border areas of the covered region.

Imperatively, the latitudes beyond 75° in the Northern and Southern Hemispheres are not covered by antenna elevation angles of higher than 5°, and as a result, their links are not reliable with acceptable quality.

- *Global Coverage by LEO Satellites*

  In LEO satellite networks, since these satellites are in motion with respect to the Earth, it is possible to adjust their orbital motions in a way to provide global coverage. If the number of such satellites is limited, then in this situation continuous coverage will not be available at certain times, and even at some occasions, it will last few weeks for a satellite to appear at a specific location of the Earth. Besides, service duration of this appearance will be less than 30 min for each satellite.

  Normally, in this network, the operation is limited to particular applications such as Earth explorations, surveying, telemetering, and remote sensing in which it is possible to store and forward the data. Also, for continuous time coverage in addition to a worldwide coverage, the number of satellites will increase greatly which will be specified according to the following issues:

- Satellite altitude from the Earth's surface
- Beamwidth of the satellite antennas
- Structure of Earth station antennas
- Traffic handling capacity

Example 3.2. If, in the previous example, we employ six circular orbits each having four equally spaced satellites, find:

1. Number of satellites
2. Type of network coverage and percentage of overlapping
3. Number of satellites having simultaneous LOS condition with every ground station

Solution. 1.

$$N_S = 4 \times 6 = 24$$

2. Noting the effective coverage area of each satellite, the total surface is equal to

$$S_t = 24 \times 1.1238 \times 10^8 = 2.696 \times 10^9\,\text{km}^2$$

$$S_E = 4\pi R_e^2 = 5.0965 \times 10^8\,\text{km}^2$$

$$S_t > S_E \quad \Longrightarrow \quad \text{Global Coverage}$$

Percentage of overlapping is equal to

$$C\% = \frac{S_t}{S_E} \times 100 \approx 500\,\%$$

3. Noting the overlapping percentage, at each instant of time, five satellites on different positions can be viewed by a ground station. It should be noted that for specifying four parameters of length, width, height, and time, it is necessary, each GPS terminal to collect suitable data from at least four satellites in each measurement. ■

3.2.2.3 Satellite Time Coverage

Time coverage is considered in case of nonstationary satellites. This type of space networks includes LEO, MEO, and HEO satellites on equatorial, polar, and inclined orbits. In LEO network, satellite situated on equatorial plane, since the orbit is circular (or near to circular), the coverage area on the Earth's surface is almost fixed and uniform, but its location is continuously changing.

In elliptical orbits, due to the altitude variation of satellite from the Earth's surface, in addition to variation of coverage location, the coverage surface also changes.

In stationary satellites, although the coverage time is assumed constant, practically their coverage is also influenced by time-varying effects such as drifts of satellite in the space, diurnal and seasonal atmospheric variations, and different sky effects at elevation angles of less than 10° for ground stations antenna.

To establish a continuous coverage over an area of the Earth's surface by LEO and MEO satellites, a cluster of satellites must be used. In this position, the number of satellites can be calculated based on dimensions of the coverage area, geographic coordinates, elevation, and radiation pattern of satellite antenna and orbital plane.

Noting the above discussion, when the coverage area is vast and takes regional or global dimensions, providing permanent time coverage becomes difficult, and the number of satellites increases. In these circumstances, the following cases are outstanding:

- Using small antennas dimensions and at some occasions omnidirectional antennas in the related ground stations
- Automatic adjustment system required for directional antennas
- Connection between satellites and handling of main traffic and control data
- Control system of the Earth station

3.2.2.4 Traffic Coverage

Traffic coverage in satellite communications is influenced mainly by the following factors:

- Number of users in each region
- Versatility of services
- Average and maximum traffic per each user
- Traffic capacity in busy hours
- Required grade of service

To meet the above requirements, traffic study is required. Then necessary calculations must be conducted based on the estimated traffic to finalize total capacity and the number of satellites.

3.2.2.5 Coverage Footprints

To display the coverage of a GEO satellite transponder, receive/transmit footprints are used. These footprints (or contours) for which a few samples are presented in Figs. 3.11 and 3.12 contain data about values of G/T and EIRP related to these transponders. For simplicity of calculations related to the satellite links, the relevant network operators of satellite space segment prepare EIRP and G/T coverage

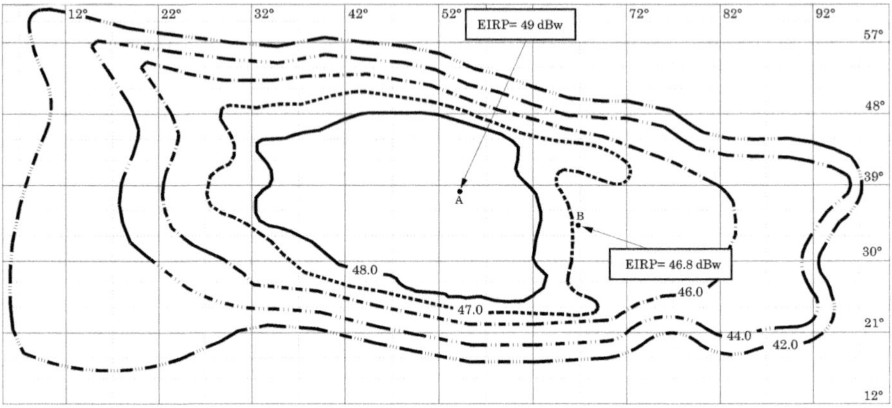

Fig. 3.11 Typical transmit footprint of satellite transponder

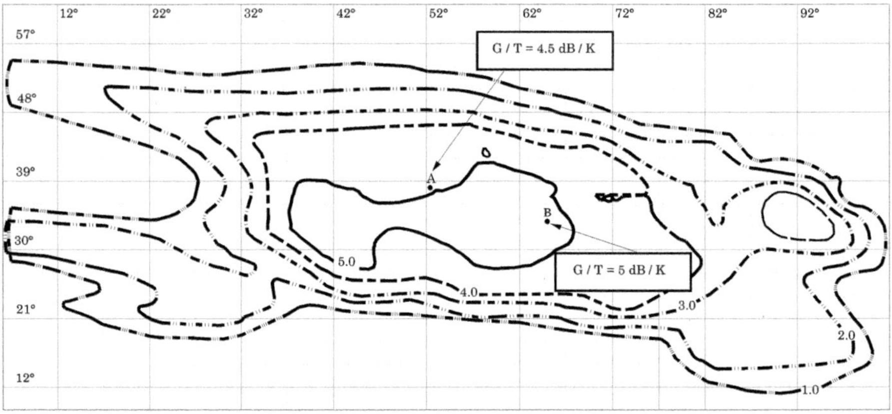

Fig. 3.12 Typical receive footprint of satellite transponder

contours related to each transponder for using them by competent experts. These contours are influenced by the following factors:

- TX power and RX sensitivity (threshold level)
- Radiation pattern of the antenna and its alignment
- Antenna gain
- Frequency band
- Satellite position

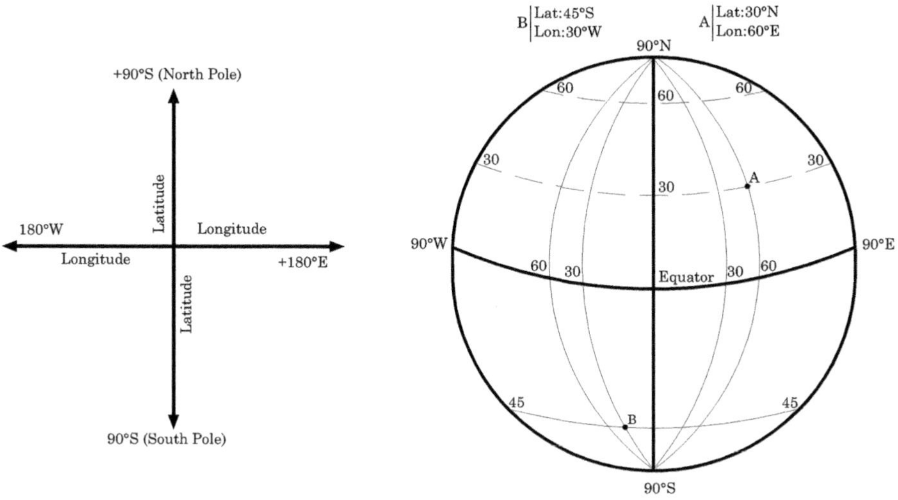

Fig. 3.13 Location coordinates (latitude and longitude)

3.2.3 Distances and Angles

3.2.3.1 Coordinates of Radio Stations

In radio network design, it is required to use a suitable coordinate system to indicate the location of a station. The most popular system is to specify each point with its latitude, longitude, and altitude. As shown in Fig. 3.13, the latitude and longitude are used to specify the precise location of points on the reference surface of the Earth.

Since the actual physical surface of the Earth is too complex for mathematical analysis, two levels of abstraction are employed in the definition of these coordinates:

- A surface which approximates the mean sea level over the oceans and its continuation under the land masses.
- To approximate the Earth by a mathematically simple reference such as a sphere.

The latitude and longitude together with some specification of altitude constitute a geographic coordinate system as detailed in the specification of ISO 19111 standard. In a radio link, reference point is the antenna center point which is specified by the following coordinates:

- "Latitude" as illustrated in Fig. 3.13 for the Northern Hemisphere is given as an angle between 0 and $+90^\circ$ or 90N measured from the Earth's equator. In the Southern Hemisphere, it is in the range of 0 to -90° or 90S.
- "Longitude" as illustrated in Fig. 3.13 is an angle measured from the reference point on the equator from 0 to 180E or 0 to 180W.

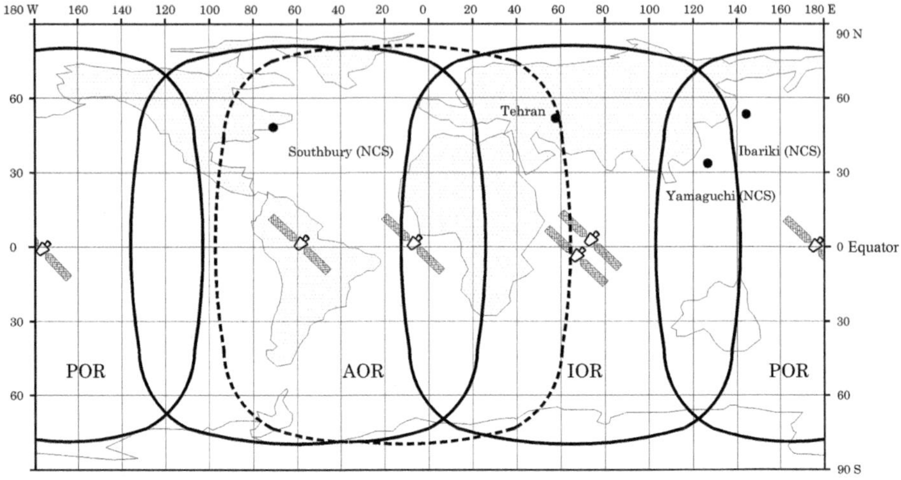

Fig. 3.14 INMARSAT global coverage

- "Altitude" which is the height of the point from a reference level measured in meters. Normally the reference level is the mean sea level for which the altitude is determined AMSL. Sometimes for special applications, above ground level or AGL is used instead of AMSL.

As an example, global coverage of INMARSAT system (the first international mobile maritime satellite system) is displayed in Fig. 3.14. Normally the topographic maps are calibrated based on geographic latitude and longitude. The AMSL heights are also specified using the available contours in those maps. Nowadays, it is possible to specify with great precision the geographic latitude and longitude of any point on the Earth using GPS receivers.

Example 3.3. INMARSAT satellite network for global coverage of year 2004 is illustrated in Fig. 3.14.

1. Which satellites can cover Turkey?
2. Specify POR satellite coverage.
3. Specify coordinates of each satellite.

Solution. 1. Turkey is covered completely by AOR satellites and also partially by IOR satellites.

2. As it can be observed from the figure, POR satellite can cover the spots located on the latitudes of 80°N to 80°S and longitudes of 105°E to 180°E and 105°W to 180°W.
3. Since all satellites are the GEO type, thus they are located in the equatorial plane with latitude of zero degree, and their longitudes are

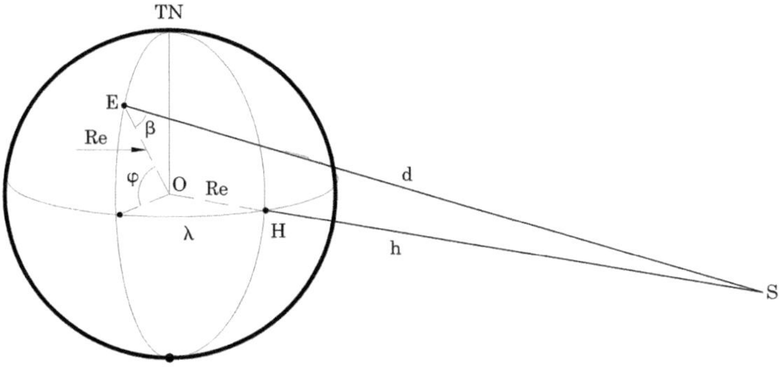

Fig. 3.15 Subsatellite point

$$\text{AOR}: \quad \text{I}: 54^\circ\text{W}, \quad \text{II}: 15.5^\circ\text{W}$$
$$\text{IOR}: \quad \text{I}: 64^\circ\text{W}, \quad \text{II}: 64^\circ\text{W}$$
$$\text{POR}: \quad \text{I}: 178^\circ\text{W}$$

■

3.2.3.2 Subsatellite Point

Figure 3.15 illustrates subsatellite point denoted by H, as the cross-point of the connecting line between the satellite reference point and the Earth's center (line SO) with the Earth's surface. Since this line is normal to the Earth's surface, it is taken as the shortest distance between the satellite and the Earth. The value of h for GEO satellites is around 36,000 km and for LEO satellites is few hundreds to few thousands of kilometers.

3.2.3.3 Distance of Two Points on the Earth

As illustrated in Fig. 3.16, the following formulas can be used for the calculations of distance of two points on the Earth and more accurately two points on the Earth's surface with zero altitude such as water, like E and H:

$$\tan\left(\frac{z}{2}\right) = \tan\left|\frac{\varphi}{2}\right| \times \cos\left[\left|\frac{\lambda}{2}\right| \times \frac{\cos\frac{\varphi}{2}}{\sin\left(\frac{L_E+L_H}{2}\right)}\right] / \cos\left[\left|\frac{\lambda}{2}\right| \times \frac{\sin\frac{\varphi}{2}}{\cos\left(\frac{L_E+L_H}{2}\right)}\right] \tag{3.10}$$

$$\widehat{EH}\ [\text{km}] = 111.12z\ \ [\text{degrees}] \tag{3.11}$$

In the above formulas:

φ: difference of latitudes of points E and H, $\varphi = (L_E - L_H)$
λ: difference of longitudes of points E and H, $\lambda = (\text{Lon } E - \text{Lon } H)$

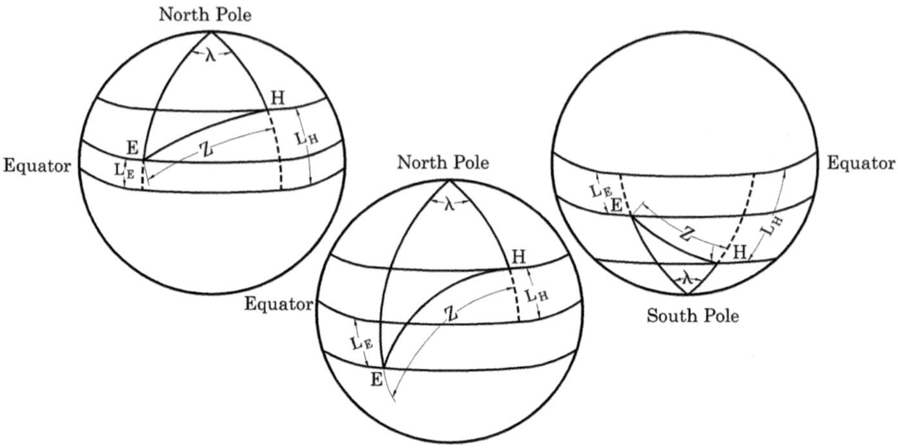

Fig. 3.16 Distance between two points on earth surface

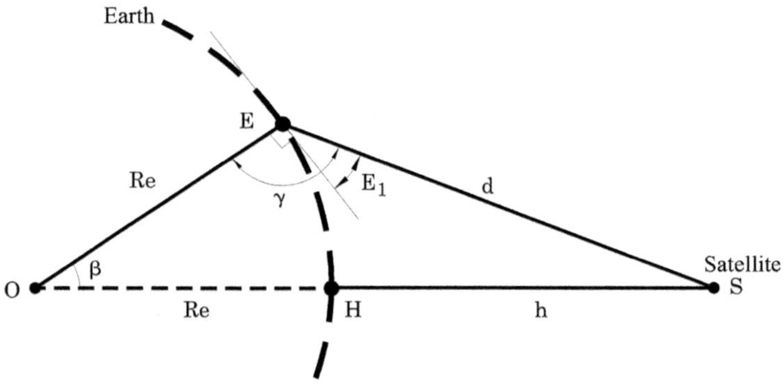

Fig. 3.17 Satellite look angle

L_H: latitude of point H
Lon H: longitude of point H
z: auxiliary angle in terms of degree
EH: curved distance of points E (Earth station) and H (subsatellite point)

According to Fig. 3.17, the look angle of the satellite link from the Earth's center is denoted as β, then

$$\beta = \widehat{EH} \quad [\text{rad}] \tag{3.12}$$

For calculation of β in a general position noting the coordinates of the Earth station (E) and subsatellite point (H) according to (3.11) and for the length of arc $\widehat{EH}$ and using the trigonometric relations, it can be concluded that

$$\beta\,[\text{rad}] = \frac{\widehat{EH}[\text{km}]}{R_e[\text{km}]} \tag{3.13}$$

In the above relation, R_e is the actual radius of the Earth and is equivalent to 6,370 km. When the satellite orbit is located in the equatorial plane, the look angle of the satellite (angle β) can be calculated from a simpler relation using the following formula:

$$\beta = \cos^{-1}[\cos\varphi \cdot \cos\lambda] \tag{3.14}$$

In the latter relation, λ is the difference of longitudes of points E and H, and the angle φ is the latitude of point E.

3.2.3.4 Distance Between Earth and Satellite Stations

Observing Fig. 3.17 and using cosines formula in the triangle SOE, the distance between Earth and satellite station is obtained from the following equation:

$$d = \sqrt{(R_e + h)^2 + R_e^2 - 2R_e(R_e + h)\cos\beta} \tag{3.15}$$

The above quantities have the same definition and unit as mentioned in the previous sections.

Example 3.4. A LEO satellite at an elevation of 830 km from the Earth's surface is located in the equatorial orbit.

1. If the effective angle of the main lobe of its antenna is 40°, calculate the satellite coverage area on the Earth at each time.
2. Specify if the points located in the latitudes of 2°N are under coverage or not and what are their minimum and maximum effective distances from the satellite.

Solution. 1. First, by noting Fig. 3.17, the maximum latitudes which can be viewed by the satellite using trigonometric relations are as follows:

$$R_e = 6,370\,\text{km}, \quad R_e + h = 7,200\,\text{km} \quad \Longrightarrow \quad \beta = 2.75^\circ$$

Calculating the coverage area using (3.9) yields

$$S_M = 2\pi R_e^2(1 - \cos\beta) = 293,458\,\text{km}^2$$

2. The maximum effective look angle is the same as the calculated value based on the previous assumptions; thus locations on latitudes of 2°N will be under coverage (in suitable times). Using (3.15), the maximum distance is also equal to

$$d_{\max} = \sqrt{(7,200)^2 + (6,370)^2 - 2 \times 6,370 \times 7,200\cos(2.75)} = 891\,\text{km}$$

The minimum effective look angle by considering the specified limitations is equal to the latitude of the point, that is, $\beta' = 2^\circ$. In this situation, the minimum distance between satellite and Earth station is equal to

$$\beta' = 2^\circ, \quad \mathrm{OS} = R_\mathrm{e} + h = 7{,}200\,\mathrm{km}, \quad \mathrm{OE} = R_\mathrm{e} = 6{,}370\,\mathrm{km}$$

$$d_{\min} = 863\,\mathrm{km}$$

■

3.2.3.5 Antenna Bearing of Earth Station

For alignment of the Earth station antenna toward a satellite, it is essential to calculate the azimuth and elevation angles of the antenna. In this case for satellites located in the equatorial plane such as GEO satellites, the following relations may be used:

- Elevation angle denoted with E_l and considering Fig. 3.17 and the fact that the antenna elevation angle is acute (less than 90°); thus, applying the trigonometric rules in the OEH triangle, we have

$$E_l = \gamma - 90 \tag{3.16}$$

$$E_l = \text{Arc cos}\left(\frac{R_\mathrm{e} + h}{d} \times \sin\,\beta\right) \tag{3.17}$$

- Azimuth of the Earth station antenna is in fact the angle of antenna axis (boresight), in the horizontal plane relative to the true north direction of TN. To calculate this angle with variation range between zero and 360° according to Fig. 3.18, there are four different cases due to the relative positions of points E and H. To calculate the exact value of the antenna azimuth indicated by A_z, it is essential to obtain the acute angle A from the following relation:

$$A = \text{Arc sin}\left(\frac{\sin|\lambda|}{\sin\,\beta}\right) \tag{3.18}$$

Then it can proceed with considering Fig. 3.18 to obtain the final value of A_z. Based on the relative position of the points of E (location of Earth station) and H (subsatellite point), it must be noted that Fig. 3.18 is treated for equatorial satellites, but it can be generalized for all situations.

Example 3.5. In case of using directional antennas in the Example 3.4, calculate the range of elevation angles for Earth station antenna.

Solution. The minimum antenna elevation angle will be obtained for satellite look angle of 2.75°, and its minimum value will be calculated for 2°. Noting Fig. 3.17, for the minimum elevation angle,

$$\beta = 2.75^\circ, \ \angle S = 20^\circ \implies \gamma = 157.25^\circ \implies (E_l)_\mathrm{m} = 67.25^\circ$$

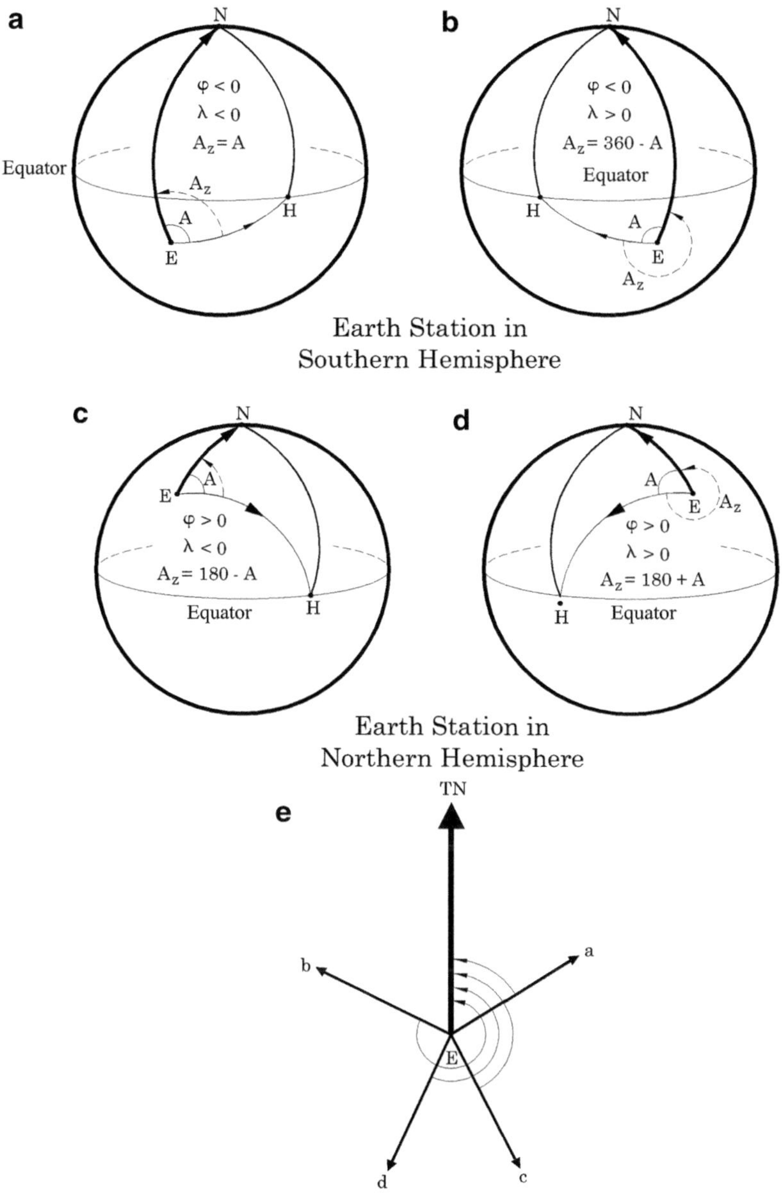

Fig. 3.18 Elevation and azimuth angles of ground station antenna

For the maximum value of the elevation angle, using trigonometric relations yields

$$\angle S' = 14.9^\circ \implies (E_l)_{\mathrm{M}} = 90 - \beta^o \angle S' = 73.1^\circ$$
$$\implies 67.25^\circ \leqslant E_l \leqslant 73.1^\circ$$

■

3.3 Main Propagation Phenomena

3.3.1 Introduction

In satellite communications, the radiowaves on a long path between the Earth and satellite stations encounter different phenomena related to the radiowaves propagation. These phenomena as depicted in Fig. 3.19 are evaluated in the four following parts:

- Local environment considerations related to the Earth station
- Effects of troposphere layer
- Effects of ionosphere layer
- Beyond of Earth or deep space effects

Considering the altitude of different layers as discussed in the first chapter, ionosphere layer has a higher altitudes and normally includes up to 500–1,000 km from the Earth's surface. Since GEO satellites are positioned at a height of around

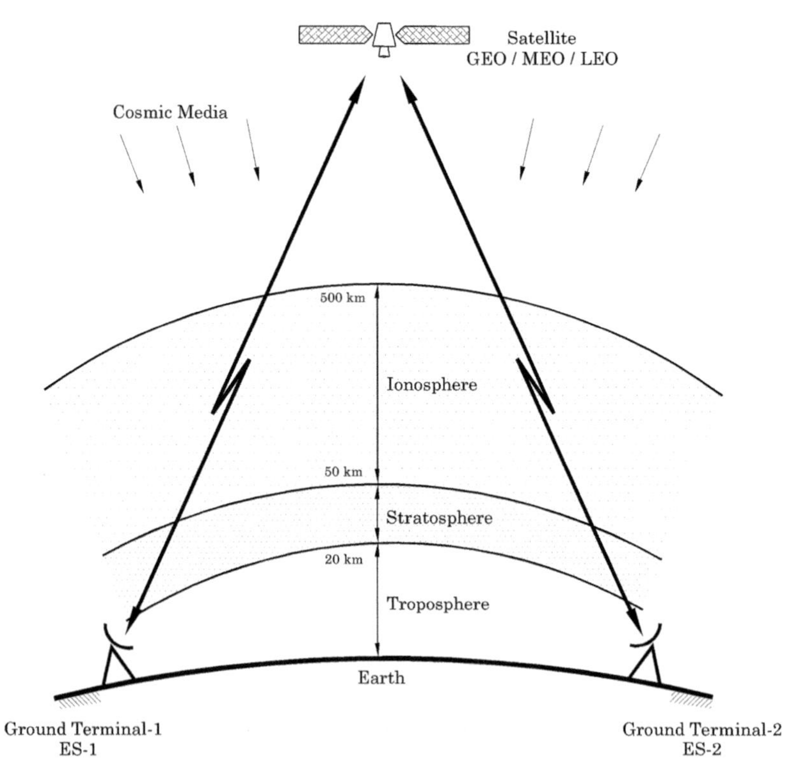

Fig. 3.19 Typical route of radiowave in satellite radio link

36,000 km, MEO satellites at a height of several thousand km and LEO satellites are positioned at the heights of more than 500 km; it can be concluded that satellite waves will be subjected to the adverse effects of all the above layers.

Because of varying nature of radiowaves propagation phenomena depending on several factors, analytical evaluations are not an easy process and their modeling are sophisticated. Radio section of International Telecommunication Union or ITU-R has conducted many research and assessment of this case which the pertinent results are presented in the form of P-series recommendations listed in Appendix B as a rich resource of data about radiowaves propagation. ITU-R website is http://WWW.itu.org which may be referred for more information.

This chapter evaluates the issues of radiowaves propagation particularly concerning satellite communications. For more details on this topic, please refer to the radiowaves propagation engineering books and/or ITU-R reports of P-series.

3.3.2 Main Phenomena

The main phenomena of satellite wave propagation can be grouped as the following:

- Attenuation in free space (clear sky)
- Attenuation of rain and other atmospheric precipitations
- Attenuation due to gases and vapors
- Attenuation from the cloud and fog
- Attenuation from dust and sand
- Scintillation and multipath effects
- Variations of the radiowaves polarization
- Group delay variation and dispersion
- Propagation delay
- Antenna gain reduction due to wave-front incoherence
- Focusing and defocusing

3.3.2.1 Effective Factors on the Phenomena

Magnitude of the above-mentioned phenomena is not fixed, and the following satellite specific parameters have outstanding effects on them:

- Satellite orbit
- Frequency band

- Type of provided services such as audio, video, data, and telemetry
- Mode of communications such as fixed satellite service FSS, mobile satellite service MSS, and TV broadcasting

3.3.2.2 Effects of Atmosphere Layers

In the satellite communications, different phenomena of radiowaves propagation depend on frequency which are nonlinear, and even some occasions relate to the third power of frequency. In addition to frequency, their magnitude is different for troposphere and ionosphere layers.

The effects of troposphere layer such as rain and dust are mostly in the form of amplitude attenuation and proportional to the RF channel frequency. The effects of ionosphere layer in addition to amplitude attenuation are in the form of phase variations and distortion caused by time delay and with inverse relation to the RF channel frequency. For this reason, in planning of satellite links, selection of frequency bands must be conducted by considering the effects of different phenomena.

3.4 Clear Sky Losses

In the absence of any atmospheric phenomena such as cloud, mist, storm, rain, snow, and hail, satellite waves undergo losses caused by the following factors which are so-called clear sky losses:

- Free-space propagation
- Antenna coupling
- Beam spreading

It must be emphasized that the above-mentioned losses exclude the energy absorption of atmosphere climatic phenomena.

3.4.1 Free-Space Loss

Since the propagation of radiowaves between satellite and ground stations is based on the LOS condition, thus as discussed in Chap. 1, the following formula is valid to calculate its free-space loss:

$$A_f = \text{FSL} = 92.4 + 20\,\log\,f + 20\,\log\,d \tag{3.19}$$

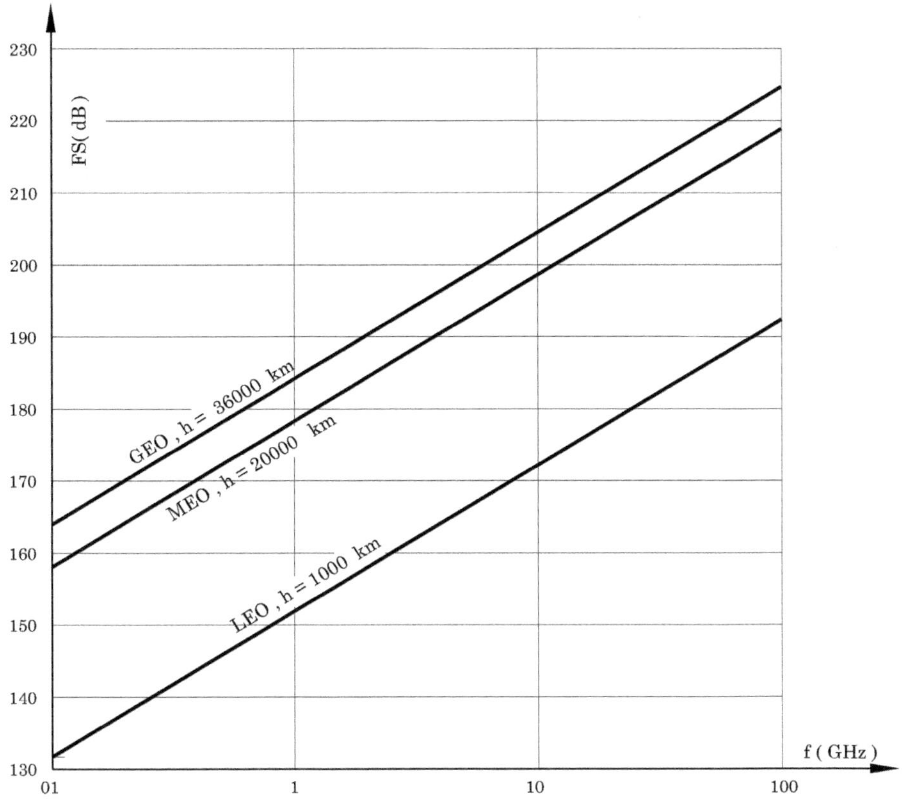

Fig. 3.20 Free space loss in satellite radio link

For proof of the above equation, refer to the first chapter of the book. As it can be noted, this loss is directly proportional to the square of frequency and distance, and thus with increase of frequency, its value also increases. In the above formula, the value of A_f is in terms of dB; f and d are in terms of gigahertz and kilometer, respectively. Figure 3.20 indicates A_f variations in terms of frequency for three typical altitudes of equatorial satellites.

Example 3.6. Calculate free-space loss for satellite radio links in the following conditions, and compare the results with Fig. 3.20.

1. INMARSAT network at L band, that is, $R = 1.5\,\text{GHz}$, $S = 1.6\,\text{GHz}$, and $d = 38,000\,\text{km}$
2. National satellite network at Ka band, $R = 20\,\text{GHz}$, $S = 30\,\text{GHz}$, and $d = 36,000\,\text{km}$
3. For reception in GPS network at frequency band of $f = 1,227\,\text{MHz}$ and distance of 20,000 km

Solution. 1.

$$\mathrm{FSL}_S = 92.4 + 20 \log 1.6 + 20 \log 38{,}000 = 188.08 \text{ dB}$$

$$\mathrm{FSL}_R = 187.5 \text{ dB}$$

Because of small difference between send-and-receive frequencies, the free-space loss at the L band for uplink and downlink is roughly the same.

2. For Ka band and considering the assumptions, it gives the following results:

$$\mathrm{FSL}_S = 213.5 \text{ dB}$$

$$\mathrm{FSL}_R = 210.02 \text{ dB}$$

3. For the GPS network using MEO satellites, it may be concluded

$$\mathrm{FSL} = 180.2 \text{ dB}$$

Comparing the results with the graphs of Fig. 3.20 is left to the reader. ■

3.4.2 *Antenna Coupling Loss*

Antenna coupling to the medium is not always perfect and matched. This is due to encountering of incoherent wave front to the RX antenna caused by small-scale irregularities in the atmospheric refractivity index.

Final impact of the antenna coupling loss is reduction in the antenna effective gain in receiving of the satellite waves. In addition, refractivity irregularities will cause abrupt changes in the received signal level which is limited and may be ignored when compared with other adverse effects.

3.4.3 *Beam Spreading Loss*

Permanent and continuous reduction of refractivity index with aboveground height results in the gradual tilting of the radiowaves, and consequently, defocusing appears at low elevation angles. The magnitude of defocusing over frequency range of 1–100 GHz is independent of frequency. In case beam spreading loss is denoted as A_{bs}, for antenna elevation angles of less than 5° at all latitudes, the following relation may be used to estimate the yearly average value of A_{bs}:

$$A_{\mathrm{bs}}[\mathrm{dB}] = 2.27 - 1.16 \log(1 + \theta_0) \tag{3.20}$$

Also, (3.20) is valid to estimate the beam spreading loss for average worst month at latitudes up to 53°.

For antenna used in a location with latitudes of more than 60°, the following relation may be used as the average worst month value:

$$A_{bs}[\mathrm{dB}] = 13 - 6.4\ \log(1 + \theta_0) \tag{3.21}$$

In the above relations, the value of θ_0 is the apparent elevation angle in terms of mrad considering the effects of the refractivity. For antenna used in a location with latitude between the two above-mentioned limits, the value of A_{bs} is derived from the linear interpolation between them, that is,

$$A_{bs} = A_{bs}(> 60^\circ) - \frac{60}{7} \times \Delta A_{bs} + \frac{1}{7} \times \Delta A_{bs} \times \gamma \tag{3.22}$$

In the latter relation, γ is the latitude of the intended point in terms of degrees and $\Delta A_{bs} = A_{bs}(> 60^\circ) - A_{bs}(< 53^\circ)$.

Example 3.7. Calculate the magnitude of beam spreading loss for antenna apparent elevation angle of 49 mrad on a point with a latitude of 55°.

Solution. Using the above relations yields

$$A_{bs}(< 53^\circ) = 2.27 - 1.16\ \log(1 + 49) = 0.299\ \mathrm{dB}$$

$$A_{bs}(> 60^\circ) = 13 - 6.4\ \log(1 + 49) = 2.127\ \mathrm{dB}$$

$$\Delta A_{bs} = 2.127 - 0.299 = 1.828\ \mathrm{dB}$$

$$A_{bs}(55^\circ) = 2.127 - \frac{60}{7} \times 1.828 + \frac{1}{7} \times 1.828 \times 55$$

$$A_{bs}(55^\circ) = 0.82\ \mathrm{dB}$$

■

3.5 Effects of Troposphere Layer

3.5.1 Introduction

Troposphere layer along with its upper layer (stratosphere) constitutes Earth's atmosphere up to heights of 50 km from its surface. Undesirable effects resulting from this layer occur in conjunction with abrupt atmosphere climatic changes such as wind, rain, snow, storm, dust, haze, cloud, and mist. Such effects for frequencies less than 1 GHz and radiowave elevation angles in excess of 10° are negligible and may be disregarded.

With the reduction of elevation angle and also increase of frequency, the value of amplitude fluctuation and angle of arrival of the antenna become significant and, particularly within high geographic latitudes, are noticeable. Also, equivalent sky-noise temperature increases due to the rain absorption and suspended aerosols in the sky and causes reduction in the received signal's carrier-to-noise ratio.

In addition to the above cases, accumulation of snow and sleet on the ground station antennas can effectively influence their radiation pattern, gain, and polarization discrimination.

The loss of radiowaves on the Earth-to-space direction (in addition to clear sky losses described previously) caused by the following factors is calculated separately, and the total loss is the summation of those factors:

A_g: loss due to atmospheric gases and vapors
A_o: loss due to oxygen molecules
A_w: loss due to water vapor molecules
A_r: loss due to the rain
A_h: loss due to the air humidity and liquid particles other than the rain
A_s: loss due to sudden atmospheric changes

Studies conducted by the ITU-R for the effects of troposphere layer are presented in the competent series of the ITU-R recommendations.

3.5.2 Atmospheric Gases and Vapors

The Earth's atmosphere gases contain permanent dipolar molecules which can resonate when they are exposed to the radiowaves. They absorb a portion of the colliding radiowaves energy at the resonance frequency. The significant resonances or energy absorption bands are depicted in Fig. 3.21 for oxygen molecules and in Fig. 3.22 for water vapor molecules.

Because of the severe loss caused at these frequencies, they must be considered in the design and utilization of satellite radio systems. Such phenomena at frequencies of 22.2, 180, and 320 GHz relate to water vapor molecules (H_2O) and at frequencies of 60 and 118.8 GHz relate to the oxygen (O_2) molecules. Resonances caused by other gases up to 300 GHz are less significant and may be ignored.

Since such phenomenon is a function of gas content constituting the atmosphere, hence this type of absorption occurs in the troposphere section only. Normally, these losses are in terms of dB/km and are devised for a reference atmosphere including one atmospheric pressure on the sea surface and relative humidity of 7.5 g/m^3 at temperature of 20 °C. Oxygen (A_o) and water vapor (A_w) absorption coefficients due to variation of the pressure, temperature, and volume of existing water vapor depend on the height from the Earth's surface. Consequently, the total value of attenuation of radiowaves related to the gases can be stated as follows:

$$A_g = \int_0^{h_0} [A_o(h) + A_w(h)] \cdot dh \qquad \text{dB/km} \tag{3.23}$$

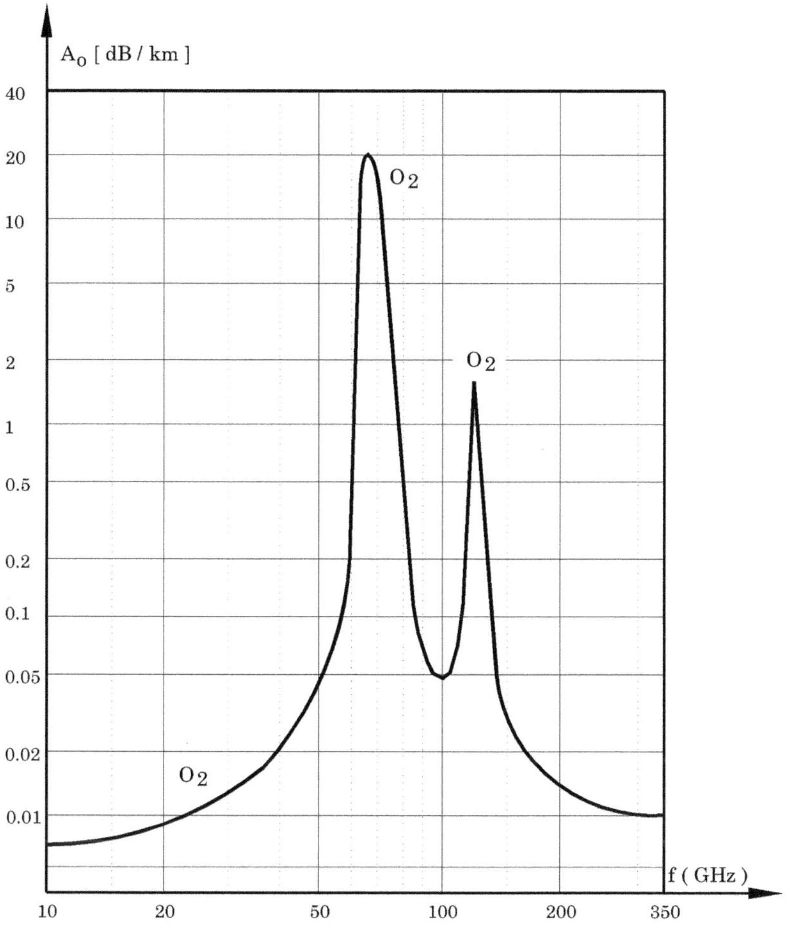

Fig. 3.21 Variations of A_o versus frequency

where h_0 is the total atmosphere height. Relation (3.23) can be approximately valid for most areas of the Earth excluding tropical regions:

$$A_g = 4\,A_o + 2\,A_w \qquad \mathrm{dB/km} \tag{3.24}$$

The values of A_o and A_w can be obtained from Figs. 3.21 and 3.22. For the routes having elevation angle (θ) more than 5°, the following approximate formula may be used:

$$A_g(\theta) = A_g(90^\circ)\ \cdot\ \mathrm{cosec}\ \theta, \quad 5^\circ \leqslant \theta \leqslant 90^\circ \tag{3.25}$$

For elevation angles less than 5°, the above approximation is unacceptable because of the atmospheric refraction and the Earth's curvature. At such an elevation angle, the route of radio beam is not a straight line and becomes curved and very

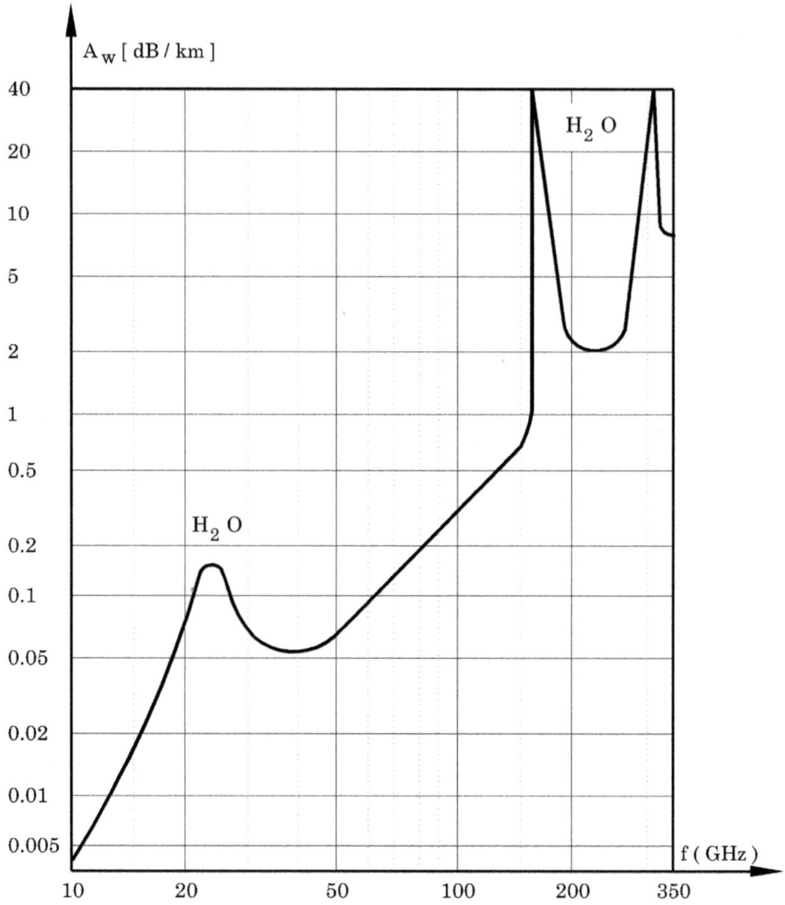

Fig. 3.22 Variations of A_w versus frequency

complex. The values of A_g which has been discussed so far are used for Earth stations which are located at the sea level, whereas for stations above sea level, the A_g value is lower due to shorter length. For frequency band of up to 20 GHz, the value of A_g is reduced nearly with a factor of 2 per each kilometer of the corresponding height. In Fig. 3.23, the variations of loss emanating from gases in the atmosphere in terms of heights from sea level are presented in the frequency range of 50–70 GHz.

Example 3.8. 1. Specify attenuation produced by oxygen molecules and water vapor at 40-GHz satellite radiowaves.

2. In case of antenna elevation angle of 20°, calculate the attenuation from the atmosphere gases.

Solution. 1. Noting Figs. 3.21 and 3.22 for $f = 40$ GHz we have

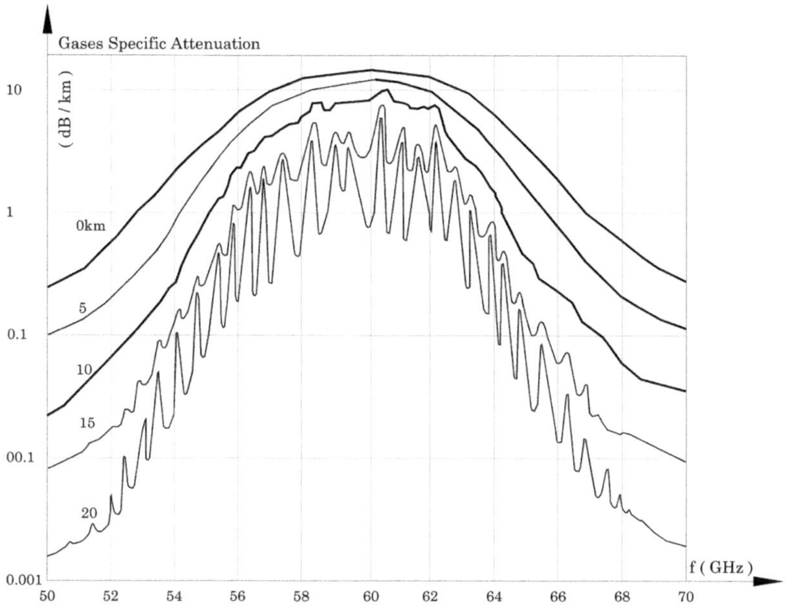

Fig. 3.23 Specific attenuation by atmospheric gases in terms of height

$$A_o = 0.02 \text{ dB/km}, \quad A_w = 0.05 \text{ dB/km}$$

2. Noting (3.24) and (3.25), we have

$$A_g(90^\circ) = 4A_o + 2A_w = 0.18 \text{ dB/km}$$

$$A_g(20^\circ) = 0.18 \times \frac{1}{\sin 20} = 0.52 \text{ dB/km}$$

■

3.5.3 *Attenuation in Worst Month*

Rain losses caused by tropospheric absorption and refraction mechanisms are created by the raindrops. Radiowaves in all frequencies are affected by such phenomena. These mechanisms are more effective only at frequencies higher than several GHz where the wavelengths are small and in the range of rain drop diameter.

Since the rain intensity varies with respect to time and geographic location, hence the losses caused by rain shall vary as well. If we consider a part of raining, because of its nonuniformity on a specified length of a route, for complete calculations of the radiowaves losses, it requires data for the following issues:

- Amount of rain intensity R (mm/h)
- Size of raindrop diameter a (mm)
- Mixed refractivity index of water drops (as a function of the temperature of the medium, velocity, and distribution of the raindrops).

Determination of these factors with a good precision is not easy and perhaps impossible. Losses emanating from the rain for an Earth to the satellite route can be obtained by the following relation:

$$A_r = \int_0^L L_0 \cdot A_{rs}(x)\, dx \quad \text{dB} \tag{3.26}$$

where A_{rs} is the specific rain attenuation (dB/km) and L in terms of kilometer is the total length which the radiowaves pass through the rain. On horizontal routes, the length of radiowaves route through the rain can reach to several hundreds of kilometers, whereby on vertical route the total length of this route is about 4 km.

It may be possible to obtain A_{rs} with the graph of Fig. 3.24 which is based on the complicated theoretical calculations considering absorption and refraction of radiowaves by raindrops, or it is more convenient to obtain from the following simple and empirical formula:

$$A_{rs} = a \cdot R^b \quad \text{dB/km} \tag{3.27}$$

where R is the rain intensity (mm/h) measured on the Earth's surface and a and b are empirical coefficients which depend on the frequency and temperature. Table 3.2 indicates the sample values for rain with a temperature of zero Celsius. Figure 3.24 and Table 3.2 can be relied on for initial estimation of radiowaves attenuation caused by rain. This shall be on the basis of the amount of raining on the Earth's surface, because to acquire the value of R statistically for a specified duration of time at a location is not difficult.

The total value of the radiowaves losses by rain can be obtained through direct measurements on the satellite link (different tests were conducted on various locations of the Earth) and/or carry out such a task based on calculation of R. There are some approaches recommended for the measurement of A_r based on the following relation:

$$A_r[\text{dB}] = A_{rs}[\text{dB}] \cdot L_e \tag{3.28}$$

where L_e is the effective length subjected to raining and it is usually a function of rain intensity (R) and elevation angle. The approximate graph of L_e, which is obtained from direct measurements in several countries, is indicated in Fig. 3.25 as a reference. This graph is sufficient for initial estimations; however, there are other estimation methods specified in competent textbooks.

Also, for calculations of rain losses, there is a detailed calculation method with the essential descriptions contributed by ITU-R recommendation P.618-7 and may be used as a basis for practical calculations.

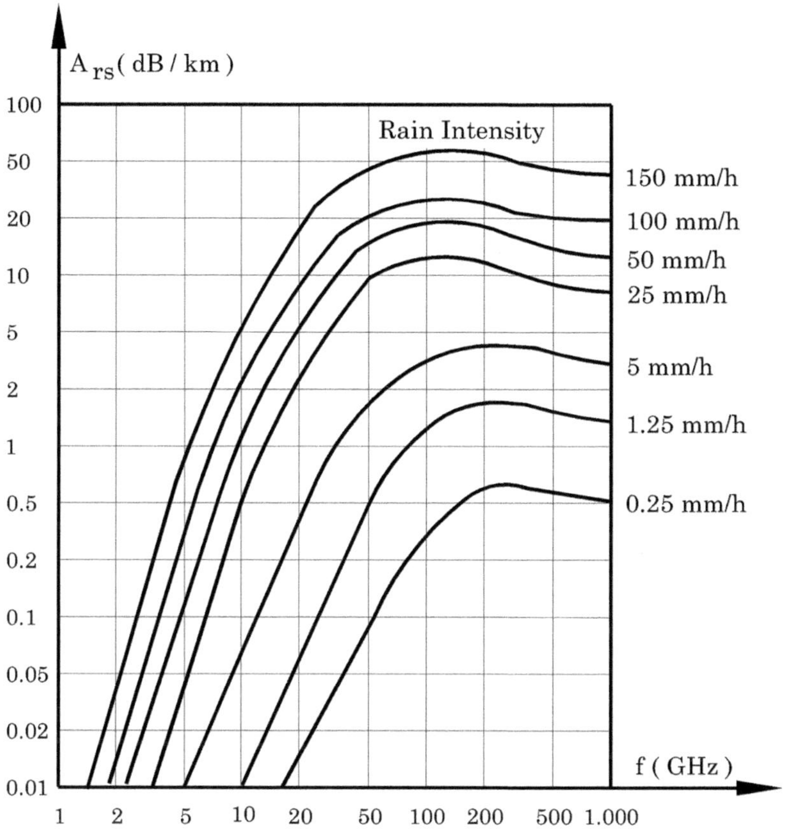

Fig. 3.24 Variations of rain specific attenuation vs. frequency and rain intensity

Table 3.2 Rain coefficients in terms of frequency and rain intensity

			A_{rs}		
GHz	a	b	$R = 10$	$R = 50$	$R = 100$
2	0.00034	0.891	0.003	0.011	0.021
4	0.00147	1.016	0.015	0.078	0.158
6	0.00371	1.124	0.049	0.30	0.657
12	0.0215	1.136	0.29	1.83	4.02
15	0.0368	1.118	0.48	2.92	6.34
20	0.0719	1.097	0.90	5.25	11.24
30	0.186	1.043	2.05	11.00	22.70

Example 3.9. Calculate rain loss with an intensity of $R = 25$ mm/h for the following satellite routes:

1. Satellite downlink at the C band ($f = 4$ GHz) and antenna elevation angle of 20°
2. Satellite uplink at the K band ($f = 14$ GHz) and antenna elevation angle of 40°

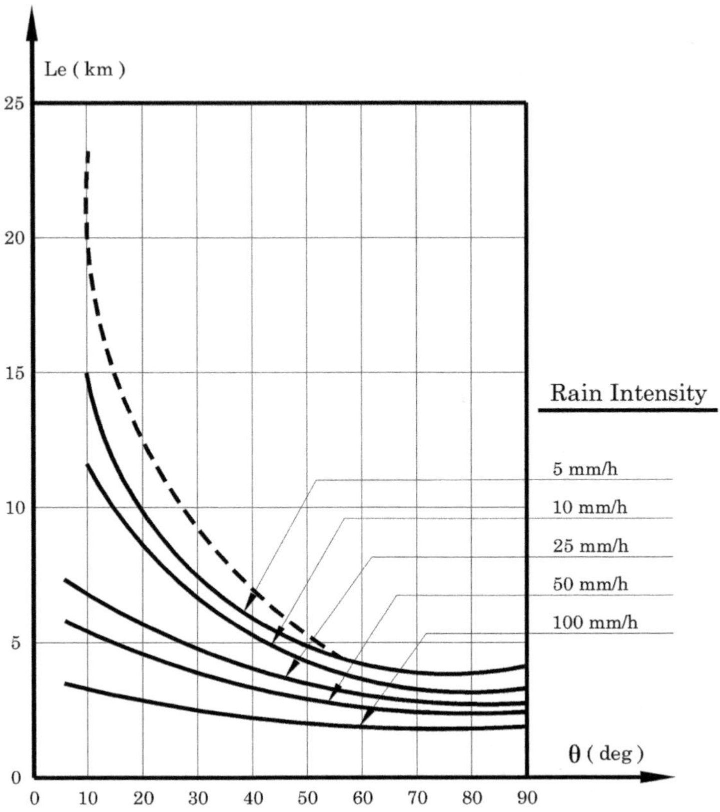

Fig. 3.25 Rain effective length in terms of elevation angles

Solution. Effective rain length for intensity of $R = 25\,\text{mm/h}$ using the graph of Fig. 3.25 is equal to

$$L_{e/C} = 6\,\text{km}, \quad L_{e/K} = 4\,\text{km}$$

Now using Table 3.2, the rain attenuation coefficients are:

- For C band,

$$f = 4\,\text{GHz} \implies a = 0.00147, \quad b = 1.016$$

$$A_{r/C} = aR^b = 0.00147 \times (25)^{1.016} = 0.0387\ \text{dB/km}$$

$$A_C = A_{r/C} \times L_{e/C} = 0.0387 \times 6 = 0.23\ \text{dB}$$

- For K band,

$$f = 14\,\text{GHz} \implies a = 0.035, \quad b = 1.122$$

$$A_{r/K} = aR^b = 0.035 \times (25)^{1.122} = 1.296\ \mathrm{dB/km}$$

$$A_K = A_{r/K} \times L_{e/K} = 1.296 \times 4 = 5.184\ \mathrm{dB}$$ ■

3.5.4 Cloud and Fog Attenuation

Specific attenuation caused by cloud and fog includes small water drops defined below:

$$A_{hs} = KM \qquad \mathrm{dB/km} \tag{3.29}$$

where M is the mass of liquid content in the cloud or fog in terms of grams per cubic meter and K is a coefficient which depends on the temperature and frequency. Graphs of K and A_{hs} are prepared for different values of frequency and M. As a sample, value of M for rainy clouds is equivalent to 0.1–5 g/m^3. Such clouds may have a height of more than 8 km, but their effective route length is shorter. For more information on this topic, refer to the P-series of ITU-R recommendations.

3.5.5 Losses Due to the Sand and Dust Storms

Sandstorms and dust storms are common in many parts of the world such as Middle East, Africa, some parts of Asia, China, and other regions of South America. This case must be considered for these regions for design of LOS radio links especially for terrestrial communications. These storms can be so vast and huge that cover an area of 400–500 km^2 and also can occur as small as around 10 km^2.

Duration of dust storms has been observed normally about 3 h but may be extended to few days which can cause negative impacts on the transmission system quality. There is distinction between sandstorms and dust storms. Sandstorm can form a cloud with very small thickness of sand at a low height (around 2 m above the ground), while dust storms consisting of small particles are very dense running at a height of several hundreds of meters above the ground and can persist for longer periods of time.

Based on the natures of dust storm and sandstorms, the former has more impacts on satellite links. Also, the attenuation caused by these types of sandstorms and dust storms depends on the existing moisture in them. When the amount of this moisture becomes larger, then its insulation coefficient will rise and the subsequent attenuation becomes larger.

These storms on high frequencies along with high moisture content of the soil can make more effect on the radiowaves propagation. Otherwise, for frequencies lower than 30 GHz and little moisture content, this factor may be ignored.

Very limited data are available about the effects of sand and dust storms on radio signals on slant-paths. At frequencies below 30 GHz, high particle concentrations and/or high moisture contents are required to produce significant propagation effects.

3.5.6 *Atmospheric Changes*

At some occasions, the radiowaves traveling in the space encounter scintillation and undergo variations of amplitude and phase and/or angle of arrival to the antenna. Such variations can happen in both of the troposphere and ionosphere layers, and its severity on the radiowaves generally is dependent on the frequency, climatic conditions, antenna size, and antenna elevation angle.

Atmospheric changes occur in the lower part of troposphere in the first few kilometers above the ground level having a high moisture content and irregular variations of refractivity index. In this situation, two factors attenuate the radiowaves. The first one is caused by defocusing, which in this case signal has little deviation from the direction of maximum antenna gain when received by the Earth station antenna. The second effect is diffusion, which in this position refractivity index related to the atmosphere is altered causing the radiowaves propagating on different orientations.

The results of the tests reveal that in attenuation related to atmospheric changes for an antenna with 30-m diameter and 5° elevation angle, the atmosphere part is 0.3 dB for 4 GHz, 0.5 dB for 6 GHz, and 2–3 dB for 30 GHz. In harsh atmospheric conditions with high humidity, these values can be much higher.

In addition to the above cases, the amplitude variations occur on the same route of satellite wave with frequencies of up to 30 GHz. Normally the range of their variations is about 2 dB, and severity of the variations has been reported to be up to 20 dB. Subsequently, it can be concluded that the attenuation caused by atmospheric changes (A_S) must be considered in the link design, particularly for sites located in regions close to the equator and with low elevation angle.

3.6 Effects of Ionosphere

3.6.1 *Introduction*

Ionosphere constitutes a part of the atmosphere which is located above troposphere and stratosphere layers and stretches up to about 500–1,000 km over the Earth's surface. It includes three sub-layers of D, E, and F having different properties. F layer is divided into F_1 and F_2 parts during the day and behaves as a unique layer during the night.

Because of solar radiations in the ionosphere, the existing gases are ionized. The total electron content (TEC) affects the radio links between satellite and Earth stations; these effects are not homogenous and also not uniform with respect to time. The ionization process includes diurnal, seasonal, and solar variations with 11-year interval and highly dependent on the geographic position and geomagnetic activities.

Considering the effective height of the ionosphere layer and positioning of telecommunications satellite of any kind (LEO, MEO, or GEO) well above it, thus the satellite radio links are deeply under influence of the ionosphere and definitely will be affected by its phenomena.

3.6.2 Main Ionospheric Phenomena

The effects of ionosphere layer on radiowaves propagation up to 12-GHz frequency are remarkable which include more effects on MEO and LEO satellites especially at frequencies less than 3 GHz. Observing these points in planning of satellite links, it is crucial to evaluate the ionospheric phenomena as summarized below:

- Polarization or Faraday rotation.
- Group delay.
- Change in the apparent direction of arrival.
- Doppler effect.
- Dispersion or distortion on group velocity.
- Scintillation with adverse effects on the amplitude, phase, and angle of arrival.

Because of the complex nature of ionosphere, it is not possible to express the above effects with simple analytical formula. The best method is to employ supplementary tables and graphs based on empirical experiments and measurements. Because of stochastic form of ionosphere irregularities, these elements can be only described with stochastic terms.

In design of the mobile satellite systems (MSS) at frequencies below 3 GHz, in addition to the above facts, the following points should be taken into account:

- Ionospheric effects are the most significant issues to be considered in the MSS system design in global scale.
- The near-surface multipath effects, in the presence of natural or man-made obstacles and/or at low elevation angles, are critical.
- Earth's obstacles.
- Temporal or location variations of the near-surface multipath effects are not dominant in the overall design of the MSS system when global-scale propagation factors are considered.
- Low antenna angles.
- Attenuation caused by hydrometeors or other aerosols is not significant compared with other effects.

Table 3.3 Typical values for main ionospheric effects (Ref.: ITU-R, P.618-9)

Effect	0.5 GHz	1 GHz	3 GHz	10 GHz
Faraday rotation	430°	108°	12°	1.1°
Propagation delay (μs)	1	0.25	0.028	0.0025
Refraction	$< 2.4'$	$< 0.6'$	$< 4.2''$	$< 0.36''$
Variation in direction of arrival (r.m.s)	$48''$	$12''$	$1.32''$	$0.12''$
Absorption (auroral and/or polar cap), dB	0.2	0.05	6×10^{-3}	5×10^{-4}
Absorption (midlatitude), dB	< 0.04	< 0.01	0.001	$< 1 \times 10^{-4}$
Dispersion, ps/Hz	0.0032	0.0004	1.5×10^{-5}	4×10^{-7}
Scintillation peak to peak, dB		> 20	≈ 10	≈ 4

3.6.3 *Frequency Dependence*

Different phenomena which were referred to in the previous section have nonlinear inverse relation with satellite wave frequency. Dependence of some of the ionospheric phenomena on the frequency is in the range of f^{-2} to f^{-3}.

Table 3.3 provides typical values of the ionospheric effects on the radiowaves for a number of frequencies.

3.6.4 *Refractivity Index*

Refractivity index of the ionosphere layer denoted as n, in case of neglecting the losses caused by electron collision and also effects of the Earth's magnetic field, can be expressed as

$$n = \left(1 - \frac{Ne^2}{\varepsilon_0 m\, \omega^2}\right)^{1/2} = \left(1 - \frac{\omega_0^2}{\omega^2}\right)^{1/2} = \left(1 - \frac{f_p^2}{f^2}\right)^{1/2} \tag{3.30}$$

In the above relation, the parameters and their corresponding units are stated below:

e: electric charge of the electron equal to 1.6×10^{-19} C
ε_0: permittivity of the medium
m: electron mass in metric system (kg)
N: electron density in terms of number per cubic meter
f and $(\omega = 2\pi f)$: frequency of the wave in Hz
f_p: plasma frequency in Hz

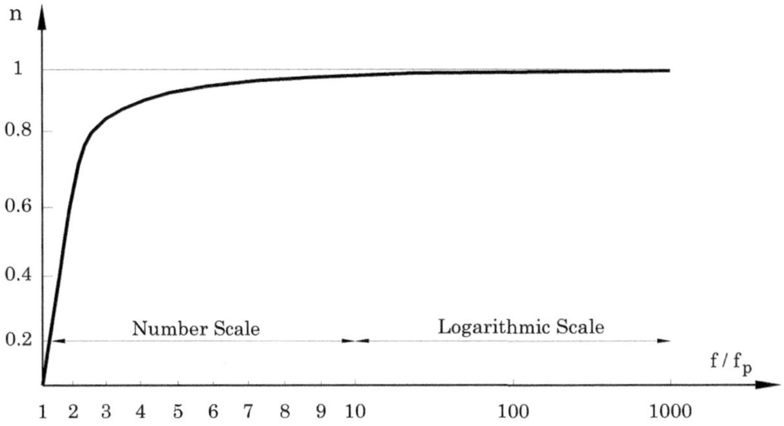

Fig. 3.26 Variations of ionospheric refractivity in terms of normalized frequency

Frequency of plasma depends on electron density according to the following relation:

$$\omega_0^2 = \frac{Ne^2}{m\,\omega^2} = (2\pi f_\mathrm{p})^2, \text{ or} \tag{3.31}$$

$$f_\mathrm{p} = \frac{e}{2\pi}\sqrt{\frac{N}{m}} \tag{3.32}$$

Inserting the mass and charge of the electron in the above relation the plasma frequency can be obtained by

$$f_\mathrm{p} = 9\sqrt{N} \tag{3.33}$$

Also, in case of absorptive medium and negligible Earth's magnetic field, then relation (3.30) is converted into

$$n = \left(1 - \frac{\omega_0^2}{\omega(\omega - jV)}\right)^{1/2} \tag{3.34}$$

In the above relation, v is the frequency of electrons collision in ionosphere layer. In case of applying the effects of the Earth's magnetic field, then the latter relation becomes very complex for which you may refer to the books on plasma physics for more information.

Evaluation of relations (3.30) and (3.34) reveals that generally the value of refractivity index in ionosphere layer is less than unity. In Fig. 3.26, the relative values of n are shown for different layers.

Example 3.10. In case the electron density in ionosphere layer is 10^{14} electrons per cubic meter, then calculate refractivity index of this layer for frequencies of 150 MHz, 1.5 GHz, and 12 GHz.

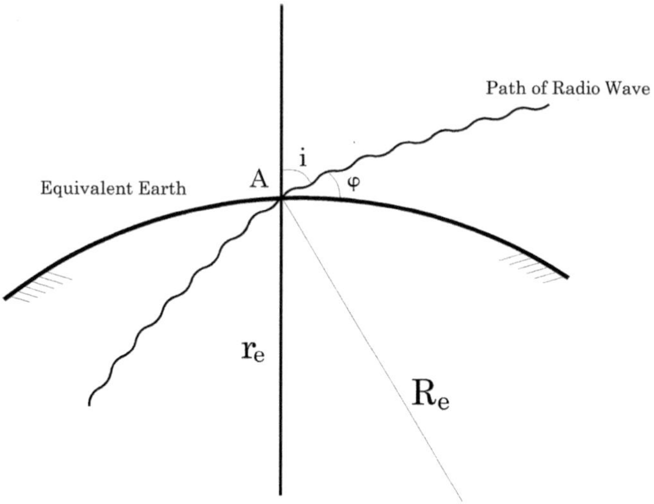

Fig. 3.27 Geometry of radiowave refraction in Earth atmosphere

Solution. First the value of plasma frequency is calculated using the relation (3.33):

$$f_p = 9\sqrt{N} \quad \Longrightarrow \quad f_p = 90\,\text{MHz}$$

Then

$$f_1 = 150\,\text{MHz} \quad \Longrightarrow \quad n_1 = \sqrt{1 - \left(\frac{90}{150}\right)^2} = 0.8$$

$$f_2 = 1.5\,\text{MHz} \quad \Longrightarrow \quad n_2 = 0.998$$

$$f_3 = 12\,\text{MHz} \quad \Longrightarrow \quad n_3 = 0.99997$$

■

Concentric levels of the Earth's atmosphere with its surface have the same refractivity index denoted as $n(r)$. The route which the radiowaves travel is a curved path with a radius of curvature denoted as p. Noting Fig. 3.27 and the Snell's law of refractivity, the following relation holds

$$n(r) \cdot r \cdot \cos\varphi = cte \tag{3.35}$$

and the radius of the radio path at point A depends on other components based on the following relation:

$$\frac{1}{p} = -\frac{\cos\varphi}{\pi} \cdot \frac{dn}{dr} \tag{3.36}$$

where

$n(r)$: refractivity index in the Earth's atmosphere at a height r from its center

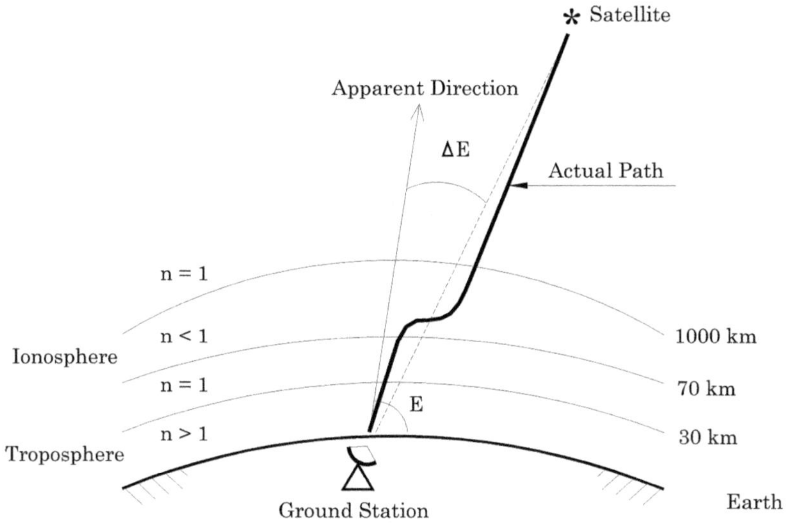

Fig. 3.28 Actual pathlength and deviation angle in satellite radio links

r: distance of point A from the Earth center
φ: angle of the wave path relative to the horizon
dn/dr: gradient of refractivity index
p: radius of wave path curvature

3.6.5 Pathlength of Radiowaves

The distance between two points in the space, like the distance between satellite and Earth terminal, is calculated by (3.15) and using geographic longitude, latitude, and the altitudes of points. However, as indicated in Fig. 3.28, it is clear that the actual wave path is longer with a magnitude of ΔL from their geometrical direct distance. This value can be calculated from the following relation:

$$\Delta l = \Delta t_t + \Delta L_{\mathrm{i}} = \int_0^S (n-1)ds \tag{3.37}$$

The value of extra length in the troposphere layer does not depend on frequency, but it is influenced by the atmosphere components and also elevation angle of the radiowave. The amount of ΔL_{t} is based on the context of recommendation ITU-R, P.834, and can be calculated by the following relation:

$$\Delta L_{\mathrm{t}} = 2.27P + \frac{1.79V}{T\ \sin\ \theta} \tag{3.38}$$

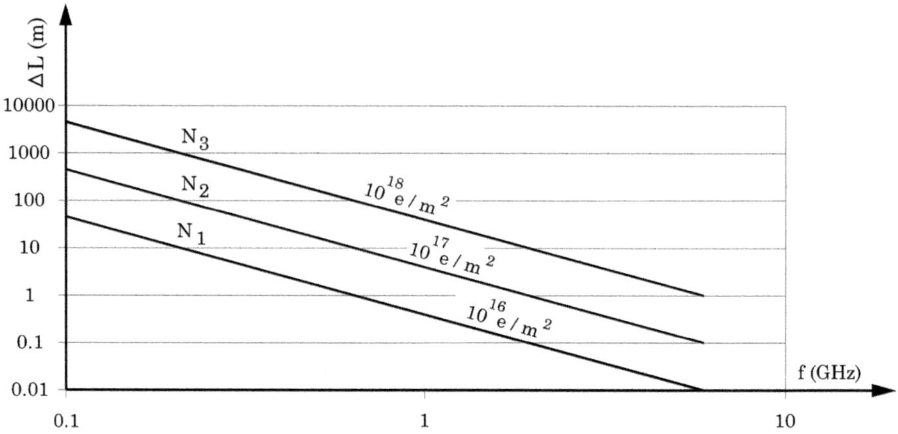

Fig. 3.29 Additional pathlength due to ionosphere layer

In the above relation, the components and units are indicated below:

ΔL_t: excess length due to the troposphere layer in meter
P: air pressure in bar ($1\ \mathrm{b} = 10^{-5}\ \mathrm{Pa}$)
T: Earth's surface temperature in Kelvin
θ: elevation angle of the wave in radians
V: amount of water vapor in kg/m^2

The value of ΔL_t is around 2.2–2.7 m from the sea level and in the direction of solar zenith. In ionosphere layer, due to dependence of refractivity index on frequency, the extra length is denoted as ΔL_i and depends on frequency, and also TEC of this layer can be expressed with the following relation:

$$\Delta L_\mathrm{i} = \frac{40}{f^2}\int_0^S N\,ds \tag{3.39}$$

In the above relation, the components and their units are:

ΔL_i: extra length caused by ionosphere layer
f: frequency of the wave in MHz
N: number of electrons per cubic meter

As an example for TEC $= 30 \times 10^{16}$ el/m^2 at 100-MHz frequency, the value of ΔL_i is about 1,200 m. In Fig. 3.29, the graph of increased length caused by ionosphere layer is presented in terms of frequency for three values of TEC.

Example 3.11. Calculate the extra path length of satellite wave with elevation angle of 10° in troposphere and ionosphere layers for frequencies of $f_1 = 1.5$ GHz and $f_2 = 6$ GHz. The air pressure is equal to 100 hectopascals, temperature 27 °C, and the existing water vapor is 10 kg/m^2 and TEC $= 10^{18}$ e/m^2.

Solution. The extra length in troposphere is independent of frequency, and observing relation (3.38), the following result is obtained:

$$P = 1\text{ mb},\quad T = 300\,^{\circ}\text{K},\quad \theta = 10^{\circ}$$

For calculation of the extra length in ionosphere layer noting Fig. 3.29, the following results are obtained:

$$f_1 = 1.5\,\text{GHz} \quad \Longrightarrow \quad \Delta L_{\text{i}1} \approx 20\,\text{m}$$

$$f_2 = 6\,\text{GHz} \quad \Longrightarrow \quad \Delta L_{\text{i}2} \approx 2\,\text{m}$$

Thus, total excess path lengths are

$$\Delta L_1 = \Delta L_{\text{t}} + \Delta L_{\text{i}1} \approx 22.6\,\text{m}$$

$$\Delta L_2 = \Delta L_{\text{t}} + \Delta L_{\text{i}2} \approx 4.6\,\text{m}$$

■

3.6.6 Satellite Antenna Deviation Angle

The path of satellite waves deviates from a straight line due to the changes of atmosphere refractivity index. In the troposphere layer, it is more than unity, while in the ionosphere layer, it is less than unity. This phenomenon, in addition to increasing the length of the path, will bend the antenna boresight with a magnitude of ΔE from the geometrical straight direction between satellite and Earth terminals. The value of ΔE based on the trigonometric relations will be

$$\Delta E = \frac{(L + r\ \sin\ E)\ \cdot\ r\ \cos\ E}{h_{\text{i}}(2r + L) + (r\ \sin\ E)^2} \times \frac{\Delta L}{L} \tag{3.40}$$

In the above relation, the components and their units are:

ΔE: antenna boresight deviation in radians
L: distance of satellite from the Earth terminal in km
r: Earth's radius in km
E: apparent elevation angle of the Earth station antenna
h_{i}: average height of ionosphere layer (300–500 km)

3.6.7 Time and Frequency Variations

Because of increased pathlength in passing troposphere and ionosphere layers along with the decreased speed of radiowaves in these layers, duration of the wave moving between satellite and ground station is more than the corresponding value in free space.

It must be also noted that the length difference in the ionosphere layer is a function of frequency and the amount of TEC. This value changes with time resulting in a slight difference of the received frequency in the Earth station with that of satellite frequency. The frequency change for satellites at the L band (1.6 GHz) is about 0.1 Hz and for satellites at the VHF band is about several hertz.

3.7 Radiowaves Polarization Variation

For optimum utilization of frequency resources, one of the prevailing techniques in radio communications is the frequency reuse. On the basis of this approach, the radiowaves are transmitted over two radio channels with identical carrier frequency but with orthogonal polarization. Generally the following orthogonal polarizations are used in this case:

- Linear orthogonal polarization (such as horizontal and vertical polarizations)
- Circular orthogonal polarization (such as right-hand circular polarization (RHCP) and left-hand circular polarization (LHCP))

The route of radiowaves propagation in satellite communications, between the Earth and space terminals, passes through ionosphere and troposphere layers.

The polarization of satellite radiowaves changes because of the following mechanisms:

- Existing ionization condition and a large TEC in the ionosphere
- Ice crystals in the upper parts of troposphere
- Atmospheric precipitation such as rain

When two orthogonal polarizations are used simultaneously, polarization variation emanating from the propagation medium causes the main radiowave to include a small component with orthogonal polarization acting as a radio noise. A similar phenomenon occurs for the second wave upon traversing the medium; it will contain a small component with orthogonal polarization, which acts as a noise. The polarization variation on the satellite waves can be originated mainly from the two following sources:

- Effect of hydrometeors such as rain and/or ice crystals, which are related to the troposphere layer
- Faraday rotation of the waves related to ionosphere layer

The following sections will provide more details concerning the above points.

Fig. 3.30 Medium effect on generation of cross polarization discrimination

Fig. 3.31 Medium effect on polarization isolation

3.7.1 Main Parameters

To express the effects of interference caused by polarization variation, the following two parameters are used:

- Cross polarization discrimination, XPD
- Polarization isolation

Cross polarization discrimination denoted by XPD is used more often to describe the polarization changes. In this case, as depicted in Fig. 3.30, the electric field of the transmitted wave has amplitude of E_1 prior to entering the medium, whereas there are two components of E_{11} and E_{12} on the RX antenna. The first one has similar polarization with the transmitted field, and the second one is an orthogonal component to it. Cross polarization discrimination, XPD, can be expressed as

$$\text{XPD} = 20 \log \frac{E_{11}}{E_{12}} \tag{3.41}$$

To explain the polarization isolation, Fig. 3.31 is presented. It illustrates a radiowave, including two orthogonal electric field components denoted by E_1 and E_2. The radiowave propagating in the medium will be converted into four components of (E_{11}, E_{12}, E_{21}, and E_{22}) whereby the fields of E_{12} and E_{21} in fact constitute the interfering fields. Since the received power is proportional to the square of electric amplitude, polarization isolation may be derived from the

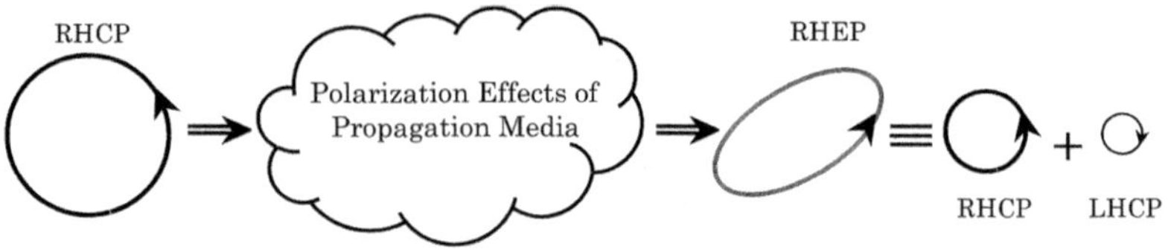

Fig. 3.32 Medium effect on RHCP radiowaves

following equations:

$$I_1 = 20 \log \frac{E_{11}}{E_{21}}, \quad I_2 = 20 \log \frac{E_{22}}{E_{12}} \tag{3.42}$$

In a position at which the amplitude of transmitted signals is equal, $E_1 = E_2$, and neglecting polarization variation of the reception system, we can write

$$I_1 = I_2 = \text{XPD} \tag{3.43}$$

The above discussions were treated for linear polarization of the radiowaves; however, it can be also true for circular polarization in a similar way. A radiowave with RHCP as depicted in Fig. 3.32, after traversing the medium, will be converted into a right-hand elliptical polarization which may be considered as combination of a RHCP plus a LHCP. Based on the said assumption, the following equations can be used for cross polarization discrimination and polarization isolation:

$$\text{XPD} = 20 \log \frac{|\text{RHCP}_2|}{|\text{LHCP}_2|} \tag{3.44}$$

$$I_1 = 20 \log \frac{|\text{RHCP}_{11}|}{|\text{LHCP}_{21}|}, \quad I_2 = 20 \log \frac{|\text{RHCP}_{22}|}{|\text{LHCP}_{12}|} \tag{3.45}$$

3.7.2 Ionosphere Polarization Variation

Polarization variations emanating from ionosphere layer relate to the existence of plasma state in this layer and being a function of TEC. In general, the value of polarization variation in this stage depends on the value of Faraday rotation, which is a function of TEC, and the Earth's magnetic field in the ionosphere region and is inversely proportional with the square of frequency.

For a radiowave with linear polarization of electric field and an intensity of E_1, the power is proportional to E_1^2. Subsequent to propagation in a medium which alters the polarization according to the Faraday rotational angle of θ, the in-phase component is E_{11} and its power is proportional to E_{11}^2. Then the polarization loss factor (PLF) in terms of decibels is

$$\text{PLF} = 20\,\log\frac{E_{11}}{E_1} = 20\,\log(\cos\,\theta) \tag{3.46}$$

The orthogonal component is equivalent to E_{12} for which the following expression is resulted for the cross polarization discrimination:

$$\text{XPD} = 20\,\log\frac{E_{11}}{E_{12}} = 20\,\log(\cot\,\theta) \tag{3.47}$$

As an example, the maximum angles of polarization rotation on the satellite C band are stated below:

- 9° for 4-GHz frequency
- 4° for 6-GHz frequency

Example 3.12. Faraday rotation at 1-GHz frequency is 108°; calculate:

1. Faraday rotation for send-and-receive frequencies of the satellite at the L, C, and Ku bands and specify in which band it may be possible to use orthogonal polarizations simultaneously.
2. Power losses due to polarization variation and also its discrimination for the waves of the L band.

Solution. 1. Considering that the Faraday rotation is inversely proportional to the square of frequency,

$$\theta_{\mathrm{r}} = 108^\circ, \quad f_{\mathrm{r}} = 1\,\mathrm{GHz}$$

$$\textbf{L band: } f_{\mathrm{LS}} = 1.6\,\mathrm{GHz}, \quad f_{\mathrm{LR}} = 1.5\,\mathrm{GHz}$$

$$\theta_{\mathrm{L/S}} = \theta_{\mathrm{r}} \times \left(\frac{f_{\mathrm{LS}}}{f_{\mathrm{r}}}\right)^{-2} = 42.19^\circ$$

$$\theta_{\mathrm{L/R}} = \theta_{\mathrm{r}} \times \left(\frac{f_{\mathrm{LR}}}{f_{\mathrm{r}}}\right)^{-2} = 48^\circ$$

$$\textbf{C band}: f_{\mathrm{CS}} = 6\,\mathrm{GHz}, \quad f_{\mathrm{CR}} = 4\,\mathrm{GHz}$$

$$\theta_{\mathrm{C/S}} = \theta_{\mathrm{r}} \times \left(\frac{f_{\mathrm{CS}}}{f_{\mathrm{r}}}\right)^{-2} = 3^\circ$$

$$\theta_{\mathrm{C/R}} = \theta_{\mathrm{r}} \times \left(\frac{f_{\mathrm{CR}}}{f_{\mathrm{r}}}\right)^{-2} = 6.75^\circ$$

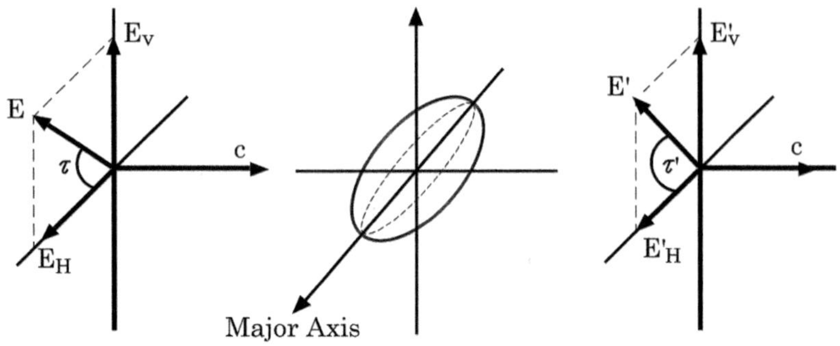

Fig. 3.33 Rain effect on radiowaves polarization

$$\textbf{Ku band}: f_{\mathrm{KuS}} = 14\,\mathrm{GHz},\quad f_{\mathrm{KuR}} = 12\,\mathrm{GHz}$$

$$\theta_{\mathrm{Ku/S}} = 0.55^\circ$$

$$\theta_{\mathrm{Ku/R}} = 0.75^\circ$$

Evaluation of the results indicates that the polarization variation on the L band is significant, on the C band has a low value, and on the Ku band it may be neglected.

2.

$$f = 1.6\,\mathrm{GHz},\quad \theta = 42^\circ$$

$$PL = 20\ \log(\cos\ 42) = -2.5\ \mathrm{dB}$$

$$\mathrm{XPD} = 20\ \log(\cot\ 42) = 0.9\ \mathrm{dB}$$

■

3.7.3 Polarization Variation Due to Rain

Raining, in addition to the attenuation of radiowaves, makes polarization variation of the satellite waves. The shape of raindrops for basic theoretical investigations is considered as a sphere which is nearly valid for small drops. In the case of big drops, it is oblate spheroid in a way that its bottom surface is flat or even has a concaved shape.

In normal cases, raindrops are in ellipsoid shape with horizontal major axis. This arrangement will change by the wind, and its major axis may be tilted downward or upward. The linear electric field, E, can be divided into two horizontal and vertical components, which for simplicity we assume the horizontal component, that is, E_{H} on the X-axis. As shown in Fig. 3.33, E_{H} makes an angle of τ with the direction of the electric field E, then in this situation, the horizontal component of the electric field will travel a longer path in the raindrop with respect to the vertical component resulting in more attenuation and phase shift. Experimental tests have shown that in such a position, the polarization variation has more effects compared to the amplitude attenuation.

Noting the relations indicated in ITU-R technical reports, the following relation is derived for the cross polarization discrimination:

$$\mathrm{XPD} = U - V \log A_{\mathrm{r}} \tag{3.48}$$

where A_{r} is the attenuation caused by the rain and U and V are coefficients which are calculated from the following relations:

$$V = 20, \quad 8 \leqslant f \leqslant 15\,\mathrm{GHz} \tag{3.49}$$
$$V = 23, \quad 15 \leqslant f \leqslant 35\,\mathrm{GHz}$$

$$U = 30 \log f = 10 \log(0.5 - 0.4697 \cos \tau) - 40 \log(\cos \theta) \tag{3.50}$$

In the above relation, f is in terms of GHz and U and V are in terms of decibels.

3.7.4 Polarization Variation by Ice Particles

When satellite waves pass through ice layer in a cloudy atmosphere, its crystals produce polarization changes in those waves. Experimental evaluations indicate that ice particles similar to the raindrops (although for different reasons) create polarization variations in the electromagnetic waves. The ice particles, in contrast to the rain, are a good dielectric and do not produce large attenuation.

These particles having sharp tips when placed irregularly, the resulting polarization variation of the waves is negligible and can be disregarded. If for any reason, such as electrical discharge and current flow, these particles are arranged regularly, then they can create abrupt increase in the cross polarization discrimination of the waves. Based on the ITU-R recommendations, the negative effects of polarization changes produced by the ice crystals can be compensated by taking into account a 2-dB loss for mild regions and 4–5-dB loss for marine regions. It must be mentioned that for time percentages less than 0.1, its effects may be ignored compared to the other factors.

3.8 Effects of Local Environment

In some fields of satellite communications such as mobile, navigational aids, and broadcasting services, the satellite receiving units may be used inside buildings, vegetation, or surrounded areas. In specific receiving locations, effects of local structures may be important. Measurement results reveal that for radiowaves attenuation, there is strong dependence on its elevation and azimuth angles.

Table 3.4 Signal attenuation inside buildings (500 MHz $< f <$ 3 GHz)

Building number	Construction	Elevation angle (degrees)	Average position: Mean loss (dB)	Average position: Standard deviation (dB)
1	Entry lobby in single-storey building, concrete tilt wall, tar roof	18	13	10
2	Office in single-storey building, block brick, tar roof	38	9	7
3	Two-storey wood frame farmhouse, metal roof, no aluminum heat-shield	33	5	4
4	Hallway and living room of two-storey wood frame house, metal roof aluminum heat shield	41	19.5	12
5	Motel room in two-storey building, brick with composite roof	37	13	6
6	Lobby of two-storey building, glass and concrete, tar roof	26	12	5

The effects of local environment are taken into account carefully in the site selection of the Earth stations, and every effort will be done to avoid such adverse effects of local environment especially for big terminals. But in the said fields, based on their application nature, receiving of non-line-of-sight (NLOS) signals in surrounded area is a crucial requirement.

Most works on this issue are experimental for which the results have been prepared in the forms of tables and graphs. In some of ITU-R recommendations including P.679-3, the effects of local environment have been evaluated.

3.8.1 Building Entry Loss

When satellite waves penetrate inside the buildings and closed areas, they undergo additional losses which at the UHF band may reach to 25 dB. When the satellite terminal is located in a suitable position, these losses can be reduced to less than 5 dB. The effective factors on the additional losses discussed so far are summarized below:

- Number of floors
- Material and chemical properties of the walls and ceilings
- Thermal insulation
- Position of the terminal
- Frequency band
- Radiowaves polarization

In Table 3.4, a typical example is given for signal attenuation at average position within buildings measured at the frequency range from 500 MHz to 3 GHz. The most important points in this case are summarized below:

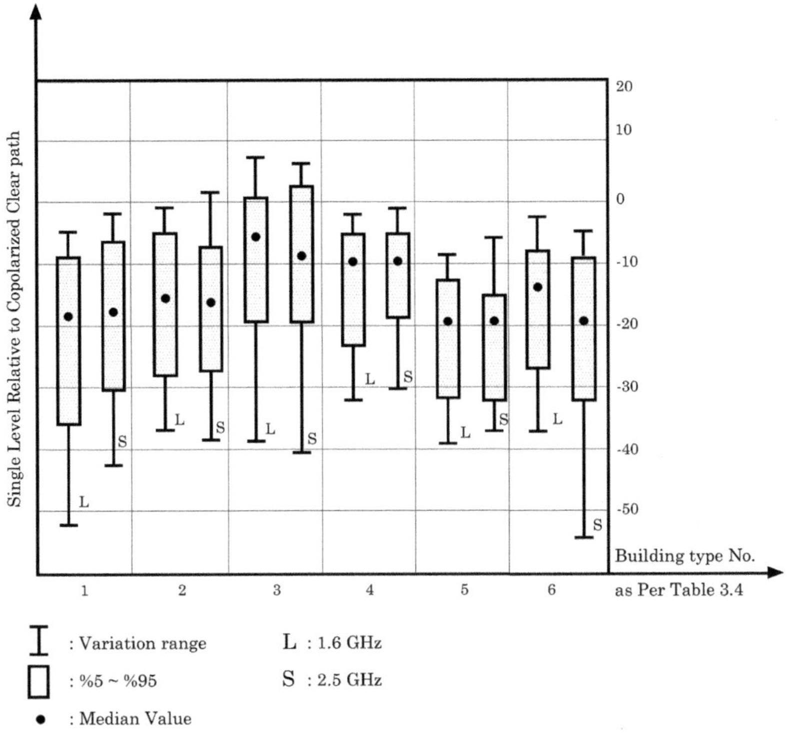

Fig. 3.34 Additional attenuation inside buildings (Ref.: ITU-R, P.679-3)

- Losses due to reflective glass door were 15 dB greater than open door.
- Effect of thermal insulator made of aluminum on the route of the waves can create additional losses of around 20 dB.
- Losses caused by brick walls are around 15–30 dB which in case of existing metallic rooftop and aluminum thermal insulator this value can rise to about 25–45 dB.
- In case of moving terminals, effects related to multipath will appear as fluctuation of the received signal level. This case has implications particularly for LEO satellites where the transmitter is moving rapidly with respect to the receiver.
- Measurement results indicate that for buildings numbered 1, 2, 4, and 6, the effect of frequency increase is around 1–3 dB/GHz and for the building No. 3 is 6 dB/GHz, while for the building No. 5, its effect is not significant. Thus, the overall excessive losses inside the buildings and closed areas are frequency-dependent mechanisms.
- Typical additional attenuation caused by buildings is given in Tables 4.6 and 4.7 of Chap. 4 for 5.1-GHz broadcasting waves at the center of corridors in the entry and inside of the building, respectively.
- Figure 3.34 is arranged based on measurement results obtained related to the six types of building specified in Table 3.4 for which the following points are outstanding:

1. Measurements performed for two frequencies at $L = 1.6\,\text{GHz}$ and $S = 2.5\,\text{GHz}$.
2. Median value of 5–95 % levels is indicated with a small bullet.
3. Extra losses through floors in dB is given by

$$L_e = 15 + 4(n - 1) \tag{3.51}$$

where n is the number of penetrated floors.
4. The effect of elevation angle of the satellite waves on 5-GHz frequency in office rooms for angles of 15° and 55° is in the form of excessive losses equal to 20 and 30 dB, respectively.

3.8.2 *Inside Vehicles*

Measurements related to the penetration of satellite signals inside the vehicles are performed in 1.5–1.6-GHz frequency band (the INMARSAT L band) at elevation angles of 8–90° for different types of antennas and vehicles and different positions of satellite terminal inside the vehicle with open windows. The obtained results have shown excessive losses with an average value of 3–8 dB. The overall considerations are summarized below:

- Signal level inside the vehicle in the NLOS condition has Rayleigh distribution.
- Type of the vehicle has limited influence on the signal penetration losses.
- Dependence of the losses on the wave elevation angle is negligible and can be disregarded.
- Radio terminal position in the vehicle has no significant effect.
- Type of antenna has influence on the extra losses.

3.8.3 *Reflection of Waves*

Subsequent to satellite waves colliding with various surfaces around the satellite terminal, they are reflected and the vector sum of the main and reflected waves is fed to the RX input. When the waves are in the opposite phase, then the main wave will be attenuated.

To determine the limits of this phenomenon, certain measurements were conducted under the supervision of ITU-R, and the conclusions were provided in the form of fluctuations of the received signal level from the TX of FM waves with circular polarization and elevation angle of around 20° as explained below:

- 15 dB for test frequency of 839 MHz
- 18 dB for test frequency of 1,504 MHz

The amount of fluctuations in urban areas for antennas with horizontal and vertical polarizations is nearly the same, and the quality of the audio signal is merely dependent on the variations of the received signal level. In suburban areas and/or rural areas, the subject of the wave reflections is a specifying factor of how to select the type of polarization since the waves having vertical polarization possess reflection coefficient of nearly zero on the Brewster angle scale whereas these conditions do not exist for horizontal polarization. Consequently, over smooth surfaces of the Earth, normally the waves with horizontal polarization have stronger reflections compared to the waves with vertical polarization, and subsequently, the received signal which is the sum of the main and reflected waves has greater maxima and minima.

3.9 Reliability of Satellite Links

The reliability of a satellite link for a period of time is normally expressed in the percentage of time which can work satisfactorily when compared to the total time. For example, a link with 99.8 % reliability means that the desirable link can be established for not less than 99.8 % of the time. Desirable communication in digital systems is in terms of bit error rate or BER and in analog systems is in terms of signal to noise ratio or SNR.

To achieve the required reliability, it is essential to consider some value to be allocated for fade margin in the link. This amount of fade margin will depend on the required reliability and also on the working frequency. Table 3.5 presents the required fade margin for the assumed reliability in terms of different frequencies.

Example 3.13. To achieve propagation reliability of 99.95 % for a satellite link, calculate:

- The fade margin required at Ka band, $f = 20\,\text{GHz}$
- The maximum annual outage for having the above reliability

Solution. 1. Noting Table 3.5 to obtain 99.95 % reliability on the Ka band, that is, nearly 20 GHz, the approximate fade margin required is FM = 20 dB.

Table 3.5 Minimum fade margin vs. link reliability

Reliability of the link (%)	Annual link outage in hours	Required fade margin in dB		
		11 GHz	20 GHz	30 GHz
99.5	44	1	3	6
99.9	8.8	3	10	20
99.95	4.4	5	20	30
99.99	0.88	15	30	–

The above values are based on the elevation angle of the Earth station antenna within the limits of 30–50° and measurements of around thirty stations at regions with temperate climate

2. With the above-mentioned reliability, the maximum outage of the link will be equal to 0.05 % for which the annual outage will be less than

$$T_0 = \frac{0.05}{100} \times 365 \times 24 \times 60 = 262.8\,\text{min}$$ ■

3.10 Satellite Link Equation

3.10.1 *Main Factors*

Main factors related to the satellite links calculations are defined and discussed in this section including the following items:

- Equivalent isotropic radiated power (EIRP)
- Merit of reception system, G/T
- Carrier-to-noise ratio of receiver

Also, some details shall be provided concerning EIRP and G/T contours of the satellite transponders.

3.10.2 *Equivalent Isotropic Radiated Power*

One of key factors in satellite link budget calculation is the equivalent isotropic radiated power, conventionally denoted as EIRP. This factor is equal to the product of TX output power and antenna gain in the desired direction. At some occasions, this parameter undergoes some additional losses caused by transmission line between TX output and the antenna input and all connections as well. The additional losses must be taken into account by radio designers in performing the relevant calculations. This parameter may be expressed in the numerical or logarithmic forms as follows:

$$\text{EIRP} = G_t \cdot P_t \tag{3.52}$$

$$\text{EIRP}[\text{dB}_W] = G_t[\text{dB}_i] + P_t[\text{dB}_W] \tag{3.53}$$

However, the logarithmic form is more often used by professional experts for technical calculations.

3.10.3 *Merit of Reception System*

Another key factor in the satellite link budget calculation is the ratio G/T, which specifies the receiving system performance.

This parameter is defined as the ratio of the receiver antenna gain (G_R) to the receiver system noise temperature (T_S) normally expressed in decibels by

$$G/T[\mathrm{dB}/K] = G_R[\mathrm{dB}_i] - T_S[\mathrm{dB}_K] \tag{3.54}$$

Example 3.14. An Earth station operating at satellite Ku band using a parabolic antenna with a diameter of 4.8 m and 55 % efficiency. Transmission and reception frequencies are assumed 14 and 12 GHz, respectively. Calculate EIRP and G/T of the Earth station if the system equivalent noise temperature and TX power are 200° K and 32 W, respectively.

Solution.

$$D = 4.8\,\mathrm{m}, \quad f_r = 12\,\mathrm{GHz}, \quad f_t = 14\,\mathrm{GHz}, \quad \eta = 55\,\%$$

Antenna gain can be calculated referring to the catalogs or the following relation for send-and-receive conditions:

$$G_t = 17.8 + 20 \log f_t \cdot D = 54.35\ \mathrm{dB}_i$$

$$G_r = 17.8 + 20 \log f_r \cdot D = 53\ \mathrm{dB}_i$$

Using relations (3.53) and (3.54) yields

$$\mathrm{EIRP} = P_t[\mathrm{dB}_W] + G_t[\mathrm{dB}_i] = 15.05 + 54.35 = 69.4\ \mathrm{dB}_W$$

$$G/T = G_r[\mathrm{dB}_i] - T_S[\mathrm{dB}_K] = 53 - 23 = 30\ \mathrm{dB}/K$$

■

3.10.4 Carrier-to-Noise Ratio

The performance of a satellite link is defined with the carrier-to-noise ratio denoted as C/N or CNR. This term is equivalent to the ratio of the received signal power and noise power

$$C/N = P_r/P_n \tag{3.55}$$

and practically is used in decibels with the following form:

$$\frac{C}{N}[\mathrm{dB}] = P_r[\mathrm{dB}_W] - P_n[\mathrm{dB}_W] \tag{3.56}$$

Replacing the values of P_r and P_n using the related equations and G/T, the following result is obtained:

$$\frac{C}{N}[\mathrm{dB}] = \mathrm{EIRP}[\mathrm{dB}_W] + G/T[\mathrm{dB}/K] - L_t[\mathrm{dB}] - K[\mathrm{dB}_{J/K}] - B_n[\mathrm{dB}_{\mathrm{Hz}}] \tag{3.57}$$

The latter relation is very significant and specifies the interrelations of the main factors on the satellite links. It requires very careful attention for the selection of units when the above relation is used. Also, by considering the equation given below,

$$\frac{C}{N_0}[\mathrm{dB_{Hz}}] = \frac{C}{N}[\mathrm{dB}] + B_n[\mathrm{dB_{Hz}}] \tag{3.58}$$

Relation (3.57) is converted into

$$\frac{C}{N_0}[\mathrm{dB_{Hz}}] = \mathrm{EIRP}[\mathrm{dB}_W] + G/T[\mathrm{dB}/K] - L_t[\mathrm{dB}] - K[\mathrm{dB}_{J/K}] \tag{3.59}$$

In the above equations, that is, (3.55) to (3.59), the parameters and relevant units are

P_r: received power in W
P_n: receiver system noise power in W
T_t: loss of receiver transmission line
K: Boltzmann constant equal to 1.38×10^{-23} W/Hz·°K
B_n: bandwidth of RF channel in Hz

Example 3.15. Distance between the Earth and satellite stations in the Example 3.14 is 37,000 km. Assuming 1.5-dB additional loss for RF feeder and antenna misalignment, then calculate the ratio of carrier to the noise power density, C/N_0 and C/N, for a transponder working on 36-MHz bandwidth. Key factors related to space segment are EIRP = 44 dB_W and G/T = 4 dB/K.

Solution. To calculate C/N_0 of the satellite, we have

$$\mathrm{FSL_u} = 92.4 + 20\log(14 \times 37{,}000) = 206.7\ \mathrm{dB}$$

$$\mathrm{EIRP_G} = 69.4\ \mathrm{dB}_W, \quad (G/T)_S = 4\ \mathrm{dB}/K$$

$$L_t = 206.7 + 1.5 = 208.2\ \mathrm{dB}, \quad K = -228.6\ \mathrm{dB}_{J/K}$$

Using relation (3.59) yields

$$\left(\frac{C}{N_0}\right)_S = \mathrm{EIRP_G} + (G/T)_S - (L_t)_S - K = 93.8\ \mathrm{dB_{Hz}}$$

$$\left(\frac{C}{N}\right)_S = \left(\frac{C}{N_0}\right)_S - B_n, \; B_n = 10\log(1.12 \times 36 \times 10^6) = 76.06\ \mathrm{dB_{Hz}}$$

$$\left(\frac{C}{N}\right)_S = 17.74\ \mathrm{dB}$$

In the downlink, we have

$$\mathrm{FSL_D} = 205.4\ \mathrm{dB}$$

$$\mathrm{EIRP} = 44\ \mathrm{dB}_W,\quad (G/T)_\mathrm{G} = 30\ \mathrm{dB}/K$$

$$L_\mathrm{t} = 205.4 + 1.5 = 206.9\ \mathrm{dB},\quad K = -228.6\ \mathrm{dB}_{J/K}$$

$$\left(\frac{C}{N_0}\right)_\mathrm{G} = \mathrm{EIRP_S} + (G/T)_\mathrm{G} - (L_\mathrm{t})_\mathrm{G} - K = 95.7\ \mathrm{dB_{Hz}}$$

$$\left(\frac{C}{N}\right)_\mathrm{G} = \left(\frac{C}{N_0}\right)_\mathrm{G} - B_\mathrm{n} = 19.67\ \mathrm{dB}$$

■

3.10.5 Coverage Footprints

To simplify the calculations related to the satellite links, the space segment operators provide coverage footprints or contours for which two samples are given in Figs. 3.11 and 3.12. The first one is dedicated to the related satellite transponder transmit coverage footprints marked by EIRP values, while the second one is dedicated to the same transponder receive coverage footprints marked by G/T values.

The application of coverage footprints is to determine all points on the Earth's surface which provide the minimum guaranteed values of either EIRP or G/T to be used in the satellite link budget calculations. At all points inside each closed contour of Fig. 3.11, the satellite power is at least equal to the indicated EIRP value. Also, at all points inside each closed contour of Fig. 3.12, the satellite receiver performance gain is at least equal to the indicated G/T value.

The extension and shape of each coverage contour are affected by the related satellite transponder parameters as stated below:

- TX power and RX sensitivity
- Radiation pattern of the antenna and its pertaining adjustment
- Antenna gain
- Frequency band
- Satellite position

Example 3.16. Using EIRP and G/T coverage footprints of a transponder presented in Figs. 3.11 and 3.12, specify the proper values of effective transmission power and satellite reception performance in the locations specified below:

A : latitude $= 40^\circ$, longitude $= 54^\circ$
B : latitude $= 35^\circ$, longitude $= 68^\circ$

Solution. Using Fig. 3.11 results in

$$48\ \mathrm{dB}_W < (\mathrm{EIRP})_A < 50\ \mathrm{dB}_W \quad\Longrightarrow\quad (\mathrm{EIRP})_A \approx 49\ \mathrm{dB}_W$$

$$46\ \mathrm{dB}_W < (\mathrm{EIRP})_B < 47\ \mathrm{dB}_W \quad\Longrightarrow\quad (\mathrm{EIRP})_B \approx 46.8\ \mathrm{dB}_W$$

Also, using Fig. 3.12 results in

$$(G/T)_A \approx 4.5 \text{ dB}/K$$
$$(G/T)_B \approx 5 \quad \text{dB}/K$$

■

3.11 Satellite Link Losses

3.11.1 Main Factors of the Losses

In satellite links as illustrated in Fig. 3.35, the overall losses are comprised of the following items:

1. Free-space loss (FSL)
2. RF feeders and RF connections loss
3. Antenna misalignment loss
4. Atmospheric constant (or clear sky) loss
5. Atmospheric variable loss

To calculate the received signal level, denoted by RSL, the overall value can be divided into uplink in Earth station toward satellite and downlink in the opposite direction. The amount of RSL in the Earth or satellite stations is not the same because of the following points:

- Significant difference between frequencies of go-and-return (send-and-receive) RF channels resulting in different values of frequency-dependent parameters such as free-space loss, antenna gain, and RF feeder loss.

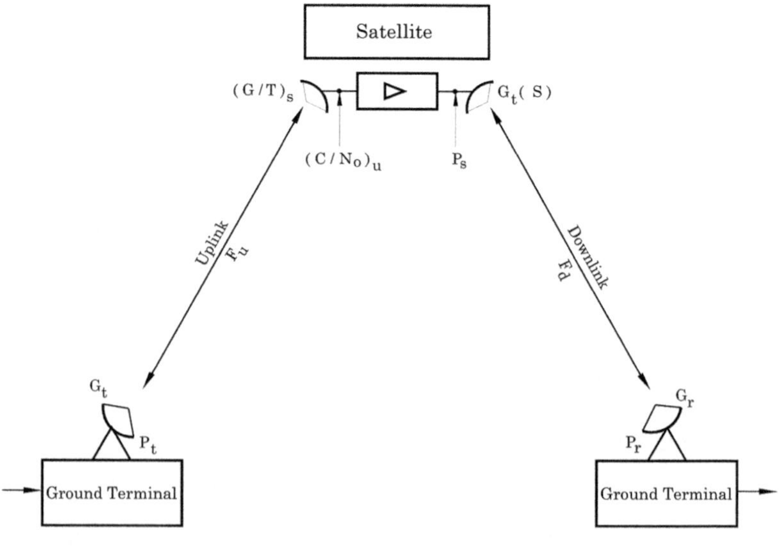

Fig. 3.35 Performance parameters of a typical satellite radio link

- Difference in the signal bandwidth, due to unsymmetrical send-and-receive capacity.
- Technical specifications of transmitter and receiver units used in the ground and space terminals.

Noting the above general issues, it will be possible to determine the total losses with the following relation:

$$L_t = \text{FSL} + L_f + L_a + L_{pf} + L_{pv} + L_m \tag{3.60}$$

In the above equation, all items are in decibels as defined below:

L_t: total path loss (L_{tu} implies L_t for uplink and L_{td} implies L_t for downlink)
FSL: free-space loss
L_f: feeder and RF connector loss
L_a: losses related to the antenna misalignment
L_{pf}: losses due to the absorbents of clear sky
L_{pv}: atmospheric variable path losses
L_m: miscellaneous losses specific to each link

3.11.2 Loss Calculation

3.11.2.1 Free-Space Loss

The free-space loss is defined and its formula derived in Sect. 1.10 of Chap. 1. This parameter in the satellite radio links for which radiowaves propagation is based on the LOS condition can be calculated by

$$\text{FSL}[\text{dB}] = 32.4 + 20 \log f[\text{MHz}] + 20 \log d[\text{km}] \tag{3.61}$$

More details about the above formula are provided in Sect. 1.10.1 of the first chapter. It must be noted that in this equation the losses emanating from absorption, dispersion, composites, and existing particles in the atmosphere are not considered in the calculations. It must be also reminded this loss constitutes the most significant loss component in the satellite links due to the long distance between the Earth and space stations which is unavoidable.

3.11.2.2 Feeder and Connector Losses

These losses include RF signals attenuation from either TX output or RX input up to the antenna including feeders, couplers, filters, and RF connections. In a complete and accurate calculation, this loss must be considered in a communication link for both Earth and space stations.

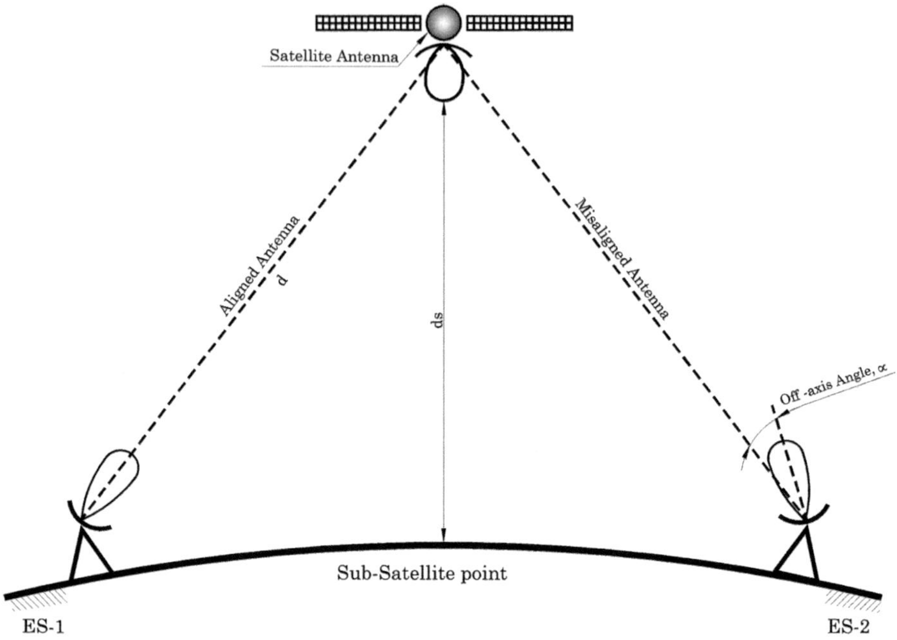

Fig. 3.36 Antenna off-axis angle in satellite link

The feeder and connector losses related to the space segment are normally included in the satellite EIRP and G/T, which the link designer must focus on the Earth station. These losses in the Earth station noting the specification of components and parts provided by manufacturers and as stated in the data sheets must be also considered in the related calculations.

3.11.2.3 Antenna Misalignment Losses

Each satellite transponder covers a vast area of the Earth. Its antenna boresight (main axis), as shown in Fig. 3.36, is directed toward a certain point on the Earth surface called subsatellite point, while the rest of the points in the covered area are located farther having the following two distinct effects:

1. More free-space loss due to longer path ($d > d_s$)
2. Reduction in satellite antenna gain due to off-axis angle α

The effects of satellite off-axis angle are included in the related EIRP and G/T coverage footprints, but if these are not available, then they should be considered in the technical calculations.

In some cases, there is antenna off-axis angle at the Earth station due to misalignment of its antenna. This mechanism will produce additional loss on the received signal levels at the both ends which is referred to as the *antenna pointing loss*. Also it may result in more loss due to the polarization misalignment.

3.11.2.4 Fixed Losses of Atmosphere

The effects of troposphere and ionosphere layers on the radiowaves are outlined in Chap. 1, while specific items related to the satellite waves and resulting losses are explained in this chapter. The atmospheric fixed losses are generally functions of operating frequency of the send-and-receive waves. These losses are caused by absorption and polarization variation phenomena which may be neglected in the satellite C band and Ku band.

3.11.2.5 Varying Losses of Atmosphere

These kinds of losses relate to the radiowaves propagation in the satellite communications as random temporal varying phenomena of the Earth's atmosphere and the surrounding space such as:

- Cloud and particles suspended in the air such as dust, soil, and other aerosols
- Atmospheric precipitations such as snow, rain, and hail
- Solar storms and variations in the atmosphere layers
- Meteor bursts

3.11.3 Received Signal Level

In satellite communications, the received signal level at the input of low-noise amplifier (LNA) can be calculated from the following equation:

$$\mathrm{RSL}[\mathrm{dB}_W] = \mathrm{EIRP}[\mathrm{dB}_W] + G_\mathrm{r}[\mathrm{dB}_i] - L_\mathrm{t}[\mathrm{dB}] \tag{3.62}$$

Based on Fig. 3.35, for calculation of RSL on each of the uplink and downlink, the corresponding values must be substituted in the above equation.

Example 3.17. The INMARSAT satellite at a position of 63° E on the equatorial orbit is used to provide services for ships sailing in the Indian Ocean and neighboring waterways. Calculate received signal levels at the both ends of the satellite link for a path length of 37,750 km with the following parameters:

The rain loss is negligible at the working frequency, and the free-space loss on the send-and-receive routes is identical. Assume other losses to be 2 dB in total.

Ship station:		Satellite station:	
• TX power:	25 W	• EIRP:	68 dB_W
• Antenna gain:	23 dB_i	• Antenna gain:	38 dB_i
• G/T:	$-4\,dB/K$	• Frequency:	$1.5 \sim 1.6$ GHz

Solution. • The received signal level at the satellite station:

$$(\mathrm{RSL})_{\mathrm{S}} = (\mathrm{EIRP})_{\mathrm{G}} + (G_{\mathrm{r}})_{\mathrm{S}} - L_{\mathrm{t}}$$

$$(\mathrm{EIRP})_{\mathrm{G}} = 10 \log 25 + 23 = 37\ \mathrm{dB}_W$$

$$\mathrm{FSL} = 92.4 + 20 \log(1.55 \times 37750) = 187.7\ \mathrm{dB}$$

$$L_{\mathrm{t}} = 187.7 + 2 = 189.7\ \mathrm{dB}$$

$$(\mathrm{RSL})_{\mathrm{S}} = 37 + 38 - 189.7 = -114.7\ \mathrm{dB}_W = -87.7\ \mathrm{dB}_m$$

• The received signal level at the Earth station:

$$(\mathrm{RSL})_{\mathrm{G}} = (\mathrm{EIRP})_{\mathrm{S}} + (G_{\mathrm{r}})_{\mathrm{G}} - L_{\mathrm{t}}$$

$$(\mathrm{RSL})_{\mathrm{G}} = 68 + 23 - 189.7 = -98.7\ \mathrm{dB}_W = -68.7\ \mathrm{dB}_m$$

■

3.12 Satellite Uplink

The satellite uplink consists of different parts that the main items are given in Fig. 3.37.

For satellite uplink, (3.59) reads

$$\left(\frac{C}{N_0}\right)_{\mathrm{u}} = (\mathrm{EIRP})_{\mathrm{u}} + (G/T)_{\mathrm{u}} - (L_{\mathrm{t}})_{\mathrm{u}} - K \tag{3.63}$$

Also, if G and S represent the Earth and satellite stations, respectively, then the above relation can be rewritten as

$$\left(\frac{C}{N_0}\right)_{\mathrm{S}} = (\mathrm{EIRP})_{\mathrm{G}} + (G/T)_{\mathrm{S}} - (L_{\mathrm{t}})_{\mathrm{S}} - K \tag{3.64}$$

In (3.63) and (3.64), the parameters are the Earth station EIRP in dB_W, satellite receiver G/T in dB/K, total free space, and miscellaneous frequency-dependent

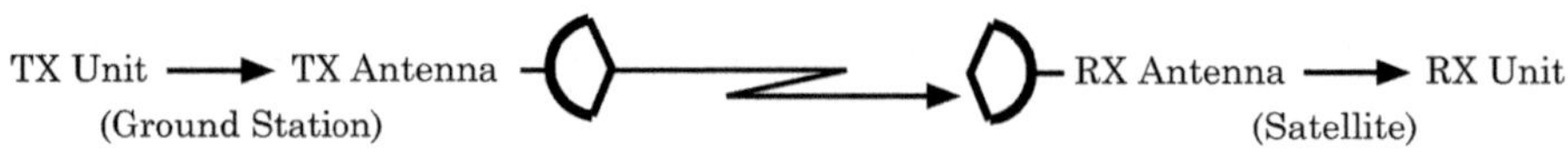

Fig. 3.37 Typical uplink in satellite communications

losses in dB and Boltzmann constant K in $\mathrm{dB}_{J/K}$. The resulting carrier-to-noise density ratio C/N_0 is in $\mathrm{dB_{Hz}}$ that appears at the satellite receiver.

3.12.1 Saturation Flux Density

In some occasions, the flux density appearing at the satellite receiving antenna is required rather than the Earth station EIRP. The flux density required at the receiving antenna to result in saturation of transmitter high-power amplifier (HPA) is called the *saturation flux density*. This parameter denoted by Ψ_S can be used to obtain the required EIRP at the Earth station.

The saturation flux density, in terms of EIRP, can be expressed by

$$\Psi_\mathrm{S} = \frac{(\mathrm{EIRP})_\mathrm{G}}{4\pi d^2} \tag{3.65}$$

In the logarithmic form, by applying mathematical operations, the following equation is concluded: (Why?)

$$\Psi_\mathrm{S}[\mathrm{dB_{W/m^2}}] = \mathrm{EIRP}[\mathrm{dB}_W] - \mathrm{FSL}[\mathrm{dB}] - 10\log\frac{\lambda^2}{4\pi} \tag{3.66}$$

with the following assumption:

$$A_\mathrm{o} = 10\log\frac{\lambda^2}{4\pi} = -21.45 - 20\log f(\mathrm{GHz}) \tag{3.67}$$

and combining it with (3.65) yields

$$\mathrm{EIRP}[\mathrm{dB}_W] = \Psi_\mathrm{S}[\mathrm{dB_{W/m^2}}] + A_\mathrm{o}[\mathrm{dB_{m^2}}] + L_\mathrm{t}[\mathrm{dB}] \tag{3.68}$$

The above equation specifies EIRP for the Earth station to generate the specified flux density on the satellite RX antenna. When other propagation losses such as the atmospheric absorption, polarization mismatch, and antenna misalignment are not included in (3.67), it implies the minimum value of EIRP and allowance must be made for these losses. Normally the saturation flux density is specified by the satellite operators, and if this value is designated by subscript S, then (3.68) is converted into the following form which can be used for the link calculations:

$$(\mathrm{EIRP_S})_\mathrm{u}[\mathrm{dB}_W] = (\mathrm{EIRP_S})_\mathrm{G}[\mathrm{dB}_W] = \Psi_\mathrm{S}[\mathrm{dB_{W/m^2}}] + A_\mathrm{o}[\mathrm{dB_{m^2}}] + L_\mathrm{t}[\mathrm{dB}] \tag{3.69}$$

3.12.2 *Input Back-Off*

When several carriers are used simultaneously, the saturation state will generate distortions resulting from intermodulations due to its nonlinear properties of the related components. Normally, the saturation flux density is declared for one carrier; in multi-carrier case, the transmit power from the Earth station must be reduced enabling the amplifier section of the satellite RX unit to operate in the linear regime of its transfer characteristic curve. This amount of power reduction of the Earth transmitter power is called input *back-off* and is indicated with IBO; therefore,

$$(\mathrm{EIRP})_{\mathrm{u}} = (\mathrm{EIRP}_{\mathrm{S}})_{\mathrm{u}} - \mathrm{IBO} \tag{3.70}$$

Combining the latter relation with the previous equations in the logarithmic system, yields

$$\left(\frac{C}{N_0}\right)_{\mathrm{u}} = \Psi_{\mathrm{S}} + A_{\mathrm{o}} - \mathrm{IBO} + \frac{G}{T} - K \tag{3.71}$$

Example 3.18. An uplink at the Ku band requires a saturation flux density of -95 $\mathrm{dB}_{\mathrm{W/m^2}}$ on the satellite transponder. With the following assumptions,

$$f = 14\,\mathrm{GHz},\ \ \mathrm{IBO} = 10\ \mathrm{dB},\ \ (G/T)_{\mathrm{S}} = -4.5\ \mathrm{dB}/K$$

$$\mathrm{FSL} = 206\ \mathrm{dB},\ \ L_{\mathrm{m}} = 1\ \mathrm{dB}$$

1. Calculate EIRP of the Earth station for the given saturation flux density of the satellite RX unit.
2. Find the carrier-to-noise density ratio at the input of satellite RX unit.

Solution. 1.

$$A_{\mathrm{o}} = -21.45 - 20\log f = -44.37\ \mathrm{dB}$$

$$L_{\mathrm{t}} = 206 + 1 = 207\ \mathrm{dB}$$

Noting (3.68) for the ground station EIRP,

$$(\mathrm{EIRP}_{\mathrm{S}})_{\mathrm{u}} = -95 - 44.37 + 207 = 67.63\ \mathrm{dB}_{W}$$

2. To calculate C/N_0, we use (3.71):

$$\left(\frac{C}{N_0}\right)_{\mathrm{u}} = -95 - 44.37 - 10 - 4.5 + 228.6$$

$$= 74.73\ \mathrm{dB}_{\mathrm{Hz}}$$

■

3.12.3 TX Output Power of Earth Station

HPA of the transmitter must be in a range enabling it to tolerate the losses of the feeder, coupler, and connectors in addition to produce the required EIRP. For this purpose, the following relation must exist in the logarithmic system:

$$(P_{\mathrm{HPA}})_{\mathrm{G}} = (\mathrm{EIRP})_{\mathrm{G}} - (G_{\mathrm{t}})_{\mathrm{G}} + (L_{\mathrm{t}})_{\mathrm{G}} \tag{3.72}$$

When several carriers are employed in the Earth station, it is essential to consider an input back-off (IBO) designated by $(\mathrm{B.O.})_{\mathrm{HPA}}$ for the power amplifier. The relation between the power P_{HPA} and saturation flux density of $P_{\mathrm{HPA,Sat}}$ can be expressed by

$$(P_{\mathrm{HPA}})_{\mathrm{G}} \leqslant (P_{\mathrm{HPA,Sat}}) - (\mathrm{B.O}_{\mathrm{HPA}})_{\mathrm{G}} \tag{3.73}$$

3.12.4 Nominal TX Power of Earth Station

When we intend to express the nominal power of TX in terms of watts, calculated in the logarithmic system for one carrier as specified in (3.72), the following equation may be used:

$$P_{\mathrm{t}}(W) = P_{\mathrm{HPA,Sat}} = 10^{[(P_{\mathrm{HPA}})_{\mathrm{G}} + (\mathrm{B.O}_{\mathrm{HPA}})_{\mathrm{G}}]/10} \tag{3.74}$$

If several carriers are transmitted from the Earth station, the nominal power formula will be converted into

$$P_{\mathrm{t}} = 10^{[\log(N_{\mathrm{C}} \cdot P_{\mathrm{C}}) + \mathrm{B.O}_{\mathrm{HPA}}/10]} = 10E[\log(N_{\mathrm{C}} \cdot P_{\mathrm{C}}) + \mathrm{B.O}_{\mathrm{HPA}}/10] \tag{3.75}$$

where

P_{t}: nominal power in terms of watts
N_{C}: number of carriers
P_{C}: output power for each carrier in terms of watts
$\mathrm{B.O}_{\mathrm{HPA}}$: output power back-off in the Earth station amplifier in terms of dB

Example 3.19. Nominal power of an Earth TX unit is allocated to transmit four satellite channels simultaneously. To prevent generation of any intermodulation products, 10-dB IBO has been considered for HPA. This transmitter is connected to an antenna with a gain of 56 dB_i, and the total path loss is 207 dB. Find nominal output power of the Earth TX unit if

$$G/T = 16\ \mathrm{dB}/K, \quad C/N_0 = 95\ \mathrm{dB_{Hz}}, \quad L_{\mathrm{f}} = 1.6\ \mathrm{dB}$$

Solution. First $\mathrm{EIRP}_{\mathrm{G}}$ is calculated using (3.64):

$$\mathrm{EIRP}_{\mathrm{G}} = 95 - 16 + 207 - 228.6 = 57.4\ \mathrm{dB}_W$$

$$P_{\mathrm{C}}[\mathrm{dB}_W] = \mathrm{EIRP}[\mathrm{dB}_W] - G_{\mathrm{t}}[\mathrm{dB}_i] + L_{\mathrm{f}}[\mathrm{dB}]$$

$$P_{\mathrm{C}} = 57.4 - 56 + 1.6 = 3\ \mathrm{dB}_W \implies P_{\mathrm{C}} = 2\,\mathrm{W}$$

Now given the values of $P_{\mathrm{C}} = 2$ W, $N_{\mathrm{C}} = 4$, and $\mathrm{B.O_{HPA}} = 10\,\mathrm{dB}$ and using (3.75), it yields

$$P_{\mathrm{t}} = 10^{[\log(4\times 2)+1]} = 10^{1.9} \approx 80\,\mathrm{W}$$ ■

3.13 Satellite Downlink

3.13.1 Received Carrier-to-Noise Ratio

As depicted in Fig. 3.38, the satellite downlink consists of the following major sections:

In this situation, (3.57) can be rewritten for the downlink as

$$\left(\frac{C}{N_0}\right)_{\mathrm{D}} = (\mathrm{EIRP})_{\mathrm{D}} + (G/T)_{\mathrm{D}} - (L_{\mathrm{t}})_{\mathrm{D}} - K \qquad (3.76)$$

In this case, if the subscripts G and S represent the Earth and satellite stations, respectively, then the above equation is changed to the following form:

$$\left(\frac{C}{N_0}\right)_{\mathrm{G}} = (\mathrm{EIRP})_{\mathrm{S}} + (G/T)_{\mathrm{G}} - (L_{\mathrm{t}})_{\mathrm{G}} - K \qquad (3.77)$$

where the value of C/N_0 is the ratio of carrier to noise density at the Earth RX input and L_{t} constitutes the total losses related to free space, feeder and connectors, rain, and other miscellaneous losses which mainly depend on the frequency of satellite downlink. To compute the carrier-to-noise ratio, the signal bandwidth is assumed same as the noise bandwidth in this calculation for which we obtain the following equation:

$$\left(\frac{C}{N}\right)_{\mathrm{G}} = (\mathrm{EIRP})_{\mathrm{S}} + (G/T)_{\mathrm{G}} - (L_{\mathrm{t}})_{\mathrm{G}} - K - B_{\mathrm{n}} \qquad (3.78)$$

Fig. 3.38 Typical downlink in satellite communications

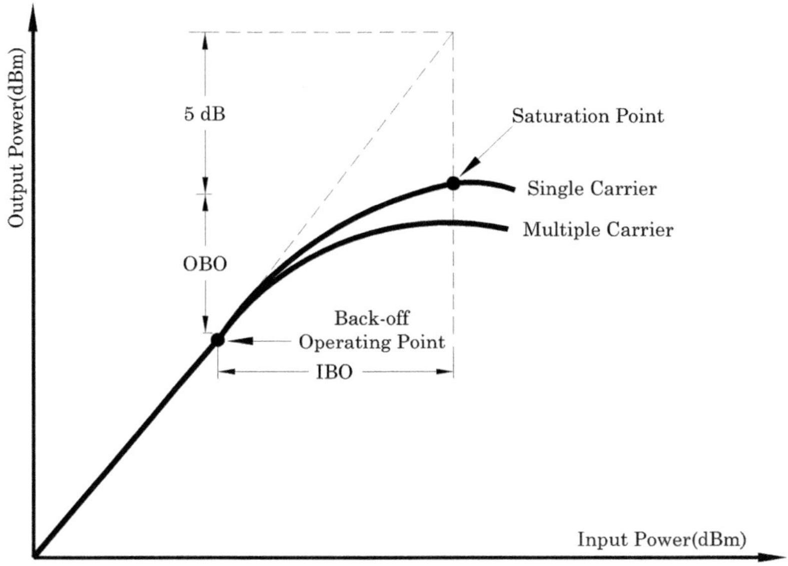

Fig. 3.39 Satellite output back-off vs. input back-off

3.13.2 Output Back-off

On the downlink, similar to the discussion of the uplink, to prevent generation of intermodulation products and performance degradation, it will also require to consider some back-off for the EIRP of the satellite station. This Output back-off denoted as OBO relates to the corresponding IBO as described in Sect. 3.12.3. Figure 3.39 illustrates the relationship of the IBO and the corresponding OBO for single and multiple carriers. As shown, this relationship is nonlinear for high powers.

Normally the saturation point is selected 5 dB below the extrapolated linear portion. Thus, the following relation is valid in the optimized condition between the OBO and corresponding IBO:

$$\mathrm{OBO[dB]} = \mathrm{IBO[dB]} - 5 \tag{3.79}$$

Denoting EIRP of satellite station by $(\mathrm{EIRP_S})_\mathrm{D}$ for saturation condition, then

$$(\mathrm{EIRP})_\mathrm{D} = (\mathrm{EIRP_S})_\mathrm{D} - \mathrm{OBO} \tag{3.80}$$

$$\left(\frac{C}{N_0}\right)_\mathrm{D} = (\mathrm{EIRP_S})_\mathrm{D} - \mathrm{OBO} + (G/T)_\mathrm{D} - (L_\mathrm{t})_\mathrm{D} - K \tag{3.81}$$

3.13.3 *Output Power of Satellite TX*

In the satellite station, the output of HPA must be adequate to tolerate the losses due to feeder and connectors in addition to the power required to produce the necessary EIRP. To meet this requirement, the output of the satellite HPA must be selected in a way that its saturation point be more than the required power by the value of OBO as a minimum, that is,

$$(P_{\mathrm{HPA}})_{\mathrm{S}} = (\mathrm{EIRP})_{\mathrm{S}} - (G_{\mathrm{t}})_{\mathrm{S}} + (L_{\mathrm{f}})_{\mathrm{S}} \tag{3.82}$$

$$(P_{\mathrm{HPA,Sat}})_{\mathrm{S}} \geq (P_{\mathrm{HPA}})_{\mathrm{S}} + \mathrm{OBO} \tag{3.83}$$

Example 3.20. Using a transponder with 36-MHz bandwidth, it is intended to exchange combined signals with a maximum bit error of BER = 10 E(-7). In case of $(C/N_0)_{\mathrm{req}} = 89\ \mathrm{dB_{Hz}}$ and G/T = 22 dB/K, $L_{\mathrm{t}} = 202$ dB and $(G_{\mathrm{t}})_{\mathrm{S}} = 30\,\mathrm{dB}_i$, then find:

1. The value of satellite EIRP.
2. Nominal power of satellite TX unit assuming OBO = 10 dB.

Solution. 1. Using (3.81) and the given data to calculate the satellite EIRP,

$$(\mathrm{EIRP})_{\mathrm{S}} = 40.4\ \mathrm{dB}_W$$

2.

$$(\mathrm{EIRP_S})_{\mathrm{S}} = (\mathrm{EIRP})_{\mathrm{S}} + \mathrm{OBO} \implies (\mathrm{EIRP_S})_{\mathrm{S}} = 50.4\ \mathrm{dB}_W$$

$$P_{\mathrm{t}}[\mathrm{dB}] = (\mathrm{EIRP_S})_{\mathrm{S}} - (9G_{\mathrm{t}})_{\mathrm{S}} + (L_{\mathrm{f}})_{\mathrm{S}}$$

$$P_{\mathrm{t}} = 22.4\ \mathrm{dB}_W \implies P_{\mathrm{t}} = 173.8\,\mathrm{W}$$

■

3.14 Total Carrier-to-Noise Ratio

As illustrated in Fig. 3.35, each satellite hop for a duplex operation consists of two ground stations and one satellite station as an active repeater. The total carrier-to-noise ratio will be discussed for the following two cases:

- Single-carrier system
- Multiple-carrier system

3.14.1 Single-Carrier System

In case of using only one carrier, the intermodulation products will not be generated. In the satellite RX unit, the received signal through the uplink will be amplified along with associated noise. This signal after required processing will be fed to the satellite RX unit for sending toward the second ground station via the downlink.

The total noises and the main signal are received at the ground station. By using proper equations and applying mathematical operations, the following results will be obtained:

$$\left(\frac{C}{N_0}\right)_{\mathrm{T}}^{-1} = \left(\frac{C}{N_0}\right)_{\mathrm{S}}^{-1} + \left(\frac{C}{N_0}\right)_{\mathrm{G}}^{-1} \tag{3.84}$$

$$\left(\frac{C}{N}\right)_{\mathrm{T}}^{-1} = \left(\frac{C}{N}\right)_{\mathrm{S}}^{-1} + \left(\frac{C}{N}\right)_{\mathrm{G}}^{-1} \tag{3.85}$$

In the above relations, the subscripts T, S, and G represent the total, satellite (uplink), and ground station (downlink), respectively. It must be noted that (3.84) and (3.85) are in the non-logarithmic system, while the carrier-to-noise ratios are normally expressed in decibels, and the necessary conversions are required to obtain final results.

3.14.2 Multiple-Carrier System

When several carriers are employed in the satellite communications, its overall performance will be influenced by special type of interference termed *intermodulation noise*. It occurs where multiple carriers pass through nonlinear devices. Both amplitude and phase nonlinearities increase intermodulation products.

Third-order component of intermodulation products is more significant. It falls on neighboring RF channels and acts as an interfering signal. In the multiple-carrier systems, the carrier to intermodulation noise denoted by $(C/N)_{\mathrm{IM}}$ should be taken into account for which (3.84) and (3.85) are extended to

$$\left(\frac{C}{N_0}\right)_{\mathrm{T}}^{-1} = \left(\frac{C}{N_0}\right)_{\mathrm{u}}^{-1} + \left(\frac{C}{N_0}\right)_{D}^{-1} + \left(\frac{C}{N_0}\right)_{\mathrm{IM}}^{-1} \tag{3.86}$$

$$\left(\frac{C}{N}\right)_{\mathrm{T}}^{-1} = \left(\frac{C}{N}\right)_{\mathrm{u}}^{-1} + \left(\frac{C}{N}\right)_{D}^{-1} + \left(\frac{C}{N}\right)_{\mathrm{IM}}^{-1} \tag{3.87}$$

In the last two equations, the subscripts T, u, D, and IM represent total, uplink, downlink, and intermodulation, respectively.

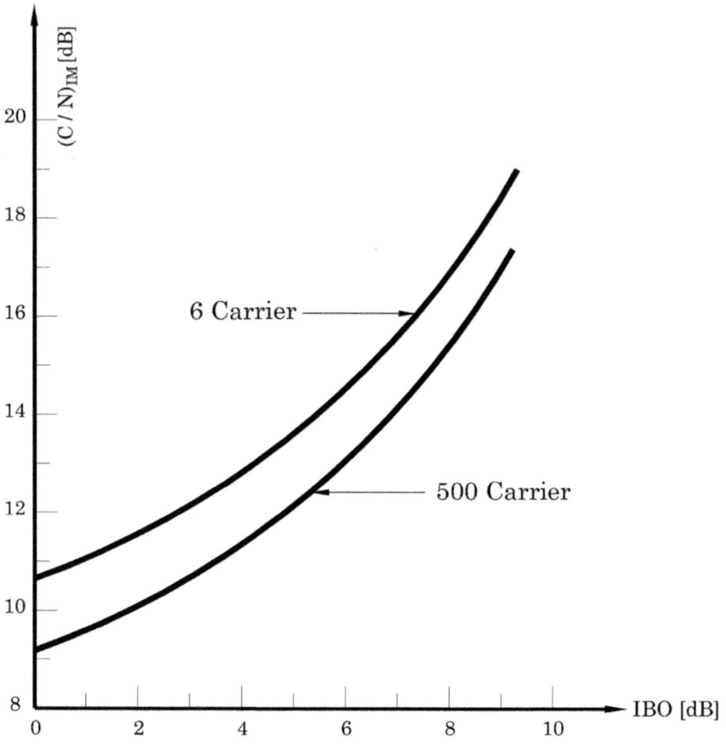

Fig. 3.40 Variation of $(C/N)_{\mathrm{IM}}$ versus input back-off

To reduce effects of the intermodulation noise, the power amplifiers must operate within their linear portion of related transfer characteristic function. To meet this requirement, it is essential to have adequate back-off in their input and output relative to the saturation position. Figure 3.40 illustrates the effects of increasing the amount of IBO for a typical satellite amplifier.

Example 3.21. The following specifications are known for a satellite link:

– Uplink		– Downlink	
• Saturation flux density:-	67 dB/W^2	• Maximum output power:	27 dB$_W$
• Input back off:	IBO = 11 dB	• Output back off:	OBO = 6 dB
• Satellite G/T:	11 dB/K	• Total loss:	196 dB
• Bandwidth:	24 MHz	• Earth station G/T:	40 dB/k
• Frequency:	6 GHz	• Frequency:	4 GHz

Find $\left(\frac{C}{N}\right)_{\mathrm{u}}$, $\left(\frac{C}{N}\right)_{\mathrm{D}}$, and $\left(\frac{C}{N}\right)_{\mathrm{T}}$ in terms of decibels.

Solution. 1.

$$\left(\frac{C}{N}\right)_{\mathrm{u}} = \left(\frac{C}{N}\right)_{\mathrm{S}} = \Psi_{\mathrm{S}} + A_{\mathrm{o}} - \mathrm{IBO} + G/T - K - B_{\mathrm{n}}$$

$$A_{\mathrm{o}} = -21.45 - 20 \log f = -37 \text{ dB}$$

$$B_{\mathrm{n}}[\mathrm{dB_{Hz}}] = 10 \log B_{\mathrm{n}} = 73.8 \text{ dB}_{\mathrm{Hz}}$$

$$\left(\frac{C}{N}\right)_{\mathrm{u}} = 28.8 \text{ dB}$$

2.

$$\left(\frac{C}{N}\right)_{\mathrm{D}} = \left(\frac{C}{N}\right)_{\mathrm{G}} = (\mathrm{EIRP_S})_{\mathrm{S}} + (G/T)_{\mathrm{G}} - L_{\mathrm{t}} - K - B_{\mathrm{n}}$$

$$\left(\frac{C}{N}\right)_{\mathrm{D}} = 19.8 \text{ dB}$$

3.

$$\left(\frac{C}{N}\right)_{\mathrm{T}}^{-1} = \left(\frac{C}{N}\right)_{\mathrm{u}}^{-1} + \left(\frac{C}{N}\right)_{\mathrm{D}}^{-1}$$

$$\left(\frac{C}{N}\right)_{\mathrm{u}} = \text{Anti log } 0.1 \left(\frac{C}{N}\right)_{\mathrm{u}} [\mathrm{dB}] = 758.6, \quad \left(\frac{C}{N}\right)_{\mathrm{u}}^{-1} = 0.00131$$

$$\left(\frac{C}{N}\right)_{\mathrm{D}} = \text{Anti log } 0.1 \left(\frac{C}{N}\right)_{\mathrm{D}} [\mathrm{dB}] = 95.5, \quad \left(\frac{C}{N}\right)_{\mathrm{D}}^{-1} = 0.01047$$

$$\left(\frac{C}{N}\right)_{\mathrm{T}}^{-1} = 0.01178 \implies \left(\frac{C}{N}\right)_{\mathrm{T}} = 84.9$$

$$\left(\frac{C}{N}\right)_{\mathrm{T}} = 10 \log \left(\frac{C}{N}\right)_{\mathrm{T}} = 19.29 \text{ dB}$$

■

3.14.3 CNR Calculations in Satellite Links

In the design phase of satellite links, in addition to considering the main and fixed parameters, the additional factors such as rain and intermodulation products must be also included in the calculations. Generally the following equations can be used to calculate $(C/N)_{\mathrm{cal}}$ value:

1. CNR for uplink with a single carrier and clear sky:

$$\left(\frac{C}{N}\right)_{\mathrm{cal}} = \left(\frac{C}{N}\right)_{\mathrm{u}} = \Psi_{\mathrm{S}} + A_{\mathrm{o}} - \mathrm{IBO} + (G/T)_{\mathrm{S}} + K - B_{\mathrm{n}} \tag{3.88}$$

2. CNR for downlink with a single carrier and clear sky:

$$\left(\frac{C}{N}\right)_{\mathrm{cal}} = \left(\frac{C}{N}\right)_{\mathrm{D}} = (\mathrm{EIRP_S})_{\mathrm{S}} + \mathrm{OBO} + (G/T)_{\mathrm{G}} - L_{\mathrm{t}} - K - B_{\mathrm{n}} \tag{3.89}$$

3. Total CNR with a single carrier and clear sky:

$$\left(\frac{C}{N}\right)_{\mathrm{cal}}^{-1} = \left(\frac{C}{N}\right)_{\mathrm{T}}^{-1} = \left(\frac{C}{N}\right)_{\mathrm{u}}^{-1} + \left(\frac{C}{N}\right)_{\mathrm{D}}^{-1} \tag{3.90}$$

4. Total CNR with multiple carrier and clear sky:

$$\left(\frac{C}{N}\right)_{\mathrm{cal}}^{-1} = \left(\frac{C}{N}\right)_{\mathrm{T}}^{-1} = \left(\frac{C}{N}\right)_{\mathrm{u}}^{-1} + \left(\frac{C}{N}\right)_{\mathrm{D}}^{-1} + \left(\frac{C}{N}\right)_{\mathrm{IM}}^{-1} \tag{3.91}$$

5. Total CNR in rainy weather:

$$\left(\frac{C}{N}\right)_{\mathrm{cal}}^{-1} = \left(\frac{C}{N}\right)_{\mathrm{R}}^{-1} = \left(\frac{C}{N}\right)_{\mathrm{T}}^{-1} \cdot \left[A_R + (A_R - 1)\frac{T_a}{T_{S,CS}}\right] \tag{3.92}$$

Equations (3.88) and (3.89) are in the logarithmic system, while (3.90)–(3.92) are in the non-logarithmic system. For definitions of each one of the above quantities, refer to their pertinent sections.

3.15 Safety Margin

3.15.1 Bit Error in Digital Communications

In digital communications, the quality of signals is evaluated by the BER. In this case, quality and performance criteria set by authorized organizations, such as ITU-R and ITU-T, should be referred based on the relevant hypothetical reference circuit. For example:

- BER $= 10\mathrm{E}(-3) = 10^{-3}$ is a criterion for out- of-service condition in digital communications, that is, when BER $> 10E(-3)$, signal quality is not acceptable and the link is considered out of service.
- BER $\leqslant 10E(-6) = 10^{-6}$ represents a good quality for telephony signals.
- BER $\leqslant 10E(-8) = 10^{-8}$ represents a good quality for data signals.

3.15.2 Minimum Carrier-to-Noise Ratio

By selecting the maximum acceptable BER for each service in the satellite communications and type of modulation, energy per bit should be calculated by

$$E_b = \frac{P_R}{R_b} = \frac{C}{R_b} = P_R \cdot T_b \tag{3.93}$$

where

P_R: average received power
R_b: transmission rate
T_b: bit period
C: carrier received power

A key parameter for satellite carrier systems is the ratio of average carrier power to the noise power density denoted by C/N_0 which is related to E_b through the following equations:

$$\frac{C}{N_0} = \frac{E_b}{N_0} \cdot R_b \tag{3.94}$$

The latter equation is rearranged in the logarithmic system by

$$\left(\frac{C}{N_0}\right)_{\text{req}} [\text{dB}_{\text{Hz}}] = \frac{E_b}{N_0}[\text{dB}] + R_b[\text{dB}_{b/s}] \tag{3.95}$$

Required E_b/N_0 in decibels for a BER range is usually prepared by manufacturer based on the used modulation type. A typical graph is given in Fig. 3.41 for BPSK/QPSK modulations.

3.15.3 Safety Margin

Safety margin refers to the difference in decibels between actual and required carrier-to-noise ratios relevant to the link of interest expressed by

$$M = \left(\frac{C}{N}\right)_{\text{acl}} - \left(\frac{C}{N}\right)_{\text{req}} \tag{3.96}$$

This margin is normally included in radio links to protect them against undesired propagation conditions resulting in the received signal fadings.

Example 3.22. For transmission of digital signals, $C/N_0 = 86\ \text{dB}_{\text{Hz}}$ is required for the maximum BER $= 10\text{E}(-6)$; find:

1. Safety margin for $C/N = 18\,\text{dB}$ and transponder bandwidth of 36 MHz.

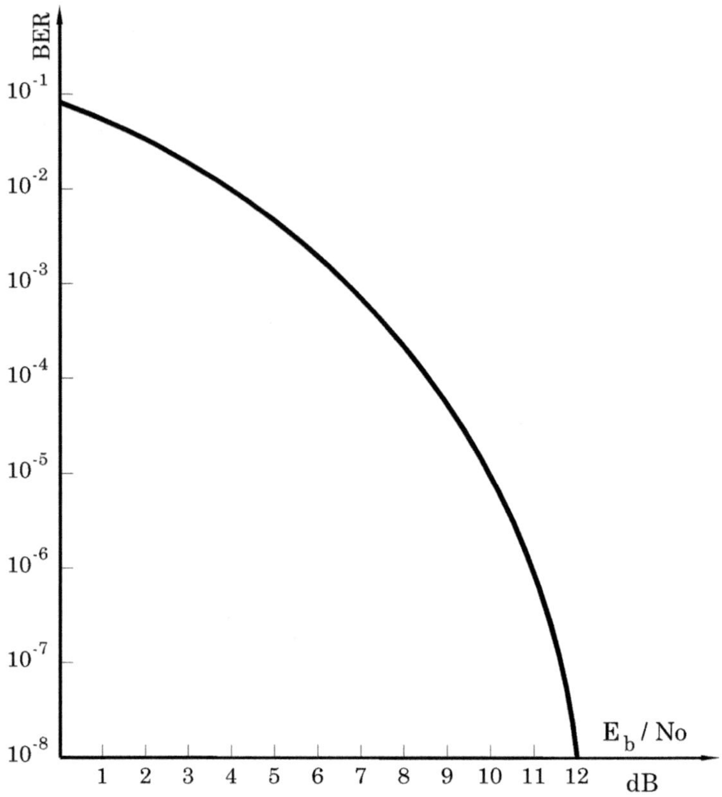

Fig. 3.41 BER variations versus E_b/N_0 (BPSK/QPSK modulation)

2. Safety margin for the link of interset if there is $(C/N_0)_{IM} = 93\ \text{dB}_{Hz}$ and transponder bandwidth of 36 MHz.

Solution. 1.

$$\left(\frac{C}{N_0}\right)_{acl} = \left(\frac{C}{N}\right)_{acl} + 10\log B_n = 18 + 75 = 93\ \text{dB}_{Hz}$$

$$M_1 = \left(\frac{C}{N_0}\right)_{acl} - \left(\frac{C}{N_0}\right)_{req} = 93 - 86 = 7\ \text{dB}$$

2.

$$\left(\frac{C}{N_0}\right)_{IM} = 93 \implies \left(\frac{N_0}{C}\right)_{IM} = 10^{-9.3} = 5\times10^{-10}$$

$$\left(\frac{C}{N_0}\right) = 93\ \text{dB}_{\text{Hz}} \implies \left(\frac{N_0}{C}\right) = 5 \times 10^{-10}$$

Total carrier to noise ratio may be found by using (3.87):

$$\left(\frac{N_0}{C}\right)_{\text{T}} = \left(\frac{N_0}{C}\right) + \left(\frac{N_0}{C}\right)_{\text{IM}} = 10^{-9}$$

$$\left(\frac{C}{N_0}\right)[\text{dB}_{\text{Hz}}] = 90\ \text{dB}_{\text{Hz}}$$

$$M_2 = \left(\frac{C}{N_0}\right)_{\text{acl}} - \left(\frac{C}{N_0}\right)_{\text{T}} = 93 - 90 = 3\ \text{dB}$$ ■

3.16 Exercises

Questions

1. Name the main applications of satellite communications, and specify the prevalent frequency bands.
2. Satellite transponders are basically used as active repeaters. Give reasons why passive repeaters are not used for this application.
3. Evaluate and specify the features of satellite communications.
4. Referring to the recommendation ITU-R, S.521, evaluate various types of hypothetical reference satellite circuits.
5. Specify the difference between various satellite bands and particularly compare them by considering free-space loss, applications, troposphere, and ionosphere effects.
6. Express the main kinds of the Earth orbits used for satellite communications. Specify their applications and the key role of GEO orbit in the global networks.
7. Explain the main features of LEO, MEO, HEO, and GEO satellites including their altitude, free-space loss, speed, TX power, total quantity for global coverage, and antenna of Earth station.
8. Define terrestrial coverage of each satellite indicating the major factors and calculation procedure.
9. List the advantages of the geosynchronous orbit (GEO) compared with the other kinds of Earth orbits.
10. In a global coverage by three GEO satellites, specify the uncovered area of the Earth's surface.
11. Define look angle of a satellite and state its formula.
12. Evaluate how the global coverage is provided (concerning geographic and time) by the LEO satellites, and specify the major factors affecting their numbers.
13. Define coverage contours of the communications satellites and their applications.

14. Define subsatellite point and its advantages in GEO satellite network.
15. Express the equations related to the distance of the Earth station from the satellite, elevation, and azimuth angles.
16. Express the main phenomena related to the satellite wave propagation.
17. Express clear sky loss and indicate its limits for LEO, MEO, and GEO satellites on L, C, Ku, and Ka bands (using graphs of Fig. 3.20).
18. Specify extra loss of troposphere layer on the satellite wave propagation.
19. State the main loss factors of ionosphere layer and their frequency dependence.
20. Explain the variation of ionospheric refractivity index for frequencies less than plasma frequency. What are its effects on the route of wave propagation?
21. What are the reasons for deviation of satellite waves from the straight path between satellite and the Earth station?
22. Define cross polarization discrimination and polarization isolation of the waves.
23. State how the polarization variations of circular waves occur, and specify the space layers having more effects on this mechanism.
24. Define the PLF and express the pertinent relation.
25. Investigate the losses of satellite waves at the areas such as wood lands, closed spaces, vehicles, and buildings, and state what factors affect them.
26. Name the main factors of satellite links such as EIRP, G/T, and C/N, and indicate their main relation.
27. Noting Fig. 3.35, write down the total losses of the waves between Earth terminals.
28. State the relation of the received signal level in terms of the main parameters such as EIRP, G/T, and L_t in a satellite link.
29. Specify the difference between C/N and C/N_0, indicating the affecting factors.
30. Define saturation flux density of a satellite and specify its main relation with EIRP of the Earth terminal and the corresponding path loss.
31. Define the back-off in transmitters of the Earth and satellite stations, and state the reasons and conditions to use them.
32. Define the following parameters and their nominal values in the satellite communications:

$$\frac{E_b}{N_0},\ \frac{C}{N_0},\ \left(\frac{C}{N_0}\right)_T,\ \text{BER},\ R_b,\ T_b$$

Problems

1. The gain of a reflecting flat plate acting as a passive repeater having an area of A, wavelength of λ, and incident angle of α is calculated from the following relation:

$$G_P[\text{dB}] = 20\ \log(4\pi A)/\lambda^2 + 20\ \log(\cos\ \alpha)$$

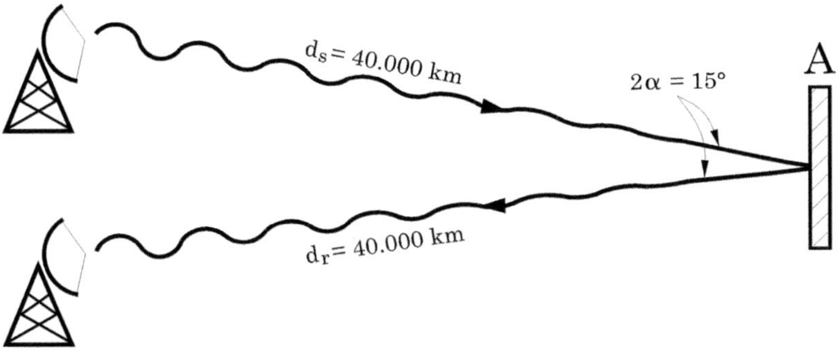

Fig. 3.42 Simple passive satellite link

Noting Fig. 3.42, calculate surface of a passive reflector at $f_1 = 12\,\text{GHz}$ and $f_2 = 4\,\text{GHz}$ in case of assuming that the return route loss under free-space conditions is compensated with the reflector gain.

2. Calculate the period and speed of GEO, MEO, and LEO satellites, on the circular equatorial orbits. Assume altitudes of 20,000 and 1,000 km from the Earth's surface for MEO and LEO satellites respectively. Compare the results using the graphs of Fig. 3.4.
3. Find the period of a LEO satellite for the following conditions:

   (a) In case of revolving on an elliptical orbit with $a = 10,000\,\text{km}$ and the Earth's center coincides with one of its focal points.
   (b) In case of revolving on a circular orbit at a height of 1,630 km above the Earth's surface.

4. For global coverage using three satellites on the GEO, obtain the related distances and angles through proper calculations. Refer to Fig. 3.6.
5. For a GEO type satellite, calculate the maximum area of coverage, and find the beamwidth of the related satellite antenna as well.
6. A LEO satellite network, with average altitude of 1,630 km from the Earth, provides continuous global coverage. Assume this coverage includes latitudes of 80°N to 80°S with 20 % overlapping coverage.

   (a) Specify the minimum number of satellites required.
   (b) If the solid angle of the antenna main lobe in suitable conditions is 18°, then specify whether the number of required satellites decreases accordingly or not.

7. (a) Calculate the number of LEO satellites situated on the equatorial orbit at an altitude of 1,650 km from the Earth's surface for continuous coverage of latitudes of ±23° for elevation angles not less than 10° for the Earth antennas.
   (b) Specify the period of these satellites.

8. MEQ communication network is assumed (satellites situated on the equatorial orbit and of MEO type) with satellites at a height of 4,630 km from the Earth's surface.

   (a) How many satellites are required to cover latitudes between 25°N and 15°S?
   (b) Points on the Earth surface with maximum received signal level.
   (c) Find the period of satellites.
   (d) How long does it take for a satellite to cover a specified point on the Earth equator during each period?

9. Solve Example 3.1 for $h = 20,000\,\text{km}$, $R_e = 6,370\,\text{km}$, and $E_l = 15°$.
10. In case of using six circular orbits in the previous problem, each orbit containing four satellites with uniform spacing, then find

    (a) Total number of satellites.
    (b) The network coverage type and percentage of overlapping.
    (c) How many satellites are in contact instantaneously by the Earth RX antenna.

11. Repeat example 3.3 for coastal terminal situated in a location with the following coordinates:

$$\text{LON} = 20°\ 15'\ 30''\ \text{W} \quad \text{and} \quad \text{LAT} = 25°\ 10'\ 40''\text{N}$$

12. A satellite of GEO type is positioned on the equatorial orbit at 63°E; find:

    (a) The angle of half-power beamwidth (3 dB, HPBW) of satellite antenna for global coverage.
    (b) The angle of half-power beamwidth for covering a surface with a diameter of 3,600 km.
    (c) Whether point A, located at (40°, 18′)N and (58°, 24′)E, is covered or not. In case of positive answer, find its distance from the satellite.

13. Calculate azimuth and elevation angles of the Earth station antenna for the previous problem.
14. Distance of a LEO satellite at an altitude of 1,200 km above the Earth's surface from the location of the Earth station is $d = 1,400\,\text{km}$; find:

    (a) Look angle of satellite link.
    (b) Angle β, suppose that the satellite is positioned on the equatorial orbit at 50°E and the Earth station is in a position of (10°, 18′)S and (53°, 24′)E.

15. In Fig. 3.43, by using a simple diagram, reduction of GEO satellite antenna gain is presented relative to its distance in terms of km from the subsatellite point.

    (a) What is the advantage of point H (subsatellite point) relative to points A and B with respect to the received signal level in terms of dB?
    (b) What is the received signal level at point K relative to point H?

Fig. 3.43 Simple sketch for normalized RSL of problem 15

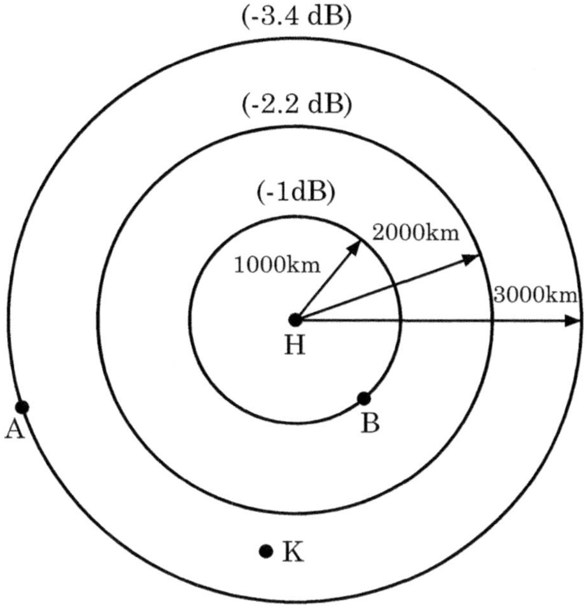

16. Calculate the free-space loss for satellite uplink and downlink at L band (S:1.6 GHz, R:1.5 GHz) and Ka band (S:30 GHz, R:20 GHz) for a GEO satellite at a distance of 40,000 km between the Earth and satellite stations.
17. Repeat the above problem for C band and Ku band.
18. Calculate radio beam spreading loss for all points situated at latitudes between 30°N and 40°N with an antenna angle of less than 5°.
19. Noting Fig. 3.24, find response to the following questions:

    (a) Suppose 3 dB as the safety margin at the Ku band, then find rain intensity to fade out the link.
    (b) Determine safety margin in case of increasing the frequency to 100 GHz for rainfall effective length of 5 km.

20. Find rain attenuation when its intensity is 50 mm/h for the following satellite routes:

    (a) Satellite uplink on the C band ($f = 6$ GHz) and elevation angle of 15°.
    (b) Downlink route on the Ka band ($f = 20$ GHz) and elevation angle of 30°.

21. Calculate the value of refractivity index of the ionosphere layer for frequencies of L band and C band with electron density variations in the range of 10E (12) $\leqslant$ N $\leqslant$ 10E (14).
22. If in Example 3.11 INMARSAT satellite (on the L band) and INTELSAT satellite (on the C band) are used, assuming $L = 40{,}000$ km and $h_{\text{i}} = 400$ km, then find:

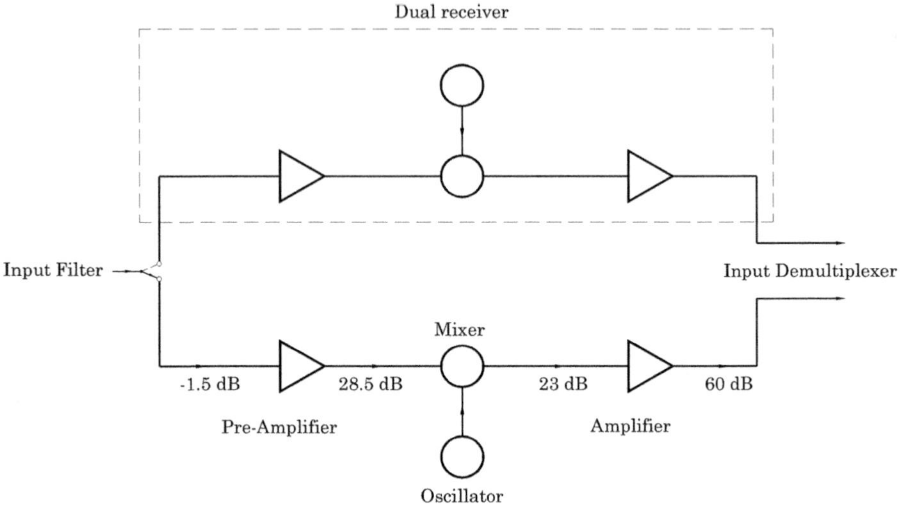

Fig. 3.44 Simple circuit diagram of satellite receiver

(a) Excess length of the total route.
(b) Deviation of antenna axis.

23. In case of assuming Faraday rotation equal to $\theta_f = 12^\circ$ at $f = 3$ GHz, then:

    (a) Plot the variations of θ_f in the range of $1\,\text{GHz} \leqslant f \leqslant 14\,\text{GHz}$.
    (b) Obtain the value of polarization rotation for send-and-receive waves on the C band.
    (c) Noting the above results, calculate the value of XPD on the RX antenna.

24. For 99.9 % reliability of satellite wave propagation, find:

    (a) Required safety margin.
    (b) Annual outage time related to the propagation unavailability.

25. Noting Fig. 3.44, which is a simple circuit diagram of a satellite RX unit, answer the following questions:

    (a) How much is the overall amplification coefficient of the RX system?
    (b) How much is the attenuation caused by the mixer?
    (c) Obtain the amplification coefficient for each of the amplifiers.
    (d) Find signal level at the RX output if its power at the reference point is 500 pW.

26. Calculate the EIRP and G/T of an Earth station with antenna diameter of 14 m and 55 % efficiency for C band using a transmitter with 2-KW output power and system equivalent noise temperature of 300 °K.
27. An antenna (Cassegrain type) with 10-m diameter and 55 % efficiency is used in an Earth station working at C band. The send-and-receive frequencies are

6 and 4 GHz, respectively. Calculate the EIRP and G/T of the Earth station assuming system equivalent noise temperature of 200° Kelvin and TX power of 500 W.

28. A distance around 38,000 km is considered between Earth and satellite stations. Calculate carrier-to-noise ratio and carrier to the noise power density under the following conditions if there is 3-dB loss for miscellaneous items:

$$(\text{EIRP})_{\text{S}} = 38\ \text{dB}_W, \quad (\text{EIRP})_{\text{G}} = 80\ \text{dB}_W$$
$$(G/T)_{\text{S}} = -0.2\ \text{dB}/K, \quad (G/T)_{\text{G}} = 27\ \text{dB}/K$$

29. On a satellite uplink at the L band, saturation flux density of the transponder is assumed to be $-90\,\text{dB}_{\text{w/m}^2}$. Based on the following data, find $(\text{EIRP})_{\text{G}}$ and $(C/N_0)_{\text{S}}$:

$$f = 1.6\,\text{GHz}, \quad \text{IBO} = 10\ \text{dB}, \quad (G/T)_{\text{S}} = -8\ \text{dB}/K$$
$$d = 38{,}000\,\text{km}, \quad L_{\text{m}} = 1\ \text{dB}$$

30. If the effective transmission power of a telecommunications satellite is $60\,\text{dB}_W$, then find satellite flux density, SFD, on the ground for the following cases:

(a) GEO satellite.
(b) LEO satellite on the altitude of 1,400 km above the ground.
(c) MEO satellite on the altitude of 5,000 km above the ground.

31. Required C/N_0 for a satellite downlink is $90\,\text{dB}_{\text{Hz}}$. Where $(G/T)_{\text{G}} = 20\,\text{dB/K}$, $L_{\text{t}} = 200\,\text{dB}$, and $(G_{\text{r}})_{\text{S}} = 32\,\text{dB}_i$, find:

(a) The value of satellite EIRP.
(b) The saturation flux density of the satellite when OBO = 20 dB.

32. On a satellite link, the carrier-to-noise ratios of the Earth terminal and space station are 18 and 26 dB, respectively; calculate:

(a) The value of total carrier-to-noise ratio.
(b) The value of total carrier-to-noise ratio if there is an intermodulation noise equal to $(C/N)_{\text{IM}} = 30\,\text{dB}$.

33. In the previous problem, it is required to consider a safety margin to compensate the rainfall effect of $(C/N)_{\text{R}} = 33\,\text{dB}$. Assuming rainfall attenuation of 1 dB, how much the value of $(C/N)_{\text{cal}}$ must be increased to satisfy these conditions?
34. In the previous problem, there is 1.5-dB loss on the uplink and 1-dB loss on the downlink due to the effects of rainfall. Assuming 1.2-dB reduction is the C/N of the downlink due to more thermal noise, then calculate the value of $(C/N)_{\text{T}}$.
35. On a satellite link in the C band with the following specifications, calculate $(C/N)_{\text{S}}$, $(C/N)_{\text{G}}$, and $(C/N)_{\text{T}}$ in the logarithmic system.

Uplink		Downlink	
• Saturation flux density:	65 dB/W^2	• Satellite TX power:	30 dB$_W$
• Input back off (IBO):	10 dB	• Output back off (OBO):	5 dB
• Satellite G/T:	10 dB/K	• Total loss:	198 dB

36. Prepare a computer program to calculate the geographic coverage of a satellite in terms of the circuit specifications and antenna radiation pattern. This program should include:

    - Calculation of the required number of satellites.
    - Calculation of the minimum elevation angle of Earth station antennas.

    Determine the minimum data required for the prepared program.
37. Prepare a computer program to calculate the antenna bearings of Earth stations for communicating with the following satellites:

    - GEO satellites
    - LEO satellites
    - LEO satellites on the equatorial orbit
    - Polar satellites
    - Satellites in inclined orbits

38. Write a computer program for calculations of satellite link budget. Based on your program, determine:

    - Input data needed.
    - Output results (products).

Chapter 4
Wireless Broadcasting

4.1 Introduction

The International Telecommunication Union (ITU) has defined services through a set of radio regulations pertinent to broadcasting as "a radio communications services for transmission of signals to be received directly by the public" which can include voice, video, and/or any type of broadcasting.

The aforementioned union has also allocated suitable frequency bands for planning and establishment of radio and television transmitter centers. In view of wide application of the broadcasting services, this chapter is devoted to the radiowaves propagation in the selected frequency bands. Generally, the radio broadcasting can be categorized into the following three groups:

- Broadcasting in frequency band less than 30 MHz (LF, MF, HF)
- Broadcasting in frequency band between 30 and 3,000 MHz (VHF, UHF)
- Broadcasting in frequency band more than 3 GHz (SHF)

Due to the importance and scope of applications of VHF/UHF bands, the major emphasis of this chapter is focused on this band. It should be reminded that the propagation of the waves in frequency bands below 30 MHz and/or satellite frequency band has been discussed in other chapters, and thus, they will be briefly discussed here. The radio section of the ITU for long-distance communications has also contributed broadcasting service (sound) in its BS series and also recommendations related to video broadcasting service (TV) in its BT series, which are quite generalized and complete.

A. Ghasemi et al., *Propagation Engineering in Radio Links Design*,
DOI 10.1007/978-1-4614-5314-7_4,

4.2 Features of Broadcasting Services

Some of the main features of the broadcasting services are as follows:

- In this service, the one way communications is from transmitting centers and broadcasting stations toward thousands and/or millions of listeners/viewers.
- The prevailing climatic and environmental conditions of the transmitter centers have major differences with those of receiver terminals especially with different types of satellites and their over horizon characteristics.
- In general, the equipments and antennas installed in the transmitter centers are high quality, sophisticated, and complete with auxiliary devices, whereas equipments and antennas of receiver terminals are relatively simple and cost effective.
- Due to the nature of one way radio communications, the arising problems from network planning and frequency assignment are mainly limited to the transmitting side.
- Because of the importance and significant influence of public broadcasting services, a great effort is made by the governments, international institutions, and regional and private agencies to improve and expand such services in different fields such as:
  - Coverage area
  - Variety of services
  - Quality of services

Considering political, military, social, cultural, and economic aspects of public broadcasting, on rare occasions, some governments and/or different groups may try to make interference and jamming against broadcasting radio signals.

4.3 Broadcasting Networks

Different types of broadcasting networks can be categorized using radiowaves. Some main types of broadcasting networks are explained in the following subsections.

4.3.1 Broadcasting with $\lambda \geq 10\,m$

Radiowaves with wavelength more than 10 m are generally recognized as MW (medium wave) or SW (short wave) operate in LF, MF and HF frequency bands. As depicted in Fig. 4.1, these waves are mostly used for voice channels and have local and long-distance remote coverage applications through the radiowaves ionospheric propagation.

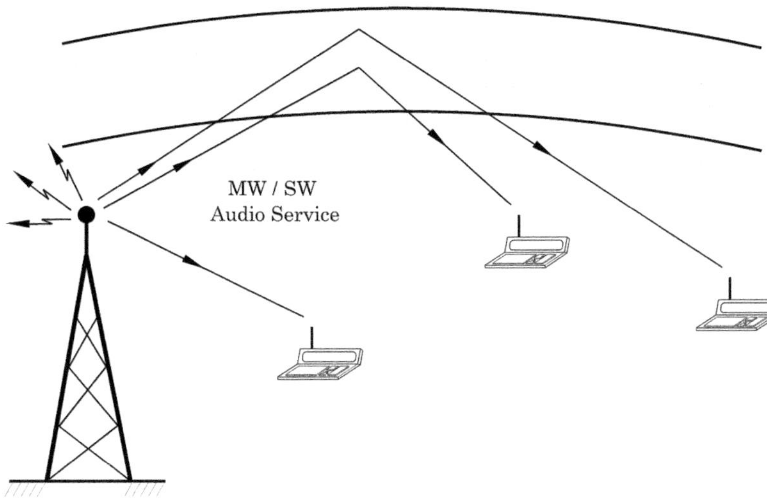

Fig. 4.1 MW and SW broadcasting radiowaves

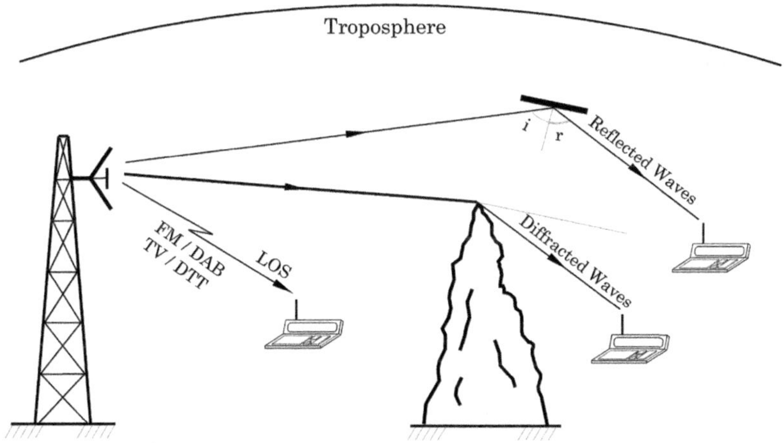

Fig. 4.2 Local broadcasting service in VHF and UHF bands

4.3.2 Broadcasting in Local Band

These types of radiowaves are basically in VHF/UHF bands and cover the frequency range of 30–3,000 MHz. This band is dedicated to broadcasting services of FM radio, analog TV, and also digital audio broadcasting (DAB) and digital terrestrial television (DTT). As indicated in Fig. 4.2, it is used in the boundaries and vicinities of the cities and ultimately for provincial and national coverage using communications infrastructures to relay the radio signals.

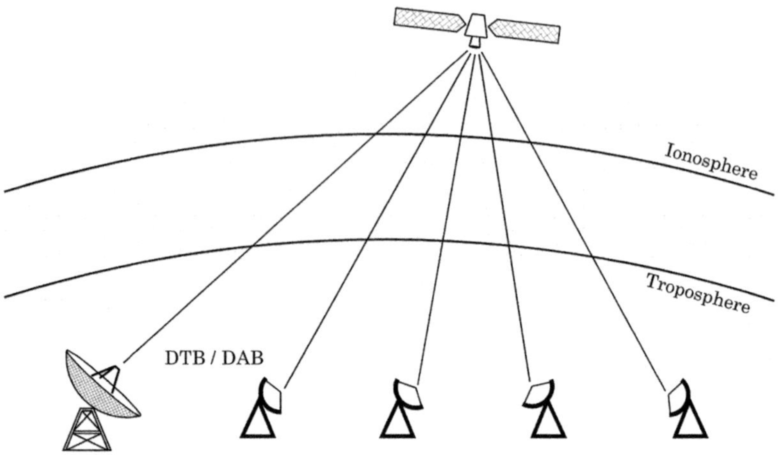

Fig. 4.3 Satellite broadcasting service in SHF and EHF bands

4.3.3 Broadcasting by Satellite

Satellite waves are used for digital audio and video broadcasting services in Ku and Ka bands. This particular service, which is generally employed by GEO satellites in equatorial orbit and at an altitude of 36,000 km above the Earth, can have global, regional, national, and local coverages (Fig. 4.3).

4.4 Frequency Bands

As it was discussed earlier, the broadcasting services from a frequency perspective are divided into three groups. The following sections will elaborate on the assignment of the bands to each one of these groups. It should be noted that based on ever-increasing demand for radio services and advancements of this particular industry and utilization of modern technologies, there are certain updates proportionate with such requirements in due times.

4.4.1 LF/MF/HF Frequency Band

The radiowaves in LF frequencies within the range of 30–300 KHz, MF within the range of 300–3,000 KHz, and HF within the range of 3–30 MHz are basically used for audio services with a low bandwidth.

Table 4.1 LF/MF/HF sub-bands for broadcasting service

	Frequency range in KHz		
Band	ITU region 1	ITU region 2	ITU region 3
LF	148.5–283.5	–	–
MF	526.5–1,606.5	525–1,705	526.5–1,606.5
	2,300–2,498	2,300–2,495	2,300–2,495
HF	3,200–3,400	3,200–3,400	3,200–3,400
	3,950–4,000	–	3,900–4,000
	3,200–3,400	4,750–4,995	4,750–4,995
	3,200–3,400	5,005–5,060	5,005–5,060
	3,200–3,400	5,900–6,200	5,900–6,200
	3,200–3,400	7,100–7,350	7,100–7,350
	3,200–3,400	9,400–9,900	9,400–9,900
	3,200–3,400	11,600–12,100	11,600–12,100
	3,200–3,400	13,570–13,870	13,570–13,870
	3,200–3,400	15,100–15,800	15,100–15,800
	3,200–3,400	17,480–17,900	17,480–17,900
	3,200–3,400	18,900–19,020	18,900–19,020
	3,200–3,400	21,450–21,850	21,450–21,850
	3,200–3,400	25,670–26,100	25,670–26,100

Noting how these waves propagate as mentioned earlier, they are used as ground waves for short distances and as ionospheric waves for long distances radio transmissions. Table 4.1 specifies the bands assignments for these services based on Article 5 of ITU radio regulations for three regions, as defined in ITU chart presented in Fig. 1.3 of Chap. 1.

4.4.2 VHF/UHF Frequency Band

The radiowaves in VHF/UHF bands which include 30–3,000-MHz-frequency range are used for audio and video broadcasting with the required bandwidth. Both types of analog and digital technologies are applicable and can be used as ground waves for local coverage. Generally, the radiowaves of this band are propagating in the form of line of sight (LOS) but sometimes can propagate non-line of sight (NLOS) using frequency-hopping phenomenon. However, this service can be used in limited intercity distances up to tens of kilometers.

Table 4.2 specifies the allocated bands for this service stipulated by Article 5 of the ITU radio regulations for three regions of the world.

The local broadcasting in this band consists of sub-bands which are shown in Fig. 4.4.

Table 4.2 VHF/UHF sub-bands for broadcasting service

	Frequency range in MHz		
Band	ITU region 1	ITU region 2	ITU region 3
VHF	47–54	–	47–50
	54–68	54–72	54–68
UHF	470–960	470–608	470–960
	–	614–890	–
	1,452–1,492	1,452–1,492	1,452–1,492
	2,520–2,670	2,520–2,670	2,520–2,670

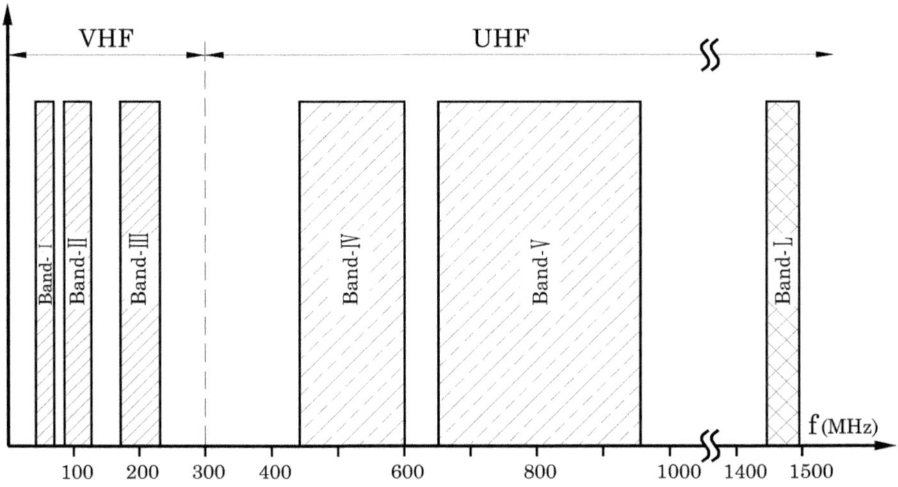

Fig. 4.4 Frequency sub-bands in local broadcasting service

Table 4.3 Frequency sub-bands for satellite broadcasting

	Frequency range in GHz		
Band	ITU region 1	ITU region 2	ITU region 3
SHF	11.7–12.5	12.2–12.7	11.7–12.5
	–	17.3–17.8	–
	21.4–22	–	21.4–22
EHF	40.5–42.5	40.5–42.5	40.5–42.5
	74–76	74–76	74–76

4.4.3 Satellite Frequency Band

Satellite services intended for broadcasting of direct to home (DTH) at current situation is generally used for satellite Ku and Ka bands. To support potential expansions, and based on advancement in the RF manufacturing and technical achievements in the telecommunication industries, more RF spectrum in higher bands is required.

This band is used for audio and video broadcasting and is based on Article 5 of ITU radio regulations. The dedicated bands are presented in Table 4.3.

4.5 Broadcasting on the LF/MF/HF Band

In the early stages of implementing the broadcasting services, the frequencies less than 30 MHz in LF, MF, and HF bands were mainly used for transmission of analog audio signals which are still valid. Nowadays because of the technical achievements in fabrication of RF devices and using of developed systems, the application of LF/MF/HF bands is restricted, for which the following factors are the main reasons:

- Improvement in the quality of radiowaves reception
- Increasing the equipment reliability
- Enhanced capability for receiving signals with lower BER in digital services and higher SNR in analog services
- Solving the wave propagation dilemma
- Providing better and more reliable coverage

4.5.1 General Phenomena

The ionospheric general propagation phenomena are presented in the most books related to the radiowaves propagation. Some of main issues affecting ionospheric propagation in LF/MF/HF bands are listed below:

- Vertical and inclined propagation in the ionosphere layer
- Optimum operating frequency
- Long distance link for broadcasting
- Day and night time frequency
- Propagation modes and their related time delays
- Sun effect and magnetic field of the Earth

4.5.2 Particular Phenomena

The frequency-specific phenomena related to radiowaves propagation in the LF/MF/HF bands and the surface waves in terms of the electrical specifications, received signal level at the receiver, and waves with vertical/horizontal polarizations are discussed in the professional books and technical papers. Also, a number of technical reports and recommendations of ITU-R are dedicated to this subject. Some of the particular phenomena related to the ionospheric propagation in the LF/MF/HF band are as follows:

- Transmission frequency through E and F layers
- Propagation modes
- Elevation angle of the radiowaves

- Field intensity of the radiowaves and received signal level
- Signal to noise ratio (SNR)
- Minimum usable frequency
- Propagation in the MF band and variation of the amplitude of the received signal and associated time delay travelling through various routes and using different modes of operation

4.6 Broadcasting Satellite System

4.6.1 Advantages of Satellite Broadcasting

The broadcasting satellite system (BSS) became popular among public and private organizations for the purpose of radio and television broadcasting once they proved their practical capabilities to provide telecommunications services. In brief, the main rationale to utilize satellites to provide such needs consists of the following requirements:

- To overcome difficulties related to providing radio line-of-sight conditions in the terrestrial communications.
- To overcome problems of ionospheric and tropospheric radiowaves propagation in the wireless communications using MF, HF, VHF, and UHF bands.
- Providing suitable and proper coverage in vast parts of the Earth.
- Providing good quality and highly reliable broadcasting services.
- Advancements of the digital technology and compression techniques of the audio and video signals.

This type of communications, implemented by the geostationary Earth orbit (GEO) satellites, have experienced a great expansion in such a way that most of the countries and/or private agencies broadcast thousands of different channels over most parts of the Earth.

4.6.2 Applications

Broadcasting audio and TV programs through satellites currently is based on digital technology and compression techniques. It is possible to transmit each channel by just using a small portion of a transponder to cover many listeners/viewers in vast areas. Therefore, to broadcast tens of TV programs with standard-definition TV (SDTV), and even high-definition TV (HDTV), quality utilizing only one transponder is a normal practice. As shown in Fig. 4.5, the application of satellites to broadcast audio and TV signals is conducted in two forms as stated below:

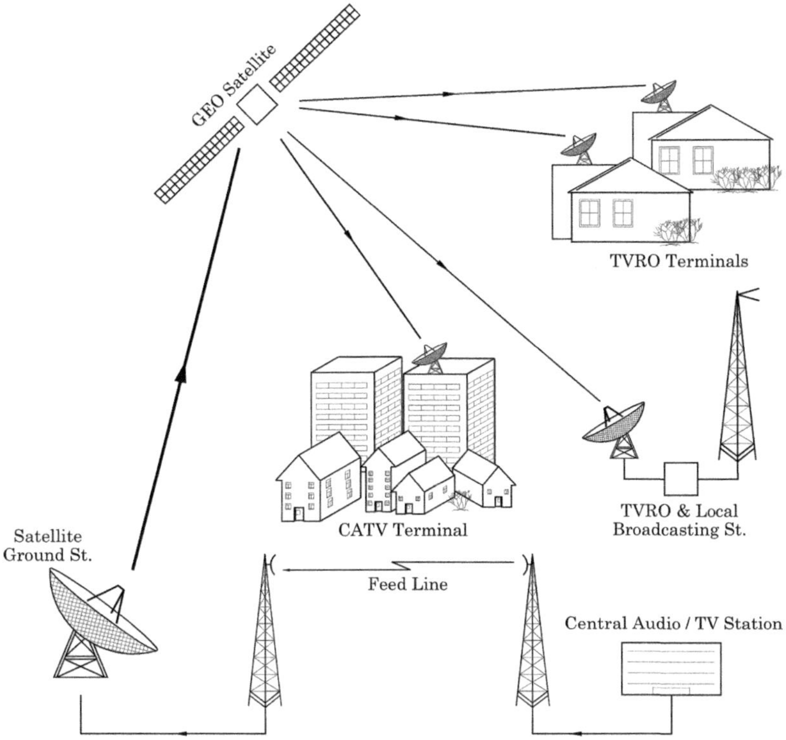

Fig. 4.5 Direct broadcasting satellite system, DBS

- Direct broadcasting satellite (DBS) system, which sometimes is referred as DTH service. In this type of broadcasting, the receivers at home and/or CATV (community antenna TV) directly receive the related radiowaves from satellite.
- Indirect broadcasting satellite service, by which the initial programs produced in the central station are transferred by satellite links to a secondary station and then broadcasted instantaneously using local transmitter in the required frequency bands.

4.6.3 Technical Parameters

In this section, a number of technical parameters related to satellite broadcasting systems are briefly discussed. For further information, the reader is referred to specialized books, technical manuals, and Internet websites dedicated to such services.

Table 4.4 Bit rates for TV channels

Pattern	Display	Pixel	Required bit rate (Mb/s) 60 frame/s	30 frame/s	25 frame/s
HDTV/1080 *i*	16:9	1,920 × 1,080	–	995	796
HDTV/720 *ρ*	16:9	1,280 × 720	885	442	334
SDTV/470 *ρ*	4:3	704 × 480	324	162	130
SDTV/480 *ρ*	4:3	640 × 480	295	148	118

4.6.3.1 Bit Rate of TV Channels

In digital TV receiver sets, without usage of compression techniques, the bit rate required is proportionate to the picture elements (i.e., pixels) for total frames per second. The mentioned bit rates are provided in Table 4.4 for HDTV and SDTV patterns conforming to the Advanced Television System Committee (ATSC) standard.

As it can be observed, noting the number of pixels and number of full pictures per second results in the bit rate required to be 118–995 Mb/s. Assuming 200-Mb/s bit rate for ordinary TV sets, it can be concluded that to transmit one TV channel, the required capacity is too large. Consequently, in the initial generations of INTELSAT satellites, a very low number of TV channels could be exchanged. In recent years, by employing the TV pictures compression techniques based on MPEG standards, the amount of bit rate required for standard color TV channels reduced drastically as stated below:

- For movie channels with a low amount of movements equivalent to 4 Mb/s
- For show channels with moderate amount of movements equivalent to 5 Mb/s
- For sport channels with fast movements equivalent to 6–15 Mb/s

Therefore, by utilization of one single transponder capable of transferring signals at 45 Mb/s, it is possible to broadcast a considerable number of compressed digital TV channels.

4.6.3.2 Transponder Capacity

The relation between frequency bandwidth and bit rate needed for satellite communications is

$$B/R = \frac{B/W \times m \times \text{FEC}}{1+\rho} \tag{4.1}$$

In the above relation, B/R is in terms of Mb/s, and B/W is in terms of MHz; ρ is the roll-off factor; FEC is the amount of forward error correction, and m is related to modulation levels (M) with the following relation:

$$m = \log_2^M \tag{4.2}$$

4.6.3.3 Error Correction

Due to large amount of compression in digital TV signals, the possibility of data iteration is very low, and thus, the bit error rate will have severe effects compared to uncompressed signals. Noting the introductory discussion provided so far, the methods of FEC with rates of 0.5, 0.75, and 0.875 are used to exchange compressed signals. To acquire a suitable BER, it is essential to have an E_b/N_0 value in excess of 6 dB; otherwise, the amount of error will be very high reducing the quality to an unacceptable level.

Example 4.1. The bandwidth of a transponder intended to broadcast television programs is 24 MHz, assuming $\rho = 0.4$ and QPSK modulation, calculate:

1. Maximum transmission capacity.
2. In case of using FEC method with FEC = 0.75, what is the useful capacity?
3. How many TV channels having desirable quality can be broadcasted with this transponder?

Solution. 1. Using relation (4.1) we have

$$M = 4 \quad \Longrightarrow \quad m = \log_2^4 = 2$$

$$B/R = \frac{24 \times 2}{1 + 0.4} = 34.4\,\text{Mb/s}$$

2. In other words, 8.6 Mb/s is spent for error correction to transfer signals with the acceptable quality:

$$\text{FEC} = 0.75 \quad \Longrightarrow \quad B/R = 34.4 \times 0.75 = 25.8\,\text{Mb/s}$$

3. Assuming that the required capacity for each color channel is 6 Mb/s, then it is possible to broadcast four channels with the said transponder. ■

4.6.3.4 Transponder Power

DBSs are divided into three groups according to their output power:

- High-power transponder, with the effective radiated power (ERP) in the range of 51 dB_w to 60 dB_w
- Medium-power transponder, with the ERP in the range of 40 dB_w to 48 dB_w
- Low-power transponder, with the ERP in the range of 33 dB_w to 37 dB_w

4.6.3.5 Frequency Bands and Polarization

Usually, the DBS satellites operate in the Ku frequency band. When using smaller antennas, it is possible to utilize higher-frequency bands as well. For a 500-MHz bandwidth selected in one of allocated frequency bands for satellite broadcasting, it is possible to assign 16 transponder each with 24-MHz RF channel spacing for one type of polarization including 3 MHz as guard band. By utilizing different polarizations such as right-hand circular polarization (RHCP) and left-hand circular polarization (LHCP) for broadcasting, the capacity will be duplicated, and 32 transponders will be available.

4.6.3.6 TVRO Terminal

Satellite TV reception terminals called TV receive only (TVRO) are used for direct reception from the DBS satellites consist of two main parts of outdoor unit (ODU) and indoor unit (IDU). The ODU part includes parabolic antenna, receiver, polarization selection device, and low-noise amplifier (LNA) or low-noise block (LNB) which following the reception of RF input in Ku band converts the signal to IF signal and transfers it to IDU part. The IDU contains tuner, demodulator, FEC, demultiplexer, and audio and video control devices.

4.6.4 Propagation of Satellite Radiowaves

4.6.4.1 Fundamental Considerations

The propagation of the satellite radiowaves relies mainly on fundamental principles and is independent of the application. However, the particular requirements in this field, where transmitters are required to cover vast areas with limited control over the receiving installation, mean that certain aspects of propagation have greater significance for public broadcasting. Generally, the fundamental considerations having major effects on the propagation of satellite radiowaves are:

- Satellite orbit and its type
- Frequency band being used
- Type of communications such as fixed or mobile
- Type of service such as voice and image
- Atmosphere dynamic variations
- Sky activities and radio noises

Broadcasting by satellite leads to propagation considerations that are not fully in compliance with those mentioned in Chap. 3 for the fixed satellite service. For proper planning of the BSSs, having appropriate propagation data and prediction methods is a crucial requirement. Additional information specific to the BSS design, based on the ITU-R recommendation No. P.679, is treated in this section. Among various issues required for more considerations specific to the broadcasting service are:

- Tropospheric effects including gaseous absorption and attenuation and depolarization by rain and other hydrometeors
- Ionospheric effects such as scintillation and Faraday rotation
- Local environmental effects including attenuation by buildings and vegetation

4.6.4.2 Tropospheric Effects

Signal impairments caused by the tropospheric layer are negligible for frequencies below 1 GHz and path elevation angles exceeding 10°. As the elevation angle decreases and/or frequency increases, the impairments become more severe, and fluctuations of the signal amplitude and the angle of arrival are significant and shall be taken into considerations for the satellite radio link design.

Sky-noise temperature increases due to precipitation resulting in further reduction of the carrier to noise (C/N) ratio of the received signal. Furthermore, snow and ice accumulation on the antenna reflector and feed horns may seriously degrade its pointing, gain, and cross polarization characteristics. Major adverse effects of troposphere on the satellite radio links are summarized below.

- *Rain Attenuation*

  For the satellite radio links employed for broadcasting, the rain attenuation exceeded for 1% of time in the worst month is usually of the greatest concern. The method for relating the worst month time percentage to annual time percentage for rain attenuation is treated in the ITU-R recommendations P.581 and P.618.
- *Atmosphere Water Vapor Attenuation*

  For most commonly used frequency bands, this kind of attenuation is not significant except within some specific bands among which 22-GHz band allocated to the broadcasting satellite service can be mentioned as an example. Figure 3.3 of Chap. 3 illustrates variation of water vapor attenuation in terms of frequency based on the ITU-R, P.676. There is a peak at 22 GHz which its magnitude may reach to 0.2 dB/km.
- *Attenuation Due to Atmosphere Gases*

  Radiowave loss caused by tropospheric gaseous absorption is not critical issue at UHF and SHF bands (300 MHz$\sim$30 GHz). As it is shown in Fig. 3.3 of Chap. 3 (extracted from ITU-R, P.618), this kind of attenuation has fluctuating nature at heights around 15$\sim$20 km above the Earth surface exceeding 10 dB/km at 60 GHz.
- *Atmosphere Precipitation and Clouds*

  Attenuation due to clouds is not serious for frequencies in the UHF and SHF (300 MHz$\sim$30 GHz) bands, while it should be taken into account for EHF band. Fog and cloud attenuation may be estimated based on its liquid water content using the procedure specified in the recommendation ITU-R, P.840.

Table 4.5 Estimated ionosphere effects on satellite radiowaves (Ref.: ITU-R, P.618-9)

Effect	Frequency dependence	0.5 GHz	1 GHz	3 GHz	10 GHz
Faraday rotation	$1/f^2$	1.2 rotation	108°	12°	1.1°
Propagation delay (μs)	$1/f^2$	1	0.25	0.028	0.0025
Refraction	$1/f^2$	$< 2.4'$	$< 0.6'$	$< 4.2''$	$< 0.36''$
Variation in direction of arrival (r.m.s)	$1/f^2$	$48''$	$12''$	$1.32''$	$0.12''$
Absorption (auroral and/or polar cap), dB	$\approx 1/f^2$	0.2	0.05	6×10^{-3}	5×10^{-4}
Absorption (midlatitude), dB	$1/f^2$	< 0.04	< 0.01	0.001	$< 1 \times 10^{-4}$
Dispersion, ps/Hz	$1/f^3$	0.0032	0.0004	1.5×10^{-5}	4×10^{-7}
Scintillation peak-to-peak, dB			> 20	≈ 10	≈ 4

- *Diurnal and Seasonal Variation of Fading*

  Signal fading depends on the time of day in the satellite broadcasting service. Faraday data obtained in various regions of the world reveals a common tendency for larger fades to occur in the afternoon and early evening hours, especially in thunderstorms conditions. Low-level fadings are more evenly distributed both seasonally and diurnally.
- *Scintillation Fading*

  Small-scale irregularities in the tropospheric refractive index called scintillation can include rapid fluctuation in signal amplitude. Scintillation adverse effects on the system performance for frequencies less than 10 GHz and radio path elevation angles above 100 are limited.
- *Depolarization Effect*

  Hydrometeors such as rain drops and ice crystals cause considerable depolarization of signals in the satellite broadcasting service at frequencies above 2 GHz.

4.6.4.3 Ionospheric Effects

Ionospheric adverse effects on the satellite radio links at ku and ka or higher-frequency bands are negligible. These effects on lower SHF band (3–10 GHz) are limited while for frequency bands less than 3 GHz, such as satellite L band and UHF band, should be taken into account for satellite radio link design of either GEO, MEO, or LEO satellites. Maximum estimated values of ionosphere effects are summarized in the Table 4.5 based on the data extracted from the ITU-R recommendation P.531 observing the following points:

- Radio link elevation angle of about 30°.
- Total elevation content (TEC) of 10E(18) electron/m^2.

- Ionospheric adverse effects on the propagation of radio links above 10 GHz are negligible.
- Scintillation values observed near the geomagnetic equator during the local early nighttime at equinox and high sunspot number.

These figures are based on measurement results presented by ITU-R, and as indicated, attenuation due to the local environments are typically in the range of 10–40 dB.

4.6.4.4 Effects of Local Environment

Different environmental and atmospheric conditions exist at the ground terminals which must be accounted for the analysis of how to receive them. In addition to the outside phenomena, basically because of the establishment of TVRO terminals at specific conditions and also their multiplicity, the specific prevailing conditions must be taken into account comprising the following cases:

- The radiowaves propagation phenomena in the building entries and surrounding areas
- Losses caused by the radiowaves encountering urban obstacles such as trees, buildings, and special structures
- Losses caused by reflection or refraction of radiowaves related to the buildings and other structures

Additional attenuation caused by local environments on the satellite radiowaves entering inside buildings typically are shown in Figs. 4.6 and 4.7.

4.7 Local Broadcasting in VHF/UHF Bands

4.7.1 Introduction

The frequency bands of the local broadcasting systems include terrestrial systems in VHF and UHF bands. Noting Table 4.2 and Fig. 4.4, the required spectrum consists of three sub-bands within VHF and two sub-bands within lower UHF (up to 1 GHz). The mentioned sub-bands comprise a specific categorization based on the international/regional agreements.

The member countries are obligated to follow these agreements for optimum usage of frequency spectrum and prevent any harmful radio interferences. These categorizations are adjusted according to different types of services and capacities as required which are accounted for by the planners for frequency management and associated channels assignments.

Radio section of ITU for the first time in 1951 prepared recommendation ITU-R, P.370 regarding waves propagation of terrestrial networks for broadcasting in

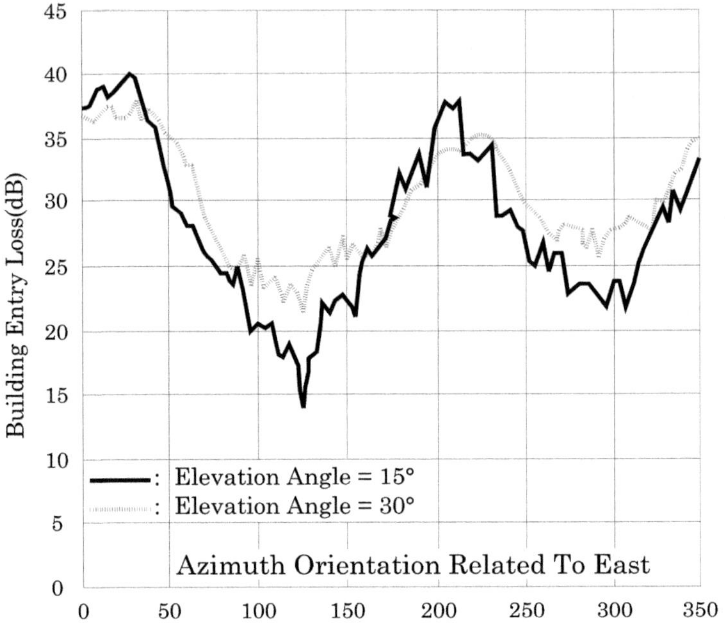

Fig. 4.6 Building entry loss in center of corridor (5.1 GHz) (Ref.: ITU-R, Rec., P.679-3)

Fig. 4.7 Inside the building loss in center of corridor (5.1 GHz) (Ref.: ITU-R, Rec., P.679-3)

VHF/UHF bands. This recommendation along with its revised and updated copies constitutes the basis of network planning for broadcasting services within the band.

Noting the development of digital systems and research and experiments conducted so far, the ITU has presented recommendation No. ITU-R, P.1546 as a replacement of previous recommendations. The latest revision of this recommendation nowadays constitutes the basis of network planning for broadcasting. Given the great importance of this recommendation, the next part is devoted to present a brief explanation, regarding P.1546.

4.7.2 Waves Propagation of Public Broadcasting in VHF/UHF

4.7.2.1 Propagation Medium

For radiowaves propagation in public broadcasting, the transmitter station normally has a fixed position, whereas the receiver stations can be located in one of the following positions:

- Fixed with antenna installed outdoors (open areas)
- Fixed with antenna installed indoors (surrounded areas)
- Mobile either handheld or moving vehicle

In general, propagation medium for the mobile radio communications consists of tough and sophisticated conditions. Specially, it includes the adverse effects of radiowave propagation in the VHF/UHF band. The radiowaves in the public broadcasting within the pertinent band as shown in the Fig. 4.8, generally, are received through multipaths which also include direct LOS and/or NLOS reception using various phenomena such as diffraction, reflection from numerous surfaces, and scattering of the radiowaves caused by troposphere propagation.

One of the prevailing issues of the radiowaves propagation in the broadcasting services is the mobility of the receivers especially audio equipments situated in the moving vehicles where the related Doppler effect and associated frequency have specific constraint on the speed of data transfer and signal quality.

The mentioned limiting factors and other complicated problems associated with the overall planning and establishment of a public broadcasting network increase the initial investment. They depend greatly on the affecting factors surrounding the radiowaves propagation en route from transmitter to mobile receivers, situated indoors and equipment installed outdoors. The broadcasting network, in addition to being undermined by general phenomena, is also influenced by trees, plants, and multipaths. These effects are due to reflection, scattering, and diffraction caused by obstacles and radio shades with their values being a function of the following items as well:

- Radio channel frequency
- The height of the receiver antenna from the ground
- Environmental conditions of the receiver location

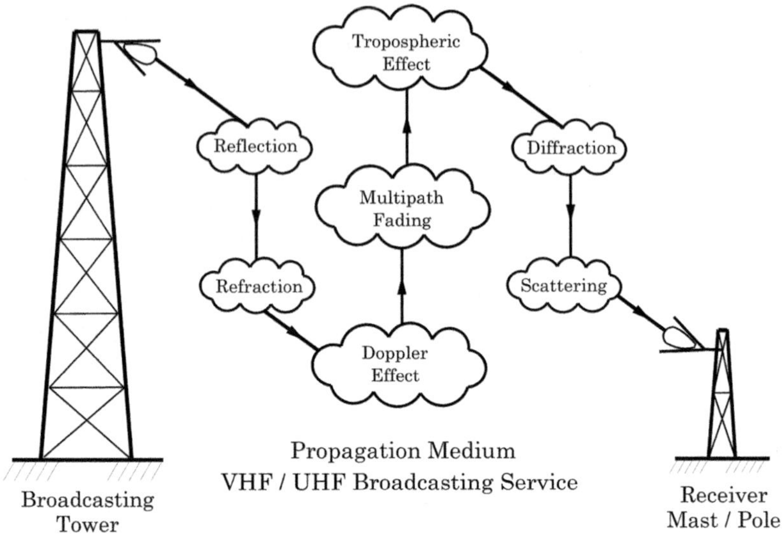

Fig. 4.8 Propagation media in VHF/UHF broadcasting service

4.7.3 Radiowaves Propagation Phenomena

The general concept of terrestrial radiowaves propagation in VHF/UHF bands is applied to the broadcasting service in the troposphere, and consequently, the overall principles and related phenomena governing the following radio links are valid for propagation of broadcasting links as well:

- Mobile radiowaves propagation
- LOS radiowaves propagation
- Propagation in surrounded areas
- Radio noises

4.7.3.1 Radio Path Attenuation

The path attenuation of the radiowaves in open area can be stated using the Feriss equation as

$$\mathrm{FSL} = 32.4 + 20\log f + 20\log d \tag{4.3}$$

In the above equation, the attenuation is in terms of decibel, frequency f in terms of megahertz, and the distance d in terms of kilometers. The actual path attenuation for terrestrial broadcasting radiowaves in VHF/UHF frequency bands which most of cases involve urban areas and also suburbs is much greater than the open area attenuation. Noting the similarities of the medium for radiowave propagation of the broadcasting with the mobile radiowaves, as a result, the distance coefficient is estimated to be higher than 20 dB in the above formula.

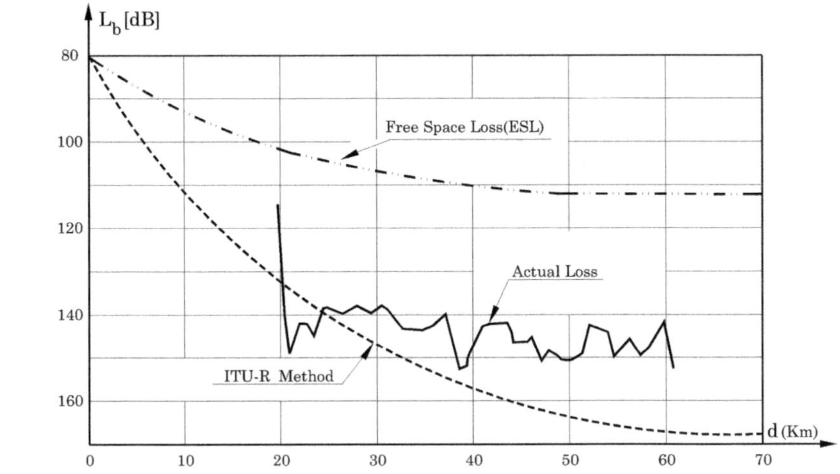

Fig. 4.9 Radiowaves path loss (ITU-R method)

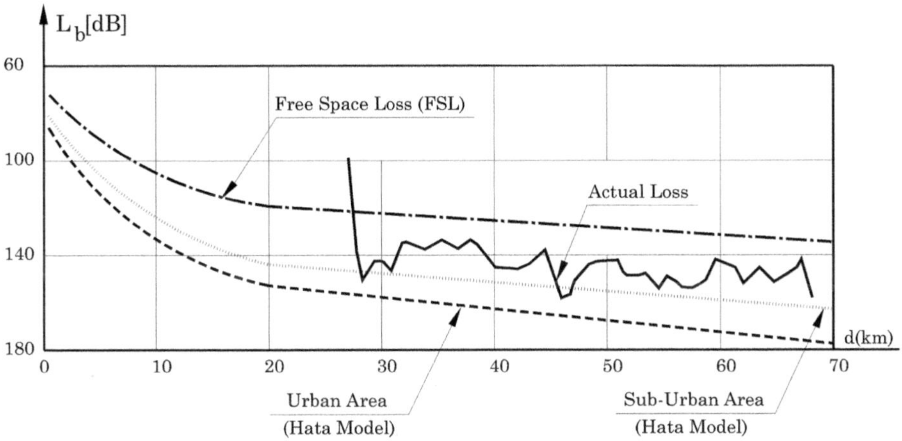

Fig. 4.10 Radiowaves path loss (Hata method)

For calculation of path attenuation, there are some alternative methods used in the mobile radio communications (Chap. 6 of this book "Propagation Engineering in Wireless Communications"). One of the popular methods is according to ITU-R, P.1546 recommendation which is discussed in the next section. In Fig. 4.9, the value of path attenuation is indicated for radio digital waves in frequency band III and its relative comparison with the measured values for open area attenuation.

Another applied method to calculate the path attenuation in the VHF/UHF bands is based on the Hata equations used for mobile radio link calculations. In Fig. 4.10, the value of path attenuation for radio digital waves in L band is based on the Hata method and provides the corresponding measured values.

To extend the application range of the (4.3), it can be written in the following form:

$$A = 32.4 + 20 \log f + n \log d \tag{4.4}$$

In the above relation, the value of n is known as the distance coefficient and medium specific which can be in the range of 20–45.

Example 4.2. 1. Calculate the free space loss for broadcasting radiowaves in L band with a path length equal to 50 km.
2. Calculate the above value using the diagram (4.10) and also the route attenuation using the Hata method in urban environments.

Solution. 1.

$$f = 1{,}500\,\text{MHz}, \quad \text{FSL} = 32.4 + 20 \log(1{,}500) + 20 \log(50)$$
$$\Longrightarrow \text{FSL} = 129.9 \text{ dB}$$

2. The following values are obtained using the related diagrams shown in Fig. 4.10:

$$d = 50\,\text{km} \overset{\text{Fig. 4.10}}{\Longrightarrow} \text{FSL} \approx 130 \text{ dB}$$

$$d = 50\,\text{km} \overset{\text{Fig. 4.10}}{\Longrightarrow} A(\text{Hata}) \approx 168 \text{ dB}$$

■

4.7.3.2 Reflection of Waves and Interference Pattern

Radiowaves encountering flat and smooth surfaces or with small roughness will reflect, and the receiver antenna collects the vector sum of the main and reflected waves. In an ideal situation, if the reflection coefficient of flat ground between transmitter and receiver is assumed to be 1, then the field intensity or the equivalent received power at any point will be a function of transmitter power, the effective gains of transmitter and receiver antennas, the path length and also the frequency of radio channel.

In Fig. 4.11, variations of received signal level in terms of the distance between the transmitter and the receiver have been plotted, assuming the rest of parameters are constant. As it can be noticed, the received field intensity at some points is a maximum of 6 dB in excess of the field intensity received from direct signal. Conversely, between any two maximum values, there is a minimum value resulting in a fluctuating pattern around the average received signal level.

Figure 4.12 illustrates the RX received signal level according to the height of antenna when the rest of parameters are constant (such as the distance between TX and RX). This case also indicates that the received field intensity (or power) relative to the RX antenna height has fluctuating variations.

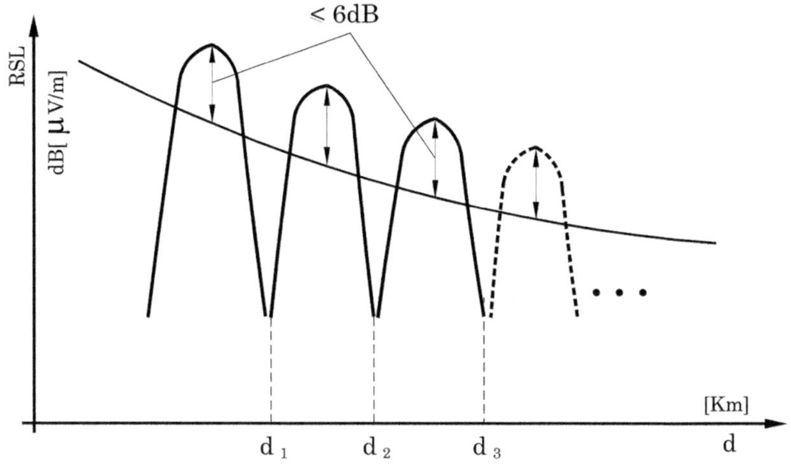

Fig. 4.11 Variations of signal level versus path length

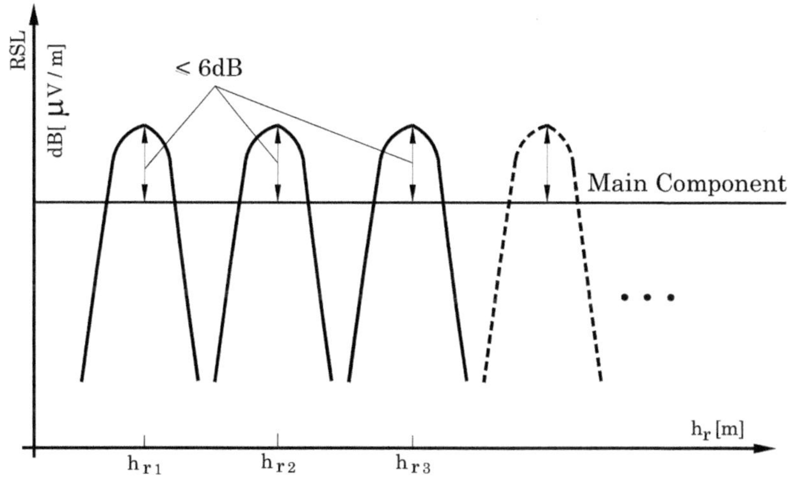

Fig. 4.12 Variations of received signal level versus RX antenna height

The main condition to meet the latter case is low roughness of the terrain between TX and RX compared with the wavelength of the radio channel. In practical terms, due to roughness of the ground, by colliding of the main wave with its surface, in addition to the formation of Snell's reflected wave, it is scattered on different directions which are also received by the RX (proportionate with the radiation pattern of the RX antenna). Reception of contrasting waves as mentioned above causes amplitude fluctuation of the received signal level. An acceptable criterion to disregard the ground roughness is to satisfy the following condition:

$$d/\lambda \leq 0.1 \sim 0.2 \tag{4.5}$$

Considering the wavelength of the VHF radiowaves (compared to those at UHF band), it can be concluded that the ground roughness for VHF band is relatively lower and the ground surface seems smoother. In other words for ground which is rough for UHF waves, there is a possibility of being regarded as smooth for VHF radiowaves.

4.7.3.3 Multipath Interference of Radiowaves

Interference of direct waves with the reflected waves was elaborated in the previous part. In an actual case, the RX unit usually receives many radiowaves including the following components:

- Main radiowave
- Reflected radiowaves
- Diffracted radiowaves
- Refracted radiowaves
- Scattered radiowaves

These waves, which are received simultaneously through different routes with unequal time delays relative to the main signal, generally cause attenuation of the main wave and can have a selective fading effect on the broadcasting wide band signals. Figure 4.13 illustrates a sample of field intensity amplitude fluctuation received by the RX unit and as a result of the effects of multipath waves.

4.7.3.4 Doppler Effect

This phenomenon is a feature of mobile radio communications caused by the relative motion of the TX and RX units. Considering a fixed broadcasting station (which is normal practice in most situations) for a fixed RX unit (not moving), the Doppler effect will be exclusively limited to the displacements within physical channel, which occurs randomly, and the amplitude of their effects is low.

The time variations related to this phenomenon depend on the velocity of the RX unit designated as V, radio frequency designated as f_0 (with wavelength of λ), and the angle between the direction of motion and the direction of radiowave propagation. Assuming identical directions for mobile displacement and radiowaves propagation, the maximum Doppler frequency can be calculated from the following relation:

$$f_{\max} = \frac{V}{C} \cdot f_0 \tag{4.6}$$

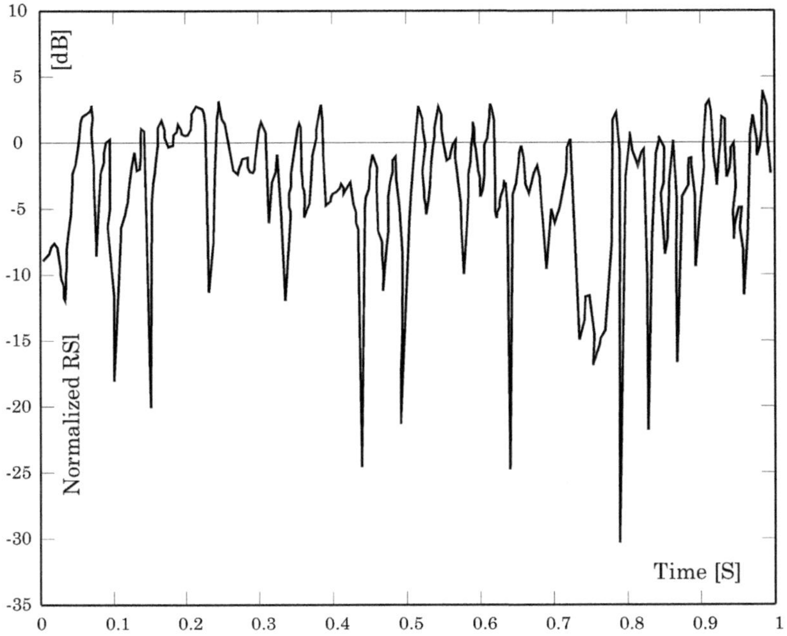

Fig. 4.13 Sample of normalized received field strength level

Table 4.6 Maximum Doppler frequency versus vehicle speed

RF channel frequency (MHz)	Doppler frequency in Hz		
	54 km/h	108 km/h	162 km/h
200	10	20	30
800	40	80	120
1,500	75	150	225

If V is in km/h and f_0 is in MHz, then the latter relation can be converted into

$$f_{\text{max}} = \frac{1}{1080} \cdot V \cdot f_0 \qquad (4.7)$$

The Doppler frequency can also be calculated from the following general equation. Where the mentioned directions are not identical:

$$f_{\text{d}} = f_{\text{max}} \cdot \cos \alpha = \frac{V}{C} \cdot f_0 \cdot \cos \alpha \qquad (4.8)$$

The sample values of the Doppler frequency for the bands III and V and L of the broadcasting service are presented in the Table 4.6. In general, the incoming frequency to the RX consists of a cluster of LOS, reflected, diffracted, and scattered waves which are received by the RX antenna on different directions. Therefore, the Doppler effect is regarded as a spectrum in place of just one Doppler frequency.

The cluster of the above-mentioned waves cause a fluctuation in the amplitude and the phase of incoming signal, and this means that it was modulated in the amplitude and phase by the physical channel. In transfer of signals with digital phase modulation, if the carrier phase changes abruptly over a period of time, T_s, then this speedy phase fluctuation can cause serious problems. The effect of Doppler phenomenon is the random oscillation of amplitude and phase having a frequency limited to f_{max}, and as a result for proper signal transmission in digital systems constituting a symbol rate time of equivalent to T_s, it is crucial to meet the following condition:

$$f_{max} \cdot T_s \ll 1 \tag{4.9}$$

4.7.3.5 Indoor Areas

Since RX units in the broadcasting networks are employed normally in indoor areas and especially the antennas of analog and digital audio systems are installed in such areas, it is essential to consider the conditions of the radiowaves penetrating into the buildings and structures. This topic will be discussed in other chapter. For radio digital systems in the band III, the mean value of incoming radiowave losses on the ground floor approximately is 8–9 dB with a deviation of 3.5 dB. For upper floors, this value decreases relatively. The loss of incoming radiowaves into the buildings in L band with frequency of 1.5 GHz is about 6–9 dB with an average value of 7.5 dB, and this value for reinforced concrete buildings is about 18–19 dB, and for metal/concrete buildings equipped with metal windows, it is equivalent to 20–30 dB.

4.7.3.6 Effect of Channel Bandwidth

Increasing bandwidth of radio and TV channels results in the increased possibility of incurring distortion in reception of signals caused by undesirable effects of multipath fading. This phenomenon incorporates a greater impact especially for signals with wider bandwidth.

Figure 4.14 illustrates the increase of fade margin to overcome the undesirable effects of multipath in radio digital broadcasting systems with channel bandwidth limited from 100 KHz to 5 MHz.

As it can be observed, the essential fade margin to eliminate adverse effects of multipath fading, in addition to channel bandwidth, will depend on the percentage probability of receiving appropriate signal. This diagram reveals that for lower bandwidth at each location coverage percentage, the variation rate of fade margin is high compared with greater bandwidths, say 1.5 MHz or higher.

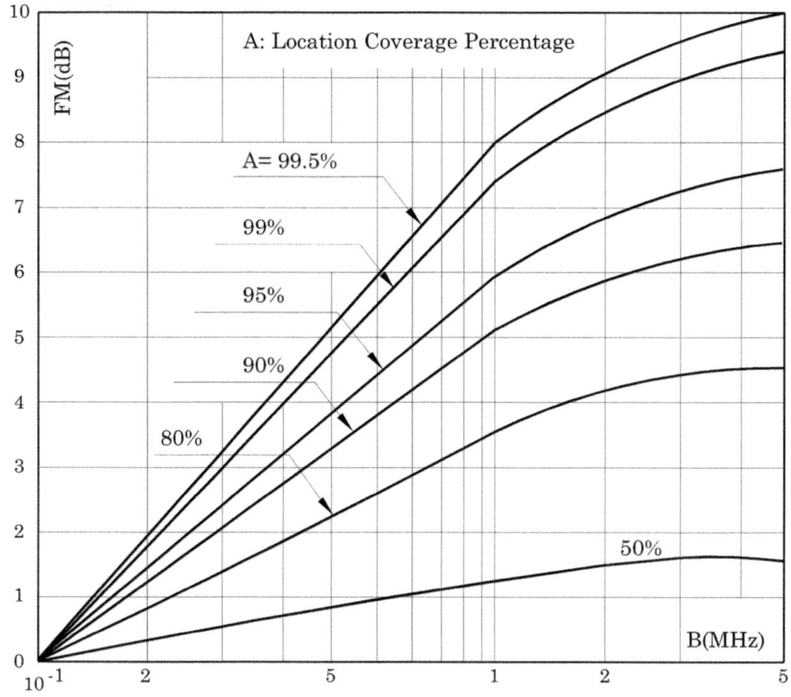

Fig. 4.14 Variations of fade margin versus bandwidth of RF channel

Example 4.3. Assume a DAB system in the L band and 2 MHz bandwidth:

1. Determine fade margin required for a desirable reception over 90% of locations and also for expanding coverage area to 99% of locations. How much should the fade margin be increased to satisfy the new conditions?
2. When a receiver is used in a vehicle moving with a velocity of 72–144 km/h, how much are the maximum and minimum Doppler frequencies?

Solution. 1. Using Fig. 4.14, it can be concluded that

$$\xrightarrow{90\%} \quad \text{FM} \approx 5.8 \text{ dB}$$

$$\xrightarrow{99\%} \quad (\text{FM})' \approx 8.4 \text{ dB}$$

$$\Delta(\text{FM}) = (\text{FM})' - \text{FM} = 2.6 \text{ dB}$$

Table 4.7 Standard minimum field strength for broadcasting services

Frequency band	II	III	IV	V
Type of service	FM radio	Digital audio DAB	Analog TV	Analog TV
RX antenna height (m)	10	1.5	10	10
Field strength edge of coverage min $dB_{\mu}V/m$	54	35	64	70

2. The minimum Doppler frequency equals to zero which is obtained for $\alpha = 90°$. In order to determine the maximum Doppler frequency given that $f_0 = 1{,}500\,\text{MHz}$ it yields

$$f_1 = \frac{1}{1{,}080} \times 72 \times 1{,}500 = 100\,\text{Hz}$$

$$f_2 = \frac{1}{1{,}080} \times 144 \times 1{,}500 = 200\,\text{Hz}$$

■

4.7.4 Coverage Prediction

4.7.4.1 Reception Criterion

In public broadcasting networks, the selection of the antenna type and height and also the receiver specification are beyond of the radio engineer's control; thus, for coverage prediction, the radio signal level in terms of electric field strength is utilized for this purpose. For this purpose, the boundary of coverage area is defined as the minimum desirable signal level which is specified by the concerning international authorities. The base values depend on the frequency band, type of service, reception quality, antenna height of the receiver, and location percentage to be covered. In Table 4.7, a sample of minimum values is provided for some types of the services for 50% location coverage.

For time and location percentages higher than median values, required field strength should be higher than the minimum value. The relation between the required and minimum values of field strength can be expressed by the following equation:

$$E_{\text{req}} = E_{\text{m}} - Q_{\text{i}}(q/100) \times \sigma \tag{4.10}$$

where:

E_{req}: Required field strength with location percentage of q in $dB_{\mu}V/m$

E_{m} : Required field strength with 50% location percentage in $dB_{\mu V/m}$

σ: Standard deviation of the Gaussian distribution of the local means in terms of dB (normally 5.5 dB)

Table 4.8 Inverse complementary cumulative normal distribution values (Ref.: ITU-R, P.1546-3)

$q\%$	$Q_i(q/100)$	$q\%$	$Q_i(q/100)$	$q\%$	$Q_i(q/100)$	$q\%$	$Q_i(q/100)$
1	2.327	26	0.643	51	−0.025	76	−0.706
2	2.054	27	0.612	52	−0.050	77	−0.739
3	1.881	28	0.582	53	−0.075	78	−0.772
4	1.751	29	0.553	54	−0.100	79	−0.806
5	1.654	30	0.524	55	−0.125	80	−0.841
6	1.555	31	0.495	56	−0.151	81	−0.878
7	1.476	32	0.467	57	−0.176	82	−0.915
8	1.405	33	0.439	58	−0.202	83	−0.954
9	1.341	34	0.412	59	−0.227	84	−0.994
10	1.282	35	0.385	60	−0.253	85	−1.026
11	1.227	36	0.358	61	−0.279	86	−1.080
12	1.175	37	0.331	62	−0.305	87	−1.126
13	1.126	38	0.305	63	−0.331	88	−1.175
14	1.080	39	0.279	64	−0.358	89	−1.227
15	1.036	40	0.253	65	−0.385	90	−1.282
16	0.994	41	0.227	66	−0.412	91	−1.341
17	0.954	42	0.202	67	−0.439	92	−1.405
18	0.915	43	0.176	68	−0.467	93	−1.476
19	0.878	44	0.151	69	−0.495	94	−1.555
20	0.841	45	0.125	70	−0.524	95	−1.645
21	0.806	46	0.100	71	−0.553	96	−1.751
22	0.772	47	0.075	72	−0.582	97	−1.881
23	0.739	48	0.050	73	−0.612	98	−2.054
24	0.706	49	0.025	74	−0.643	99	−2.327
25	0.674	50	0.00	75	−0.674		

$Q_i(q/100)$: Inverse complementary cumulative normal distribution per different values of $q\%$ are given in Table 4.8. Other percentages can be found using interpolation between subsequent lower and upper limits

Example 4.4. Table 4.9 indicates the required field strength for proper reception of radio digital radiowaves in the L band with 99% location coverage by the vehicle antenna at an altitude of AGL $= 1.5$ m, find:

1. The minimum required field strength and in case of decreasing the incoming field intensity to $62\,\text{dB}_{\mu\text{V/m}}$. Then determine location coverage percentage. Assume the probability distribution function to be logarithmic normal distribution with a standard deviation of $\sigma = 5.5$ dB.
2. The increase of the RX antenna gain for using at a distance of 70 km within a medium of $n = 30$. Assume the TX radio coverage is designed for a coverage radius of 50 km.

Table 4.9 Standard minimum field strength for DAB service

Reception condition	Field strength band III	Field strength L-band
Mobile RX, $h_r = 1.5$ m/AGL, $q = 99\%$	58 $dB_{\mu V/m}$	69 $dB_{\mu V/m}$
Mobile RX, indoor, $q = 95\%$	63 $dB_{\mu V/m}$	80 $dB_{\mu V/m}$
Fixed RX, $h_r = 10$ m/AGL, $q = 95\%$	37 $dB_{\mu V/m}$	57 $dB_{\mu V/m}$

Solution. 1. Table 4.9 indicates that for the given system the minimum required field strength is 69 $dB_{\mu V/m}$. Noting Table 4.8, the following result can be concluded:

$$q = 99\% \xrightarrow{\text{Table 4.8}} Q_i = -2.327$$

$$E_m = E(99\%) + Q_i \cdot \sigma \implies E_m = 56.2\ dB_{\mu V/m}$$

$$Q_i'(q'/100) = -\frac{E(q') - E_m}{\sigma} \implies Q_i'(q'/100) = -1.05$$

$$Q_i'(q'/100) = -1.05 \xrightarrow{\text{Table 4.8}} q' = 85.5\%$$

2. The additional loss due to the longer distance for $n = 30$ will result in:

$$L_e = 30\ \log(70/50) = 4.38\ \text{dB}$$

This additional loss should be compensated by increasing the receiving antenna gain. ■

According to the European Broadcasting Union (EBU), the minimum values for DAB under different RX conditions are provided in Table 4.9.

As an example in band III, the minimum reception required for a mobile RX unit with a 99% location coverage according to the mentioned table will be 58 $dB_{\mu V/m}$. However, it must be noted that if the RX unit is located at a height of more than 1.5 m above the ground level, then it will receive a field intensity greater than the above value.

The base values usually are provided for 50% locations. If higher location percentage (normal practice for most cases) is intended, assuming that its probability distribution function is log-normal, then by using relation (4.10) the value of field intensity required must be determined and to be used as the basis of the calculations.

4.7.4.2 Minimum Signal Level

Table 4.7 contains the base values for the received field strength in some of the broadcasting services. As an example for TV channels in band IV, the base value is equal to 64 $dB_{\mu V/m}$. This objective is fulfilled by using an ordinary roof mounted antenna at a height of AGL = 10 m to receive the desirable images. In order to determine the minimum received field strength, the SNR must be specified which its value on the output image of an ordinary TV set is approximately a minimum of 30–33 dB, and for vestigial side band (VSB) TV receivers, it is about 38–41 dB. The specified points may be expressed by

$$\mathrm{SNR} = \mathrm{S/N} = 20\ \log(V_\mathrm{r}/V_\mathrm{n}) \tag{4.11}$$

where V_r is the receiver input voltage level and V_n is the equivalent voltage level of noise temperature at the receiver input which has a relation in the logarithmic scale as stated below:

$$V_\mathrm{n}[\mathrm{dB}_{\mu V}] = 10\ \log[kT_0BR] + F + 120 \tag{4.12}$$

In this equation, parameters and associated units are:

K: Boltzmann constant equal to 1.38×10^{-23}
T_0: Reference temperature equal to 290°K
B: Bandwidth of the receiver in Hz
R: Input impedance of the receiver in ohms
F: Receiver noise figure in dB

The (4.12) can be rewritten in the following form:

$$V_\mathrm{r}[\mathrm{dB}_{\mu V}] = (\mathrm{SNR})_{\mathrm{min}} + V_\mathrm{n}[\mathrm{dB}_{\mu V}] \tag{4.13}$$

Also, the relation of electrical field strength with the input voltage of the receiver using the isotropic antenna is given by

$$E_\mathrm{r} = V_\mathrm{r} + L - G - 33.6 + 20\ \log\ f \tag{4.14}$$

The same formula relative to dipole antenna is converted to the following equation:

$$E_\mathrm{r} = V_\mathrm{r} + L - G_\mathrm{d} - 32 + 20\ \log\ f \tag{4.15}$$

In these relations, parameters and their pertinent units are:

E_r: Received electrical field strength in $\mathrm{dB}_{\mu V/m}$
V_r: Voltage at the input port of receiver in $\mathrm{dB}_{\mu V}$
L: Feeder loss in dB
G: Antenna gain of receiver relative to isotropic antenna in dB_i
G_d: Antenna gain of receiver relative to dipole antenna in dB_d
f: Radio channel frequency in MHz

Example 4.5. To receive VSB television signals, the minimum value of SNR = 41 dB is necessary. If the required bandwidth is 5 MHz, RF channel frequency is 500 MHz, the antenna input impedance is 75 ohms, and the receiver noise figure is 6 dB, calculate:

1. The input noise voltage
2. The minimum received voltage of the main signal and electrical field strength having 10 dB_d antenna gain along with 3 dB feeder loss

Solution. 1. Using the following values and relation (4.12), it yields

$$(k = 1.38 \times 10^{-23},\ T_0 = 290^\circ\text{K},\ B = 5 \times 10^6\text{Hz},\ R = 75\ \Omega,\ F = 6\ \text{dB})$$

$$\Longrightarrow\ V_\text{n} = 7.76\ \text{dB}_{\mu\text{V/m}}$$

2. To calculate E_r using the (4.15) yields

$$E_\text{r} = V_\text{r} + L + G - 32 + 20\ \log\ f$$

$$\Longrightarrow\ E_\text{r} = 63.74\ \text{dB}_{\mu\text{V/m}}$$

■

4.7.4.3 Coverage and Interference Regions

The coverage area of each broadcasting transmitter implies all points by which the signal level of radiowaves is equal or higher than the base value. In an ideal position as it is illustrated by Fig. 4.15, this level includes a circle with a radius of R_c. The covering radius is the maximum distance that is obtained according to the position, height, and power of the TX unit and also the conditions of the radiowave propagation. In practical terms, due to various reasons such as the composition of the ground, heterogeneous phenomena of the radiowave propagation in different directions and the radiation pattern of the TX antenna result in a covered area in the form of a closed contour, normally not a circle.

In terrestrial broadcasting networks, the planning requires utilization of numerous TX units in different positions to cover vast areas. Outside the covered area, the radiowaves are detected lower than the base value but with a poor quality. As indicated in Fig. 4.15, the interference region implies the points in which the level of radiowaves is less than the base value, but it is higher than a predetermined value denoted by E_p. Noting the above-mentioned diagram and assuming that the received signal is equivalent to E_r, the three distinct regions can be distinguished as:

$$E_\text{r} \geq E_\text{b} \qquad \text{Coverage region} \tag{4.16}$$

$$E_\text{b} > E_\text{r} \geq E_\text{p} \qquad \text{Interference region} \tag{4.17}$$

$$E_\text{r} < E_\text{p} \qquad \text{Uncovered or free region} \tag{4.18}$$

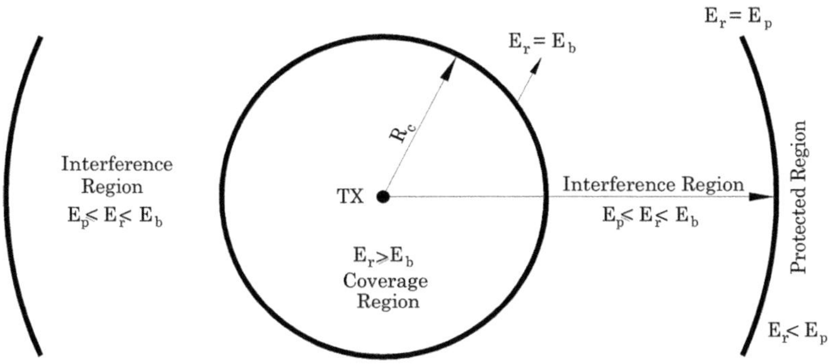

Fig. 4.15 Coverage, interference, and protected regions of a broadcasting station

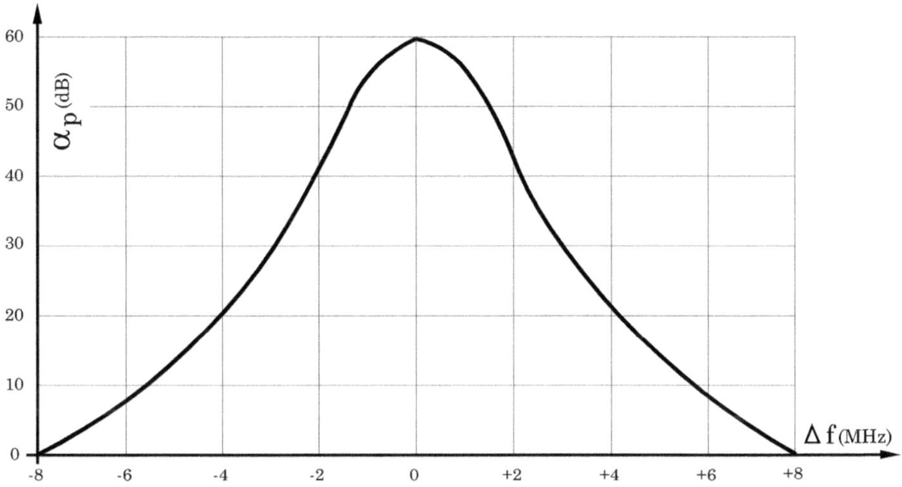

Fig. 4.16 Variations of protection ratio (p)

The interference region is a function of the following factors:

- Effective transmitting power of the broadcasting station (EIRP)
- Receiver and its antenna specifications
- Position and the specifications of the TX and RX units
- Ground composition and the conditions of radiowave propagation
- Radio channel frequency and the type of service
- Protection ratio, α_p

The protection ratio (α_p) according to its definition in the logarithmic scale and given in the Fig. 4.16 is equal to

$$\alpha_p = E_b - E_p \tag{4.19}$$

In the above relation, E_b and E_p are the desirable signal level and interfering signal level, respectively, in terms of $\mathrm{dB}_{\mu V/m}$. The value of this ratio depends on the level of impairment and is specified by the concerning authorities. The impairment levels are defined by the ITU-R in the following grades:

- Grade 5: Imperceptible
- Grade 4: Perceptible but not annoying
- Grade 3: Slightly annoying
- Grade 2: Annoying
- Grade 1: Very annoying

Obviously in any service, the value of α_p with grade 5 of impairment level has the greatest value. The significant point is the amount of α_p which is a function of frequency separation of two radio channels denoted by Δf. As shown in Fig. 4.16, when $\Delta f = 0$, that is, similar channels are used, the interference is of co-channel type, and the amount of α_p is maximum and equals to 60 dB, while for $(\Delta f) >$ 8 MHz, this ratio reaches the least value.

In practice, due to terrain topography and man-made structures and buildings, the limit of coverage area is not exactly a static circle, but it is a dynamic closed curve as given in Fig. 4.17 for predetermined conditions including some contours each one dedicated to a specific value of E_r.

Example 4.6. Assuming that for the analog TV service in the band IV, the minimum value of protection ratio is 27 dB:

1. Specify the minimum signal level within coverage area.
2. If the maximum covering distance is 30 km and the distance coefficient related to the path loss is 35, calculate the minimum distance of interference.

Solution. 1. Noting Table (4.7), it can be concluded that

$$E_{\min} = 64\ \mathrm{dB}_{\mu V/m}$$

2.

$$\alpha_\mathrm{p} = E_\mathrm{b} - E_\mathrm{p} = 27\ \mathrm{dB}$$

$$27 = 35\ \log(R_I/R_C) \implies R_I/R_C = 5.9$$

$$R_I = 5.9 \times 30 = 177\,\mathrm{km}$$

■

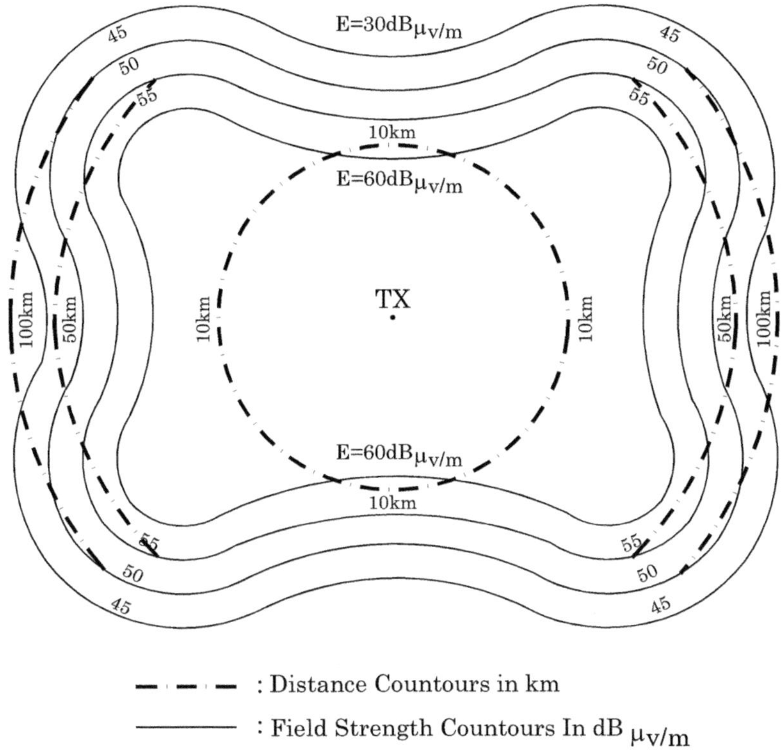

Fig. 4.17 Terrestrial broadcasting coverage contours

4.8 Path Loss of Radiowaves in VHF/UHF Bands

4.8.1 ITU-R Recommendations

To calculate the path loss of radiowaves broadcasting service, the ITU had contributed the principles, relations, and curves concerning mobile radio communications in 30–3,000-MHz-frequency band in the context of ITU-R, P.370 recommendation. Because of similarity in the nature of radiowaves propagation in broadcasting and mobile radio services, the above-mentioned recommendation was used for designing of radio broadcasting networks.

In recent years, this recommendation along with some amendments and modifications has been completed and presented as ITU-R, P.1546. The first revision was prepared in 2001, and the second revision was presented in 2005. Because of great importance of this recommendation and its application for calculation of the path loss of broadcasting radiowaves, it is briefly outlined here.

The recent recommendation and its curves are dedicated to the three following bands:

- VHF band in the frequency range of 30–300 MHz with reference frequency of 100 MHz covering the sub-bands of I, II, and III for public broadcasting services
- Lower UHF band in the frequency range of 300–1,000 MHz with reference frequency of 600 MHz which includes the sub-bands of IV and V for public broadcasting services
- Upper UHF band in the frequency range of 1,000–3,000 MHz with reference frequency of 2,000 MHz covering the L band for public broadcasting services

Due to the close similarities of relations and for the sake of brevity of this chapter, only a set of the propagation curves from 300- to 1,000-MHz-frequency range have been presented here, and therefore, the examples are generally provided for this particular band. For design of the broadcasting radio links, readers are addressed to the complete set of the curves and related software presented in the ITU-R recommendation P1546.

4.8.2 The Propagation Curves

The mentioned recommendation includes a set of curves similar to those given in the Figs. 4.18–4.25 indicating received field strength in terms of distance based on some fixed parameters. The curves are valid for typical terrain structure such as land and sea paths and the following ranges:

- Frequency range 300–1,000 MHz
- Height of TX antenna from 10 to 1,200 m
- Time percent of coverage 1%, 10%, and 50%
- Distance range 1–1,000 km

The figures include eight series of curves which are prepared to represent the plots of 300–1,000 MHz. For proper use of the curves, the following key points should be taken into account:

- ERP of the TX in terms of EIRP is assumed 1,000 w.
- Reference frequency is 600 MHz.
- Reference TX antenna height denoted by h_1 is given for 10, 20, 37.5, 75, 150, 300, 600, and 1,200 m.
- The first three plots relate to land (onshore) routes with a location probability reception of 50% and time probabilities of 50, 10, and 1% respectively.
- The following five plots relate to sea (offshore) routes with a location probability reception of 50% and different time probabilities.

In the proceeding parts, a brief discussion is presented based on the procedure set by ITU-R.

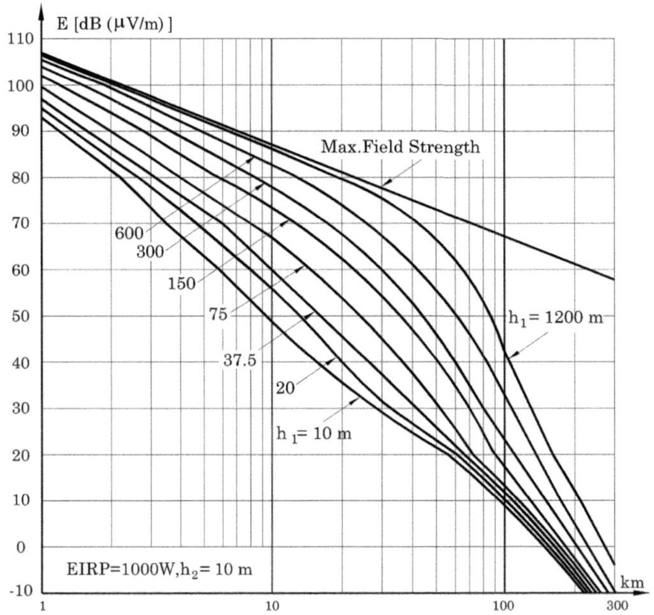

Fig. 4.18 Field strength versus distance for land path (for 600 MHz, 50% time, 50% locations) (Ref.: ITU-R, P.1546-3)

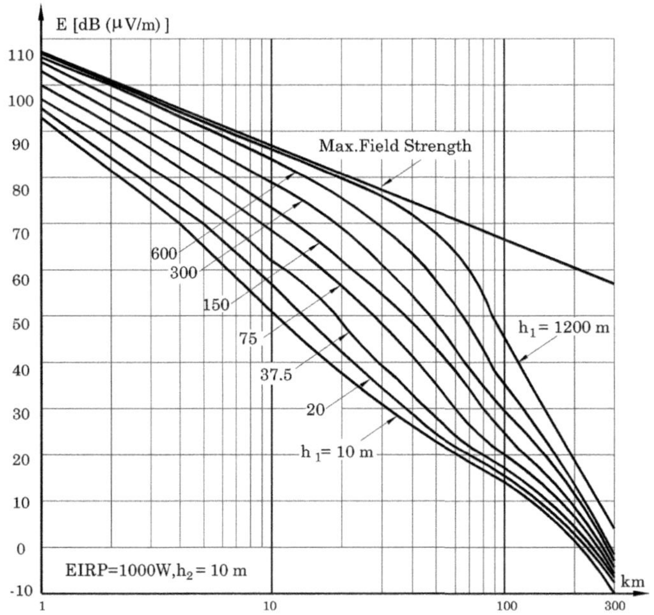

Fig. 4.19 Field strength versus distance for land path ($f = 600$ MHz, 10% time, 50% locations) (Ref.: ITU-R, P.1546-3)

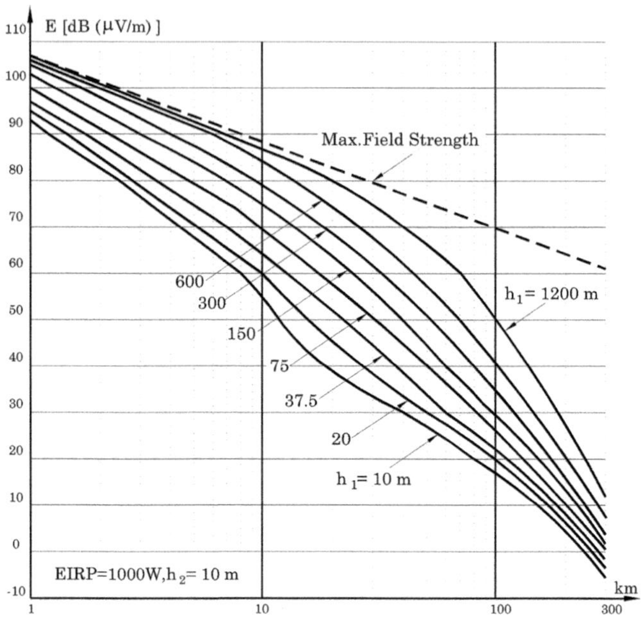

Fig. 4.20 Field strength versus distance for land path ($f = 600$ MHz, 1% time, 50% locations) (Ref.: ITU-R, P.1546-3)

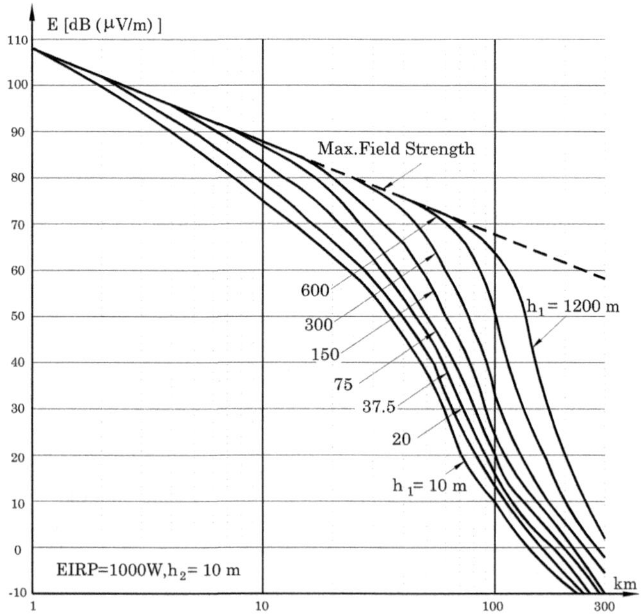

Fig. 4.21 Field strength versus distance for sea path ($f = 600$ MHz, 50% time, 50% locations) (Ref.: ITU-R, P.1546-3)

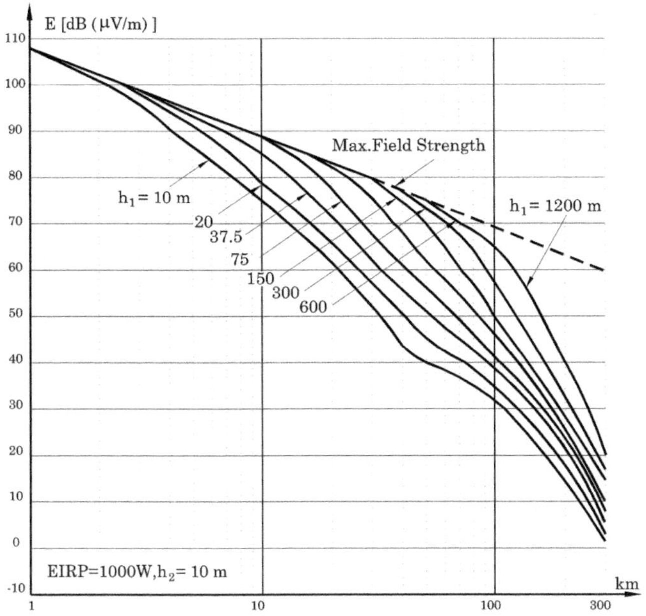

Fig. 4.22 Field strength versus distance for warm sea path ($f = 600\,\text{MHz}$, 10% time, 50% locations) (Ref.: ITU-R, P.1546-3)

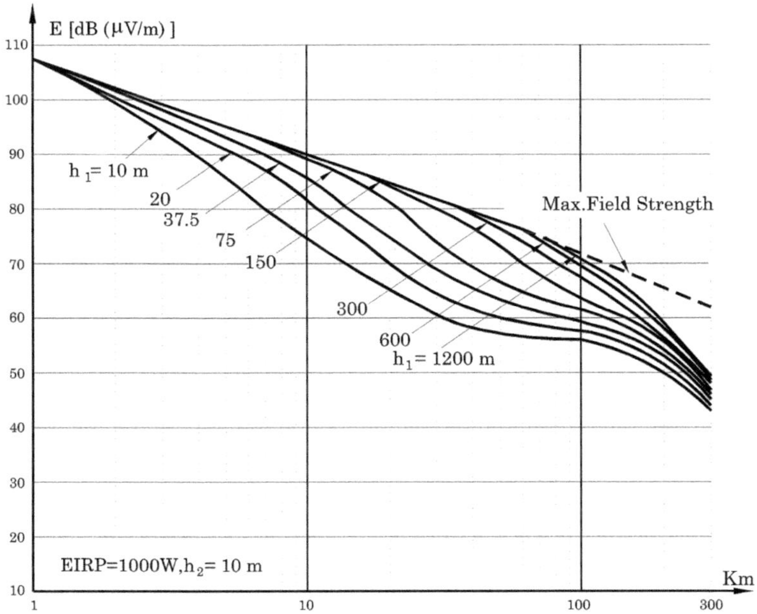

Fig. 4.23 Field strength versus distance for warm sea path ($f = 600\,\text{MHz}$, 1% time, 50% locations) (Ref.: ITU-R, P.1546-3)

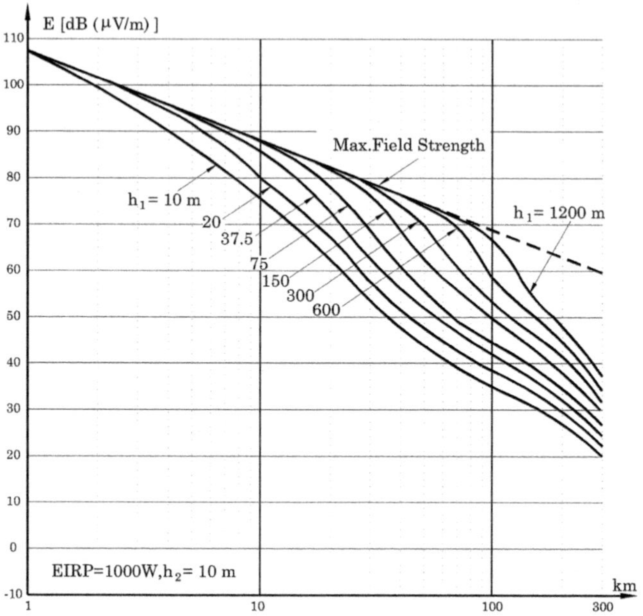

Fig. 4.24 Field strength versus distance for cold sea path ($f = 600$ MHz, 10% time, 50% locations) (Ref.: ITU-R, P.1546-3)

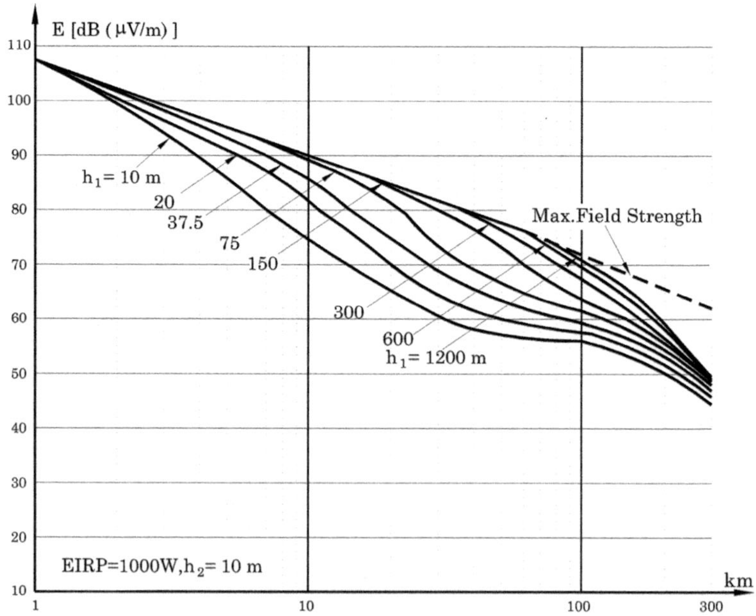

Fig. 4.25 Field strength versus distance for cold sea path ($f = 600$ MHz, 1% time, 50% locations) (Ref.: ITU-R, P.1546-3)

4.8.3 Basic Principles

4.8.3.1 Maximum Received Field

The maximum value of electrical field strength (intensity) at any point derived from calculations must not exceed the value of $E_{\max}$ conforming the following definition:

$$E_{\max}[\mathrm{dB}_{\mu\mathrm{V/m}}] = E_{\mathrm{fs}}, \quad \text{For land paths} \tag{4.20}$$

$$E_{\max}[\mathrm{dB}_{\mu\mathrm{V/m}}] = E_{\mathrm{fs}} + E_{\mathrm{se}}, \quad \text{For sea paths} \tag{4.21}$$

In the above relations, E_{fs} is free space field strength given by

$$E_{\mathrm{fs}}[\mathrm{dB}_{\mu\mathrm{V/m}}] = 106.9 - 20 \log d \tag{4.22}$$

And the value of E_{se} is the pertinent component for the reflection of the waves from the sea surface and calculated by the following equation:

$$E_{\mathrm{se}}[\mathrm{dB}] = 2.38[1 - \exp(-d/8.94)] \cdot \log(50/t) \tag{4.23}$$

where d is the distance in terms of kilometer and t is the time percentage. In principle, any correction factor discussed later must be limited in a manner that the value of field strength at a distance of d does not exceed the above values.

4.8.3.2 TX Antenna Height

Noting the Fig. 4.26 in the recommendation ITU-R, P.1546, three kinds of heights are defined by the ITU-R. Figure 4.26 has been prepared in order to realize these concepts.

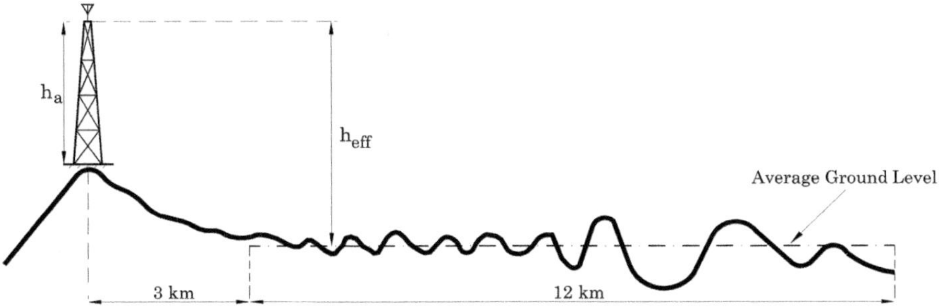

Fig. 4.26 Concept of effective antenna height

h_a: The actual antenna height from the station ground in meters which in fact is regarded as the height of antenna tower (if the antenna is installed on the top.)

h_{eff}: The effective antenna height in meters which is defined as the antenna height from average ground level. The average ground level is taken as the distance of 3–15 km from the TX station toward the RX station.

h_1: The transmitting antenna height in meters used in the curves and relations. The height is defined and calculated as follows:

$$h_1 = h_a, \quad d \leq 3\,\text{km} \tag{4.24}$$

$$h_1 = h_a + (h_{eff} - h_a)(d-3)/12, \ \ 3\,\text{km} < d \leq 15\,\text{km} \tag{4.25}$$

$$h_1 = h_{eff}, \quad 15\,\text{km} < d \tag{4.26}$$

In the above relations, h_1, h_a, and h_{eff} are in terms of meters, and d is in terms of kilometers. Noting the above definitions is intended for specifying the TX antenna height as a distinct criterion in practice.

Example 4.7. 1. For sea paths, prove that $h_1 = h_a$.
2. If the altitude of TX station is 1,250 m above mean sea level (AMSL), calculate the value of h_1 at a distance of 9 km. Assume the height of antenna tower as 30 m.

Solution. 1. For sea path conditions, referring to the Fig. 4.26, $h_a = h_{eff}$ and under such circumstances all kinds of TX antenna height defined by (4.24)–(4.26) can be assumed $h_1 = h_a$.
2. Applying the assumptions given in this example, it yields

$$h_a = 30\,\text{m}$$

$$h_{eff} = h_d + (1{,}250 - 800) = 480\,\text{m}$$

Since the distance of RX station from TX station is within 3–15 km, thus using the relation (4.25), we have

$$h_1 = 30 + (480-30)(9-3)/12 = 255\,\text{m}$$ ■

In general, the height h_1 is equivalent to the antenna height from the average level of the radiowave path at a distance of 20–100% of the path, (i.e., 0.2d up to 1.0d). For instance, for a 10-km path length, the average ground elevation is determined by considering the average value of elevations from 2 to 10 km while h_1 is the actual antenna height from this average level. The height h_1 will be equal to the actual antenna height from this average level.

4.8.3.3 Equivalent Basic Transmission Loss

In order to calculate the equivalent basic transmission loss denoted as L_b, the following equation can be used:

$$L_\text{b}[\text{dB}] = 139 - E - 20 \log f \tag{4.27}$$

In the above equation, each of the parameters and their units are:

E: Field strength in $\text{dB}_{\mu\text{V/m}}$, produced by the TX unit with 1 kW ERP
f: RF channel frequency in MHz

Example 4.8. Antenna height at TX station is $h_1 = 37.5$ m, and the radiowave path at land is intended for time percentages of 10–50%. Calculate the equivalent basic transmission loss at a distance of 20 km for RF channel frequency of 600 MHz.

Solution.

Figure (4.18) ⇒ for (600 MHz frequency) + (50% time) + (land path)
Figure (4.19) ⇒ for (600 MHz frequency) + (10% time) + (land path)

Among the different curves of the above-mentioned figures, the curve related to $h_1 = 37.5$ is selected, and the value of E is read at a distance of 20 km:

$$E(10\%) \approx 48\ \text{dB}_{\mu\text{V/m}}$$

$$E(50\%) \approx 46\ \text{dB}_{\mu\text{V/m}}$$

Now using the (4.27), the following results are obtained:

$$L_\text{b}(10\%) = 139 - 48 + 20 \log 600 \implies L_\text{b} = 146.4\ \text{dB}$$

$$L_\text{b}(50\%) = 139 - 46 + 20 \log 600 \implies L_\text{b} = 148.4\ \text{dB}$$

■

4.8.3.4 Radio Path Clearance

In radio communications, a route is regarded as cleared and unobstructed where all points (taking into account the atmospheric K-factor and also all natural Earth bulges) of the radiowaves path to be above all of the Earth bulges with a measure of 60–100% of first Fresnel radius. For point-to-area radio communications in the VHF and UHF bands with antennas having the heights of h_r and h_1 for a route without any high or low lands, ITU-R has set the following approximate equation to determine the path length which just achieves a clearance of 60% of the first Fresnel zone over a smooth curved Earth:

$$D_{0.6} = \frac{D_\text{f} \cdot D_\text{h}}{D_\text{f} + D_\text{h}}, \quad \text{km} \tag{4.28}$$

$$D_{\mathrm{f}} = 3.89 \times 10^{-5} \cdot f \cdot h_{\mathrm{t}} \cdot h_{\mathrm{r}} \tag{4.29}$$

$$D_{\mathrm{h}} = 4.1(\sqrt{h_{\mathrm{t}}} + \sqrt{h_{\mathrm{r}}}) \tag{4.30}$$

Each parameter in the above equation and its relevant unit is:

$D_{0.6}$: The maximum distance between TX and RX in kilometer (km) in which the minimum waves path is a measure of 0.6 of first Fresnel radius above the ground
D_{f} : Frequency dependent term in km
D_{h}: Asymptotic term defined by horizon distances in km
f: Radio frequency in MHz
h_1, h_{r}: TX and RX antenna heights above smooth Earth in m

Example 4.9. For TX and RX antennas with actual heights of $h_{\mathrm{t}} = 36\,\mathrm{m}$, $h_{\mathrm{r}} = 4\,\mathrm{m}$ and working at 450 MHz, find the following:

1. Value of $D_{0.6}$
2. Maximum first Fresnel radius for the radio path

Solution. 1.

$$f = 450\,\mathrm{MHz} \implies \lambda = 0.67\,\mathrm{m}$$

$$D_{\mathrm{f}} = 3.89 \times 10^{-5} \times 450 \times 36 \times 4 = 2.52\,\mathrm{km}$$

$$D_{\mathrm{h}} = 4.1(\sqrt{4} + \sqrt{36}) = 32.8\,\mathrm{km}$$

$$D_{0.6} = \frac{2.52 \times 32.8}{2.52 + 32.8} = 2.34\,\mathrm{km}$$

2. Fresnel radius is maximum at middle of the path:

$$r_1 = \sqrt{\frac{(D_{0.6})^2}{4\,D_{0.6}}} = 19.8\,\mathrm{m}$$

■

4.8.4 *Generalized Equations*

To determine the field strength when the values of frequency, TX antenna height, and time percents coincide with the figures indicated for the propagation curves, then the required field strength may be obtained directly from the plotted curves or the associated tabulations.

To generalize the equations to cover other values, the required field strength should be interpolated or extrapolated from field strengths obtained from two adjacent curves, (i.e., superior or inferior curves).

4.8.4.1 Transmitter Antenna Height

If the value of h_1 coincides with one of eight basic heights, namely, 10, 20, 37.5, 75, 150, 300, 600, or 1,200 m, the required field strength will be obtained directly from the given curves or the associated tabulations. Otherwise, the required field strength will be calculated by the following equation:

$$E[\mathrm{dB}_{\mu V/m}] = E_i + (E_s - E_i) \cdot \log(h_1/h_i)/\log(h_s/h_i) \tag{4.31}$$

where

$$\left| \begin{array}{ll} h_i = 600\,\mathrm{m} & \text{if } h_1 > 1{,}200\,\mathrm{m} \\ h_i = h_1 & \text{if } h_1 < 1{,}200\,\mathrm{m} \end{array} \right.$$

$$\left| \begin{array}{ll} h_s = 1200 & \text{if } h_1 > 1{,}200\,\mathrm{m} \\ h_s = h_1 & \text{if } h_1 < 1{,}200\,\mathrm{m} \end{array} \right.$$

h_i and h_s are the adjacent (the nearest) nominal effective inferior and superior heights:

E_i = Field strength value for h_i at required distance

E_s = Field strength value for h_s at required distance

The field strength resulting from extrapolation for $h_1 > 1,200$ m should be limited (if necessary) such that it does not exceed the maximum value set in the (4.20) or (4.21).

Example 4.10. For a radio broadcasting transmitter with an effective antenna height of 400 m operating at 600 MHz and the following assumptions: EIRP = 1,000, 50% locations and 50% time, $h_2 = 10$ m over land at a distance $d = 20$ km, find:

1. h_i and h_s
2. E_i and E_s
3. Field strength at receiving station

Solution. 1.

$$h_1 = 400\,\mathrm{m} \quad \Rightarrow \quad h_i = 300\,\mathrm{m} \quad \text{and} \quad h_s = 600\,\mathrm{m}$$

2. Based on the assumptions and using curves of the Fig. 4.18 the following values are concluded:

$$h_i = 300\,\mathrm{m} \ \Rightarrow \ E_i = 68\ \mathrm{dB}_{\mu V/m}$$

$$h_s = 600\,\mathrm{m} \ \Rightarrow \ E_s = 75\ \mathrm{dB}_{\mu V/m}$$

3. Using (4.31) for E, yields:

$$E = 68 + 7 \times \log(400/300) : \ \log(600/400)$$

$$E = 72.97\ \mathrm{dB}_{\mu V/m}$$

■

4.8.4.2 RF Channel Frequency

Field strength value for any frequency other than nominal frequencies, that is, 100, 600, and 2,000 MHz, should be obtained by interpolation of nominal values. In the case of frequencies below 100 MHz or above 2,000 MHz, the interpolation must be replaced by extrapolation from two nearest nominal frequencies. For most paths, the mentioned procedure can be used, but for some paths over sea, when the required frequency is less than 100 MHz, it is necessary to use an alternative method.

For land paths and sea paths where the RF channel frequency is greater than 100 MHz, the field strength should be calculated by the following equation:

$$E = E_{\mathrm{i}} + (E_{\mathrm{s}} - E_{\mathrm{i}}) \times \log(f/f_{\mathrm{i}}) : \log(f_{\mathrm{s}}/f_{\mathrm{i}}) \tag{4.32}$$

In (4.32), f_{i} and f_{s} are adjacent below and above nominal frequencies respectively. For cases which meet the following conditions:

- Sea path
- Frequency less than 100 MHz
- $d \le d_f$

The following equations may be used to estimate the field strength:

$$E[\mathrm{dB}_{\mu\mathrm{V/m}}] = E_{\max} \qquad d \le d_f \tag{4.33}$$

$$E[\mathrm{dB}_{\mu\mathrm{V/m}}] = E_{df} + (E_{d600} - E_{df}) \times \times \log(d/d_f) : \log(d_{600}/d_f) \quad d > d_f \tag{4.34}$$

In the above equations, each term and related units are:

f: The required frequency in MHz
f_{i}: Lower nominal (inferior) frequency in MHz
f_{s}: Higher nominal (superior) frequency in MHz
E_{i}: Field strength value for f_{i} in $\mathrm{dB}_{\mu\mathrm{V/m}}$
E_{s}: Field strength value for f_{s} in $\mathrm{dB}_{\mu\mathrm{V/m}}$
$E_{\max}$: Maximum field strength at the required distance in $\mathrm{dB}_{\mu\mathrm{V/m}}$
E_{df}: Maximum field strength at a distance equal to d_f in $\mathrm{dB}_{\mu\mathrm{V/m}}$
d_{600}: Distance for which the path has $0.6r_1$ clearance at the 600 MHz calculated as $D_{0.6}(600, h_1, 10)$ based on method explained in (4.8.3.4) in km
d_f: Distance for which the path has $0.6r_1$ clearance at the 600 MHz calculated as $D_{0.6}(f, h_1, 10)$ based on method explained in (4.8.3.4) in km
E_{d600}: Field strength at distance d_{600} and the required frequency calculated using (4.32) in $\mathrm{dB}_{\mu\mathrm{V/m}}$

Example 4.11. Repeat example (4.10) for $f = 420$ MHz assuming $E = 73.97\,\mathrm{dB}_{\mu\mathrm{V/m}}$

Solution.

$$f = 420\,\text{MHz} \quad \Rightarrow \quad 100\,\text{MHz} < f < 600\,\text{MHz}$$

As calculated in the example (4.10), the value of E_s is equal to $72.97\,\text{dB}_{\mu\text{V/m}}$, thus using (4.32) yields:

$$E = 73.97 - \log(420/100) : \ \log(600/420)$$

$$E = 69.95\ \text{dB}_{\mu\text{V/m}}$$

■

4.8.4.3 Percentage of Time

Field strength for a given percentage of time between 1 and 50% time should be calculated by interpolation between nominal values 1 and 10% or between the nominal values 10 and 50% of time by the following equation:

$$E = E_s(Q_i - Q_t)/(Q_i - Q_s) + E_i(Q_t - Q_s)/(Q_i - Q_s) \tag{4.35}$$

where:

t: Required percentage time
t_i: Lower nominal (inferior) percentage time
t_s: Upper nominal (superior) percentage time
E_i: Field strength value for time percentage t_i
E_s: Field strength value for time percentage t_s
Q_t: $= Q(t/100)$
Q_i: $= Q(t_i/100)$
Q_s: $= Q(t_s/100)$

It should be noted that $Q(q/100)$ is the inverse complementary cumulative normal distribution function. Values of Q for different values of q are given in Table 4.8.

Example 4.12. Find the received field strength with 40% percentage time for path over cold sea with the following assumptions:
$d = 20\,\text{km}$, $f = 600\,\text{MHz}$, $h_1 = 37.5\,\text{m}$, $h_2 = 10\,\text{m}$, 50% locations.

Solution.

$$Q_i = Q_{10} \xrightarrow{\text{Table 4.8}} Q_{10} = 1.282$$

$$Q_s = Q_{50} \xrightarrow{\text{Table 4.8}} Q_{50} = 0.0$$

$$Q_t = Q_{40} \xrightarrow{\text{Table 4.8}} Q_{40} = 0.253$$

$$E_{\mathrm{i}} \xrightarrow[d=20,\ h_1=37.5]{\text{Fig. 4.22}} E_i = 71.5\ \mathrm{dB}_{\mu\mathrm{V/m}}$$

$$E_{\mathrm{s}} \xrightarrow[d=20,\ h_1=37.5]{\text{Fig. 4.22}} E_s = 70.5\ \mathrm{dB}_{\mu\mathrm{V/m}}$$

Using (4.35) yields:

$$E = 70.7\ \mathrm{dB}_{\mu\mathrm{V/m}}$$

■

4.8.4.4 Mixed Paths

A particular method has been developed by ITU-R for radio paths where the terrain is mixed of land and sea portions. This method uses $E_l(d)$ and $E_{\mathrm{s}}(d)$ to represent the field strength at distance d from the TX antenna at the representative clutter height, R, for all-land and all-sea paths respectively. Interpolation or extrapolation may be used to generalize the method for antenna height h_1, frequency, and percentage time when they are other than nominal values.

The following steps should be followed to determine the field strength of any path with a mixture of land and sea parts. If the path contains both warm sea and cold sea portions, the warm sea curves should be selected for calculation of $E_{\mathrm{s}}(d)$. The value of h_1 should be calculated using the procedure introduced in previous sections. Normally, this value of h_1 will be used for both E_l and E_s. However, if h_1 is less than 3 m, it should still be used for E_l, but a value of 3 m should be used for E_s:

Step 1: Calculate the total length of path that lies over land, d_{l}.

Step 2: Calculate the quantity Δ:

$$\Delta = d_{\mathrm{l}}[E_l(1\,\mathrm{km}) - E_{\mathrm{s}}(1\,\mathrm{km})], \quad \text{if } d_{\mathrm{l}} < 1\,\mathrm{km} \tag{4.36}$$

$$\Delta = E_l(d_{\mathrm{l}}) - E_{\mathrm{s}}(d_{\mathrm{l}}), \quad \text{if } d_{\mathrm{l}} > 1\,\mathrm{km} \tag{4.37}$$

Step 3: Calculate the mixed path value at RX antenna distance, d_{t}:

$$E_{\mathrm{mix}}(d_{\mathrm{t}}) = E_{\mathrm{s}}(d_{\mathrm{t}}) + \Delta \tag{4.38}$$

Limit $E_{\mathrm{mix}}(d_{\mathrm{t}})$ such that $E_l(d_{\mathrm{t}}) \leq E_{\mathrm{mix}} \leq E_{\mathrm{s}}(d_{\mathrm{t}})$

Step 4: Calculate the difference, ΔE, between the mixed path and land path field strengths at the required total path distance, d_{t}:

$$\Delta E = E_{\mathrm{mix}}(d_{\mathrm{t}}) - E_l(d_{\mathrm{t}}) \tag{4.39}$$

Step 5: Calculate an interpolation factor to take account of the long range effect of land on propagation using d_{l} and the TX antenna height, h_1:

$$\chi = \alpha + (1-\alpha)\exp[-\beta d_1^{2.42-0.0003527h_1}] \quad (4.40)$$

where $\alpha = 0.3$ and $\beta = 0.0001$

Step 6: Finally, calculate the field strength for the mixed path:

$$E = E_l(d_t) + \Delta E \cdot \chi \quad (4.41)$$

4.8.4.5 Radio Path Length

In Figs. 4.18–4.25, field strength is plotted in terms of distance in the range of 1–1,000 km. Interpolation is not needed in the case of reading the field strength directly from these graphs. For greater precision and for computer implementation, field strengths should be obtained from the associated tabulations prepared by ITU-R. In this case, unless d coincides with one of the tabulation distances given by ITU-R, the field strength, E, should be linearly interpolated for the logarithm of the distance using

$$E = E_i + (E_s - E_i) \cdot \log(d/d_i) : \log(d_s/d_i) \quad (4.42)$$

where:

d: Required distance
d_i: Nearest tabulation distance less than d
d_s: Nearest tabulation distance greater than d
E_i: Field strength value for d_i
E_s: Field strength value for d_s

4.8.5 Correction Factors

To increase the accuracy of calculations, applying some correction factors is needed. These correction factors as specified in ITU-R recommendation P.1546 consist of the following items:

- RX antenna height
- Urban and suburban paths
- Terrain clearance angle
- Location variability

4.8.5.1 Correction for RX Antenna Height

Increasing of RX and TX antenna heights will decrease the overall path loss associated with radiowaves. The field strength values given by the land curves prepared by ITU-R are for a reference RX antenna at a height, R(m), representative of the height from the ground cover surrounding the RX antenna, subject to a minimum height of 10 m.

In real cases, RX antenna height may be higher or less than the specified minimum value. Examples of reference heights are:

- 30 m for dense urban areas
- 20 m for urban areas
- 10 m for suburban areas
- 10 m for dense sea paths

Where the receiver antenna is adjacent to land, the elevation angle of the arriving wave should be taken into account by calculating a modified representative clutter height R'(m), given by

$$R' = (1{,}000d \cdot R - 15h_1)/(1{,}000d - 15) \tag{4.43}$$

where h_1, R', R are in meters, distance d in km and

$$h_1 < 6.5d + R \quad \Longrightarrow \quad R' \approx R \tag{4.44}$$

In case the receiver antenna located in an urban environment, the correction factor is then given by

$$\alpha_h[\text{dB}] = 6.03 - J(V), \qquad h_2 < R' \tag{4.45}$$

$$= K_{h2} \cdot (\log h_2/R'), \qquad h_2 \geq R' \tag{4.46}$$

For calculating of α_h, the following auxiliary equations may be used:

$$J(V) = \left[6.9 + 20\log\left(\sqrt{(V-0.1)^2+1} + V - 0.1\right)\right] \tag{4.47}$$

$$V = K_n\sqrt{h_d \cdot \theta_c} \tag{4.48}$$

$$h_d\,[\text{m}] = R' - h_2 \tag{4.49}$$

$$\theta_c(\text{degrees}) = \arctan(h_d/27) \tag{4.50}$$

$$K_{h2} = 3.2 + 6.2\log f \tag{4.51}$$

$$K_n = 0.0108\sqrt{f} \tag{4.52}$$

α_h is negative for $h_2 < R'$ and positive for $h_2 > R'$, also the following points should be taken into account when the above equations are applied:

- Where the receiver antenna is adjacent to land in a rural or open area, the correction factor is given by (4.46) for all values of h_2.
- Where the receiver antenna is adjacent to sea for $h_2 \geq 10$ m, the correction factor should be calculated using (4.46) with R' set to 10 m.

- Where the receiver antenna a is adjacent to the sea for $h_2 < 10\,\text{m}$, the following alternative method must be applied based on the path lengths at which 0.6 of the first Fresnel zone (r_1) is just clear of obstruction by the sea surface.

$$\alpha_h[\text{dB}] = 0.0 \qquad \text{for } d \le d_h \tag{4.53}$$

$$\alpha_h[\text{dB}] = C_{10}\log(d/d_{h2})/\log(d_{10}/d_{h2}), \quad \text{for } d_{h2} < d < d_{10} \tag{4.54}$$

In the above equation, each parameter is defined as follows:

C_{10}: Correction for the required value of h_2 at distance d_{10} using (4.46) with R' set to 10 m

d_{10}: Distance at which the path just has $0.6r_1$ clearance for $h_2 = 10\,\text{m}$ calculated as $D_{0.6}(f,\ h_1,\ 10)$.

d_{h2}: Distance at which the path just has $0.6r_1$ clearance for the required value of h_2 calculated as $D_{0.6}(f,\ h_1,\ h_2)$.

It should be noted that the above procedure is not valid for receiver antenna heights, h_2, less than 1 m when adjacent to land or less than 3 m when adjacent to sea.

Example 4.13. Calculate correction factor at a distance of 15 km, $R = 10\,\text{m}$, for receiver antenna heights 5 and 15 m. Assume $h_1 = 75\,\text{m}$ and $f = 400\,\text{MHz}$.

Solution.

$$R' = 9.93\,\text{m}$$

$$h_2 = 5\,\text{m} \implies h_2 < R' \implies \alpha_h = 6.03 - J(V)$$

The following parameters are calculated using (4.47)–(4.52):

$$K_n = 0.216, \quad h_d = 4.93, \quad \theta_c = 10.35^\circ, \quad V = 1.54$$

$$J(V) = 6.9 + 20\ \log(3.19) = 16.98\ \text{dB}$$

$$\alpha_h(h_2 = 5\,\text{m}) = 6.03 - 16.98 = -10.95\ \text{dB}$$

The above result reveals that there is 10.95 dB more attenuation where $h_2 = 5\,\text{m}$, while for $h_2 = 15\,\text{m}$ there will be 3.4 dB gain:

$$h_2 > R' \implies \alpha_h = K_{h2} \cdot \log(h_2/R')$$

$$K_{h2} = 3.2 + 6.0\ \log 400 = 19.3$$

$$\alpha_h = 19.3\ \log(15/9.93) = 3.46\ \text{dB}$$

■

Table 4.10 Typical C-value (height gain factor)

Area	VHF band	UHF band
Rural (open)	4 dB	4 dB
Suburban	5 dB	6 dB
Urban	6 dB	8 dB

For α_h, ITU-R has presented the following simple rough relation:

$$\alpha_h = \frac{C}{6} \times 20 \log(h_2/10) \quad \text{for } 1.5\,\text{m} \leq h_2 \leq 40\,\text{m} \tag{4.55}$$

The C-parameter is given in the Table 4.10.

4.8.5.2 Correction for Urban/Suburban Area

For urban or suburban radio links of length less than 15 km and covered with buildings of identical heights over flat terrain, a correction factor for reduction of field strength due to building clutter should be considered. The urban/suburban correction factor, α_u in dB, is expressed by the following equation:

$$\alpha_u = -3.3(\log f)(1 - 0.85 \log d)[1 - 0.46 \log(1 + h_a - R)] \tag{4.56}$$

where:

h_a: Antenna height in m/AGL
R: Reference height of the ground cover surrounding receiver (or transmitter) antenna as defined in (4.8.5.1)

This correction factor only applies under the following conditions:

$$d < 15\,\text{km}, \quad h_1 - R < 150\,\text{m}, \quad h_a > R \tag{4.57}$$

4.8.5.3 Correction for Terrain Clearance Angle

For radio links over land, and when the receiver antenna is on a land section of a mixed path, for predicting the field strength for reception conditions in specific areas (e.g., small reception area), a correction factor may be made based on a terrain clearance angle. The terrain clearance angle θ_{tca} as shown in Fig. 4.27 is defined by

$$\theta_{tca} = \theta - \theta_r \tag{4.58}$$

$$\theta_r = \arctan\left(\frac{h_{2s} - h_{1s}}{1{,}000d}\right) \tag{4.59}$$

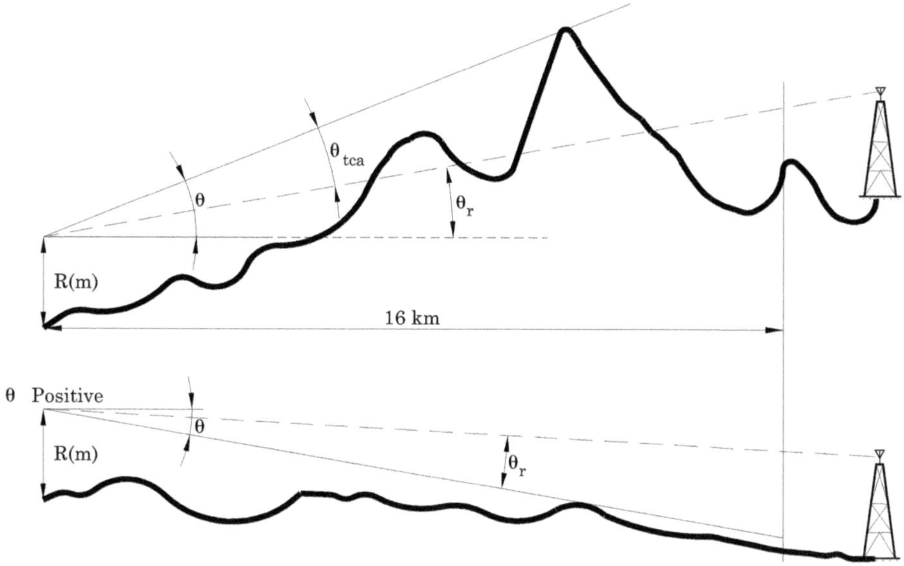

Fig. 4.27 Concept of terrain clearance angle

For a known terrain clearance angle, the correction factor to be added to the field strength is calculated using

$$\alpha_c = J(V') - J(V) \tag{4.60}$$

where $J(V)$ is defined by (4.47) and

θ_{tca}: Terrain clearance angle in degrees
f: Frequency in MHz
θ_r: Reference angle in degrees
h_{1s}: Height of transmitter antenna in m
h_{2s}: Height of receiver antenna in m

$$V' = 0.036\sqrt{f} \tag{4.61}$$

$$V = 0.065 \cdot \theta_{tca} \cdot \sqrt{f} \tag{4.62}$$

Figure 4.28 illustrates the correction factor of terrain clearance angle in terms of nominal frequencies. The correction factor as defined by (4.60) is valid for $-0.8^\circ \leq \theta_{tca} \leq +40^\circ$, and beyond this limit, it should be the same as $\theta_{tca} = -0.8^\circ$ for lower values and $\theta_{tca} = +40^\circ$ for upper values.

It should be noted that the land field strength curves given in Figs. 4.18–4.25 take account of losses due to typical shielding of the receiver antenna by gently rolling terrain. Thus, the terrain clearance angle correction factor is zero (in logarithmic scale) at a small positive angle typical for receiver antenna positions.

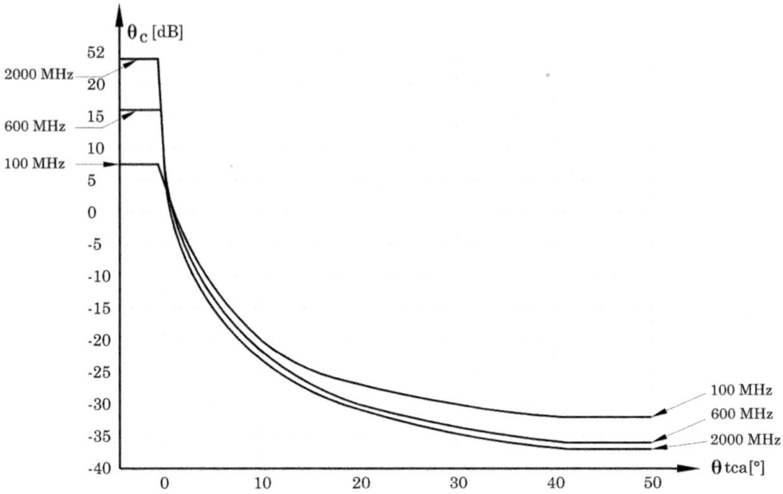

Fig. 4.28 Correction factor versus terrain clearance angle (Ref.: ITU-R, P.1546-3)

Example 4.14. Calculate α_c for $\theta_{tca} = 10°$ and $f = 100\,\text{MHz}$ using the given equations and compare the result with α_c value determined using Fig. 4.28

Solution.

$$V = 0.065 \times 10 \times \sqrt{100} = 6.5$$

$$J(V) = 6.9 + 20\,\log(12.8) \quad \Longrightarrow \quad J(V) = 28.4\text{ dB}$$

$$V' = 0.035 \times \sqrt{100} = 0.36$$

$$J(V') = 6.9 + 20\,\log(1.29) \quad \Longrightarrow \quad J(V') = 8.4\text{ dB}$$

$$\alpha_c = J(V') - J(V) \quad \Longrightarrow \quad \alpha_c = -20\text{ dB}$$

Using Fig. 4.28 and selecting the curve related to $f = 100\,\text{MHz}$, for $\theta_{tca} = 10°$, α_c is found to be approximately $-20\,\text{dB}$ which is in conformity with the calculated value. ■

4.8.5.4 Location Variability

In broadcasting networks, the received signal level in the receiver will vary continually or randomly with location changes. Area coverage prediction methods are required to provide the statistics of reception conditions for a given area rather than at any particular point. Usually affecting factors are categorized in the following major groups:

- *Multipath Fading*
  Signal variations will occur over scales of the order of wavelength due to phasor summation of multipath effects, for example, reflections from the ground, man-made structures, and buildings.
- *Local Ground*
  Received signal level will vary due to obstruction by ground cover in the local vicinity, for example, buildings and trees, over scales of the order of the sizes of such objects. Normally, local ground cover effects are more than multipath variations.
- *Terrain Structure*
  Signal variations will also occur due to changes in the terrain structure and geometry along the entire propagation path, for example, the presence of hills, mountains, and lakes. For all except very short paths, the scale of these variations will be significantly larger than other type of variations.

In broadcasting service at VHF and UHF bands, location variations are usually measured in a square area with 100–200-m side. Signal level distribution in urban area has almost log-normal distribution. Also, since multipath fading has frequency selective nature, identifying the signal effective bandwidth is necessary for further analysis.

Field strength at the receiving location, E_q, for $q\%$ of time can be calculated by

$$E_q\ [\mathrm{dB}_{\mu\mathrm{V/m}}] = E_\mathrm{m}[\mathrm{dB}\ \mu\mathrm{V/m}] + Q_\mathrm{i}(q/100)\cdot\sigma_\mathrm{L}\ [\mathrm{dB}] \tag{4.63}$$

where:

Q_i: Log-normal distribution coefficient according to Table 4.8
σ_L: Standard deviation in dB
E_m: Median received signal level in $\mathrm{dB}_{\mu\mathrm{V/m}}$

For calculating the standard deviation used for digital signals with bandwidth less than 1 MHz and also analog signals, the following equation may be used:

$$\sigma_\mathrm{L}\ [\mathrm{dB}] = K + 1.6\log f\ [\mathrm{MHz}] \tag{4.64}$$

In the above equation, value of constant K is selected as follows:

- $K = 2.1$ for wireless systems in urban areas
- $K = 3.8$ for wireless systems in suburban areas
- $K = 5.1$ for analog broadcasting systems

For digital systems with bandwidth equal or greater than 1 MHz, the standard deviation is equal to 5.5 dB for all frequencies. Table 4.11 shows the standard deviation values for different frequency bands.

Example 4.15. A digital radio broadcasting network at UHF band is designed to provide 90% location coverage within a suburban area. To increase the coverage area to 95%, how much the antenna gain should be added?

Table 4.11 Location variability standard deviation (Ref.: ITU-R, P.1546-3)

Frequency range (MHz)	30–300 MHz	300–1,000 MHz	1,000–3,000 MHz
Nominal frequency (MHz)	100	600	2,000
Broadcasting/analog	8.3	9.5	–
Broadcasting/digital	5.5	5.5	5.5

Table 4.12 Time variability standard deviation (σ_t) in dB (Ref.: ITU-R, P.1406-1)

Frequency band	Terrain (km) structure	Distance in km			
		50	100	150	175
VHF	Land or sea	3 dB	7 dB	9 dB	11 dB
UHF	Land	2 dB	5 dB	7 dB	–
	Sea	9 dB	14 dB	20 dB	–

Solution.

$$\xrightarrow[\text{DAB/UHF}]{\text{Table 4.11}} \sigma_L = 5.5 \text{ dB}$$

$$\text{FM}(90\%) = Q_i(90\%) \times \sigma_L = 7.05 \text{ dB}$$

$$\text{FM}(95\%) = Q_i(95\%) \times \sigma_L = 9.05 \text{ dB}$$

$$\Delta(\text{FM}) = 2 \text{ dB}$$

So the new antenna gain should be 2 dB greater than the existing antenna gain. ■

4.8.5.5 Time Variability

Received signal level in addition to location variability includes some time variations, which have statistical nature and are mostly due to climate condition variability and motion. The time variability standard deviation value is represented by σ_t which is a function of the following parameters:

- Distance between the transmitter and receiver
- Terrain structure along radio path
- Working frequency band

In Table 4.12, σ_t value is given for VHF and UHF bands in the mobile radio communications.

4.8.5.6 Total Location and Time Variability

Considering the correlation coefficient between location variability and time variability, as ρ then combined standard deviation value is calculated from the following equation:

$$\sigma = \sqrt{\sigma_L{}^2 + \sigma_t{}^2 + 2\rho\sigma_L\sigma_t} \tag{4.65}$$

According to the low correlation between these random processes, ρ is usually neglected resulting in the following equation:

$$\sigma = \sqrt{\sigma_L{}^2 + \sigma_t{}^2} \tag{4.66}$$

Example 4.16. In a broadcasting network, the maximum predicted coverage distance is 25 km. For operating at VHF (160 MHz) and UHF (420 MHz) frequency bands find:

1. Value of σ_L and σ_t for both channels
2. Fade margins for 95% location coverage and 90% time coverage

Solution. 1.

$$\sigma_L = 3.8 + 1.6 \log 160 = 7.33 \text{ dB} \quad \text{for VHF band}$$
$$\sigma'_L = 3.8 + 1.6 \log 420 = 8 \text{ dB} \quad \text{for UHF band}$$

$$\sigma_t \xrightarrow[d<50\,\text{km, VHF}]{\text{Table 4.12}} 3 \text{ dB}$$

$$\sigma'_t \xrightarrow[d<50\,\text{km, UHF}]{\text{Table 4.12}} 2 \text{ dB}$$

2. Standard deviations for total location and time variabilities are

$$\sigma = \sqrt{\sigma_L^2 + \sigma_t^2} = 7.92 \text{ dB} \qquad \text{for VHF band}$$
$$\sigma' = \sqrt{\sigma_L'^2 + \sigma_t'^2} = 8.25 \text{ dB} \qquad \text{for UHF band}$$
$$q = 95\% \xrightarrow{\text{Table 4.8}} Q_i(95/100) = -1.645$$
$$\text{FM(VHF)} = -Q_i(95/100) \cdot \sigma = 13 \text{ dB}$$
$$\text{FM}'\text{(UHF)} = -Q_i(95/100) \cdot \sigma' = 13.57 \text{ dB}$$

■

4.9 Exercises

Questions

1. State the main frequency bands allocated to propagation of radiowaves in the broadcasting service.
2. Specify the basic features of the broadcasting services and indicate their main differences with those related to telephony communications.
3. Study article 5 of the latest version of ITU radio regulations and assess whether or not there are any modifications in each of the LF, MF, HF, VHF, UHF, and satellite frequency bands compared to those indicated in Tables 4.1–4.3.
4. Study about audio and video channels in the local broadcasting service within the VHF/UHF spectrum and prepare a report discussing the result of your study.
5. Specify the reasons for using EHF satellite frequency bands and its major differences with SHF and UHF satellite frequency bands.
6. Explain the radiowaves propagation phenomena related to the broadcasting in LF, MF, and HF bands.
7. Explain the advantages of utilizing the satellite technology for the audio and video broadcasting services.
8. Describe the reasons for installation of LNB amplifier close to TVRO antennas.
9. Specify the reasons for using digital technology in the video/audio broadcasting systems.
10. Describe the main phenomena affecting satellite radiowave propagation in the broadcasting services.
11. Assess the dependence of the radiowaves path loss to the distance in VHF/UHF bands for the broadcasting service and determine the differences between ITU-R and Hata methods.
12. State the features of the propagation medium associated with radiowaves in the local broadcasting services.
13. Explain Doppler effect in the broadcasting service and indicate for which kind of various audio/video services it has more effects.
14. State the application of diagrams (4.18) up to (4.25) for selecting the locations of TX broadcasting centers.
15. Describe how the path loss and the received signal level are calculated based on the method presented in the recommendation ITU-R, P.1546.
16. Specify the main effective parameters on the received signal level in the broadcasting services.
17. State major factors in the effective TX antenna height at VHF/UHF bands.

Problems

1. Using diagrams (4.9) and (4.10), calculate the approximate value of radiowaves path loss in broadcasting service at distances of 20 and 40 km from the TX station in the band III.
2. If the path loss is defined with the following general equation:

$$\text{FSL} = 32.4 + 20 \log f + n \log d$$

   Using the curves indicated in the diagrams (4.9) and (4.10), calculate the value of n for three distances of 10, 30, and 50 km for open area, based on the ITU-R method and also according to the Hata method, for urban and suburban areas.
3. If the values of n obtained from the above problem is generalized for the broadcasting path loss in the VHF and lower UHF bands for TV service, then calculate and sketch similar curves for these bands.
4. The coverage range of a DAB network using ordinary RX antenna is 50 km for 95% locations, find:
   (a) The coverage range in case of using RX antenna with a 2-dB tolerance of the effective gain. (Assume the signal bandwidth of 1.5 MHz).
   (b) The coverage range in case of using RX antenna with 2 dB gain in open areas at L band.
5. Assume a DAB system operating at the L band with a 1.5 MHz bandwidth.
   (a) Specify the required fade margin for appropriate reception of the radiowaves in more than 80% of locations.
   (b) Calculate the increase in TX output power for expanding the reception probability of 95%, locations.
   (c) If the TX power is increased to provide 90% location coverage, then how much the RX antenna gain must be increased in order to obtain 95% coverage?
6. Signals of FM and digital radio in the L band are received in a vehicle with speed in the range of 100–150 km/h, calculate:
   (a) Limits of the Doppler frequency when radiowaves propagation and driving directions are opposite.
   (b) The Doppler frequency where the angle between radiowaves propagation and driving directions is 60°. Indicate the ideal condition for proper reception of the radio signals.
7. Repeat the example (4.4) for the following conditions:

$$\text{SNR} = 33\text{ dB},\quad B = 6\,\text{MHz},\quad f = 495\,\text{MHz},$$

$$N = 7\text{ dB},\quad R = 50\,\Omega$$

8. For a system with $V_n = 10\,dB_{\mu V}$ and SNR $= 40\,$dB, calculate:
   (a) The minimum value of V_r
   (b) The received field intensity by a matched antenna with a $8\,dB_d$ gain and loss of transmission line equal to 4 dB at 495 MHz
9. Assuming the RF channel spacing equal to 6 MHz, find:
   (a) The required minimum protection ratio to prevent interference between them
   (b) The ratio of interference region to coverage area in smooth Earth where the distance factor is $n = 30$
10. For continuous reception of radio digital signals in the band III, it requires a minimum field intensity equal to $40\,dB_{(\mu V/m)}$ for a 95% location coverage. If it is assumed that the probability distribution function is in the form of log-normal distribution, find:
    (a) The median value of the field intensity with a standard deviation of $\sigma = 6\,$dB
    (b) The amount of system gain increase in terms of decibel (by adjusting antennas and TX/RX parameters), for expanding the location coverage to 99%
11. Using the assumptions given below,

$$d = 15\,\text{km}, \quad f = 600\,\text{MHz}, \quad h_1 = 20\,\text{m}$$

$$h_2 = 10\,\text{m}, \quad 50\%\ \text{Locations}$$

    Find the radiowave path loss on an inland route for time percentages of 10 and 50.
12. Repeat the problem No. 11 for frequency $f = 100\,$MHz, 50% location coverage and for time percentages of 10 and 50.
13. The calculated height of TX antenna is 60 m working at 600 MHz, using the following assumptions:

$$50\%\ \text{location coverage}, \quad 10\%\ \text{time coverage}, \quad \text{EIRP} = 1{,}000\,w$$

$$d = 15\,\text{km}, \quad \text{land path}, \quad h_2 = 10\,\text{m}$$

    (a) Calculate the field strength.
    (b) Repeat the case for $f = 100\,$MHz.
14. Using the same parameters as in the previous problem except $f = 400\,$MHz:
    (a) Find the field strength.
    (b) Calculate the path loss.

15. For a radio link having the following specifications:

$$d = 10\,\text{km},\ h_1 = 20\,\text{m},\ h_2 = 10\,\text{m},\ \text{land path},\ 50\%\ \text{location coverage}$$

(a) Find the field strength for a TX having an effective radiation power of 1 kw and time percentages of 5, 20, 30, and 40, respectively.
(b) Sketch the curve of E in terms of time percentage in the interval from 1 to 50%.

16. Find the radio link path loss with the following specifications:

$$h_2 = 10\,\text{m},\ \ d = 15\,\text{km},\ \ f = 420\,\text{MHz},\ \ h_1 = 50\,\text{m},\ \ \text{land path},$$
$$50\%\ \text{location coverage and at}\ 40\%\ \text{time coverage}$$

17. In a suburban broadcasting service at $f = 220$ MHz, the receiver is used in one of the following modes of operation:

(a) Inside a vehicle with an antenna height of 2 m
(b) In a fixed station with an antenna height of 20 m

Assuming that $R = 10$ m, calculate the correction factor of the RX for each case.

Chapter 5
Trans-Horizon Radiowaves Propagation

5.1 General Concept

5.1.1 Introduction

This chapter is dedicated to the trans-horizon (also called over-horizon) radiowaves propagation through troposphere layer at the high altitudes of several kilometers above the ground. As it was elaborated on the first chapter, the radio horizon of a TX which has an antenna installed at a height of h_t from the ground level, assuming that all the path elevations and obstructions are ignored and only the smooth Earth's curvature is accounted for, can be expressed by the following relation:

$$R_t = \sqrt{2KR_e h_t} \tag{5.1}$$

As an example for standard climatic conditions with $K = 1.33$ and $h_t = 30\,\text{m}$, this distance will approach around 22.6 km. In order to be able to transmit the radiowaves directly to distances beyond the radio horizon without the need for any terrestrial repeater stations, certain techniques must be used. One of the distinguished techniques with a desirable performance as depicted in Fig. 5.1 is the troposcatter radio communications which will be discussed in this chapter.

5.1.2 Trans-Horizon Communications

Radiowaves transmission and reception methods, using long hops in the order of hundreds of kilometers and through over-horizon radio communications, can be summarized as follows.

A. Ghasemi et al., *Propagation Engineering in Radio Links Design*,
DOI 10.1007/978-1-4614-5314-7_5,

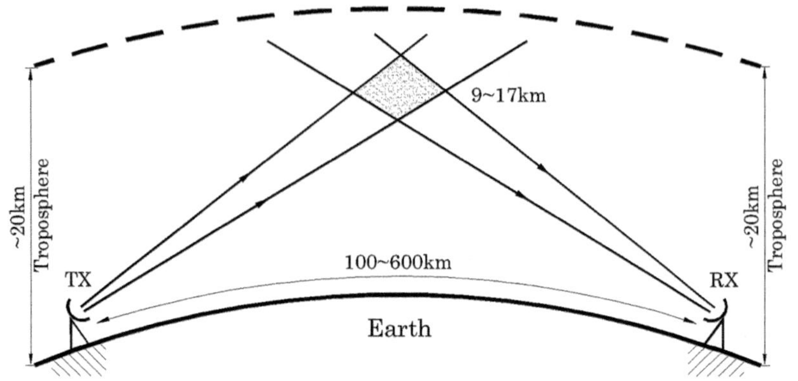

Fig. 5.1 Typical troposcattering propagation

5.1.2.1 MF and HF Communications

In this method, the refraction and reflection phenomena are used for transmitting the radiowaves to distances of many hundreds/thousands of kilometers. The average bandwidth of the transmission is limited to one or two telephone channel(s) including restriction of using these sub-bands due to frequency dependence of such communications to the diurnal time (i.e., daytime and night time). This method was widely used for maritime, broadcasting, and long-distance services prior to the application of satellite communications.

5.1.2.2 Ionospheric Scatter

This method utilizes the scattering properties of the radiowaves in the ionosphere layer (a phenomenon similar to troposcatter) and constitutes hops of up to several thousands of kilometers in the frequency bands of VHF up to 100 MHz. The average bandwidth in this method is very limited, and due to the restrictions caused by fading effect, this method is not a common practice for professional communications, and its application is very limited.

5.1.2.3 Meteorites

This method employs the reflections produced by the ionized trails of meteorites which always persist on the upper layers of atmosphere. Due to the physics of this phenomenon, the continuity of transmission is not guaranteed, and consequently, the radiowaves must be transmitted in the form of bursts. This method is under study and currently is not practical for common applications.

5.1.2.4 Troposcatter

Tropospheric scatter simply called troposcatter, which is the topic of this chapter, provides transmission of many telephone channels through long hops in the order of several hundreds of kilometers. This technology at some occasions is regarded as a suitable and cost-effective solution for local networks.

5.1.2.5 Diffraction

This method facilitates the transmission of many telephone channels over short distances beyond the horizon. This phenomenon is used in the mobile communications and broadcasting networks in VHF/UHF bands, which is discussed in other chapters with more details.

5.1.2.6 Satellite Communications

It is the most appropriate method for very long hops (e.g., intercontinental communications), but because of high cost and limited capacity, it is not an effective alternative for over-horizon networks. There are detailed descriptions concerning the satellite communications provided in Chap. 3.

5.2 Trans-horizon Communications via Troposphere

5.2.1 Types of Tropospheric Propagation

When trans-horizon propagation is limited to the troposphere layer, the only mechanisms for radio propagation beyond the horizon which occur permanently for frequencies in VHF and UHF bands are:

- Diffraction at the Earth's surface
- Scattering from troposphere irregularities

This type of propagation is defined by ITU in the recommendation No.R.310 as:

> Tropospheric propagation between points close to the ground, the reception point being beyond the radio horizon of the transmission point.

Trans-horizon propagation may be due to a variety of tropospheric mechanisms such as diffraction, scattering, and reflection from troposphere layer (excluding ducting effect), while tropospheric scatter propagation is referred only to tropospheric propagation due to scattering from many inhomogeneities and discontinuities in the refractive index of the atmosphere.

Because of dissimilarity of the two mechanisms, it is necessary to consider them separately. In practice, diffraction-based propagation is used in the point-to-area radio links such as mobile radio and broadcasting networks. For point-to-point trans-horizon radio links, tropospheric scatter propagation is used.

5.2.2 Position of Troposcatters

Nowadays, the microwave and satellite communications have developed on a massive scale, while the use of troposcatters is limited and dedicated for some specific applications. For instance, the length of a single hop in the tropo-links is longer than those in microwave links and can reach $500 \sim 600$ km which is a good and impressive reason for the significance of this type of communications.

The capacity and quality of tropo-communications have a better and improved status compared to MF/HF communications, in such a way that the troposcatters are able to transfer over 60 digital or over 300 analog audio channels and/or one TV channel.

Due to narrow beamwidth in tropo-links, the security and anti-interference capability are better than the satellite communications. For radio links with equal length, the initial establishment and maintenance costs are lower compared to microwave communications. It can be stated that even compared to the leased satellite lines, the cost of each audio channel when the span of tropo-communication is less than 400 km is lower; therefore, it requires fewer number of personnel, while the security of related sites are provided very conveniently.

The tropo-link bandwidth is about a hundred times the bandwidth in meteorite trail communications. The security of troposcatter communications is more than the satellite communications and also public telephone networks. Therefore, as a conclusion, troposcatter communications can be recommended for some areas such as desert, swamp, forest, islands, and densely populated remote and scattered areas.

The tropo-communications can be also considered as a competitive communications method to establish communication links in the offshore oil fields. The sun spots, magnetic storms, and nuclear explosions have no effect on the troposcatter propagation, and as a result, they are suitable for military communications in nuclear warfares. Troposcattering equipments are capable of transferring digital telephone, fax, and image data, and they can be also used for telemetry, remote measurements, TV, and data transfer.

5.3 Main Aspects and Applications

5.3.1 Main Aspects

The main aspects of troposcatter systems can be summarized as follows:

- Long routes up to several hundred kilometers
- Very high path loss
- High-power transmitter
- High-gain parabolic antennas
- Low-noise and very sensitive receivers
- Frequency range of 200–5,000 MHz
- Frequency modulation techniques
- Traffic capacity up to 120 telephone channels and more
- Limited bandwidth
- Antenna space diversity to improve the quality of radiowaves reception

5.3.2 Main Applications

Establishing long hops is the most outstanding characteristic of troposcattering links in radio communications. These radio links do not require any intermediary repeaters and cover distances much longer than the line-of-sight (LOS) microwave links. This property may be useful, specifically in cases where there are certain communications problems due to natural obstructions and limitations such as:

- Communications within or beyond big desert or forest.
- Communications between a remote island and the mainland and/or with other islands.
- Military networks which must be safeguarded against possible sabotage and intrusion.
- Communications of a remote offshore oil platform with onshore offices and industrial plants.

The troposcatter systems are able to provide telephony, fax, image, telemetry, and TV services. By employing error correction techniques, troposcatter may be used for data transfer as well. These systems have ad hoc applications for establishing secret links with specific objectives such as military communication trunks with low capacity and/or medium capacity for tactical requirements. In addition, it is compatible with Integrated Services Digital Network (ISDN) and operates as a communication means between two points in air defense systems.

5.3.3 Major Advantages

Major advantages of troposcatter systems may be outlined as follows:

- The possibility of establishing link on a long path in a manner that each hop can cover a path length of hundreds of kilometers, that is, ten times more than the path lengths provided by an ordinary LOS link.
- The possibility of straightforward and simple planning for rough terrains. The significant issue in planning a troposcatter system is the selection of sites in a way to minimize the effects of terrain topography on the system design. To do this, tilt angles of antennas should be very low and a bit downward.
- Covering very vast areas which can be contemplated by few hops.
- No need for repeater due to excessive length of the hop.
- Reduction in the number of stations and optimum utilization of frequency spectrum, resulting in the increased length of the hop.
- Less problems of repair and maintenance because of limited number of active operating stations.
- Increasing safety factor of the system due to diminished number of operating stations.
- Troposcatter system is a suitable method for linking remote offshore oil platforms with a 2–8 Mbps bit rate and intended for distances of 100–600 km.
- Low initial investment for buildings, roads, electrical power facilities, spare parts, test instruments, and maintenance personnel.
- Great safety against interception due to using antennas with very narrow beamwidth.
- Since tropospheric propagation is relatively immune against adverse effects of sun spots, magnetic storms, and nuclear explosions, they are very useful and suitable for military communications in nuclear warfare.
- In some occasions and applications, the deployment expenses for a troposcatter radio system are low and acceptable compared to its benefits.

5.4 Troposphere Radiowaves Propagation

Troposphere is the lowest layer of the Earth's atmosphere in which normally increasing the altitude will result in the temperature reduction. The span of this area stretches to an altitude of 9 km over the Earth's poles and 17 km over the equator. The variations of temperature, pressure, and humidity such as clouds and rain can have adverse effects on the radiowaves propagation from one point to another point. Ionization of the gases in the atmosphere can be neglected, but at altitudes of 60–1,000 km, the existence of these ions is impressive. These layers constitute the ionosphere, which have a considerable effect on radiowaves at frequencies below 40 MHz. For higher frequencies, the following phenomena must be considered:

- Radiowaves dispersion based on the fluctuations of refractivity index concentrated at a location in the troposphere.
- Reflection due to the variations of refractivity index at horizontal layers.
- Ducting caused by large negative gradients in the refractivity index.

All of these mechanisms can transfer a considerable amount of energy over-horizon and create interference between different radio paths. Any further reflections affect radiowaves in the range of 30–1,000 MHz, and ducting occurs mainly on frequencies above 1,000 MHz. Duct formation does not occur on the Earth's highlands, and most of them exist above the sea surface.

In addition, the variations of refractivity index proportional to the altitude cause bending of the radiowaves particularly over the span of optical trans-horizon. This phenomenon can be significant for small elevation angles for all frequencies.

Radio propagation, apart from the effects of refractivity index, on frequencies well above 3 GHz and on the presence of heavy raining may be affected severely, and on 15 GHz and higher, the radiowaves attenuation due to the existing oxygen and water vapor in the air can be significant. In addition to the attenuation caused by rain and atmosphere gases, it creates an equivalent amount of noise temperature.

The above-mentioned effects of the Earth are of distinct importance, and in frequencies above 30 MHz, the presence of the hills and terrain structure has a great impact on the amount of transmitted energy through trans-horizon propagation. At higher frequencies, the buildings and other obstacles have a considerable effect due to diffraction, scattering, and direct reflection mechanisms, when the wavelength is small compared to the dimensions of the obstacle.

5.5 Geometry of Troposcattering Propagation

5.5.1 Introduction

The geometry of radiowaves path in the troposcattering propagation has a major role in the design calculations including the parameters required for estimation of losses and distortions of the path. Definition of geometric parameters and the main formula related to them will be discussed in this section. Initially, it will assess the basic information required for troposcattering link design including definitions of parameters and also some of the supplementary relations.

5.5.2 The Radiowaves Path Profile

The troposcatter link path can be specified by its path profile. These profiles are similar to radio link profile shown in Fig. 5.2. These sketches are usually plotted using generalized orthogonal coordinate system, and the acquired curve is adjusted

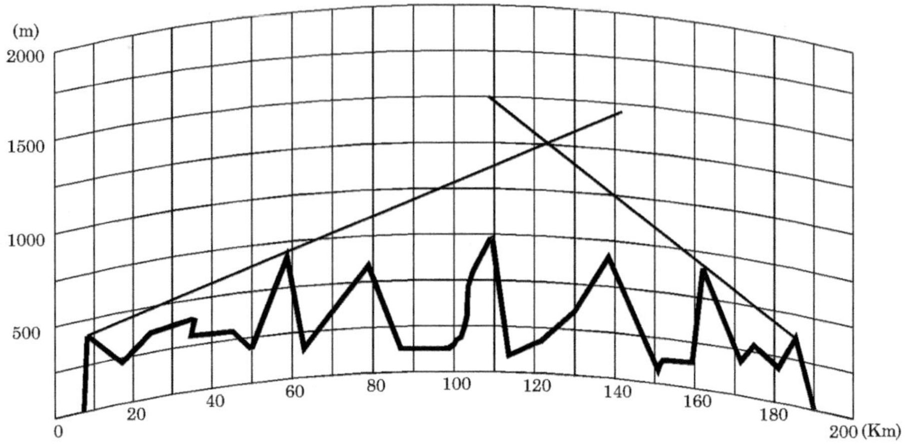

Fig. 5.2 Typical path profile for troposcattering propagation

in a manner that the electromagnetic ray appears as a straight line related to the equivalent Earth's curvature. Consequently, an effective radius denoted as R'_e is defined for the Earth which depends on the actual Earth's radius (with a value of 6,370 km) and corresponding K-factor, that is, $R'_e = KR_e$.

The value of K-factor depends on the atmosphere refractivity index and has a standard value of 1.33. In most practical cases for the purpose of planning calculations, the value of K-factor is considered less than the standard value.

The radio profiles are usually plotted for $K = 1.33$ and also specific ratio of horizontal to vertical units. However, plotting a complete and precise profile for troposcatter path is not necessary, but instead it is essential to specify the distance and elevation angles of the radiowaves at both sides. Such data are acquired from the maps or measurements in the pertaining sites. In some instances, the calculations are conducted using the exact value of K-factor in place of standard value noting the climatic conditions and geographical position of the TX and RX stations. The path profiles are used to determine the related parameters and also conduct proper calculations.

Figure 5.3 indicates a typical troposcatter route also showing the antenna beams. However, the following features must be also considered by noting Fig. 5.4:

- Antennas shall be directed toward the horizon (or to be more precise the radio horizon due to the variation of the Earth's curvature which can form a different horizon). The radio horizon can be seen from the radar sites with a vertical elevation or a takeoff which can be negative, zero, or positive. The beam oriented toward radio horizon in most of the pertinent literature is mistaken with the antenna axis. The actual case in practical situations is somewhat the same, as indicated in Fig. 5.5 which shows a 3 dB beam is slightly obstructed by the horizon. Thus, it can be concluded that a path beamwidth must be considered in the calculations.

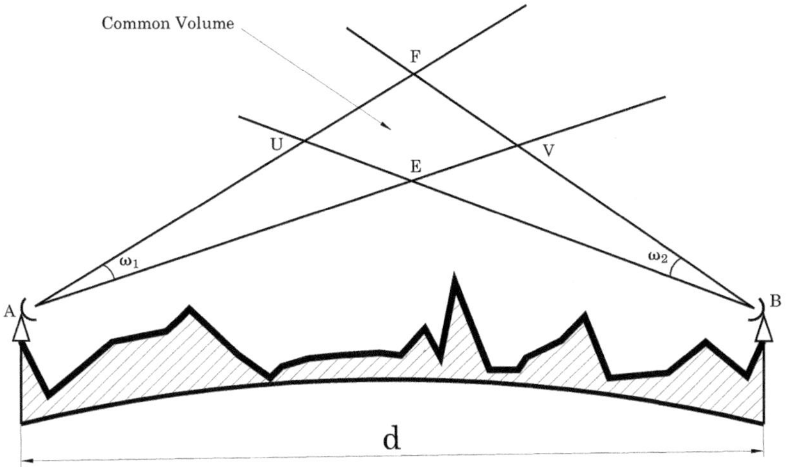

Fig. 5.3 Route of troposcattering radiowaves

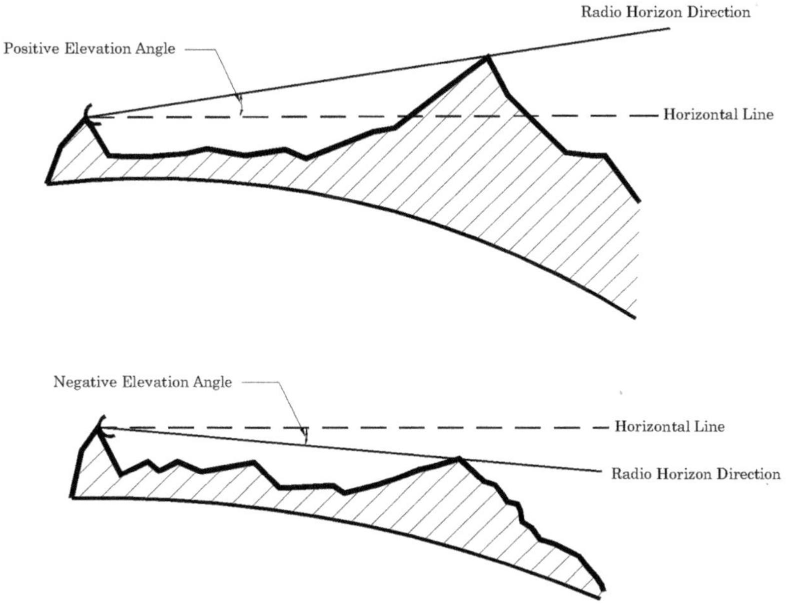

Fig. 5.4 Elevation angle of radiowaves

- Radiowaves within 3 dB beamwidth of TX and RX antennas will include a portion of troposphere marked with VFUE in Fig. 5.3 which is called common volume.
- There is a dark region between the two horizons.

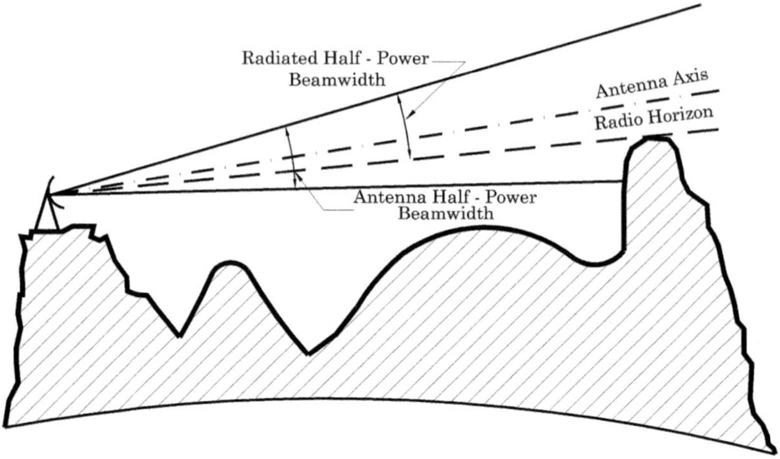

Fig. 5.5 Radiated beamwidth

- The beams of one antenna oriented to another direction such as AFB and AEB possess different lengths which result in various transmission delays from different paths and consequently will cause signal distortion.

  The study of profile parameters will entirely start with geometric specifications. There are a certain number of basic parameters which must be initially recognized in the process. But some others emanate from these basic parameters. Other significant factors such as the beamwidth of the antennas and multipaths will be considered in the following sections.

5.5.3 Geometrical Parameters

5.5.3.1 Basic Data

The following parameters known as the basic data must be defined prior to the path calculations:

- The path length, d.
- The heights of h_1 and h_2 from the sea surface relating to the TX and RX antenna locations, that is, the height of the two antenna center lines from the sea surface.
- The heights of h'_1 and h'_2 relating to the two horizons from the sea surface (these parameters are not considered when the sea itself is taken as the horizon).
- The distances d_1 and d_2 from each site to its relating horizon (when the sea itself is taken as the horizon, and then these parameters are obtained through calculations).

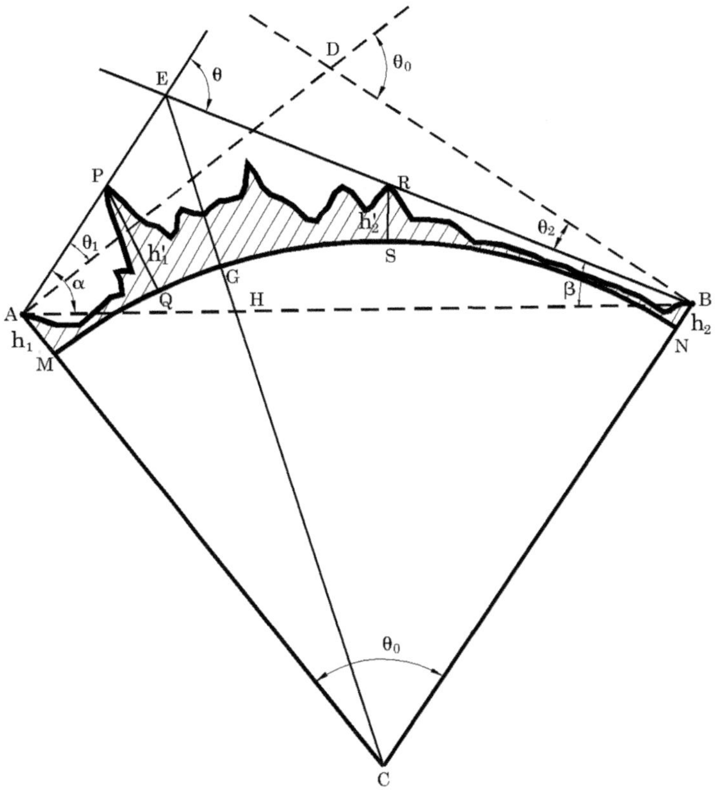

Fig. 5.6 Geometrical parameters in troposcatter link

- The correction factor of refractivity index of the air (K) for the Earth's radius R_e (the standard value of R_e is equal to 6,370 km and for K is equal to 1.33).

Other geometric parameters of the path can be calculated using these parameters. In some instances, the elevation of the horizon can be used instead of the height and distance from the horizon. The prevalent values related to the path length are limited from 100 to 500 km. The heights of transmission and reception sites and also the horizons from the sea surface may reach to several hundred meters. The distance of one site from its pertaining radio horizon usually includes a range of few kilometers up to tens of kilometers.

5.5.3.2 Angular Parameters

Normally the angles which are used in the tropospheric communications are so small that their values in terms of radian may replace their sine or tangent values in the formulas with an acceptable accuracy. Noting Fig. 5.6, the useful angular parameters are specified below:

$$\theta_0 = \frac{d}{KR_e} \tag{5.2}$$

For example, for a route of 500 km, angular length of θ_0 and $K = 1.33$, the above value becomes 59 mrad (3.4°).

The elevation angles θ_1 and θ_2 pertaining two horizons with respect to the horizontal plane in each site are given with the following relations:

$$\theta_1 = \frac{h'_1 - h_1}{d_1} - \frac{d_1}{2KR_e} \tag{5.3}$$

$$\theta_2 = \frac{h'_2 - h_2}{d_2} - \frac{d_2}{2KR_e} \tag{5.4}$$

When $h'_2 = h'_1 = 0$, that is, the horizon positioned over the sea, the above formulas are simplified to:

$$\theta_1 = -\frac{d_1}{KR_e} \tag{5.5}$$

$$\theta_2 = -\frac{d_2}{KR_e} \tag{5.6}$$

Distances:

AM: Height of antenna 1 from the sea surface, h_1
BN: Height of antenna 2 from the sea surface, h_2
EG: Height of intersection point of the beams (h_0) from the sea surface
EH: Height of intersection point of the beams (h) from the line AB
PQ: Height of horizon obstacle A from the sea surface (h'_1)
RS: Height of horizon obstacle B from the sea surface (h'_2)
AB: Path length (d) equal to the arc of MN
MQ: Distance d_1 from horizon of station A
NS: Distance d_2 from horizon of station B
$CM = CG = CN$: Effective radius of the Earth (R'_e)
AD: Horizontal line from site A (AD $\perp$ AC)
BD: Horizontal line from site B (BD $\perp$ BC)
MG: Distance of sub-intersection point of the beams (E) from site A
NG: Distance of sub-intersection point of the beams (E) from site B

Angles:

$M\widehat{C}N = \widehat{D} = \theta_0$: Angular length
$\widehat{E} = \theta$: Radio path deviation angle
$E\widehat{A}D = \theta_1$: Elevation angle of radio horizon at site A (is positive since it is above AD)
$E\widehat{B}D = \theta_2$: Elevation angle of radio horizon at site B (is negative since it is below BD)
$\alpha = E\widehat{A}B$: Elevation angle of radio horizon at A with respect to the line AB
$\beta = E\widehat{B}A$: Elevation angle of radio horizon at B with respect to the line AB

For all small angles: $\sin x = \tan x = x, \cos x = 1 - x^2/2$

Ratios:

$s = \alpha/\beta$: The path symmetry factor (if $\alpha/\beta > 1$, then $s = \beta/\alpha$)

These angles can be also measured in the TX and RX sites; however, when the horizon is the sea itself, this is not a good advice, since the error created by refraction may be a big mistake and unacceptable.

The elevation angles of radio horizons at sites A and B (i.e., angles of θ_1 and θ_2) practically are limited to $-1.7° \sim +2°$ equivalent to -30 to $+35$ mrad. For example, the elevation angle of the peak of a mountain with a height of 3,000 m and located 60 km apart from a sea-level station is equal to -27 mrad ($-1.52°$). It must be explained that the optical elevation angle and radio angle of the horizon are slightly different. The angles of α and β between the radio horizons and line AB connecting two sites are indicated with the following relations:

$$\alpha = \theta_1 + \frac{d}{2KR_e} + \frac{h_1 - h_2}{d} = \theta_1 + \frac{1}{2}\theta_0 + \frac{h_1 - h_2}{d} \tag{5.7}$$

$$\beta = \theta_2 + \frac{d}{2KR_e} - \frac{h_1 - h_2}{d} = \theta_2 + \frac{1}{2}\theta_0 - \frac{h_1 - h_2}{d} \tag{5.8}$$

- The scatter angles or the angular distance θ, as an angle between the two beams, can be expressed as:

$$\theta = \theta_0 + \theta_1 + \theta_2 \tag{5.9}$$

$$\theta = \alpha + \beta \tag{5.10}$$

The scatter angles are valid in the range of few mrad up to 80 mrad (4.6°) but practically do not exceed 35 mrad (2°).

- The symmetry factor denoted as "s" has a value less than unity and is defined below:

$$s = \begin{cases} \alpha/\beta, & \alpha/\beta \le 1 \\ \beta/\alpha, & \alpha/\beta > 1 \end{cases} \tag{5.11}$$

5.5.3.3 Length Parameters

The length parameters which are useful in the calculations of troposcatter radio links are given below:

- The radar horizon of antenna center of each one of the stations is specified by the following relations:

$$d_{1h} = \sqrt{2KR_e h_1} \tag{5.12}$$

$$d_{2h} = \sqrt{2KR_e h_2} \tag{5.13}$$

- The lengths of r_0 and s_0 indicate the radio path length from the sites up to the location of their intersection and are expressed by the following equations:

$$r_0 = EA = d\frac{\beta}{\theta} \tag{5.14}$$

$$s_0 = EB = d\frac{\alpha}{\theta} \tag{5.15}$$

- The height of intersection point of radio beams from the sea level shown with h_0 is specified with the following equations:

$$h_0 = h_1 + \frac{d^2}{2KR_e}\left(\frac{\beta}{\theta}\right)^2 + d\left(\frac{\beta}{\theta}\right)\theta_1 \tag{5.16}$$

$$h_0 = h_2 + \frac{d^2}{2KR_e}\left(\frac{\alpha}{\theta}\right)^2 + d\left(\frac{\alpha}{\theta}\right)\theta_2 \tag{5.17}$$

This value also determines the approximate height of the common volume base which varies from hundreds of meters to thousands of meters.

- Δd as the difference between the actual radio path length with that of the straight distance between two sites can be calculated by the following relation:

$$\Delta d = r_0 + s_0 - d \approx \frac{\alpha\beta d}{2} \tag{5.18}$$

- The height h related to the location of radio horizon beam intersection from the interconnecting line of the two sites can be calculated by the following relation:

$$h = \frac{d \cdot s \cdot \theta}{(1+s)^2} \tag{5.19}$$

- $\Delta d_{1,2}$ as the difference between the length of AFB and AEB routes in Fig. 5.3 for $\omega = \omega_1 = \omega_2 = \mathrm{HBW}_1 = \mathrm{HBW}_2$, that is, the difference between the longest and shortest paths along the radio route, can be derived as:

$$\Delta d_{1,2} = \frac{d}{2}\left(\omega^2 + \omega\theta\right) \tag{5.20}$$

The variations of this parameter for actual cases are within the limits of 10–200 m.

5.5.3.4 Geometric Parameters of the Antennas

The half-power beamwidth of an antenna, defined as a 3 dB beamwidth, depends on particular circumstances. The 3 dB beamwidth for a parabolic antenna with a proper approximation in degrees can be calculated from the following equation:

$$\mathrm{HBW} = 70\frac{\lambda}{D} = \frac{21{,}000}{f \cdot D} \tag{5.21}$$

In which λ is the wavelength (in meters), D is the diameter of parabolic reflector (in meters), and f is the working frequency (in megahertz). This beamwidth can change in the practical troposcatter band from a few degrees (e.g., 8° for small antennas on lower frequency range) to a fraction of a degree (e.g., 0.8° for bigger antennas on higher frequencies).

5.5.3.5 Earth's Effective Radius

The Earth's effective radius can be calculated from the following equation:

$$R'_{\mathrm{e}} = KR_{\mathrm{e}} \tag{5.22}$$

in which R_{e} is the actual radius of the Earth (6,370 km) and K is the atmosphere refractivity with a standard value of 1.33 and 1.18 for microwave and light waves, respectively.

5.6 Received Signal Power

For calculation of the received signal power in a troposcatter radio link, the following equation is used:

$$P_r = 1.8 \times 10^{32} \cdot P_t \cdot k_0^{-5/3} \cdot g_t \cdot g_r \cdot C_n^2 \cdot d^{-17/3} \cdot \theta_{1/2}^3 \cdot \cos^2\theta \tag{5.23}$$

The parameters used in the above equation are defined as:

P_r, P_t: RX and TX power in W
k_0: Wave number
g_t, g_r: Gain of TX and RX antennas
C_n: Structure coefficient for refractive index
d: Distance between TX and RX in m
θ: Radiowave deviation angle in mrad
$2\theta_{1/2}$: Antenna half-power beamwidth angle (3 dB) in mrad

Noting that on long hops, the common volume of troposcatter is situated at higher altitude from the Earth's ground level, thus, the value of C_n^2 will be smaller, and there will be more reduction in the received power due to the increase of distance d. In order to determine the effects of altitude from ground level on the structure coefficient, normally the following relation is used:

$$C_n^2 = 4.2 \times 10^{-14} h^{-1/3} e^{-h/h_0} \quad m^{-2/3} \tag{5.24}$$

Where $h_0 = 3,200$ m and h is the height of antenna center line from the Earth's surface. The range of d is equal to $2\sqrt{2KR_e h}$; thus, the above equation will be modified to:

$$C_n^2 = 1.68 \times 10^{-13} d^{-2/3} R_e^{1/3} e^{-d^2/8KR_e h} \tag{5.25}$$

As a result, it is expected that the rate of decrease in the average received power to have an inverse relation with $d^{6.3}$. The following example will reveal an application of typical values of the mentioned parameters:

Example 5.1. Specifications of a troposcatter link are indicated below:

- TX power: 2 kW
- Antennas gain: 10^5
- Antennas diameter: 14 m
- Radio frequency: 3 GHz
- Structure coefficient: $C_n = 10^{-8}$ $m^{-1/3}$
- Distance between stations: 400 km

1. Calculate the received signal level.
2. Determine how much the received signal is attenuated compared to the signal in the free space.

Solution. 1. $f = 3\,\text{GHz} \implies \lambda = 0.1\,\text{m}$

Noting the relation (5.21), the amount of half-power beamwidth is:

$$\theta_{1/2} = \text{HBW} = 0.5^\circ = 8.6 \times 10^{-3}\,\text{rad}$$

Using equation (5.23) for $\theta = \frac{d}{R_e} << 1$ and $\cos\,\theta \approx 1$, hence:

$$P_r \approx 4 \times 10^{-12}\ \text{W}$$

2. The received signal under free-space conditions is determined by using the formula stated in Chap. 1:

$$P_{rf} = \frac{P_t \cdot g_t \cdot g_r}{(4\pi)^2 \cdot d^2} \times \lambda^2 = 7.92 \times 10^{-3}\ \text{W}$$

$$A_{\text{tropo}} = \frac{P_r}{P_{rf}} = \frac{4 \times 10^{-12}}{7.92 \times 10^{-3}} \approx 5 \times 10^{-10}$$

$$A_{\text{tropo}}[\text{dB}] = 10\ \log \frac{P_r}{P_{rf}} = -93\,\text{dB}$$

The last result shows that troposcatter link has an additional attenuation of about 93 dB. ■

Example 5.2. In the above example assuming 1 MHz bandwidth for the signal and an equivalent system noise temperature of 600 K, then calculate the signal-to-noise ratio.

Solution.

$$P_n = kTB$$

$$P_n = 1.38 \times 10^{-23} \times 10^6 \times 600 = 8.28 \times 10^{-15}\ \text{W}$$

$$\text{S/N} = \frac{P_r}{P_n} = \frac{4 \times 10^{-12}}{8.28 \times 10^{-15}} = 4.8 \times 10^2$$

$$\text{S/N[dB]} = 10 \log \frac{P_r}{P_n} \implies \text{S/N} = 26.8\,\text{dB}$$ ■

In the above example, the approximate value of the signal-to-noise ratio is roughly 26.8 dB. To overcome the fading adverse effect which may occur occasionally, it is essential to increase the transmission signal power in order to increase the signal-to-noise ratio to S/N = 40 dB which obviously by decreasing the bandwidth it is also possible to reduce the received noise power level and as a result to increase S/N ratio.

In addition to free-space loss, there is more attenuation for troposcatter radio link which may be derived by dividing the (5.23) by value of $P_t \cdot g_t \cdot g_r \lambda^2/(4\pi)^2 d^2$:

$$L_{ex} = \frac{7.2 \times 10^{32} k_0^{1/3} \theta_{1/2}^3 C_n^z}{d^{11/3}} \tag{5.26}$$

5.7 Climatic Classification of Earth

To obtain more precise values of radiowaves losses in the troposcatter communications corresponding to the prevailing geographical and climatic conditions according to the Fig. 5.7, the Earth is divided into nine regions as described below.

5.7.1 *Equatorial Regions*

These areas are situated within the boundaries in the latitudes of 10° north to 10° south, and their corresponding climate has mild variations, with intensive heats, periodic heavy raining, and permanent humidity. The annual mean value of refractivity index is about 360 N-units, and annual range of variations is 0–30 N-units.

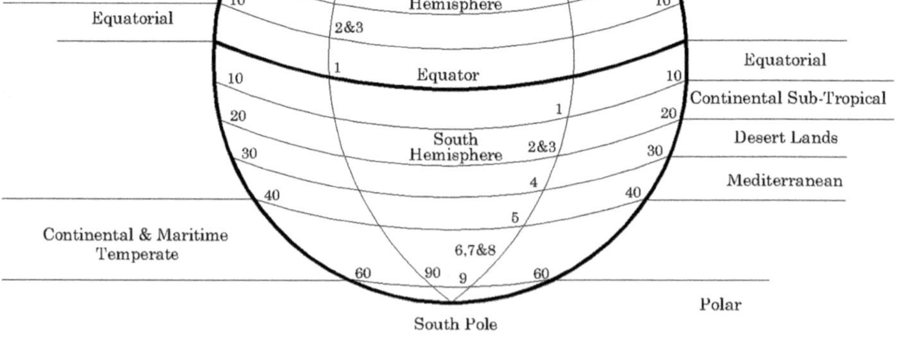

Fig. 5.7 Climatic classification chart

5.7.2 Continental Subtropical Regions

These regions are bounded in the limits of 10–20° latitudes and have a climate characterized by dry winters and rainy summers. In these regions the radiowaves propagation has a significant diurnal and annual variations and possesses minimum loss in the rainy seasons. In this area, radio ducts exist on the dry lands for a long period of time during the year. The mean value of N_S is about 320 N-units and has a variation limited from 60 to 100 N-units during the year.

5.7.3 Maritime Subtropical Regions

These areas similar to the latter one are situated within the limits of 10–20° latitudes, and most of them are located in low-altitude lands and neighboring the seas. These areas are severely affected by seasonal winds which blow from the sea toward the land in the summer, transferring a great amount of humidity from the sea into the land. Despite the existing radiowaves losses at the beginning and ending of the season, these winds are weak, and at the mid-time of the season, the Earth's atmosphere up to high altitudes is uniformly humid, and even with a high N_S value, the amount of radio losses is significant. The mean annual value of the N_S is about 370 N-units, and the magnitude of its variation is from 30 to 60 N-units.

5.7.4 Desert Lands

Desert lands correspond to two land areas which are roughly situated between 20 and 30° latitudes. Throughout the year there are semiarid conditions and extreme diurnal and seasonal temperature variations. These conditions are regarded as unsuitable for troposcatter radiowaves propagation during the summer. The mean value of refractivity index on the land per year, that is, N_S, is roughly 280 N-units, and magnitude of variations on the monthly basis is in the range of 20–80 N-units.

5.7.5 Mediterranean Regions

These areas are situated at the northern/southern parts of desert regions and limited to 30–40° latitudes and neighboring the seas. The climate of this region is of relatively high temperature, which decreases at vicinities of the seas and has summers which are almost without any rain. The radiowaves propagation over the seas includes a very high variation, and radio ducts are formed in most of the time during the summer.

5.7.6 Continental Temperate Regions

These areas are limited to 30–60° latitudes of northern and southern hemisphere. The climate is mild, and the changes of radiowaves propagation conditions are immense. The western parts of these regions are deeply subjected to the effects of oceans in a way that temperature variations are mild and there is a possibility of raining throughout the year and at any time. In the eastern areas of this region, the temperature fluctuation and raining during winter are decreased. The conditions of radiowave propagation are desirable during summer, and their annual variations are relatively high. The mean annual value of N_s is equal to 320 N-units with mean monthly variations of 20–40 N-units.

5.7.7 Maritime Temperate, Overland

These areas are also situated in 30–60° latitudes, and the prevailing winds carry the sea humidity into dry lands. Some of typical parts of this area can be named as the Northern Europe, Northern America, and also northwestern coasts of Africa. The mean annual value of N_s is 320 N-units with relatively low variations of 20–30 N-units during the year.

5.7.8 Maritime Temperate, Oversea

These regions are limited to 30–60° latitudes and include coastal and overseas climates. The radio duct formation for low percentages of time occurs frequently.

5.7.9 Polar Regions

These regions consist of lands bounded beyond 60° latitudes up to both poles in the northern and southern hemispheres. The climate constitutes low temperature and also with relatively small precipitation.

5.8 Calculations of Troposcatter Radiowaves Loss

5.8.1 Introduction

For planning troposcattering links, it requires a proper estimation of how the radiowaves propagate in this type of communications. The main components in this case must be identified and their interrelations to be characterized accordingly. In addition to this requirement, the obtained results or implemented experiences and empirical experiments must have desirable adaptations. In troposcatter links, the received signals consist of two types of variations:

- Slow variations which are related to refractivity index of the Earth's atmosphere and are defined by the hourly median distribution functions. These functions have log-normal nature with a standard deviation of 4–8 dB where their values are functions of environmental and climatic conditions.
- Rapid variations due to the movement of irregularities on a small scale close to Rayleigh distribution and acting for a short period of time which is about 5 min.

Since the radio channel frequency in troposcatter links is limited to 200–5,000 MHz, the losses generated by rainfall and also gases, vapor, fog, and clouds etc., are negligible. Most of the losses in addition to free-space losses include those items which are stated in the following section.

5.8.2 Radiowaves Propagation Loss

According to Fig. 5.8, the radiowaves propagation loss between TX and RX can be calculated using a suitable method as described by the ITU-R recommendation No.P.617. It must be mentioned that the validity of this method is limited to the frequency range of 200–5,000 MHz:

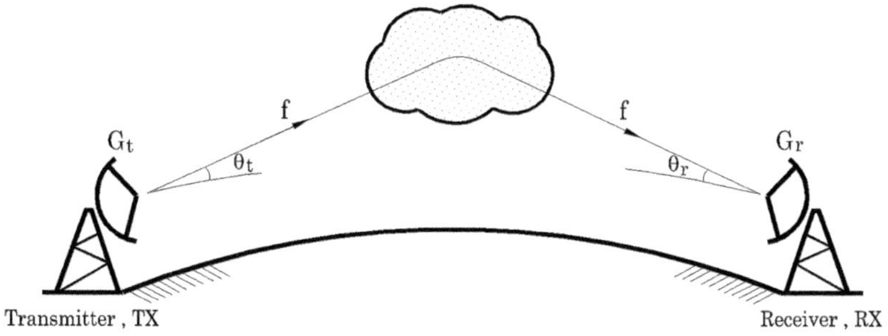

Fig. 5.8 Typical troposcatter link

Table 5.1 Atmospheric structure and meteorological parameters (Ref.: ITU-R, P.617-1)

Area	1	2	3	4	5	6	7	8	9
M (dB)	39.60	29.73	19.30	38.50	38.50	29.73	33.20	26.00	33.20
$\gamma(\text{km}^{-1})$	0.33	0.27	0.32	0.27	0.27	0.27	0.27	0.27	0.27

Step 1: Among the nine typical regions, the most suitable one is selected for the intended link.

Step 2: Using Table 5.1 and in accordance with the selected region in the step 1, the following parameters are determined:

- Parameter M related to meteorological structure
- Parameter γ related to atmospheric structure

Step 3: Calculate the scatter angle θ from the following relation:

$$\theta = \theta_c + \theta_t + \theta_r \quad \text{(mrad)} \tag{5.27}$$

θ_t and θ_r are the elevation angles of the transmitted and received radiowaves in terms of milliradian (mrad), and θ_c in terms of mrad is calculated according to the relation given below:

$$\theta_c = d \times 10^3 / KR_e \tag{5.28}$$

In the above relation, each of the parameters together with their units is indicated below:

d: Distance of radiowaves path in km.
R_e: Actual Earth's radius in km (6,370 km).
K: K-factor related to the Earth's refractivity index which has a value of 1.33 unless measured more accurately.

Step 4: Estimate the transmission loss dependence on the height of troposphere common volume; LN is calculated from the following relation:

$$LN = 20\,\log(5 + \gamma H) + 4.34\,\gamma h \tag{5.29}$$

where

$$H[\text{km}] = 10^{-3} \cdot \theta \cdot d/4 \tag{5.30}$$

$$h[\text{km}] = 10^{-6} \cdot \theta^2 \cdot K \cdot R_e/8 \tag{5.31}$$

In the above relations, θ is the angular distance in terms of mrad, and γ is the Earth's atmospheric structure which is selected from Table 5.1.

Step 5: Calculate the conversion factor $Y(q)$ which can be derived from the following relation for time percentages other than 50%:

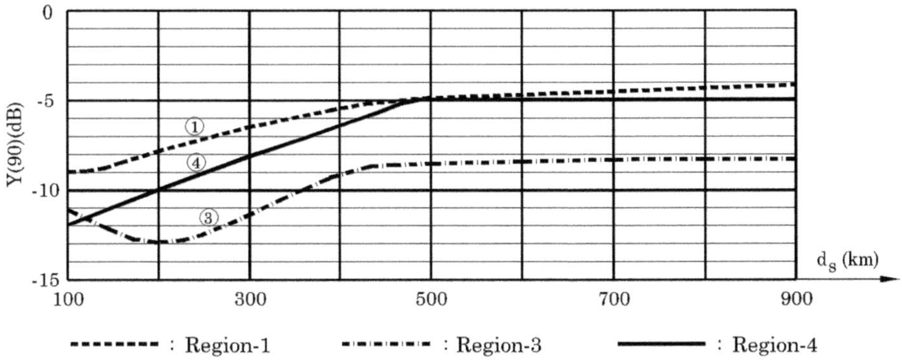

Fig. 5.9 Conversion factor, $Y(90)$ for regions 1, 3, and 4 (Ref.: ITU-R, P.617-1)

Table 5.2 Typical values of C(q) (Ref.: ITU-R, P.617-1)

q	50	90	99	99.9	99.99
$C(q)$	0	1	1.82	2.41	2.90

$$Y(q)[\text{dB}] = C(q) \cdot Y(90) \tag{5.32}$$

In the above relation, $Y(90)$ is the conversion factor for $q = 90\%$ which are determined in terms of dB for the climate of each region according to the following relations:

- For regions 2, 6, and 7 in terms of dB:

$$Y(90) = -2.2 - (8.1 - 2.3 \times 10^{-4} f) \ \exp\ (-0.137\ h) \tag{5.33}$$

- For region 8 in terms of dB:

$$Y(90) = -9.5 - 3\ \exp\ (-0.137\ h) \tag{5.34}$$

Also, for regions 1, 3, and 4, the $Y(90)$ may be found according to Fig. 5.9 in which d_s is approximately calculated from the following relation:

$$d_s[\text{km}] = \theta \cdot K \cdot R_e/1{,}000 \tag{5.35}$$

In the above equation, f is in terms of MHz, and the height h is obtained from (5.31) and is in terms of km. The conversion factor $C(q)$ is obtained for common percentages of time according to Table 5.2. Also for other percentage of time, the value of $C(q)$ may be determined using Table 4.8 given in Chap. 4.

Step 6: Calculate the coupling loss of TX and RX antennas to the medium using the following equation:

$$L_c[\text{dB}] = 0.07\ \exp\ [0.055(G_t + G_r)] \tag{5.36}$$

where G_t and G_r are the both ends antenna gains in dB.

Step 7: Estimate the average annual transmission loss not exceeding $q\%$ of the time by:

$$L(q)[\mathrm{dB}] = M + 30\ \log(\theta \cdot f \cdot \sqrt[3]{d}) + L_N + L_c - G_t - G_r - Y(q) \tag{5.37}$$

Example 5.3. For a troposcatter link between two points located on the northern hemisphere of the Earth latitudes 30–40° with a distance of 400 km and Mediterranean climate using common volume height at 8.25 km from the ground level, find:

1. Transmission loss with antennas $G_t = G_r = 45\ \mathrm{dB}_i$ and time percentages of $q = 99\%$ and $\theta_t = \theta_r = 10\ \mathrm{mrad}$.
2. Received signal level on the RX input. Assume the TX output power is equal to 2 kW and the total miscellaneous losses as 3 dB.

Solution. 1. *Link position* $\Longrightarrow$ Region 5

$$(\text{Region 5}) \xrightarrow{\text{Table 5.1}} M = 38.5\,\mathrm{dB},\ \gamma = 0.27\,\mathrm{km}^{-1}$$

$$K = 1.33 \Longrightarrow \theta_c = 47\,\mathrm{mrad}$$

$$\theta = \theta_c + \theta_t + \theta_r \Longrightarrow \theta = 67\,\mathrm{mrad}$$

Using (5.30), (5.31), and (5.29) yields:

$$H = 6.7\,\mathrm{km},\ h = 4.77\,\mathrm{km},\ L_N = 22.25\,\mathrm{dB}$$

$$d_s = \theta \cdot K \cdot R_e/1{,}000 = 569.5\,\mathrm{km} \xrightarrow{\text{Fig. 5.9}} Y(90) = -5\,\mathrm{dB}$$

$$q = 99\% \Longrightarrow C(q) = 1.82$$

Equations (5.36) and (5.37) result in:

$$L_c = 0.07\ \exp[0.055 \times 90] = 9.88\,\mathrm{dB}$$

$$L(99\%) = 38.5 + 30\ \log(67 \times 3{,}000 \times \sqrt[3]{400})$$

$$+ 22.25 + 9.88 - 45 - 45 + 9.1$$

$$L(99\%) = 174.8\,\mathrm{dB}$$

2.

$$P_t = 2\,\mathrm{kW} \Longrightarrow P_t = 10\ \log(2 \times 10^6) = 63\,\mathrm{dB}_m$$

$$P_r = P_t - L(99\%) - L_m = 63 - 174.8 - 3 = -114.8\,\mathrm{dB}_m$$

■

5.8.3 Worst Month Loss

In the previous section, the average transmission loss was calculated and related to the mean value of transmission for time percentages in normal months. In order to calculate the average worst month transmission loss, the following procedure must be followed:

Step 1: Calculate the average annual distribution for the required non-exceeding of time percentage (say 90, 99, 99.9, etc.) in the region of interest using step-by-step procedure specified in Sect. 5.8.2.

Step 2: Determine the basic transmission loss difference between the average annual and worst month distributions from Fig. 5.10 for the required percentages of time and region of interest.

Step 3: Add the extra value found in step 2 to the corresponding average annual value obtained in step 1. The result is the average worst-month transmission loss for the required non-exceeding time percentage.

Example 5.4. 1. Find the average basic transmission loss for the worst month considering a troposcatter link with the assumptions specified in Example 5.3.
2. How much the calculated loss is in excess of free-space loss?

Solution. 1. Extra loss for the worst month using proper graph of Fig. 5.10 for d = 400 km and q = 99% is:

$$d_{\mathrm{WM}} = 4.35\,\mathrm{dB}$$

$$L(99\%)_{\mathrm{WM}} = L(99\%) + \Delta_{\mathrm{WM}} = 174.8 + 4.35 = 179.15\,\mathrm{dB}$$

2.

$$\mathrm{FSL} = 92.4 + 20\,\log\, f \cdot d \quad \Longrightarrow \quad \mathrm{FSL} = 154\,\mathrm{dB}$$

These calculations reveal that troposcattering propagation loss in the worst month may exceed free-space loss by:

$$L_e = L(99\%) + G_{\mathrm{t}} + G_{\mathrm{r}} - L_C - \mathrm{FSL} + \Delta_{\mathrm{WM}} = 104.27\,\mathrm{dB}$$

■

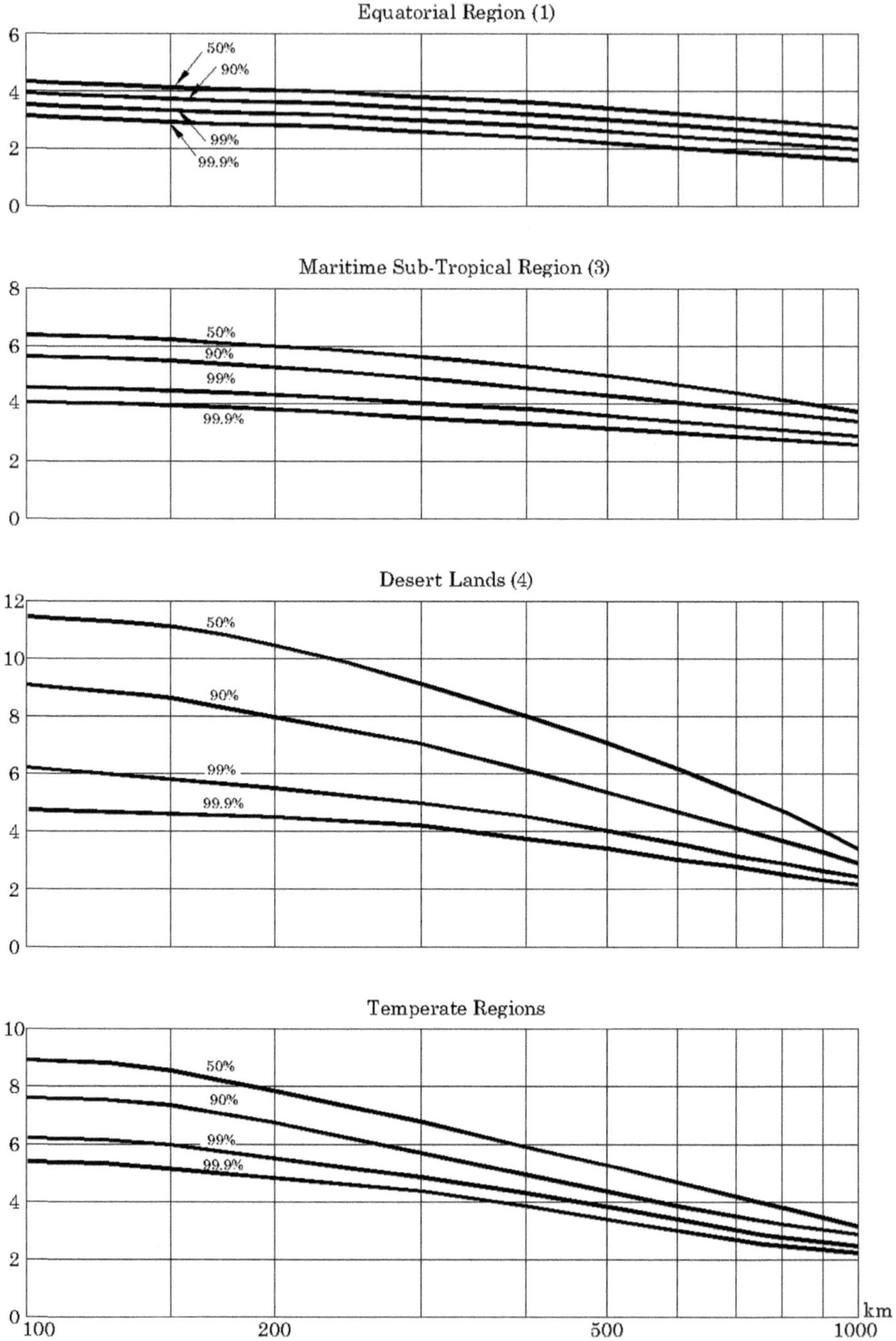

Fig. 5.10 Extra basic transmission loss for worst month(Ref.: ITU-R, P.617-1)

5.9 Exercises

Questions

1. Name the over-horizon communication methods.
2. Specify the most suitable frequency range for troposcattering propagation.
3. Apart from the radiowaves troposcattering propagation, what kinds of other mechanisms may cause the over-horizon propagation through troposphere layer?
4. Determine the major reasons for the distortion of received signal in a troposcatter link, and state the most suitable polarization for this kind of radio communications.
5. What are the main characteristics of troposcatter systems to be considered in the planning phase, and specify characteristics that mainly affect the system performance.
6. Name the major reasons for limitation of using troposcatter links.
7. One of the most important applications of troposcatter system is in the military field; name some of the reasons for using this method for military purposes.
8. Noting that satellite communications are better choice in place of over-horizon communications, why at some instances is troposcatter system preferred to the satellite links?
9. One of the features of troposcatter communications is the use of high-power TX(s) and high-gain antennas. What are the advantages and disadvantages of this feature for radio propagation compared to other systems?
10. Summarize advantages and disadvantages of troposcatter communications.
11. Noting Fig. 5.6 of this book, determine the length and angular parameters in tropospheric communications.
12. What are the angular distance and its pertinent relations?
13. For the purpose of troposcatter communications, define the geographical division of the world, given the Fig. 5.7.
14. How do gases, water vapor, rain, snow, and cloud affect troposcatter radiowaves propagation?
15. Express the difference between the annual median and the worst month losses, and determine its limits.

Problems

1. If the diameter of a parabolic antenna is 10 m and its working frequency is 1.8 GHz, then:

   (a) Specify the 3 dB beamwidth.
   (b) Calculate the antenna gain assuming the antenna efficiency of 0.6.

2. Assume a troposcatter link with the following specifications:

   - Transmission power: 10 kW
   - Antenna gain: 40 dB
   - TX/RX distance: 300 km
   - Wavelength: 10 cm

   (a) Assume the center of common volume at the height of 6 km from the Earth's ground level, and then calculate the values of C_n and the received signal power.
   (b) If this distance is doubled, calculate the reduction in the received signal level in terms of dB.
   (c) Assuming the bandwidth BW = 2 MHz and equivalent noise temperature of $T = 600$ K, calculate the received noise power and signal-to-noise ratio for both cases. Is this signal-to-noise ratio acceptable? If not, how you can increase it?

3. For a link similar to Example 5.3 located in continental temperate region at northern latitudes of $40 \sim 45°$, calculate:

   (a) The median value of the path loss for a year.
   (b) The median value of the path loss for worst month.
   (c) The received power at RX location for both of the above-mentioned positions.

4. To establish a link from an onshore location situated in a temperate region with an offshore oil platform at a distance of 300 km away, a troposcatter station is used at 2.5 GHz using antennas with 43 dB$_i$ gain. In case of $\theta_t = 14$ mrad and $\theta_r = 10$ mrad, find:

   (a) M and γ parameters.
   (b) The median value of annual transmission loss.
   (c) Free-space loss.
   (d) How much is added to the above values on worst month period.
   (e) In case of the nominal TX power equal to 1 kW, calculate the received power for normal and also for the worst month conditions.

5. Prepare a computer program for calculations of troposcatter links to obtain the minimum path loss, the length and angular parameters of the link, and the received power and electric field strength levels at the RX location. For this program:

   (a) What are the input data?
   (b) Using the prepared program, proceed with Examples 5.3 and 5.4, and compare the results with the calculated values in the mentioned examples.

Chapter 6
Propagation of Radar Waves

6.1 Introduction

6.1.1 Definition of Radar

The word RADAR stands for *radio detection and ranging* and a radar network constitutes all transmitter and receiver units (usually at the same location), related antenna(s) along with pertinent devices, accessories, and software. As illustrated in Fig. 6.1, the radar scenario involves with a target at range R and a radar wave in electromagnetic sinusoidal form that travels the round-trip between the radar and related target.

The radar signals are generated by the TX unit, amplified, and at time t_1 are transmitted through radar antenna and proceed with receiving and detecting the reflected waves from the target body at the same point (or corresponding instruments at another point) at time t_2. The reflected electromagnetic energy returned to the radar receiver indicates the presence of a target, and by comparing and processing echo signals, its location can be determined along with additional information. Radar networks may perform their functions under adverse conditions such as darkness, rain, snow, hail, haze, and fog situations where optical or infrared sensors fail to operate. Its ability to measure distance with high accuracy in all weather conditions has found many applications in military, navigation, and traffic control fields.

The radiowaves encounter various phenomena traveling from TX antenna to RX antenna. The main objective of this chapter is to focus on this subject. Obviously, the general and basic principles such as EM waves and their pertinent properties and troposphere and ionosphere radiowave propagation phenomena discussed in the previous chapters are valid for radar waves as well. Analysis of the radar echoes requires deep knowledge of target reflectivity, natural clutter, clutter artifacts, and their susceptibility to the radio interference and noise. Effective use of the radar networks for remote sensing and space research activities requires further

A. Ghasemi et al., *Propagation Engineering in Radio Links Design*,
DOI 10.1007/978-1-4614-5314-7_6,

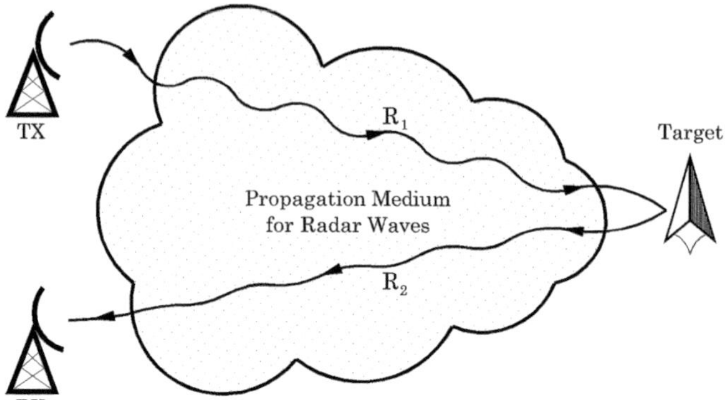

Fig. 6.1 Typical radar radio link

understanding of the propagation media. For this purpose, it is required to study unwanted reflections from sea and land areas, precipitation and chaff, thermal noise, and jamming as well.

6.1.2 Brief History

Maxwell equations were presented for electromagnetic waves in the *nineteenth* century, and it followed with the generation and propagation of radiowaves by Hertz. Then, many experts and scientists were involved with the application of radiowaves for detection and tracking of fixed and moving target.

The preliminary radars in the early twentieth century (1904) were produced by Christian Huelsmeyer in Germany for transmission of radiowaves and detecting their reflections after colliding with ships and trains. The radars employed prior to the Second World War were implemented in VHF frequency band.

In the late 1930s and early 1940s, modern radars were designed and manufactured in microwave band, and with the emergence of oscillators and magnetron amplifiers, at present time, most of radars still operate in this particular band. In recent years for the purpose of detection and locating the position of targets, the over-horizon radars in HF band have been developed by the experts in this field.

As mentioned, radars were initially employed for military purposes and applications. In the later stages, in addition to the advancement and evolution of this system for military use, it was utilized for public and peaceful applications such as transportation, meteorology, and recognition of the composition of the atmosphere layers.

6.1.3 *Applications*

The major applications of radars include, but not limited to, the following fields:

- Military applications including control networks, reconnaissance, and tactical
- Operational applications such as civil aviation radars to control airports and radars installed on ships, oil tankers, etc.
- Traffic control applications including aviation/maritime/terrestrial
- Reconnaissance applications for detection and separation of targets
- Special services such as meteorology radars and ground-penetrating radars

In the 1980s, the radar technology reached mature state, and its applications increased. Currently, radar systems have acquired many applications in different fields to meet operational and technical requirements.

6.1.4 *Categorization of Radars*

Radars can be categorized in different ways based on their main aspects. A few examples of their types are indicated below:

- Application such as military, traffic control, and meteorological
- Frequency bands such as S-band or X-band radars
- Type of station
- Operational requirements

6.1.5 *Fixed and Mobile Radars*

When the local position of a radar station is fixed relative to the ground and in other words its topology does not change, this radar is a fixed type which is generally used in sites for covering a particular region. Radars intended to cover airport areas, control of harbors, meteorological or military radar sites are some typical examples of fixed radars. For fixed or stationary radars, to specify its site location and preparing coverage plots is a prime concern which are explained in Sect. 6.6.5. Generally this type of radar is equipped with a very powerful TX, large and high-gain antenna, auxiliary equipments to improve the safety margin, and also back-up electric power system.

Mobile radar stations are used in great numbers compared to the fixed type. The local position of the radar is altered relative to the ground, and its topology is not fixed. This type of radar is generally utilized in planes, ships and military weaponry. These radars possess simpler antennas with a less powerful TX and consequently their covering range is limited.

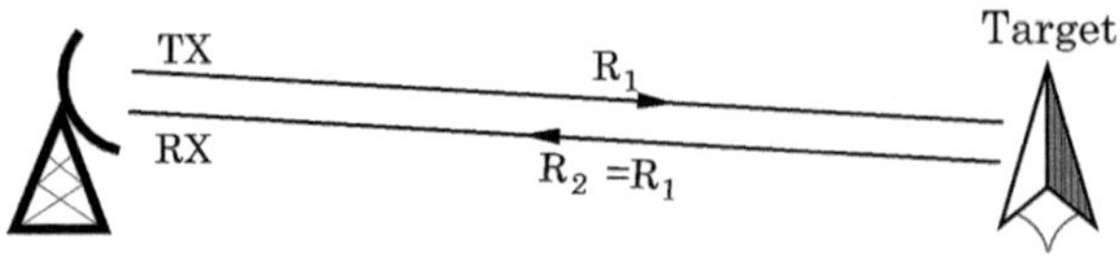

Fig. 6.2 Monostatic radar network

6.1.6 Radar Network Architecture

The major configurations of radar networks include three different combinations as described below:

- Monostatic radar
- Bistatic radar
- Multistatic radar

6.1.6.1 Monostatic Architecture

As shown in Fig. 6.2, the radar equipment in this configuration are positioned in a single location and employ a single antenna for both transmission and reception of the radar waves. If the distance between TX and RX units from the target is denoted by R_1 and R_2, respectively, and the distance between them is denoted by d, then in such a structure,

$$R_1 = R_2, \quad d = 0 \tag{6.1}$$

6.1.6.2 Bistatic Architecture

As indicated in Fig. 6.3, the transmission and reception equipments in this structure are situated in different locations, and they utilize different antennas dedicated to TX and RX units.

For bistatic architecture using the same notations, we have

$$R_1 \neq R_2, \quad d \neq 0 \tag{6.2}$$

6.1.6.3 Multistatic Architecture

As illustrated in Fig. 6.4, in a multistatic architecture, there are several TX units employed for transmission of the radar waves toward the target,and several RX units

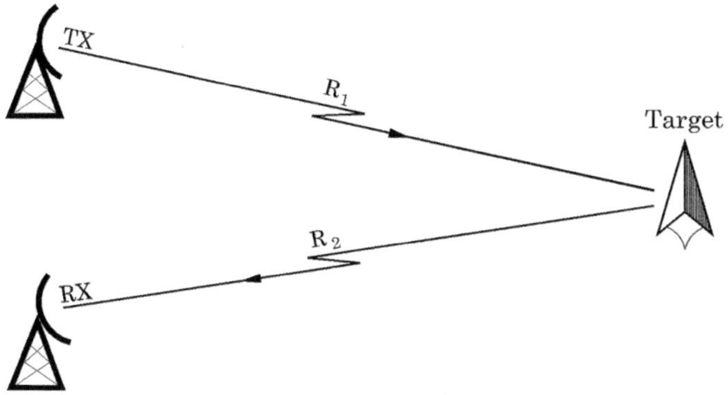

Fig. 6.3 Bistatic radar network

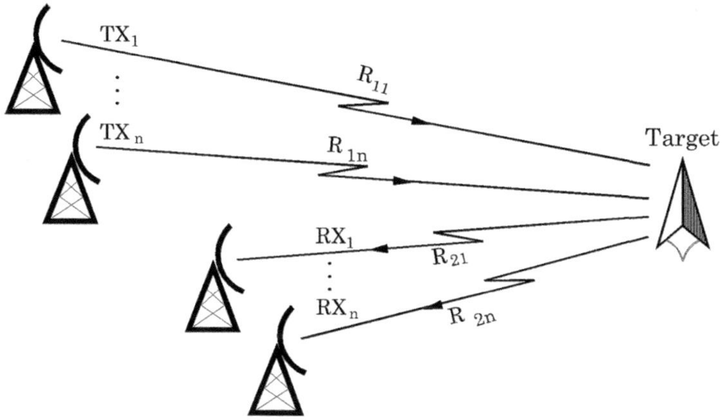

Fig. 6.4 Multistatic radar network

are used for reception of the reflected waves from the target. Generally in such a position, the following relation holds true:

$$R_{1i} \neq R_{2i}, \quad d_i \neq 0 \tag{6.3}$$

6.1.7 Basic Measurements by Radar

Among a variety of basic measurements conducted by radar networks, the following are more significant and popular:

- Distance, position, and the velocity of the target
- Identification, separation, and sizing of the target

For the above-mentioned measurements, radar(s) sends powerful radiowaves in the form of continuous wave, CW, or pulsed waves and receives the subsequent reflected waves from the target and proceed with processing these data for indicating the required parameters.

The changes observed in the returned signal provide some information regarding the target position and its properties. Simply, the delay time of the received signal includes data on the range. The frequency yields data on the target velocity, while the antenna pointing direction with maximum received signal level contains azimuth and elevation angles of the target relative to the radar. Also, by further processing of the returned signal by suitable software, more information regarding the target such as its composition and trajectory of movement may be provided.

6.2 Main Aspects

This section will present some of the main parameters related to frequency bands, average and maximum powers, effective transmitting power, radar equation, radar cross section, and signal-to-noise ratio.

6.2.1 *Frequency Bands*

Radio frequency spectrum was introduced in the Chap. 1 of this book. However, for dedicated radar bands, a distinct classification was presented during the Second World War which still is used. Table 6.1 indicates the range of each radar frequency bands and their pertaining symbols.

Table 6.1 Radar frequency classification

No.	Band	Frequency limit	Unit	No.	Band	Frequency limit	Unit
1	HF	$3 \sim 30$	MHz	7	X	$8 \sim 12$	GHz
2	VHF	$30 \sim 300$	MHz	8	Ku	$12 \sim 18$	GHz
3	UHF(P)	$300 \sim 1{,}000$	MHz	9	K	$18 \sim 27$	GHz
4	L	$1 \sim 2$	GHz	10	Ka	$27 \sim 40$	GHz
5	S	$2 \sim 4$	GHz	11	V	$40 \sim 75$	GHz
6	C	$4 \sim 8$	GHz	12	W	$75 \sim 120$	GHz

6.2.2 Radar Transmission Power

6.2.2.1 Maximum Transmitter Power

When the TX output signal voltage in the following form is applied into a matched load such as radar antenna (R_a), then maximum power will be transferred to the antenna.

$$S(t) = a(t) \cdot \cos(\omega_0 t + \varphi) \tag{6.4}$$

Since the voltage amplitude is variable with time, thus the instantaneous value of power is given by the following equation:

$$(P_t)_i = \frac{S^2(t)}{4R_a} = \frac{a^2(t)}{8R_a}[1 + \cos(2\omega_0 t + 2\varphi)] \tag{6.5}$$

It is noted that the mentioned maximum power is itself a sinusoidal function of time and its average value designated as P_t is equal to:

$$P_t = \frac{a^2}{8R_a} \tag{6.6}$$

6.2.2.2 Average Transmitter Power

For radars operating in continuous wave transmission, (CW mode), clearly the following relation exists between the average power P_{av} and the maximum power P_t:

$$P_{av} = P_t \tag{6.7}$$

For radars transmitting pulsed signals in periodic form with the time interval between two consecutive pulses denoted by T_R and pulse duration of τ, the following relation exists:

$$P_{av} = \frac{1}{4R_a \cdot T_R} \int_0^{T_R} S^2(t) \cdot dt \tag{6.8}$$

$$P_{av} = \frac{\tau}{T_R} \cdot P_t \tag{6.9}$$

6.2.2.3 Effective Transmission Power

To increase the effective transmission power of radar systems, high-gain antennas with parabolic reflectors are used to concentrate the transmission power in the direction of antenna main axis. If the antenna gain is denoted as g_t, then the effective transmission power will be equal to

$$(P_t)_e = P_t \cdot g_t \tag{6.10}$$

In the above relation, P_t and $(P_t)_e$ have identical unit. Normally in radio calculations, the effective transmission power is denoted by EIRP, and antenna gain is evaluated with respect to an isotropic antenna, then

$$\text{EIRP}(W) = P_t(W) \cdot g_t \tag{6.11}$$

$$\text{EIRP}(\text{dB}_W) = P_t(\text{dB}_W) + G_t(\text{dB}_i) \tag{6.12}$$

Example 6.1. A radar system in the C-band is connected to an antenna with a gain of $G_t = 47\ \text{dB}_i$ and input impedance of 50 ohms. The output signal of TX in terms of KV is specified by $S(t) = 10\cos(\omega_0 t + \pi/4)$, find:

1. The maximum and average output power of the radar if $\tau = 80\,\mu\text{s}$ and $T_R = 1$ ms
2. The effective radar transmission power

Solution. 1.

$$P_t = \frac{a^2}{8R_a} = \frac{(10^4)^2}{8 \times 50} = 2.5 \times 10^5\,\text{W} = 250\,\text{KW}$$

$$P_{av} = \frac{\tau}{T_R} \times P_t \implies P_{av} = 20\,\text{KW}$$

2.

$$G_t = 47\ \text{dB}_i \implies g_t = \text{Antilog}\left(\frac{G_t}{10}\right) = 50{,}000$$

$$\text{EIRP} = P_t \cdot g_t = 250 \times 50{,}000 = 12.5\,\text{GW}$$

The calculation may be performed in the logarithmic form as follow:

$$\begin{aligned}\text{EIRP}[\text{dB}_W] &= P_t(\text{dB}_W) + G_t(\text{dB}_i)\\ &= 10\log(250 \times 10^3) + 47 = 54 + 47 = 101\ \text{dB}_W\end{aligned}$$

■

6.2.3 Radar Cross Section

Basically the ability of radar targets to reflect the radiowaves toward radar RX antenna is evaluated by a parameter called radar cross section or RCS. This component is designated as σ; its unit is square meter and defined below:

$$\sigma \equiv \lim(4\pi R^2 \cdot \frac{|E_s|^2}{|E_i|^2}) \tag{6.13}$$

In the above relation, each one of the components is described below:

R : Distance of target to the radar in meters
E_s : Intensity of scattered electric field in V/m
E_i : Intensity of incident electric field in V/m

The average power densities of incident and scattered radar waves are denoted by S_i and S_s respectively and expressed by the following formulas:

$$S_i = \frac{1}{2}\,\varepsilon_0 C|E_i|^2 \quad (\mathrm{W/m^2}) \tag{6.14}$$

$$S_s = \frac{1}{2}\,\varepsilon_0 C|E_s|^2 \quad (\mathrm{W/m^2}) \tag{6.15}$$

Besides the incident power intensity in terms of watt per steradian denoted by Φ_s is equal to:

$$\Phi_s = S_s \times R^2 = \frac{1}{2}\,\varepsilon_0 C|E_s|^2 \cdot R^2 \tag{6.16}$$

Combining the above relations gives the following result:

$$\sigma = 4\pi \frac{\Phi_s}{S_i} \tag{6.17}$$

In the radar equation and analysis of target data, the radar cross section is a major parameter. There is some correlation between RCS and physical dimension of the target, but other factors such as shape, aspect angle, and wavelength also affect this value.

To calculate the radar cross section or RCS of different objects reference is made to the books about electromagnetic fields and waves. Using such sources, the RCS of some of the simple structures is provided in Table 6.2.

Basically, the radar cross section of composite structures such as aircrafts and ships is very diverse and is a function of horizontal and vertical angles, frequency, and polarization of the wave. A simple and straightforward expression for a composite structure is

$$\sigma = 4\pi \cdot \frac{A^2}{\lambda^2} \tag{6.18}$$

where A is orthogonal projection of the structure on a surface perpendicular to the radiation direction of radar waves. As shown in Fig. 6.5, if the line perpendicular to the conducting plate and the direction of the main wave having a linear polarization

Table 6.2 RCS of basic simple structures

No.	Shape	RCS (radar cross section)	Remarks
1	Sphere with radius of a	πa^2	$a >> \lambda$
2	Cylinder with a length of L and base radius of a	$\frac{a\lambda}{2\pi} \times \frac{\cos\theta \cdot \sin^2(KL\sin\theta)}{\sin^2\theta}$	θ is deviation angle from the main axis of radiation
3	Cylinder with a length of L and base radius of a where $\theta = 0$	$(2\pi a L^2)/(\lambda), \theta = 0$	θ is deviation angle of the cylinder axis from the direction of radiation
4	Conductive plate with a surface A	$(4\pi A^2)/(\lambda^2)$	Direction of radiation is perpendicular to the plate surface
5	Dihedral	$(8\pi a^2 \cdot b^2)/(\lambda^2)$	a and b width and height of the dihedral
6	Triangular trihedral	$(4\pi a^4)/(3\lambda^2)$	a small crest of the trihedral
7	Square trihedral	$(12\pi a^4)/(\lambda^2)$	a small crest of the trihedral

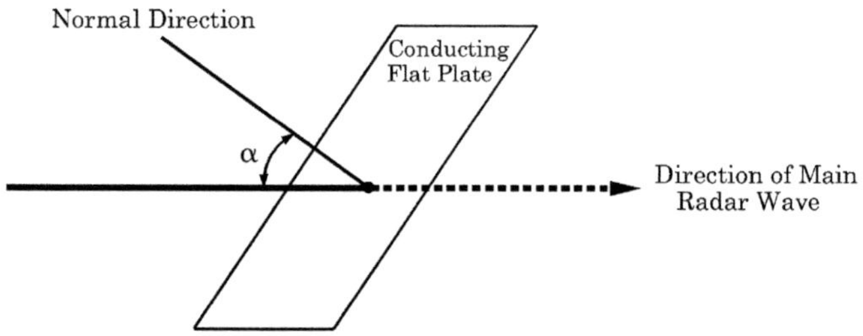

Fig. 6.5 Radar cross section of oblique plate

makes an angle α, then the radar cross section is calculated by the following relation:

$$\sigma = 4\pi \cdot \frac{A^2}{\lambda^2} . \cos^2\alpha \tag{6.19}$$

Example 6.2. Normally passive reflector in radio relay networks is made of metallic conductive flat plates with specific geometrical shapes such as rectangular, circular, elliptical, or rhombic. Assuming that the maximum gain of these reflectors in the direction of the main axis is determined by $G_p[\text{dB}] = 20 \log 4\pi A/\lambda^2$, then find:

1. The relation of reflector gain with its radar cross section
2. If the reflector is positioned in a way that makes an angle α with the direction of incident wave, what is its radar cross section and gain and how much the gain will decrease if $\alpha = 60°$.

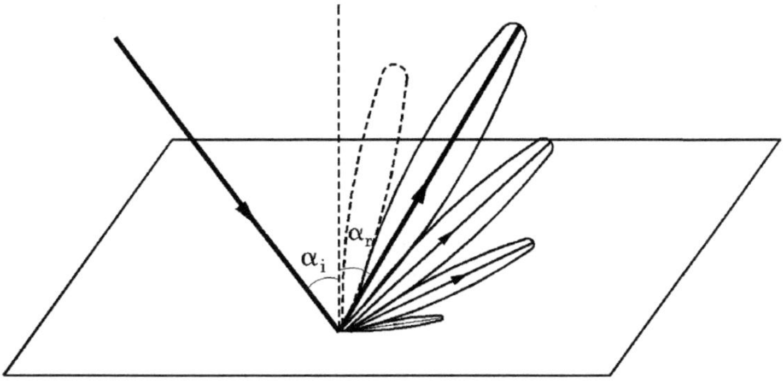

Fig. 6.6 Scattering of radiowaves

Solution. 1. Since (6.18) is related to the radar cross section of conductive flat plate with an area of A in the nonlogarithmic system, thus it is converted to the reflector gain by

$$g_P = \text{Antilog}(G_P/10) = \left(\frac{4\pi A}{\lambda^2}\right)^2 = \frac{4\pi}{\lambda^2} \cdot \frac{4\pi A^2}{\lambda^2}$$

$$\Longrightarrow \; g_P = \frac{4\pi}{\lambda^2} \times \sigma$$

2.

$$G'_P[\text{dB}] = 20 \log \frac{4\pi \cdot A \cdot \cos \alpha}{\lambda^2} = G_P[\text{dB}_i] + 20 \log(\cos \alpha)$$

$$\alpha = 60^\circ \; \Longrightarrow \; G'_P = G_P - 6 \; \Longrightarrow \; \Delta G = 6\,\text{dB}$$

Thus, with 60° incident angle, the reflector gain is decreased by 6 dB. ■

6.2.4 Scattering of Radar Waves

When a radar beam encounters a conductive flat surface according to Fig. 6.6, in addition to the reflected wave based on the Snell's rule ($\hat{i} = \hat{r}$), there will be some reflected waves with an angle different from the incident angle ($\hat{i} \neq \hat{r}$). There are certain components of the wave which propagate in different directions based on phenomenon called wave scattering.

The scattered waves that are received by the radar RX as unwanted signals play a negative role in the process of detecting the desired targets. The amount of scattered wave intensity at each direction depends on the radiation pattern of the plate.

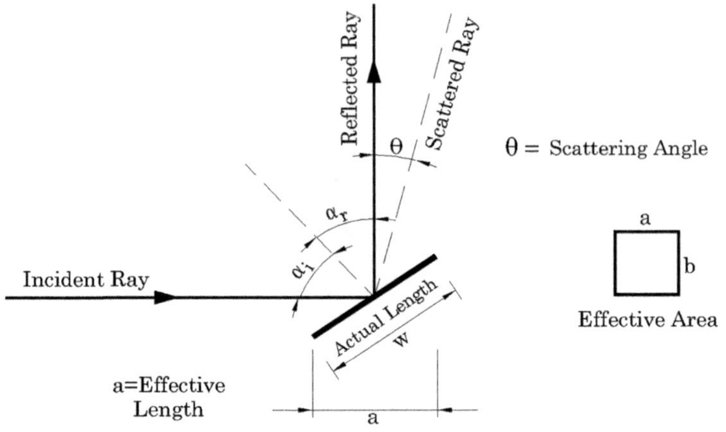

Fig. 6.7 Scattering of incident radiowaves from flat rectangular plate

In order to derive the relationships between the radiation pattern and shape of the plate, reference is made to the advanced electromagnetic engineering books. Due to the significant effects of scattering mechanism in the radar networks, we will examine the following simple cases including their illustrations and formulas.

6.2.4.1 Rectangular Flat Plate

The architecture of a rectangular plate is shown in Fig. 6.7, including reflected beams and also scattered waves in the direction of the angle θ.

Given the context and quantities indicated in the figure, it is possible to derive the following relation for the field intensity of scattered wave in the direction of θ:

$$\bar{E} = \mathrm{j}\frac{\mathrm{e}^{-\mathrm{j}\frac{2\pi r}{\lambda}}}{\lambda \cdot r} \cdot \hat{\alpha_\theta} \cdot a \cdot b \cdot E_0 \frac{\sin(\frac{\pi a}{\lambda}\sin\theta)}{\frac{\pi a}{\lambda}\sin\theta} \quad (6.20)$$

$$a = W \cdot \cos \alpha_\mathrm{i} \quad (6.21)$$

$$a \cdot b = W \cdot b \cdot \cos \alpha_\mathrm{i} = A \cos \alpha_\mathrm{i} \quad (6.22)$$

Assuming the relations given below, the normalized value of the radar wave's power with respect to the same power in the Snell's direction is stated below:

$$u = \frac{\pi a}{\lambda}\sin\theta \ , \ \lambda = \frac{c}{f} \quad (6.23)$$

$$P(u) = \left(\frac{\sin u}{u}\right)^2 \quad (6.24)$$

$$E/P = 20 \log u \quad (6.25)$$

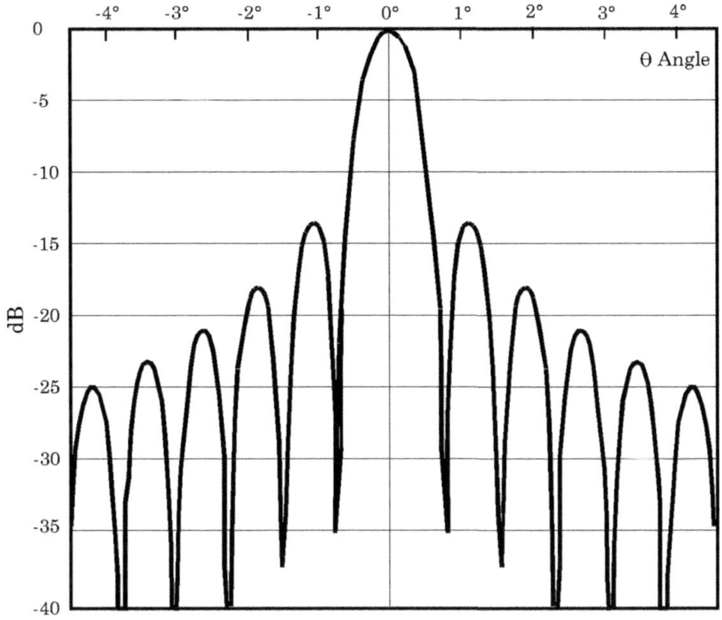

Fig. 6.8 Radiation pattern of flat rectangular conducting plate

In the above relations, W is the real length of rectangular plate, a is the effective length, b is the width of the plate, α_i and α_r are incident and reflected angles, λ is the wavelength, E is the vector of electric field intensity (E_0 is its value for incident wave), $P(u)$ is the normalized radiation power, and E/P is the envelope of peak point of the radiation side lobes.

The radiation pattern of conductive rectangular plate on the main propagation plane ($\varphi = 0$) and for $\alpha_i = 0$ is plotted according to Fig. 6.8. It must be noted here $\alpha_i = 0$ implies the main axis of incident wave and the direction of reflected wave are the same.

6.2.4.2 Elliptical Plate

The structure and geometry of a conductive elliptical plate and incident and reflected beams including the scattered beam in the direction of angle θ is illustrated in Fig. 6.9. Noting the values indicated in this figure, it is possible to state the electric field intensity due to the scattering of the wave in the direction of θ as:

$$\bar{E} = \frac{\pi D^2 E_0}{2} \cdot \hat{\alpha}_\theta \cdot \frac{J_1(\frac{\pi D}{\lambda} \sin\theta)}{\frac{\pi D}{\lambda} \sin\theta} \tag{6.26}$$

$$\frac{\pi D^2}{4} = A_e = \frac{\pi \cdot a \cdot b}{4} \cos\alpha_i \tag{6.27}$$

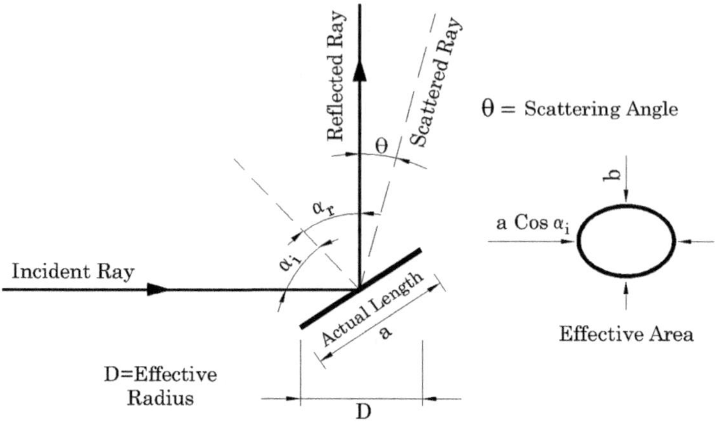

Fig. 6.9 Scattering of incident radiowaves from flat elliptical conducting plate

With the following assumption, the normalized value of strength of the radar wave's power with respect to the same power in the Snell's reflection direction is given by

$$u = \frac{\pi D}{4} \sin\,\theta\ ,\quad \lambda = \frac{C}{f} \tag{6.28}$$

$$P(u) = [\Lambda_i(u)]^2 \tag{6.29}$$

$$\Lambda_i(u) = 2/u \cdot J_1(u) \tag{6.30}$$

$$J_1(u) = \frac{u}{2! \times 0! \times 1!} - \frac{u^3}{2^3 \times 1! \times 2!} + \frac{u^5}{2^5 \times 2! \times 3!} \tag{6.31}$$

In the above relations, a and b are the lengths of major and minor axis of the ellipse, D is the equivalent diameter of the circle, α_r and α_i are the incident and reflected angles, λ is the wavelength, E is the electric field intensity in the direction of θ (E_0 is its value for incident wave) and $P(u)$ is the normalized radiating power. The radiation pattern of elliptical conductive plate is plotted according to Fig. 6.10. It must be noted that $\alpha_i = 0$ implies the main axis of the incident wave is exactly in the direction of reflected wave. (For more details of these relations, please refer to the books and technical manuals related to radio passive reflectors.)

6.2.5 Basic Radar Equation

For processing and obtaining the essential conclusions, the radar equipment must be capable of proper reception of the returned signals from the target. Along with the

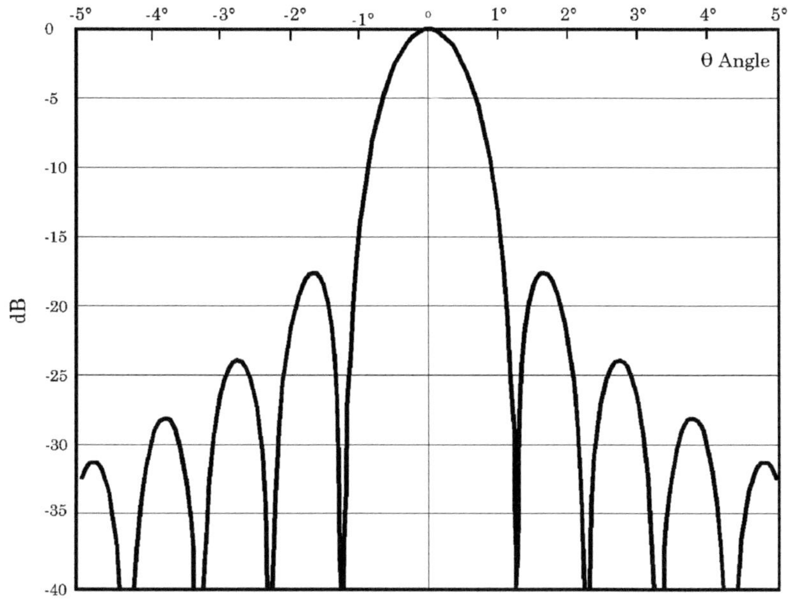

Fig. 6.10 Radiation pattern of flat elliptical conducting plate

main signals, there are noise and other impairments. Thus, higher received signal levels will result in better functioning of the radar.

The target's range, bearing, and velocity can be measured accurately subject to a good signal-to-noise ratio (SNR), while these parameters may be obtained within a certain range at somewhat lower SNR. To evaluate and analyze the received signals, the radar equation is a main tool and source of information on the expected returned signal and SNR.

It is required for this purpose to determine the relation between the received signal and the affecting components. Noting Fig. 6.11, the radar equation can be generally expressed on the basis of different position of TX and RX antennas relative to the target and simply expand it for monostatic structure.

Assuming that the first radar transmits a pulse with maximum carrier power P_t through an antenna of gain g_t, then the power flux density at the target distance R_1 will be

$$S'_r = \frac{P_t \cdot g_t}{4\pi R_1^2 . L_t \cdot L_{p1}} \quad (\mathrm{W/m^2}) \tag{6.32}$$

Now assume that the target reflects back all the power intercepted by its RCS, then power reflected isotropically is given by

$$S'_t = S'_r \cdot A_e = S'_r \cdot \sigma \ (W) \tag{6.33}$$

$$S'_t = \frac{P_t \cdot g_t}{4\pi R_1^2 \cdot L_t \cdot L_{p1}} \times \sigma \ (W) \tag{6.34}$$

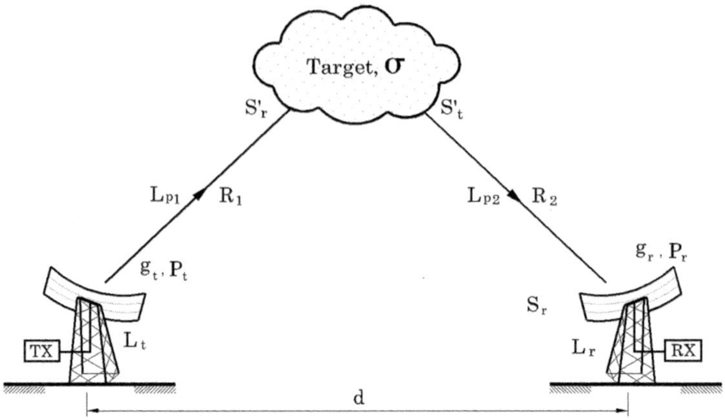

Fig. 6.11 Typical radio link for radar equation

The reflected radar waves will be received by the second receiving antenna with gain g_{r}, thus the received signal power is

$$S_{\mathrm{r}} = \frac{P_{\mathrm{t}} \cdot g_{\mathrm{t}} \cdot \sigma}{4\pi R_1^2 \cdot L_{\mathrm{t}} \cdot L_{p1}} \times \frac{1}{4\pi R_2^2 \cdot L_{p2}} \quad (\mathrm{W/m^2}) \tag{6.35}$$

$$P_{\mathrm{r}} = S_{\mathrm{r}} \cdot \frac{g_{\mathrm{r}} \cdot \lambda^2}{4\pi \cdot L_{\mathrm{r}}} \quad (W) \tag{6.36}$$

$$P_{\mathrm{r}} = \frac{P_{\mathrm{t}} \cdot g_{\mathrm{t}} \cdot g_{\mathrm{r}} \cdot \lambda^2 \cdot \sigma}{(4\pi)^3 R_1^2 \cdot R_2^2 \cdot L_{\mathrm{t}} \cdot L_{p2} \cdot L_{p2} \cdot L_{\mathrm{r}}} \tag{6.37}$$

The latter relation which in fact provides the value of received power level in terms of other components is known as the basic radar equation. The following conditions exist in monostatic case:

$$R_1 = R_2 = R \ , \ g_{\mathrm{t}} = g_{\mathrm{r}} = g \ , \ d = 0$$
$$L_{\mathrm{t}} = L_{\mathrm{r}} = L \ , \ L_{p1} = L_{p2} = L_p \tag{6.38}$$

Thus, the radar equation (6.37) can be simplified to the following:

$$P_{\mathrm{r}} = \frac{P_{\mathrm{t}} \cdot (g)^2 \cdot \lambda^2 \cdot \sigma}{(4\pi)^3 \cdot R^4 \cdot L^2 \cdot L_p^2} \tag{6.39}$$

In the above relations, the components and their units are

P_{t} : TX power in terms of watts
S'_r : Power density in the target location in terms of watt per square meter
$g_{\mathrm{r}}, g_{\mathrm{t}}$: TX and RX antennas gain ratios

R_1, R_2 : Distance of target from radar TX and RX antennas in terms of meter
L_r, L_t : RX and TX feeder cable losses
L_{p1}, L_{p2} : Excess loss ratios of the send and receive paths of the radar waves between TX/RX antennas and the target excluding free space loss (without unit)
σ : Radar cross section of target in terms of square meter
λ : Wavelength in terms of meter
S_t' : Reflected power from the target in terms of watt
S_r : Received power density at the input of RX antenna in terms of watt per square meter
P_r : Received power at the RX input in terms of watt

Example 6.3. A monostatic radar with 500-kW output power encounters an aircraft with a radar cross section of $4.5\,\text{m}^2$ at a distance of 30 km. The sum of additional losses in the send and receive routes of the waves reduces the reception power to 25 %. Radar frequency is 2.5 GHz and its antenna gain is 33 dB_i, calculate:

1. The received signal level
2. The radar range for a target with a radar cross section of $10\,\text{m}^2$ assuming the threshold level of the RX is equal to 5×10^{-12} W

Solution. 1.

$$f = 2.5\,\text{GHz} \implies \lambda = 0.12\,\text{m}$$

$$G = 33\ \text{dB}_i \implies g = 2{,}000$$

$$P_r = \frac{5 \times 10^5 \times (2{,}000)^2 \times (0.12)^2 \times 4.5}{(4\pi)^3 \times (30{,}000)^4 \times 4} \approx 2 \times 10^{-11}\,W$$

2.

$$R^4 = \frac{5 \times 10^5 \times (2{,}000)^2 \times (0.12)^2 \times 10}{(4\pi)^3 \times 4 \times 5 \times 10^{-12}}$$

$$R^4 = 7.2566 \times 10^{18} \implies R = 5.19 \times 10^4\,\text{m}$$

$$\implies R \approx 50\,\text{km}$$

■

Since it is prevalent to conduct radio calculations in logarithmic system for ease and convenience of operations, hence, the radar equation in logarithmic form is described below:

$$P_r[\text{dB}_W] = P_t[\text{dW}] + 2G[\text{dB}_i] + 20\ \log\ \lambda + 10\ \log\ \sigma - 30\ \log(4\pi) - 40\ \log\ R - 2L[\text{dB}] - 2L_p[\text{dB}] \qquad (6.40)$$

In the above equation, the symbols inside brackets are the logarithmic units of the pertinent quantities. The units of the rest of quantities are given below:

λ : Wavelength in terms of meter
σ : Radar cross section of the target in terms of square meter
R : Distance of target from radar in terms of meter

Simpler form of (6.40) is expressed below:

$$P_r[\text{dB}_W] = -33 + P_t[\text{dB}_W] + 2G[\text{dB}_i] - 2L[\text{dB}] - 2L_p[\text{dB}] + 10 \log\left(\frac{\sigma\lambda^2}{R^4}\right) \tag{6.41}$$

By using the antenna effective area (A_e), the relation of antenna gain with the wavelength and efficiency (η_e) is given as:

$$g = \frac{4\pi A_e}{\lambda^2} \times \eta_e \tag{6.42}$$

In that case, the radar equation is converted into the following formula:

$$P_r = \frac{P_t \cdot g \cdot A_e \cdot \eta_e \cdot \sigma}{(4\pi)^2 \cdot R^4 \cdot L^2 \cdot L_p^2} \tag{6.43}$$

6.2.6 Signal-to-Noise Ratio

Evaluation of the radar received signals may be performed using the SNR as a function of the range R. To develop this form of radar equation, it is required to discuss about received noise.

The ability to detect radar returned signals received through the RX unit is determined by a parameter called signal-to-noise ratio denoted by SNR.

In addition to the radio noises, the existing thermal noises in any electronic device such as radar receiver are unavoidable. This parameter is specified by a factor called the *noise figure*, NF in the RX, and a great effort is done to select premium quality electronic components and parts in order to limit this value as low as possible (say few decibels).

For proper reception of signal in the receiver, it is crucial to have the signal-to-noise ratio not less than a specified value. The radio noise power density is given by the following relation:

$$\rho_n = kT \tag{6.44}$$

In the above relation, each one of the components and their pertinent units are

ρ_n : Noise density in terms of joule per hertz
k : Boltzmann constant equal to 1.38×10^{-23}
T : System temperature in terms of degree Kelvin

Assuming the noise bandwidth equal to B, then the radio noise power will be

$$P_n = kTB \tag{6.45}$$

Noting that in pulsed radars, the value of $B = 1/\tau$ and assuming the correction factor of input filter equal to C_B, then the value of radar P_n can be expressed by

$$P_n = kTC_B \frac{1}{\tau} \tag{6.46}$$

It is also possible to define SNR in the following form, using the above relations:

$$\text{SNR} = \frac{P_r}{P_n} \tag{6.47}$$

Using relations (6.39) and (6.46), the following formula is obtained:

$$\text{SNR} = \frac{P_t \cdot g^2 \cdot \lambda^2 \cdot \sigma \cdot \tau}{(4\pi)^3 \cdot R^4 \cdot kTC_B \cdot L^2 \cdot L_p^2} \tag{6.48}$$

The above relation is converted into logarithmic system as indicated below:

$$\begin{aligned} \text{SNR[dB]} &= -33 + P_t[\text{dB}_W] + 2G[\text{dB}_i] + 10 \log\left(\frac{\sigma\lambda^2\tau}{R^4}\right) \\ &\quad - 2L[\text{dB}] - 2L_p[\text{dB}] - 10 \log(kTC_B) \end{aligned} \tag{6.49}$$

Example 6.4. A radar system operating in the S-band with 200-kW output power is connected to an antenna with 40-dB$_i$ gain. The operating frequency is 3 GHz and the radar cross section of the target is 1 dm^2, and time duration of each pulse is 100 ns. Assuming the following parameters,

$$R = 25\,\text{km}, \;\; T = 580^\circ\text{K}\,, \;\; C_B = 1\,, \;\; L = 2.5\,\text{dB}$$

Then calculate the SNR value in terms of decibel.

Solution.

$$P_t = 200\,\text{kW} \implies P_t[\text{dB}_W] = 53\ \text{dB}_W$$

$$f = 3\,\text{GHz} \implies \lambda = 0.1\,\text{m}$$

$$\tau = 100\,\text{ns} \quad \Longrightarrow \quad \tau = 10^{-7}\,\text{s}$$

$$10\ \log\left(\frac{\sigma \cdot \lambda^2 \cdot \tau}{R^4}\right) = -286\,\text{dB}$$

$$10\ \log(KTC_B) = -201\,\text{dB}$$

$$\text{SNR} = -33 + 53 + 2 \times 40 - 286 - 5 + 201 = 10\,\text{dB}$$

■

6.3 Propagation of Radar Waves

Propagation of radar waves in any medium is subject to a number of phenomena and mechanisms which influence the coverage range of a radar. These effects may extend the radar range significantly or reduce it drastically. Thus, it is important to account for the Earth's environment for prediction of radar performance.

In addition to the free space attenuation, the following phenomena affect the radar performance:

- Refraction of radar waves due to the variations in the air refractivity index which results in the bending of the waves including subrefraction and superrefraction.
- Forward scattering or reflection of radar waves from the Earth's surface resulting in the periodic increase and decrease of the radar path gain.
- Backscattering from land and sea resulting in received signal impairments.
- Diffraction of radar waves by the Earth's bulge and natural or man-made obstructions. By this mechanism, propagation beyond radio horizon is available.
- Duct formation which causes the radar waves to be trapped in surface or elevated air ducts resulting in extending of radar range beyond control which gives misleading data.
- Radio noises produced by ground surface, atmosphere, sky, Sun, radio stars, and other galactic origin. Also, man-made noises due to electric machineries, lightning, and rainfall should be taken into account.
- Radar clutters (excluding atmospheric precipitation) producing unwanted echoes and impairments which make it difficult to detect the desired signals.
- Atmospheric precipitation which results in additional attenuation and more equivalent noise temperature.

We will discuss main issues related to the radar waves propagation in this section. Radar-specific items in the propagation of radar waves include medium, radio frequency, type of transmission, effects of troposphere and ionosphere layers, K-factor and radar horizon, sea and land clutters, Doppler effect, waves polarization, and noise.

General phenomena related to troposphere and ionosphere layers are presented in Chaps. 3 and 4 of the book *Propagation Engineering in Wireless communications* A. Ghasemi et al., 2011 Springer.

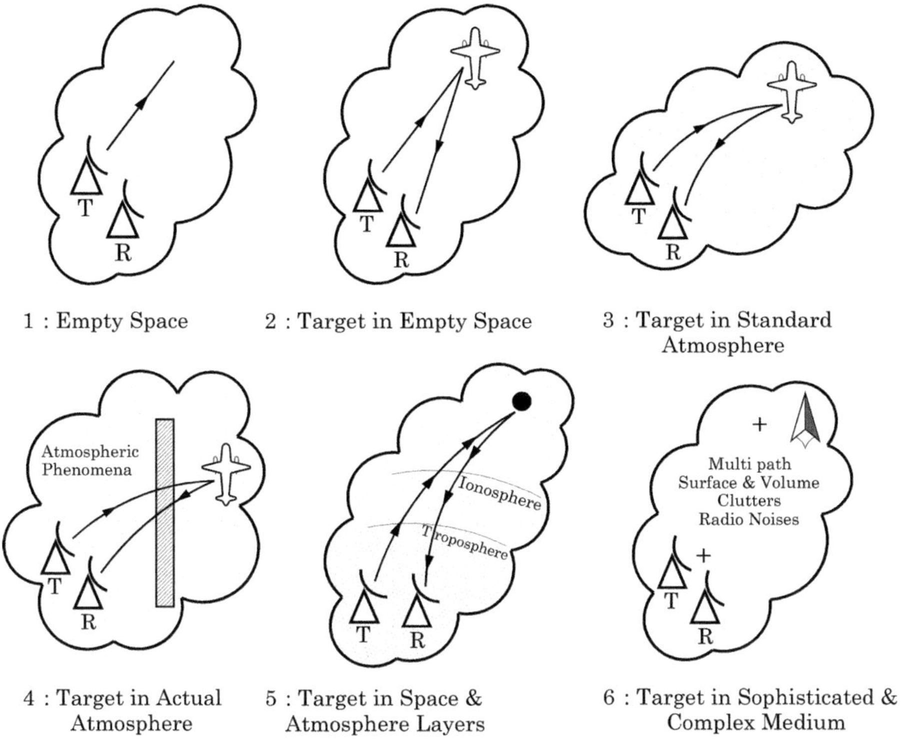

Fig. 6.12 Propagation media for radar waves

6.3.1 Propagation Medium

The propagation medium of the radar waves in UHF and SHF bands and higher given in the Fig. 6.12 can be classified as follows:

1. The simplest form of radar waves propagation which is regarded as hypothetical case relates to a vacuum medium. Under such condition, the waves propagate as straight geometrical line, and there is no return wave, and the waves attenuation is limited which will include only the free space losses.
2. In the next stage, there is a target with a noticeable radar cross section in the vacuum medium. In such a state, a part of the transmitted waves which are traveling as a straight geometrical line upon the collision with the target are reflected back to the RX antenna.
3. In this situation, there is a target in the standard atmospheric medium of the Earth with a noticeable radar cross section. The standard atmosphere includes some humidity and natural gases, and imperatively additional losses occur proportionate to the wave frequency. Under these conditions, the waves path is

also deviated from straight geometrical path due to the variation of atmosphere refractivity index.

4. In the proceeding stage, in addition to the above issues in item (3), different atmospheric phenomena occur such as rain, snow, hail, fog, wind, and dust creating undesired losses which depend on the radio channel frequency.
5. In the next stage, the radar waves are directed toward other distant planets or artificial objects outside the Earth's atmosphere. In this situation, the radar waves traverse troposphere and ionosphere layers and are subjected to the effects of particular phenomena existing within and beyond these layers.
6. Finally, the most complex condition occurs where in addition to the previous factors, other adverse phenomena of the wave propagation such as multipath fading, surface and volume clutters, and radio or thermal noises are added to the process.

6.3.2 Effects of Frequency

Operating frequency bands in radar systems were listed in Sect. 6.2.1 of this chapter and summarized in Table 6.1. In general, most phenomena and effects on the radar waves propagation are dependent on the selected frequency band. These issues are discussed extensively in different textbooks and recommendations relevant to the radiowaves propagation. Some phenomena concerning radar waves propagation depending on frequency are listed below:

- Free space loss according to the relations specified in the first chapter.
- Losses due to rain at frequencies lower than the X-band are negligible and become noticeable at higher bands.
- Losses due to water vapor and gas content in the atmosphere at frequencies below K-band are negligible, while they become more significant as the frequency increases.
- Fresnel radius according to the formula given in the other chapters which depends on radio waves frequency.
- Ionosphere phenomena are inversely dependent on f^2 and f^3 as summarized in the Table 4.5 of the fourth chapter.

In a homogenous medium, the route of radar waves is a geometrical straight line, but in the real atmosphere where it is not homogenous, the radio path between the radar antenna and related target will bend slightly due to the variations of atmosphere refractivity index.

Frequency dependence of losses related to the atmospheric gases, rain, and water vapor is illustrated in the graphs of Fig. 6.13 for the frequency range of the radar waves.

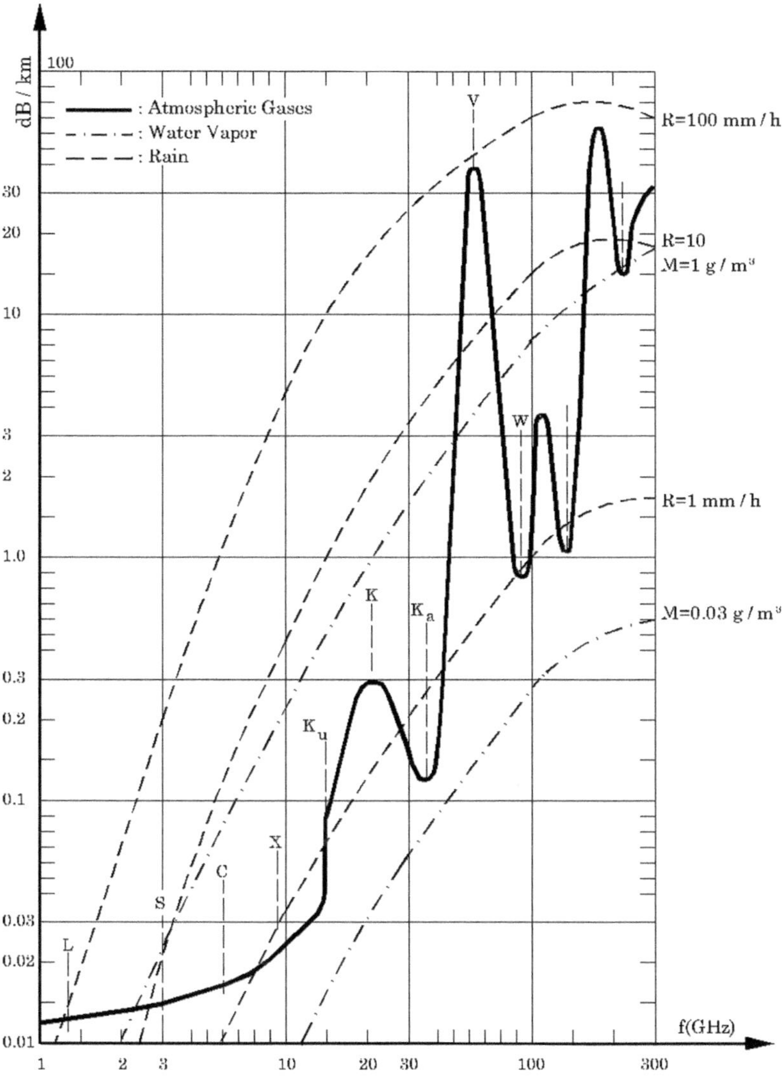

Fig. 6.13 Round-trip attenuation of radar waves in atmosphere

6.3.3 *Types of Radar Links*

Generally the most basic radar links are interchanged between radar and the target based on the line of sight or LOS conditions. Therefore, the general principles governing line of sight communications are valid for this type of radar links, and we can refer to the pertinent principles governing LOS radio links. Furthermore,

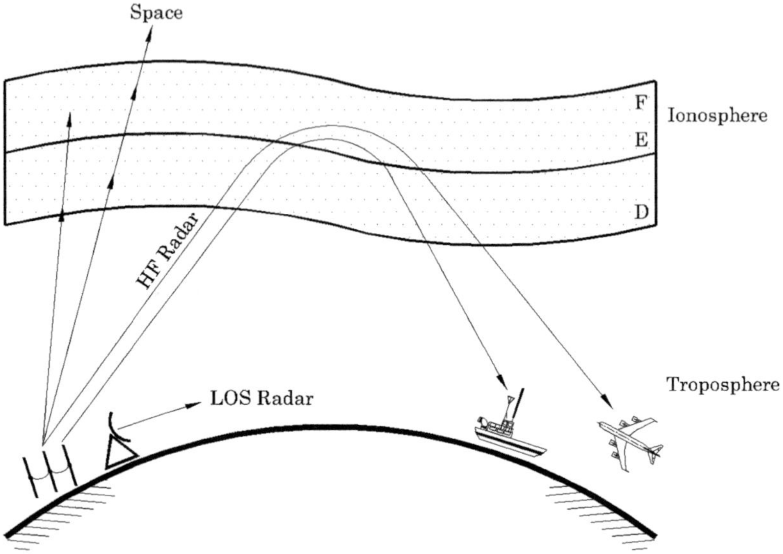

Fig. 6.14 Long range radar in HF band

as depicted in Fig. 6.14, propagation in HF band based on refraction theory in the ionosphere layer may be used for long range and over-horizon radar waves.

For more details regarding ionospheric propagation in the HF band, reference is made to other books of radiowave propagation including Chaps. 4 and 5 of *Propagation Engineering in Wireless Communications*.

6.3.4 Effects of Atmosphere Layers

The effects of troposphere layer on the radar waves are summarized below, while more details are provided in other reference books.

- Environmental conditions including fixed and variable compositions of the Earth's atmosphere such as temperature, humidity, oxygen, nitrogen, and water vapor
- Effects of atmospheric phenomena such as wind, hurricane, dust, thunderstorms, rain, snow, and hail
- Sky effects such as magnetic storms and the sunspots and diurnal or seasonal variations
- Artificial cases such as obstacles related to high towers, chemical substances, and steams existing in urban utilities and huge industrial plants

The effects of the ionosphere layer on the radar waves are also summarized below. More details are provided in other reference books:

- Faraday rotation
- Propagation delay
- Refraction/reflection
- Scattering
- Absorption
- Abrupt changes
- Antenna arrival angle variations
- Dispersion and distortion because of the radar wave phase and group delay

6.3.5 Effects of Earth

Terrain under the radar waves including Earth surface structure such as mountains, hills, surface roughness, lakes, seas, oceans, rivers, forests, and even man-made structures and their composition have a great influence on attenuation or producing echoes which impair the returned signals. Among different effects of Earth, land and sea clutters are significant which have been discussed in Sect. 6.4.

6.3.6 Free Space Loss

Considering that free space propagation is a fundamental reference for the engineering of a radar link, ITU-R assembly through recommendation. P.525 decided the following method be used for the calculation of attenuation in free space which represents a special case because the radar wave is subjected to a loss while propagating both from the transmitter to the target and in the opposite direction up to radar receiver.

For radars using a common antenna for both TX and RX, the free space basic transmission loss, L_{br}, can be written as

$$\mathrm{FSL} = L_{\mathrm{br}}[\mathrm{dB}] = 103.4 + 20\log f + 40\log d - 10\log\sigma \qquad (6.50)$$

where

σ : Radar cross section of target in m^2
d : Distance from radar to the related target in km
f : Radar operating frequency in MHz

6.3.7 Doppler Effect

One of the radar applications is the measurement of the target velocity which is based on the Doppler effect. This phenomenon is caused by relative motion of the

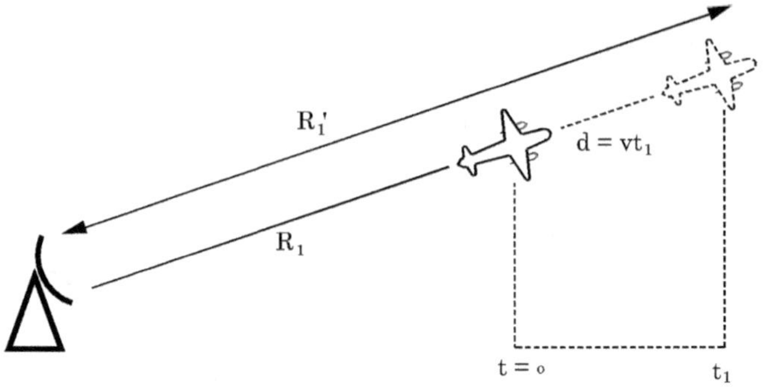

Fig. 6.15 Doppler effect concept

radar antenna and the target. Doppler effect is based on the difference between the frequency of transmitted and received waves. The difference of transmitted frequency denoted by f and received frequency denoted by f_r comprises the Doppler frequency denoted by f_d:

$$f_d = f_r - f \tag{6.51}$$

To extract the relation of velocity of the relative movement of the target with Doppler frequency, Fig. 6.15 is used to illustrate the case. Therefore, while the initial distance of the target from the radar is R_1, but the length of round-trip of radar waves will be $2R_1'$ instead of $2R_1$ because of the relative movement of the target. Assuming that departing velocity of the target away from the radar is positive, then

$$R_1' = R_1 + d = R_1 + V \cdot t_1 \tag{6.52}$$

This difference of distance results in the increase of received signal wavelength by λ_d' (for outgoing path):

$$\lambda_d' = \frac{-C\lambda}{V} \tag{6.53}$$

$$f_d' = \frac{-V}{\lambda} \tag{6.54}$$

The return waves from the target toward the RX antenna also undergo a similar frequency change which consequently the TX and RX frequencies of the radar will have a difference equal to f_d.

$$f_d = 2f_d' = -2\frac{V}{\lambda} \tag{6.55}$$

When direction of the target movement has an angular difference of θ with the direction of radar wave, then the only component of the target velocity in the same direction of radar wave will influence Doppler frequency and the following result can be concluded:

$$f_d = -2(V/\lambda) \cdot \cos\theta \tag{6.56}$$

Noting all the above-mentioned facts and assuming the following expression for the transmitted signal by the radar in the following general form,

$$S_t(t) = a(t) \cdot \cos[\omega t + \theta(t) + \varphi_0] \tag{6.57}$$

then the radar received signal in its general form will be

$$S_t(t) = K \cdot a(t - t_R) \cdot \cos[(\omega + \omega_d)(t - t_R) + \theta(t - t_R) + \varphi_0] \tag{6.58}$$

$$\omega_d = 2\pi f_d \ , \ t_R = \frac{2R_1'}{C} = \frac{2(R_1 + Vt_1)}{C} = \frac{2R_1}{C - V} \tag{6.59}$$

where the total losses due to the propagation adverse mechanisms are included in K.

Example 6.5. A radar station operating at f = 5,540 MHz detects a target moving with a speed of 720 km/h toward its antenna. Assuming 100-km distance from the target, calculate:

1. Doppler frequency shift
2. Round-trip time of radar signal

Solution. 1.

$$f_d = \frac{-2v \cdot f}{C} \quad \Longrightarrow \quad f_d = +7.35\,\text{KHz}$$

2. Using the relation (6.59) yields

$$t_R = \frac{2 \times 100 \times 10^3}{3 \times 10^8 - 200} \approx 0.67\,\text{ms}$$ ■

6.3.8 K-Factor and Radar Horizon

The K-factor is presented in Sect. 1.8.5 of the first chapter. In radar waves propagation and to cover big distances, the value of K-factor in the radar S-, C-, and X-bands are approximately in the range of 1–1.33. For higher frequencies, lower values of K-factor are taken as the basis of calculations. In different regions of the world, the effective and the minimum values of K-factor due to the environmental

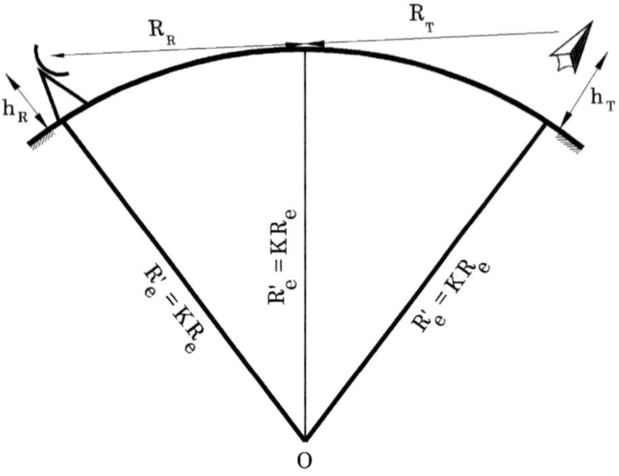

Fig. 6.16 Geometry of radar horizon

and geographical conditions are different; therefore, it is essential to take its suitable values based on measurements or experimental data. The variation of the Earth's radius from its actual value of R_e to the effective value of R'_e depends on the K-factor coefficient according to the following relation:

$$R'_e = KR_e \tag{6.60}$$

Noting Fig. 6.16, the radar horizon which in fact is the maximum distance of perceiving the target by the radar in LOS condition, under normal conditions (i.e., $K = 1.33$), is calculated by the following relation:

$$R_{RH} = \sqrt{2R'_e h_r} + \sqrt{2R'_e . h_t} \tag{6.61}$$

$$R_{RH} = \sqrt{2KR_e}(\sqrt{h_r} + \sqrt{h_t}) \tag{6.62}$$

Each of the components and their pertaining units for the above relation are given below:

R_{RH} : Radar horizon in meter
R_e : Earth's radius in meter
K : K-factor (without unit)
h_r : Height of radar antenna in meter
h_t : Height of the target in meter

Example 6.6. Find the minimum radar perceiving distance for targets higher than 20 m where radar antenna height is 12 m. The effective value of K in the location is assumed 0.9.

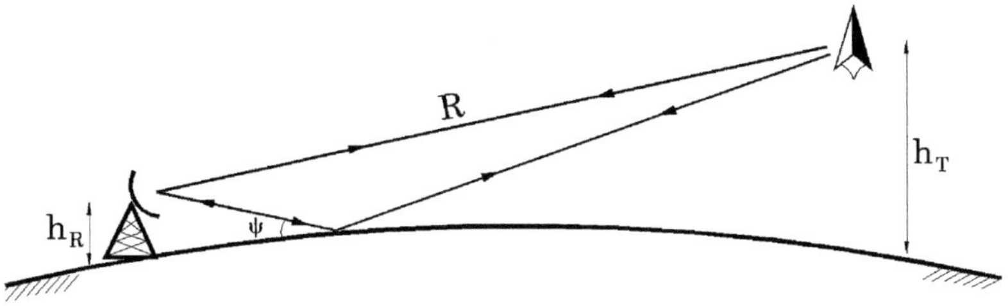

Fig. 6.17 Geometry of multipath in radar link

Solution.

$$K = 0.9\ ,\ \ h_r = 12\,\text{m}\ ,\ \ h_t = 20\,\text{m}$$

$$R_{RH} = \sqrt{2\times 6,370\times 10^3\times 0.9}\times(\sqrt{12}+\sqrt{20})$$
$$= 26,873.3\,\text{m} = 26.87\,\text{km}$$

■

6.3.9 *Multipath by Reflection*

One of the negative phenomena in radio propagation and particularly for radar waves is multipath reception due to the reflection of the waves from the Earth's surface. The basic principles and relations of this function for radiowaves are discussed in the Chap. 3 of the book *Propagation Engineering in Wireless Communications, A. Ghassemi, Springer 2011* and finally led to the definition of a new parameter called *path gain factor* denoted as F in below:

$$F = \frac{\bar{E}_t}{\bar{E}_d} = 1 + \rho e^{j\Delta\Phi} \tag{6.63}$$

The value of $|F|$ in terms of pertaining components is given below:

$$|F| = 2|\sin(k\cdot h_t\cdot h_r/R)| \tag{6.64}$$

Noting the path geometry of radar wave propagation according to Fig. 6.17 and the following conversions:

$$k = 2\pi/\lambda \tag{6.65}$$

Path gain factor (6.64) is converted to

$$|F| = 2|\sin(\frac{2\pi h_r\cdot h_t}{\lambda\cdot R})| \tag{6.66}$$

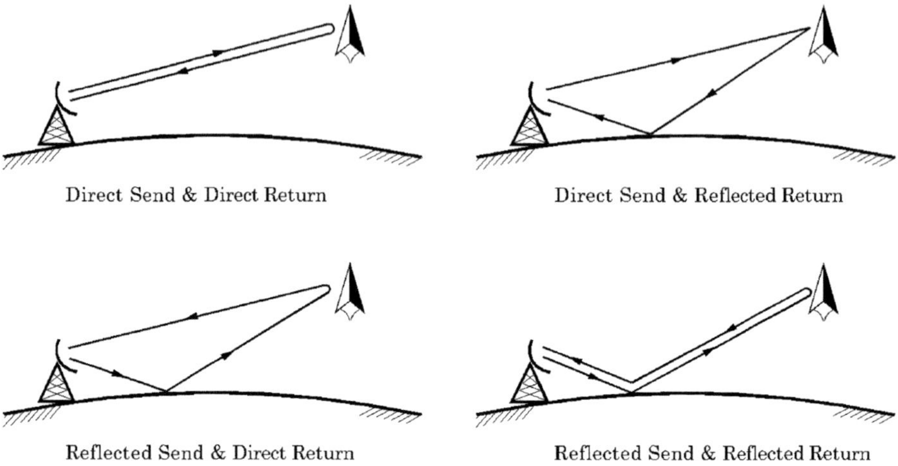

Fig. 6.18 Direct and reflected waves in radar link

The above relations are valid only for good reflection coefficient ($\rho = -1$), small h_r and h_t compared to the distance R, and the reflection point is located in the half-power beamwidth (3 dB) of radar antenna main lobe. With the following assumption for practical situations,

$$k \cdot h_r \cdot h_t << 1 \tag{6.67}$$

relation 6.66 is simplified to

$$|F| = 2 \cdot k \cdot h_r \cdot h_t/R \tag{6.68}$$

Considering that the received power is proportional to the square of electric field magnitude, it can be concluded that

$$P_r \propto |F^2| \tag{6.69}$$

For each transmitted radar wave, there are a great number of returned waves among which four cases are significant as illustrated in the Fig. 6.18.

In this position, the waves combination relative to the direct wave is multiplied to $|F^2|$ factor once on the radar to the target and once more on the return route (target to RX) is multiplied to $|F^2|$ factor and finally multiplied by $|F^4|$ factor, that is,

$$P_r = P_d \cdot |F^4| \tag{6.70}$$

$$|F^4| = 16 \sin^4\left(\frac{2\pi h_r \cdot h_t}{\lambda R}\right) \tag{6.71}$$

$$|F^4| \approx 16\left(\frac{2\pi h_\mathrm{r} \cdot h_\mathrm{t}}{\lambda R}\right)^4 \approx (\Delta\Phi)^4 \tag{6.72}$$

Example 6.7. A radar system is assumed to use an antenna with a height of $h_\mathrm{R} = 10\,\mathrm{m}$. At a distance of 20 km, calculate:

1. The path gain factor and also the ratio of the received power to direct wave power at $f = 5$ GHz when target height is 45 m
2. The radar blind height
3. The value of $|F^4|$ to observe targets at a height of 1 km

Solution. 1.

$$f = 5\text{ GHz} \implies \lambda = 0.06\,\mathrm{m}$$

$$h_\mathrm{R} = 10\,\mathrm{m}\ ,\ h_T = 45\,\mathrm{m}\ ,\ R = 20{,}000\,\mathrm{m}$$

$$|F| = 2\ \sin\left(\frac{2\pi \times 10 \times 45}{0.06 \times 20{,}000}\right) = 1.4$$

2. At blind height, the value of F should be zero:

$$\frac{2\pi \cdot h_\mathrm{R} \cdot h_T}{\lambda R} = n\pi \implies h_T = \frac{n \times 1{,}200}{20}$$

$$h_T = 60n\ ,\ n = 1 \implies h_T = 60\,\mathrm{m}$$

3. In this position, the radar antenna is pointed upwards and the reflected waves may be disregarded, thus

$$|F| = 1 \implies |F^4| = 1$$

■

6.3.10 Polarization Losses

Polarization of radar waves is subject to change during propagation through the troposphere and ionosphere layers for various reasons such as the Earth's magnetic field resulting in some losses in the received power. Polarization mismatching of the RX antenna with the polarization of the received waves for any reason creates some extra losses.

In radar links, in addition to the effective factors of the medium, the targets also make some changes in the waves polarization. For example, to receive the waves with circular polarization through a linearly polarized antenna will result in at least 3 dB extra losses. The polarization losses due to reception of the radar waves with elliptical polarization through an antenna with linear polarization can be calculated by the following relation:

$$L_P = 0.5(1 + K \cos 2\theta) \tag{6.73}$$

$$K = \frac{a_K^2 - 1}{a_K^2 + 1} \tag{6.74}$$

The parameters of the above relation are indicated below:

L_P : Polarization losses
a_K : The ratio of major axis to the minor axis of the ellipse
θ : The angle between linear polarization direction and boresight of the ellipse

Relation (6.73) can be also stated in the logarithmic system as

$$L_P = -3 + 10 \log(1 + K \cos 2\theta) \tag{6.75}$$

If the polarization variation does not lead to the alteration of its type but the polarization axis rotates, then the polarization losses are given by

$$L_P = 0.5(1 + \cos 2\theta) = \cos^2 \theta \tag{6.76}$$

$$L_P[\text{dB}] = 20 \log(\cos \theta) \tag{6.77}$$

In addition to the phenomena pertaining to propagation medium, the radar target is also capable to change its waves polarization. This factor called "target polarization loss" will decrease the effective radar cross section by

$$\sigma = \sigma_s \times \rho_P \tag{6.78}$$

where

σ : Effective radar cross section of the target
σ_s : Actual radar cross section of the target
ρ_P : Target polarization efficiency

In fact, the polarization efficiency is a number smaller than unity which is an indication of a portion of the waves power being scattered by the target itself.

Example 6.8. The radar waves with elliptical polarization are received by an antenna having horizontal polarization. Assuming that $a_K = 2$ and $\theta = 30°$, then calculate:

1. The polarization loss factor
2. The polarization loss if elliptical polarization is replaced by circular or linear polarizations
3. Total loss in dB if the polarization efficiency of the target is 50%

Solution. 1.

$$a_K = 2 \implies K = \frac{3}{5} = 0.6$$

$$L_{\text{P}} = 0.5(1 + 0.6\cos 60^\circ) = 0.65 \implies L_{\text{P}}[\text{dB}] = -1.87\,\text{dB}$$

2.

$$\text{Cirular Polarization} \implies a_K = 1 \implies K = 0$$

$$L_{\text{P}} = 0.5 \implies L_{\text{P}}[\text{dB}] = -3\,\text{dB}$$

$$\text{Linear Polarization} \implies L_{\text{P}} = 3/4 \implies L_{\text{P}}[\text{dB}] = -1.25\,\text{dB}$$

3.

$$L_{\text{P}} = 0.65 \quad , \quad \rho_{\text{P}} = 0.5$$

$$L_{\text{P/t}} = L_{\text{P}} \cdot \rho_{\text{P}} = 0.325$$

$$L_{\text{P/t}}[\text{dB}] = 10\,\log(0.325) = -4.88\,\text{dB}$$

■

6.3.11 Target Losses

Targets change radar waves in different ways resulting in more attenuation. Among different factors affecting radar waves are shape, material, motion, and rotation of the target which make the following effects or losses:

- Absorption of the radar waves
- Relative motion of the target and radar location (Doppler effect)
- Radar waves scattering
- Losses related to the polarization change of the radar waves
- The change in target effective radar cross section

6.4 Radar Clutter

6.4.1 Introduction

In the radar technology, *clutter* implies unwanted echoes mostly from natural environment and sometimes from artificial structures. It is a significant factor in the propagation of radar waves acting as an adverse effect. Radar clutters either in surface or volume forms decrease its coverage range. Major sources of radar clutters include, but not limited to, the following items:

- Land and sea backscattering
- Atmospheric aerosol such as dust and chaff
- Climatic precipitation like rain, snow, and hail
- Groups of flying birds and insects

Large clutter echoes create radar signal impairment limiting its capability for proper detection of the desired signals. Radar echoes due to the propagation process are not always undesired like in the case of atmospheric weather observation, remote sensing and in cases when the clutter itself is considered as a radar target.

6.4.2 Types of Clutters

The radar clutters are divided into the following types:

1. Surface clutter for which echoes from land and sea surface are examples of this type
2. Volume clutter for which rain, snow, hail, dust storm, and artificial chaffs are typical examples

For evaluation of clutter effects, it is common to employ normalized radar cross section (NRCS) of the clutter. This value is clutter cross section per unit area or volume.

When surface clutters are considered, the magnitude of the echo from distributed surface clutter is proportional to the area illuminated by the radar antenna (within its half-power beamwidth). To measure the clutter echoes, its NRCS denoted by σ^o is commonly used which is defined by the following relation:

$$\sigma^o = \frac{\sigma_c}{A_c} \tag{6.79}$$

where σ_c is the radar cross section of the surface clutter covering an area A_c, σ^o is a dimensionless quantity, and both σ_c and A_c have the same units (like m^2/m^2).

When volume clutter is of interest, the magnitude of its echo is proportional to the volume illuminated by the radar antenna (within its half-power beamwidth). Similar to surface clutter, to measure the volume clutter echoes, it is common to express related NRCS denoted by η and defined by the following relation:

$$\eta = \frac{\sigma_c}{V_c} \tag{6.80}$$

where σ_c is the radar cross section of the clutter with total volume V_c. The unit of η is m^{-1} and sometimes called reflectivity of the volume clutter.

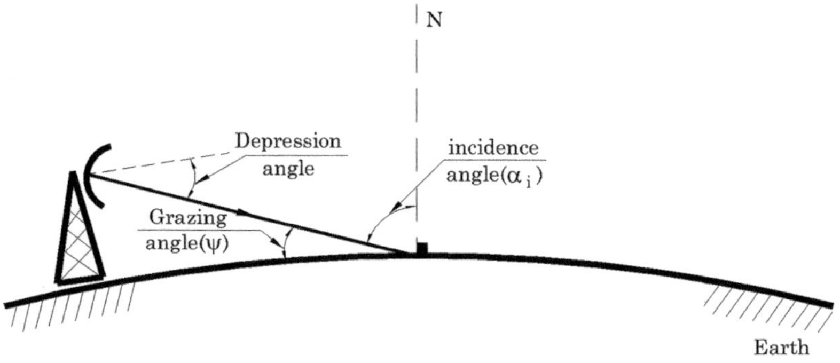

Fig. 6.19 Angles in surface clutter

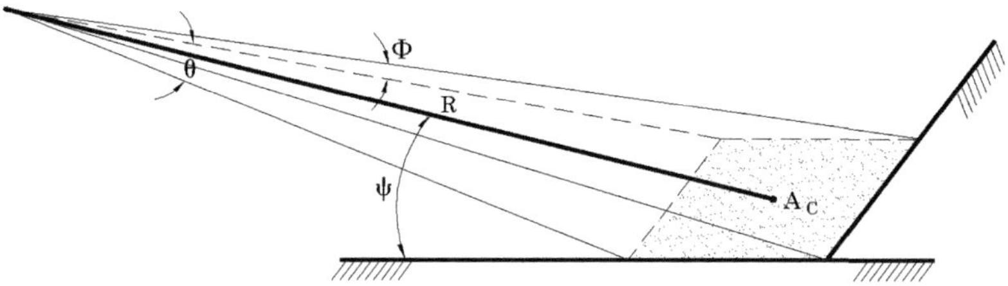

Fig. 6.20 Geometry of radar surface clutter

6.4.3 Surface Clutter

The surface clutter and its typical samples along with normalized radar cross section σ^o was introduced in the previous section. Detection of the radar signals in the surface clutter condition is limited, and basic radar equation (6.48) is not valid for evaluation where the receiver sensitivity is subject to the noise as the only limiting factor.

Surface clutter will be discussed for different grazing angles of the radar waves along with incident and depression angles which are shown in the Fig. 6.19.

To calculate the illuminated area by a radar antenna, the required concept and geometry are depicted in Fig. 6.20 in terms of parameters related to the incident waves, that is, θ for elevation and Φ for azimuth half-power beamwidths, respectively.

In both cases of low and high grazing angles, the following relations can be utilized with an acceptable approximation.

- Low Grazing Angle

  For low grazing angle where it meets the following criteria:

$$c\tau >> R\theta / \sin\,\psi \tag{6.81}$$

The illuminated area can be approximated by

$$A_c \approx R\Phi \cdot \frac{c\tau}{2\cos\,\psi} \tag{6.82}$$

where θ and Φ are in terms of radian, R in meters, A_c in square meters, c is the velocity of propagation in m/s, and τ is the pulse width in s.

Considering the above results and based on the basic radar equation, then the echo received power P_c will be

$$P_c = \frac{P_t \cdot g_t \cdot A_e \cdot \sigma}{(4\pi)^2 R^4} \tag{6.83}$$

where:

P_t : Transmitter power in W
g_t : Radar TX antenna gain
A_e : Receiving antenna effective area in m^2
R : Radar distance from clutter in m
σ : Radar cross section of clutter in m^2

In clutter condition $\sigma = \sigma_c = \sigma^o \cdot A_c$, then using (6.82) yields

$$P_c = \frac{P_t \cdot g_t \cdot A_e \cdot \sigma^o \cdot \Phi(c\tau/2)}{(4\pi)^2 \cdot R^3 \cdot \cos\,\psi} \tag{6.84}$$

Equation (6.84) reveals that in surface clutter condition at low grazing angles, the returned power from clutter is inversely proportional to the cube of range rather than fourth power.

- High Grazing Angle

For high grazing angle where it meets the following criteria:

$$c\tau << R\theta/\sin\,\psi \tag{6.85}$$

The illumination area can be approximated by

$$A_c \approx \frac{\pi R^2}{4} \times \frac{\theta \cdot \Phi}{\sin\,\Psi} \tag{6.86}$$

Similarly, the echo received power, P_c, will be

$$P_c = \frac{P_t \cdot g_t \cdot A_e \cdot \sigma^o \cdot \theta \cdot \Phi}{64\pi R^2 \cdot \sin\,\psi} \tag{6.87}$$

Equation (6.87) reveals that in the surface clutter condition at high grazing angles, the returned power from clutter is inversely proportional to the square of range rather than fourth power as in the basic radar equation.

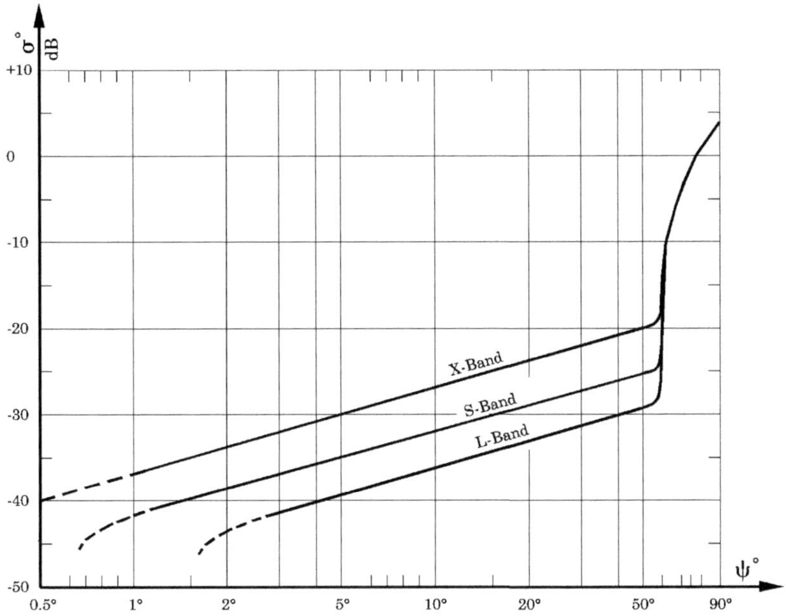

Fig. 6.21 Radar cross section of sea clutter versus grazing angle

6.4.4 Sea and Land Clutters

The land and sea clutters are typical surface clutters affecting the operations of radar systems.

6.4.4.1 Sea Clutters

Backscattering from surface of the sea limits the performance of radar systems, used for different purposes in the oceans, seas, and lakes. For all cases, the normalized radar cross section of the related surface clutter should be taken into account for radar range and performance evaluation.

The NRCS of sea clutter is a function of frequency, polarization, grazing angle, wind speed, and air ducts. As an example, variations of σ^o in terms of grazing angle are depicted in Fig. 6.21, for L-band, S-band, and X-band radar systems. Also, wind speed on the sea influences the clutter σ^o around 10 dB because of surface roughness

One of outstanding mechanism in dealing with sea clutter is continuous changes of its surface roughness with time. However, the radar echoes are mostly returned from those parts of the sea with roughness equal to the radar wavelength. On the other hand, the sea clutter is related to the wind speed over the sea which may reach more than 10 m. The wind speed on the sea will influence its normalized radar cross section, σ^o around 10 dB because of surface roughness.

Table 6.3 Sea states based on WMO classification

Sea state	*Wave height (m)*	*Description*
0	0	Calm, glassy
1	0 ~ 0.1	Calm, rippled
2	0.1 ~ 0.5	Smooth, wavelets
3	0.6 ~ 1.2	Slight
4	1.2 ~ 2.4	Moderate
5	2.4 ~ 4.0	Rough
6	4.0 ~ 6.0	Very rough
7	6.0 ~ 9.0	High
8	9.0 ~ 14.0	Very high
9	Over 14	Phenomenal

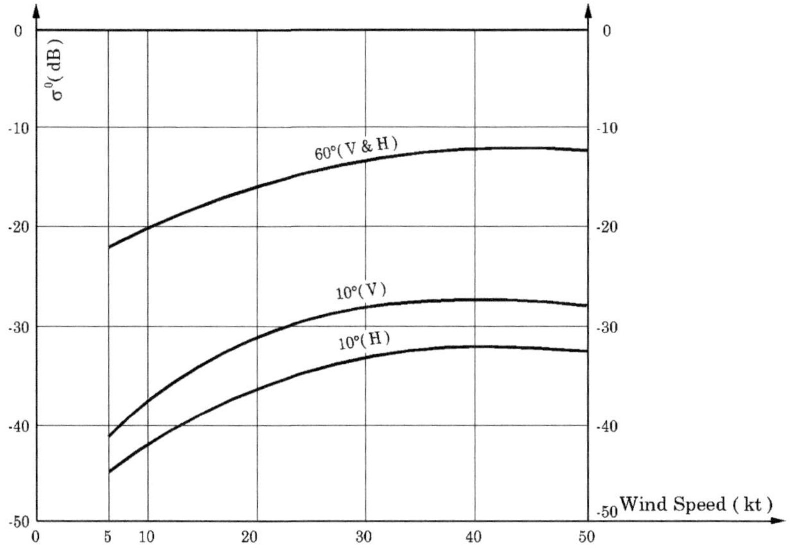

Fig. 6.22 Effect of wind speed on sea clutter

The sea state defined by the World Meteorological Organization is given in the Table 6.3. Also the effect of wind on sea clutter at grazing angles of 10 and 60° for X-band is shown in the Fig. 6.22.

6.4.4.2 Land Clutter

Analysis of land clutters is more complex than sea clutters because of the following facts:

- The variety of land material and their electrical characteristics
- The land backscatter amplitude distribution
- The moisture content of the soil or surface cover

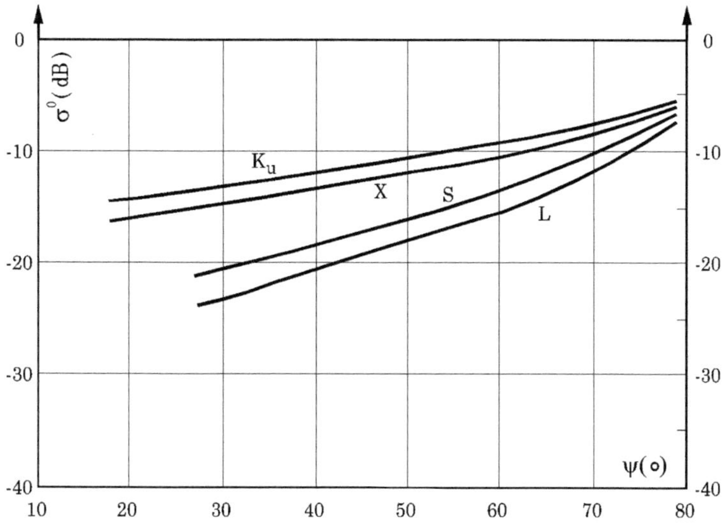

Fig. 6.23 Variation of land clutter σ^o versus grazing angle ψ for L, S, X and Ku bands

In the most references, land clutter is explained at low, medium, and high grazing angles.

To describe theory of clutter and backscattering from land, some approaches are selected, and many experiments have been conducted by experts using empirical models. In principle, backscatter of surface can be analyzed by applying Maxwell's equations, but in practice, the land properties and terrain structure make it too complicated.

To study land clutter, its normalized radar cross section is a key factor which is a function of many factors. Figure 6.23 illustrates roughly dependence of σ^o on grazing angle at L, S, X, and Ku frequency bands. Also, variation of σ^o in terms of the grazing angle, ψ, is shown in Fig. 6.24 for a number of typical structures of the terrain.

Experimental measurements reveal that the value of σ^o at medium grazing angles ($8^\circ < \psi < 70^\circ$) is less sensitive to polarization of radar waves. With a good approximation, it is linearly proportional to the sin Ψ which gives the following result:

$$\gamma = \frac{\sigma^o}{\sin\,\psi} \tag{6.88}$$

γ is a specific land clutter parameter expressed in dB (m^2/m^2). The average value of γ in terms of frequency and terrain structure is given in Fig. 6.25. Sometimes the following equation is used to express γ as a function of frequency:

$$\gamma = \gamma_0 + 5\,\log(f/f_0) \tag{6.89}$$

where γ_0 and f_0 are reference values (as an example $\gamma_0 = -10\,\text{dB}$ and $f_0 = 10\,\text{GHz}$).

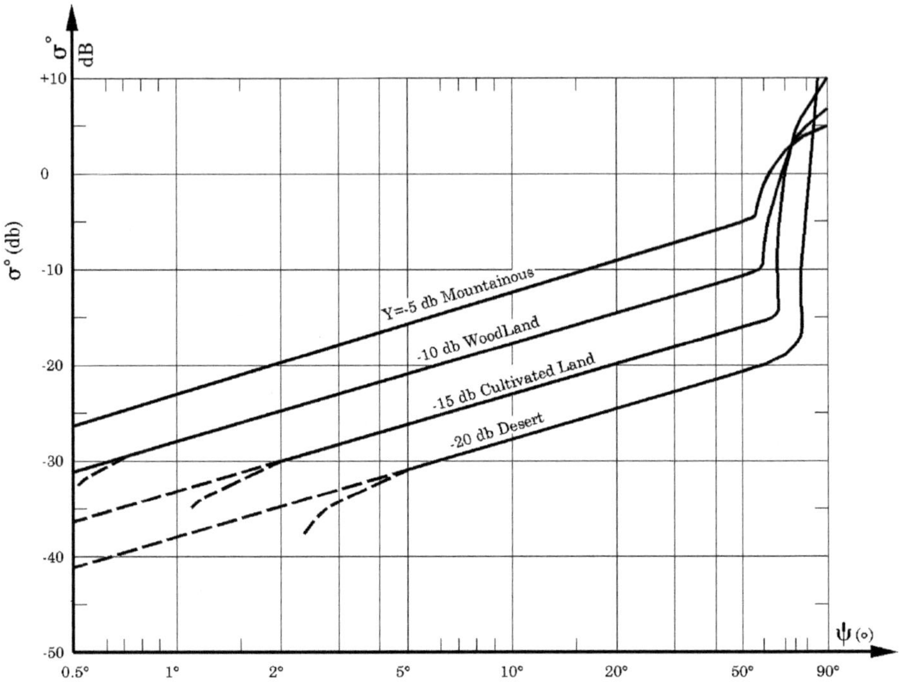

Fig. 6.24 NRCS of land surface clutter versus grazing angle ψ

Fig. 6.25 Rain attenuation versus frequency

6.4.5 Volume Clutters

When a radar target is located in a place having a large number of undesired scatterers such as raindrops, snowflakes, hail stones, or chaff dipoles, the summation of the returned signals from these particles may impose a powerful adverse effect on the radar reception system similar to the environment noise which is commonly called "radar volume clutter."

Most of volume clutters are related to the meteorological phenomena such as rain, snow, fog, hail, dust and cloud. A large portion of radiated signals are reflected by raindrops, hail, and snowflakes. Climatic precipitation affects radar waves with the following mechanisms:

- Attenuation which may be neglected for frequencies less than 10 GHz. The main atmospheric precipitation is rain for which attenuation is a function of its rate and radar frequency. Figure 6.26 provides some typical rates of rain attenuation in terms of frequency.
- Backscattering phenomenon which dominates the detection and tracking of radar systems for frequencies as low as L-band.

The backscatter spectrum from precipitation and chaff is broadened by the wind shear, vertical fall rate, and air turbulence. These mechanisms limit the ability of the Doppler radars to distinguish targets from clutter.

To study volume clutter, its normalized radar cross section denoted by η is a key factor. The Fig. 6.26 illustrates rainfall rate for some frequencies common for radar waves.

To calculate the volume clutter radar cross section, it is necessary to obtain V_c of the volume clutter which may be estimated by

$$V_c \approx \frac{\pi}{4} \cdot R^2 \cdot \theta \cdot \Phi \cdot \frac{c\tau}{2} \tag{6.90}$$

where R = range, θ = horizontal half-power beamwidth, Φ = vertical half-power beamwidth, c = propagation velocity, and τ = pulse duration.

Now to derive the received power by the antenna due to the volume clutter, a new parameter called *radar reflectivity factor* denoted by z is defined as follows:

$$z = a\, r^b \tag{6.91}$$

where r is rainfall in mm/h and a and b are empirical constants. For rain, the (6.91) is simplified to the following equation by radar experts:

$$z = 200\, r^{1.6} \tag{6.92}$$

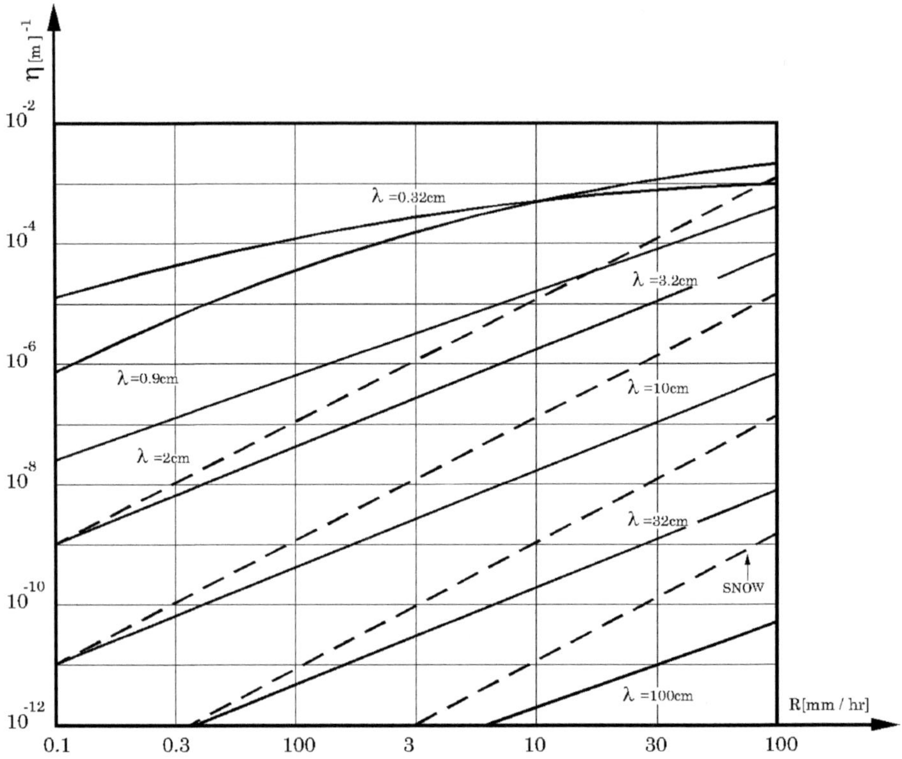

Fig. 6.26 Typical volume normalized RCS versus rainfall rate

Similar expression for radar reflectivity factor related to snowfall (with equivalent water content) is

$$z = 2{,}000\, r^2 \tag{6.93}$$

Considering the above equations, the received power level due to the volume clutter can be expressed by the following equation:

$$P_c = C = \frac{K_1 \cdot P_t \cdot g_t \cdot \tau \cdot z}{R^2 \cdot \lambda^2}, \quad K_1 = 1.2 \times 10^{-10} \tag{6.94}$$

In addition to the above-mentioned atmospheric clutters, there are some other volume clutters in reality among which the following are more popular:

- Group of flying birds
- Group of flying insects
- Air turbulence

For more details regarding characteristics of the above clutters, reference is made to radar textbooks.

6.5 Radar Equation in Real Conditions

The basic radar equation for the monostatic configuration was derived in Sect. 6.2.5. This is accurate for ideal situations where there is only free space and basic transmission attenuation including atmospheric absorption and RF feeder loss as well. Since in real situations where adverse conditions in the propagation of radar waves are crucial, they should be taken into account for proper evaluation of the received signal level and performance.

The actual conditions in the propagation of radar waves result in more sophisticated expressions for the radar equation. Among a number of adverse conditions, the following are studied in this section:

- Noisy media
- Reflective media and multipath reception
- Clutter
- Interference and jamming

6.5.1 *Noisy Media*

Effective noise in the radar links originates from two main components of radio noises (during the waves propagation) and receiver thermal noise. These noises exist in the real cases and cannot be ignored. The RX thermal noise is expressed by its noise figure denoted as N_F and defined by

$$N_F = \frac{1}{G} \cdot \frac{N_O}{N_i} = \frac{S_i}{S_O} \cdot \frac{N_O}{N_i} \tag{6.95}$$

In the above relation, G is numerical RX system gain, N_i and N_O are the input and output noise powers, and S_i and S_O are the input and output signal power related to the radar RX unit. Applying the following modification to the noise figure yields

$$\mathrm{SNR} = \mathrm{S/N} \tag{6.96}$$

$$N_F = (S_i/N_i)/(S_O/N_O) = \frac{(\mathrm{SNR})_i}{(\mathrm{SNR})_O} \tag{6.97}$$

The input noise power can be expressed as follows:

$$N_i = P_n = kTB_n \tag{6.98}$$

In the latter relation, each of the components and their pertaining units are stated below:

k : Boltzmann constant equal to 1.38×10^{-23}
T : Equivalent noise temperature in degrees Kelvin
B_n : Equivalent noise bandwidth in Hz

In case of rewriting (6.97) in terms of input signal power, then it yields

$$S_i = kTB_n \cdot N_F \cdot S_O/N_O \tag{6.99}$$

To produce the required signal-to-noise ratio (at the IF output), a certain amount of S_O/N_O is needed. In such circumstances, the minimum input signal power is

$$S_i = S_{min} = kTB_n \cdot N_F \cdot S_O/N_O \tag{6.100}$$

Considering the input power fed to the RX according to (6.43) and assuming $L_t = L^2 \cdot L_p^2$, then

$$P_r = S_i \tag{6.101}$$

$$\frac{P_t \cdot g_t \cdot \sigma}{(4\pi)^2 \cdot R^4 \cdot L_t} \times A_e = kTB_n \cdot N_F \cdot \frac{S_O}{N_O} \tag{6.102}$$

$$(\mathrm{SNR})_O = \frac{S_O}{N_O} = \frac{P_t \cdot g_t \cdot A_e \cdot \sigma}{(4\pi)^2 \cdot R^4 \cdot L_t \cdot kTB_n \cdot N_F}, \quad \text{or} \tag{6.103}$$

$$(\mathrm{SNR})_O = \frac{S_O}{N_O} = \frac{P_t \cdot g_t \cdot g_r \cdot \lambda^2 \cdot \sigma}{(4\pi)^3 \cdot R^4 \cdot L_t \cdot kTB_n \cdot N_F} \tag{6.104}$$

Finally for a minimum signal-to-noise ratio equal to S_{min}, the maximum radar range can be obtained by

$$R_{max}^4 = \frac{P_t \cdot g_t \cdot A_e \cdot \sigma}{(4\pi)^2 \cdot L_t \cdot kTB_n \cdot N_F \cdot S_{min}} \tag{6.105}$$

Example 6.9. A radar system with a TX output power of 50 kW at 3-GHz frequency is connected to an antenna with 40-dB$_i$ gain. In case of 5-dB receiver noise figure and all miscellaneous losses equal to 20 dB, ambient temperature of 27°C and effective surface of RX antenna being 8 m^2 and B_n = 1 MHz, find:

1. The output signal-to-noise ratio for an elevated target at h = 6 km with radar cross section of 2 m^2 at a distance of 30 km
2. The maximum radar range if the minimum required SNR is 10 dB

Solution. 1. Noting that an elevated target is assumed with $E_l >> 0$, then $|F^4| = 1$, and

$$G_t = 40\ \mathrm{dB}_i \implies g_t = 10{,}000$$

$$N_F[\mathrm{dB}] = 5 \implies N_F = 3.16$$

$$L_t[\mathrm{dB}] = 20 \implies L_t = 100$$

$$kTB_n = 1.38 \times 10^{-23} \times 300 \times 10^6 = 4.14 \times 10^{-15}$$

Using (6.103) gives the following result:

$$(\mathrm{SNR})_0 = \frac{(50\times 1{,}000)\times 10^4 \times 8\times 2\times 1}{(4\pi)^2\times (30{,}000)^4\times 100\times 4.14\times 10^{-15}\times 3.16} = 47.8$$

$$\Longrightarrow \quad (\mathrm{SNR})_0 = 16.8\,\mathrm{dB}$$

2. For $(\mathrm{SNR})_0$ = 10 dB and applying (6.105) yields

$$(\mathrm{SNR})_0[\mathrm{dB}] = 10\,\mathrm{dB} \quad \Longrightarrow \quad (\mathrm{SNR})_0 = 10$$

$$R_{\max} = \left[\frac{(50\times 1{,}000)\times 10^4\times 8\times 2\times 1}{(4\pi)^2\times 100\times 4.14\times 10^{-15}\times 3.16\times 10}\right]^{1/4}$$

$$R_{\max} = 44.4\,\mathrm{km}$$

■

6.5.2 *Reflective Media on Multipath Reception*

In reflective medium as explained before, in addition to the direct route for LOS transmission between the radar antenna and the target, there are reflective routes making multipath condition. To evaluate multipath effects, the path gain factor denoted by F was introduced with a magnitude range of zero to 2 by the following formula:

$$|F| = 2\left|\sin\left(\frac{2\pi h_{\mathrm{R}}\cdot h_{\mathrm{t}}}{\lambda R}\right)\right| \tag{6.106}$$

where h_{R} and h_{t} are the heights of radar antenna and target, respectively, λ is wavelength of RF channel frequency, and R is the distance between radar and related target.

In a radar reflective medium, final received signal level (P_{r}) is related to $|F|$ and direct received power level at LOS condition (P_{d}) by the following formula for a round-trip route:

$$P_{\mathrm{r}} = P_{\mathrm{d}}\cdot |F^4| \tag{6.107}$$

Then, the basic radar equation expressed by (6.39) will be converted to the following equations in the reflective medium and multipath conditions:

$$P_{\mathrm{r}} = \frac{P_{\mathrm{t}}\cdot g_{\mathrm{t}}\cdot g_{\mathrm{r}}\cdot \lambda^2\cdot \sigma\cdot |F^4|}{(4\pi)^3\cdot R^4\cdot L^2\cdot L_p^2} \tag{6.108}$$

$$P_{\mathrm{r}} = \frac{P_{\mathrm{t}}\cdot g_{\mathrm{t}}\cdot A_{\mathrm{e}}\cdot \sigma\cdot |F^4|}{(4\pi)^2\cdot R^4\cdot L^2\cdot L_p^2} \tag{6.109}$$

If we consider the thermal noise effect in a multipath environment and substitute $L^2 \cdot L_p^2$ by L_t, then the following equations shall be applied for signal-to-noise ratio and maximum radar range:

$$(\mathrm{SNR})_0 = \frac{P_t \cdot g_t \cdot A_e \cdot \sigma \cdot |F^4|}{(4\pi)^2 \cdot R^4 \cdot L_t \cdot kTB_n \cdot N_F} \tag{6.110}$$

$$(\mathrm{SNR})_0 = \frac{P_t \cdot g^2 \cdot \lambda^2 \cdot \sigma \cdot |F^4|}{(4\pi)^3 \cdot R^4 \cdot L_t \cdot kTB_n \cdot N_F} \tag{6.111}$$

$$R_{\max}^4 = \frac{P_t \cdot g \cdot A_e \cdot \sigma \cdot |F^4|}{(4\pi)^2 \cdot L_t \cdot kTB_n \cdot N_F \cdot S_{\min}} \tag{6.112}$$

Example 6.10. In case of using the radar system specified in the previous example in a reflective condition where $|F|$ varies from 0.6 to 1.5, find:

1. Radar basic range in ideal situation without any kind of impairment for $P_{th} = 8 \times 10^{-14}$ W
2. Maximum radar range in the worst case
3. Radar range tolerance for the specified conditions

Solution. 1. For $g_t = 10{,}000$, $L_t = 100$, $\sigma = 2\,\mathrm{m}^2$, $P_t = 50\,\mathrm{kW}$, $\lambda = 0.1$ m, and using (6.39) results in

$$R_1^4 = \frac{50 \times 10^3 \times 10^8 \times (0.1)^2 \times 2}{(4\pi)^3 \times 8 \times 10^{-14} \times 100}$$

$$\Longrightarrow R_1^4 = \frac{10^{11}}{64\pi \times 8 \times 10^{-11}} \quad \Longrightarrow R_1 \approx 50\,\mathrm{km}$$

2. Assumptions for the worst case are

$$|F| = 0.6,\ \ \mathrm{SNR} = 10,\ \ kTB_n = 4.14 \times 10^{-15},\ \ N_F = 3.16,\ \ A_e = 8\,\mathrm{m}^2$$

Now by applying (6.112) the following result is concluded:

$$R_{\max}^4 = \frac{50 \times 10^3 \times 10^4 \times 8 \times 2 \times 0.6^4}{(4\pi)^2 \times 100 \times 4.14 \times 10^{-15} \times 3.16 \times 10} = \frac{80 \times 10^8 \times 0.6^4}{0.2066 \times 10^{-8}}$$

$$R_{\max} = 26.6\,\mathrm{km}$$

3. For the best condition, where there is no noise and $|F| = 1.5$, then using equation (6.39) yields

$$R_i = R_1 \times |F| \quad \Longrightarrow \quad R_i = 50 \times 1.5 \approx 75\,\mathrm{km}$$

■

6.5.3 Interference and Jamming Condition

Jamming is a radar-specific terminology which implies RF interference that is produced intentionally. It is used to mislead or disable enemies' receiver by noise or false information. During a jamming condition, the basic radar equation specified by (6.39) is not valid and should be modified accordingly. A simple way to include this modification is to consider the target radiates high-power radiowaves as a noise source with the following density:

$$S_J = \frac{P_J \cdot g_j}{4\pi R_J^2} \ \mathrm{W/m^2} \tag{6.113}$$

where

P_J : Jammer power in W
g_j : Jammer antenna gain (including RF feeder losses)
R_J : Jammer distance to the search radar

Assuming that the relative bandwidth of the jammer B_J is greater than the bandwidth of the radar receiver, then the RX total noise power will be

$$N = kT_s B_N + \frac{P_J \cdot g_j \cdot B_N \cdot A_e}{4\pi R_J^2 \cdot B_J \cdot L_r \cdot L_a} \tag{6.114}$$

where

A_e : Radar antenna effective aperture area
L_r : RF feeder losses of the radar receiver
L_a : Atmospheric loss for jamming radiowaves

It is common to consider that the jamming power is strong enough to make troubles for proper receiving of radar signals, that is, the jamming noise density must be greater than the RX system noise density. Then assuming $L_t = L^2 \cdot L_p^2$, the radar basic equation for the effective range in jamming condition can be written as

$$R_J^2 = \frac{P_t \cdot g_t \cdot B_J}{4\pi B_N \cdot (S/J) \cdot L_t \cdot L_r \cdot L_a} \times \left(\frac{\sigma}{P_J \cdot g_j}\right) \tag{6.115}$$

Example 6.11. A jammer is radiating 1 KW power in a 500 MHz bandwidth with $g_j = 10$ to interfere with a radar system specified below:

$$P_t = 1\,\mathrm{MW},\ f = 3\,\mathrm{GHz},\ G_t = 2\times 10^4,\ \sigma = 1.5\,\mathrm{m^2},\ B_n = 1\,\mathrm{MHz}$$

Considering $(S/J)_{\min} = 10$, then calculate:

1. Effective range of radar for proper detection of the target in jamming condition
2. How much the maximum range will be changed for targets with $3\,\mathrm{m^2}$ cross section

Solution. 1. Applying (6.115) yields

$$R_j^2 = \frac{10^6 \times 2 \times 10^4 \times 500 \times 10^6 \times 1.5}{(4\pi) \times 10^3 \times 10 \times 10^6 \times 10} \implies R_j \approx 3.45\,\text{km}$$

2. For the target with RCS = 3 sqm,

$$R'_j = R_j \times \sqrt{3/1.5} \implies R'_j \approx 4.89\,\text{km}$$

■

6.5.4 Reception in Clutter Condition

To extract radar equation in the clutter condition, it will be treated in the following cases:

- Low grazing surface clutter
- High grazing surface clutter
- Volume clutter

In all of the above cases, it is assumed that other reception impairments such as radio noise, RX thermal noise, multipath reception, radio interference, and jamming are negligible or they are small when compared with clutter echoes.

- Low Grazing Angle
  As discussed before, the received power related to a surface clutter at low grazing angles can be expressed by (6.83) while for power of the direct received signal, the equation (6.39) is used ($L \cdot L_p = 1$). The ratio of these power levels simply yields

$$\frac{P_r}{P_c} = \frac{S}{C} = \frac{2\,\sigma_t \cdot \cos\,\psi}{\sigma^o.R \cdot \theta \cdot (c\tau)} \tag{6.116}$$

To obtain the maximum range of radar in this situation, the minimum signal-to-clutter ratio, $(\frac{S}{C})_{\min}$, should be selected, then

$$R_{\max} = \frac{2\,\sigma_t \cdot \cos\psi}{(S/C)_{\min} \cdot \sigma^o \cdot \theta \cdot (c\tau)} \tag{6.117}$$

- High Grazing Angle
  When the radar reception is subject to surface clutter at high grazing angles, the received power of the returned echoes can be expressed by (6.87) while for power of the direct received signal, the equation (6.39) is used ($L \cdot L_p = 1$). The ratio of these power levels simply results in

$$\frac{P_r}{P_c} = \frac{S}{C} = \frac{4\pi \cdot \sigma_t \cdot \sin\,\psi}{R^2 \cdot \theta \cdot \Phi \cdot \sigma^o} \tag{6.118}$$

To obtain the maximum range of radar in this situation, the minimum signal-to-clutter ratio, $(S/C)_{\min}$, should be selected, then

$$R_{\max}^2 = \frac{4\pi \cdot \sigma_t \cdot \sin\psi}{(S/C)_{\min} \cdot \theta \cdot \Phi \cdot \sigma^o} \quad (6.119)$$

- Volume Clutter
  If a radar target is subject to the rain or snow volume clutter, the equation (6.94) can be used to calculate the power level of direct received signal specified by (6.39) assuming $(L \cdot L_p = 1)$, then

$$\frac{P_r}{P_c} = \frac{S}{C} = \frac{P_t \cdot g_t \cdot \lambda^4 \cdot \sigma_t}{K_1 (4\pi)^3 \cdot R^2 \cdot z \cdot \tau}, \quad K_1 = 1.2 \times 10^{-10} \quad (6.120)$$

To determine the maximum range at which a radar target is detectable in the presence of volume clutter, we should accept the minimum signal-to-clutter ratio, $(S/C)_{\min}$, then

$$R_{\max}^2 = \frac{K_2 \cdot g_t \cdot \lambda^4 \cdot \sigma_t}{(S/C)_{\min} \cdot z \cdot \tau}, \quad K_2 = 4.2 \times 10^6 \quad (6.121)$$

6.6 Radar Coverage

6.6.1 Introduction

This section outlines some practical issues pertaining to the fixed radar networks including:

- Coverage diagrams
- Principles of site selection for fixed radar stations
- Safety against electromagnetic radiations

The above points may be slightly different based on the type and application of the radar systems, but we will focus on the common and general principles, and system-specific issues are left to other references.

6.6.2 System of Coordinates

Different systems of coordinates are used in the analysis of the radar waves propagation. For coverage diagrams, spherical coordinates are more popular with radar antenna acting as a focal point along with the following components:

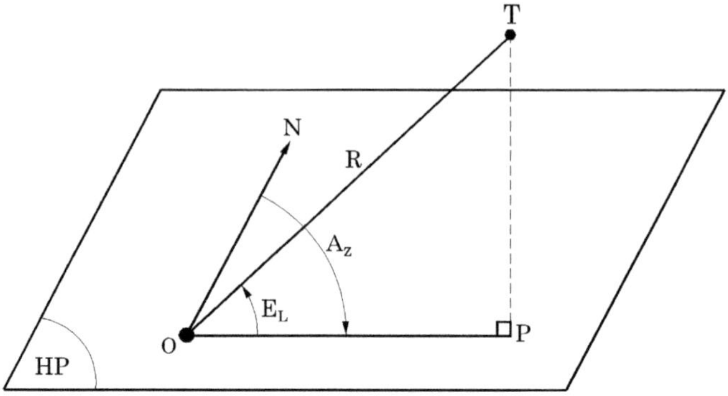

Fig. 6.27 3-dimensional coordinate system

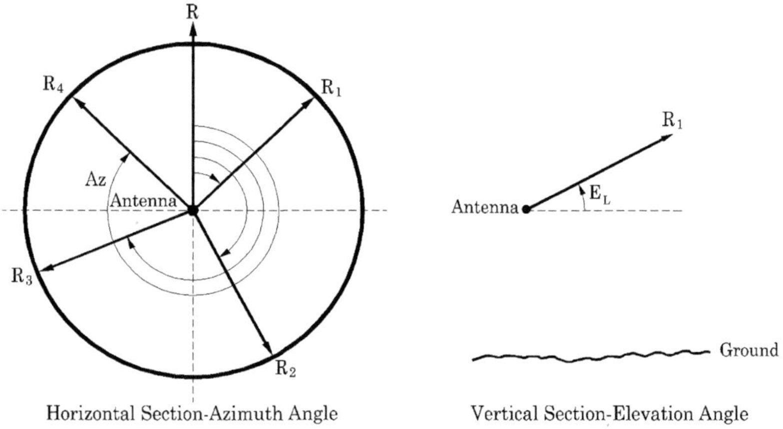

Fig. 6.28 Elevation and azimuth angles

- Distance denoted by R
- Elevation angle denoted by E_l
- Azimuth (or horizontal) angle denoted by A_z

Noting Fig. 6.27 and also HP plane which is a flat horizontal plane traversing the point O, each one of the above-mentioned components are defined below:

R : Distance between the origin O from the target location T, R = OT
E_l : Elevation angle between the line OT and its image on the HP plane
A_z : Azimuth angle between OT projection on the HP plane and true north direction

As shown in Figs. 6.27 and 6.28, the whole space around a radar station can be determined by R, E_l, and A_z. The limits of the mentioned angles are expressed below:

$$-90^\circ \leq E_l \leq 90^\circ \quad , \quad 0 \leq A_z \leq 360^\circ \tag{6.122}$$

Table 6.4 Main factors in radar coverage

Factors due to radar system characteristics	*Factors due to the structure of the propagation medium*
Antenna gain	Land and sea clutters
Altitude of antenna centerline	K-factor and propagation phenomena
Antenna A_z and E_l angles	Multipath reception
Transmitter output power	Terrain roughness
Type of radar and its application	Radio interferences/jamming
Coordinates of radar antenna	Radar cross section of targets

6.6.3 Main Types of Radar Diagrams

To analyze the area covered by each radar site, a number of diagrams are used, among which the following are common:

- Coverage area diagram in terms of the distance and azimuth angle (elevation angle would be a parameter)
- Range diagram in terms of azimuth for a fixed elevation angle
- Elevation angle diagram in terms of azimuth for a fixed distance
- Above ground level, AGL, coverage area diagram in terms of A_z for fixed E_l
- Mean sea level, MSL, coverage diagram in terms of A_z for fixed E_l

The coverage diagrams are colored charts organized on the topographic data of surrounding points around a radar site. These data along with radar-specific data as given in the Table 6.4 and a competent computation algorithm are used to produce a coverage diagram. The next section will provide some details about different types of the mentioned diagrams.

6.6.4 Coverage Diagrams

6.6.4.1 Covered Area

Two types of covered area diagrams are presented in Figs. 6.29 and 6.30. These diagrams are prepared for all points located in the range up to 150 km and 250 km, respectively, from the radar site for different elevation angles. The blank parts include all points covered by the radar when its antenna is adjusted for the specified elevation angle while dotted parts contain noncovered points.

Example 6.12. Refer to Fig. 6.29 and assume that the height of radar antenna and point A from sea level are 1,000 and 1,200 m, respectively, find:

1. Coordinates of point A
2. Determine altitudes at point A which are definitely under the radar coverage

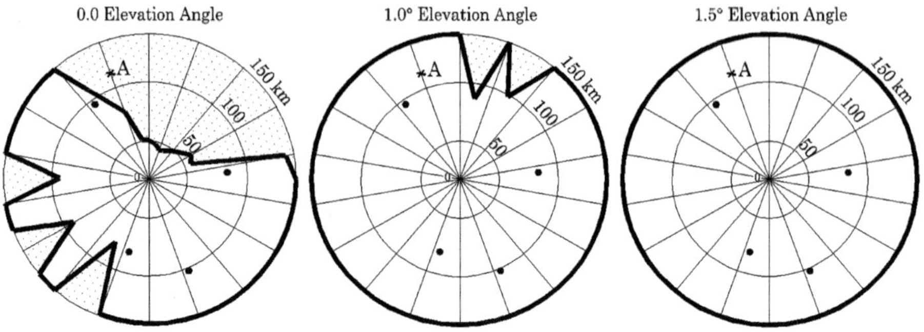

Fig. 6.29 Radar coverage in a 150-km range

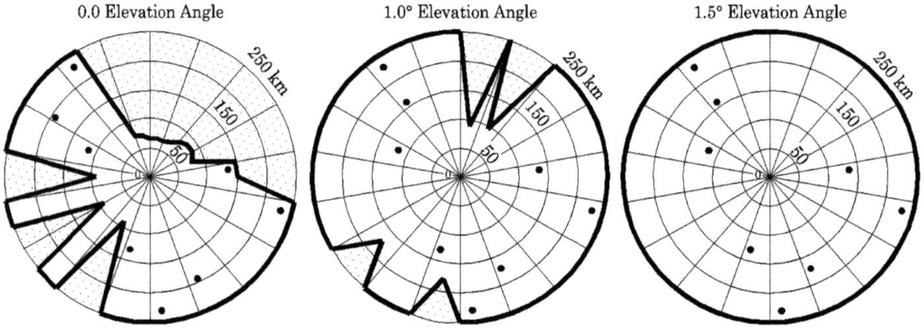

Fig. 6.30 Radar coverage in a 250-km range

Solution. 1. Noting Fig. 6.29, it yields

$$R = 120\,\text{km}\ ,\ E_l = 1^\circ\ ,\ A_z = 340^\circ$$

2. Noting the mentioned graph, all points with elevation angles more than 1° relative to point O are covered by the radar, thus,

$$\Delta h = 120\ \tan^{-1}(1^\circ)\ \implies\ \Delta h = 2{,}094\,\text{m}$$

$$h_A = (1{,}000 + 2{,}094) - 1{,}200 = 1{,}894\,\text{m}$$

■

6.6.4.2 Radar Range Diagram

A sample of this type of linear diagram is presented in Fig. 6.31. The horizontal axis includes azimuth angles in the range between 0 and 360°, and vertical axis indicates a distance of up to 250 km from the radar site.

The diagram is prepared for the antenna elevation angle adjusted for 0° to indicate the maximum coverage distance in the desired direction. For example, all points in the direction of an $A_z = 30^\circ$ are under coverage for a distance of up to 200 km, but the same value for an azimuth angle of 160° is limited to 50 km.

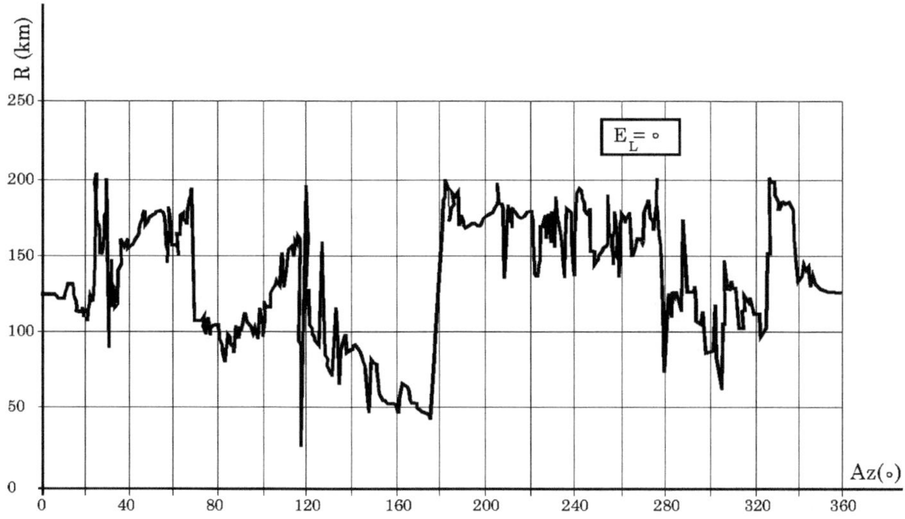

Fig. 6.31 Radar range diagram

6.6.4.3 Radar Angle Diagram

A sample of this type of linear diagram is shown in Fig. 6.32. Similar to the previous diagram, the horizontal axis includes azimuth angles in the range of 0–360°, vertical axis determines radar antenna elevation angles, and the required coverage range is considered as parameter of the diagram. Obviously by alteration of covering range, the variations of this diagram will be different.

As an example for a distance parameter of 100 km, the direction with $A_z = 100^\circ$ will be covered with elevation angles exceeding 6°. In other words, by elevation angle less than 6°, the 100-km range cannot be covered in the said direction.

Example 6.13. Using the coverage diagrams, determine:

1. The minimum covered distance for azimuth angles of 180–200° when the elevation angle of the antenna is set to 1°
2. The minimum antenna elevation angle for a coverage of 100 km for azimuth angles of 40–60°

Solution. 1. Considering the diagram in Fig. 6.31, the following range is concluded:

$$180^\circ \leq A_z \leq 200^\circ \implies R = 170\,\text{km}$$

2. Noting the diagram in Fig. 6.32 for the assumed parameters,

$$40^\circ \leq A_z \leq 60^\circ \ , \ R = 100\,\text{km} \implies E_l = 1^\circ$$

■

Fig. 6.32 Radar angle diagram

6.6.4.4 AGL Diagram

A sample of covered area diagram with specified altitudes from the ground surrounding the radar site called AGL diagram for 3 typical altitudes of 500, 1,000, and 1,500 m, and maximum covering range of 200 km is displayed in Fig. 6.33.

In fact, these diagrams determine the coverage status of a set of points with a specified heights from the ground surrounding the radar site up to a certain radius.

For radar systems which have a fixed antenna elevation angle, there is only one set of coverage diagrams available. Whereas for radars with variable elevation angle, a set of coverage diagrams may be prepared for every elevation angle of the radar antenna to facilitate selection process of the optimized coverage of interest.

These types of diagrams may be employed to specify meteorological radar coverage and to analyze the climate conditions of the cities, cultivated lands, and airports.

6.6.4.5 MSL Diagram

These types of diagrams with some difference are similar to the AGL diagrams which by using them, it is possible to determine the coverage status of points located in the space surrounding the radar with a fixed height relative to the mean sea level MSL. A sample of this diagram is displayed in Fig. 6.34 for altitudes of 2,000, 3,000, and 4,000 m and covering range of up to 200 km.

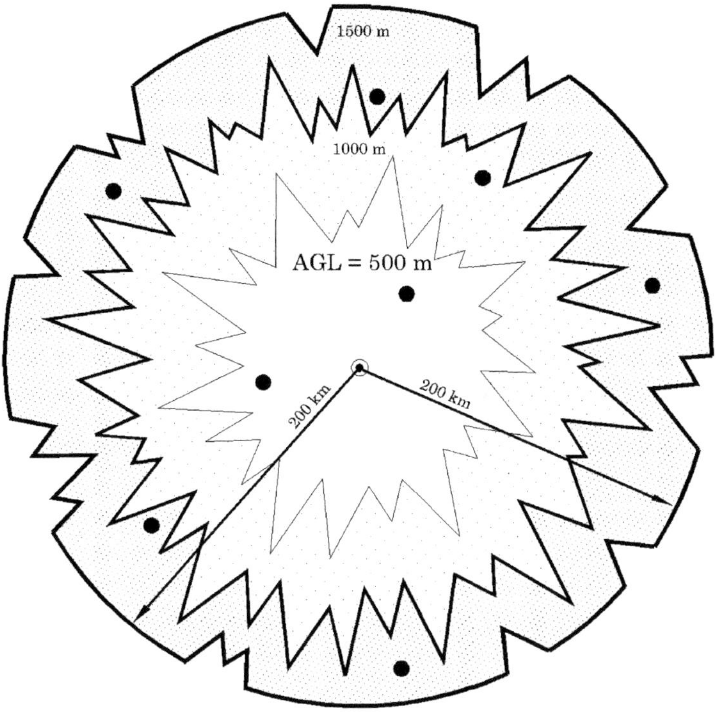

Fig. 6.33 Above ground level diagram (for 500, 1,000, and 1,500 m)

6.6.5 Radar Site Selection

Site selection for fixed radar stations is one of the major issues in design of radar networks having significant role in its operations and maintenance. The main steps in site allocation for a radar-fixed station that must be considered by the radar design engineers are indicated below:

- Specifying the objectives, data, and products of radar system.
- Evaluation and specifying initial alternatives using local studies and collecting relevant topographic data and maps.
- Producing and assessing coverage diagrams of each alternative and specifying superior ones.
- Study and specify how the pertaining station can be accessed.
- Local inspection, evaluation, and measurements.
- How to provide main and emergency electric power.
- How to provide communication channels for the purpose of connecting to the main telecom networks.

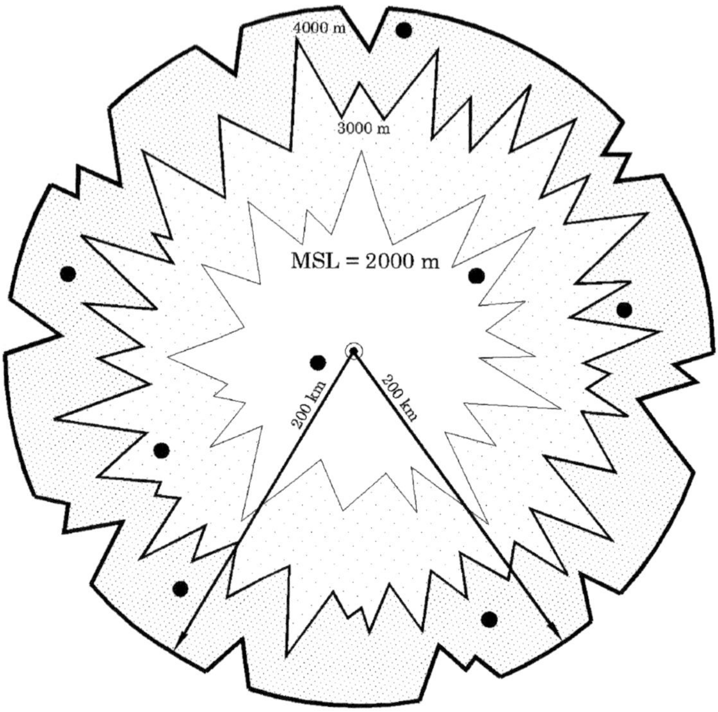

Fig. 6.34 MSL diagram for 2,000, 3,000, and 4,000 m altitudes

- Study the structure of station ground regarding geology, soil-bearing capacity, and other data required for civil works.
- Evaluating security and subversion issues.
- How to provide fuel and supply water for the station.
- Considering civil works required for construction of buildings and erection of radar tower.
- Requirement study for remote control and supervisory equipments.
- Study of overlapping with the neighboring radar sites for large networks to provide proper coverage.
- Evaluating electromagnetic pollutions and safety of radio emissions for humans and surrounding medium.
- Taking into account the RF interferences with different stations such as telecom centers, audio and TV transmitting stations, satellite terminals, and other neighboring radar sites.
- Considering issues related to the land acquisition of the radar site.
- Comparing the superior alternatives in terms of technical, operational, and economic aspects.
- Determination of optimized solution.

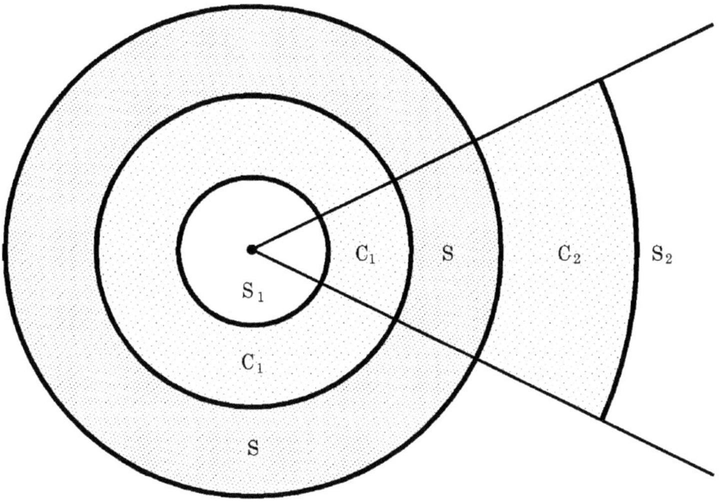

Fig. 6.35 Hazardous area classification around radar antenna

6.6.6 Radar Emissions

Considering a lot of losses created on the path of radar waves between the TX and RX units, it is essential to use a high-power transmitter and a large directional antenna. As an example, with TX power of 250 kW in C-band and antenna gain equal to 46 dB_i, the amount of effective transmission power is given by

$$\mathrm{EIRP} = 10 \log P_t + G_t = 54 + 46 = 100\,\mathrm{dB}_W$$

$$\mathrm{EIRP} = \mathrm{Antilog}[\frac{\mathrm{IIRP}}{10}] = 10^{10}\,\mathrm{W}$$

Thus, the effective transmission power is in the order of few hundred millions to few tens of billions watts potentially being able to make adverse effects described below:

- Generation of undesired electromagnetic waves interfering with other radio stations under specific conditions
- Causing harmful signals for human health

To counteract such undesired effects of radio interference, the pertaining calculations must be conducted on the basis of working frequency of radio channels, radiation patterns of antennas, TX(s) power, and RX sensitivity. Then, based on the obtained results, suitable methods must be considered in the design stage.

From safety perspective the radio emissions of radar stations given in Fig. 6.35, three zones can be identified and defined as below:

- Hazardous zone denoted as H
- Controlled zone denoted as C

- Safe zone denoted as S

The overall radius of each zone is influenced by the following factors:

- Radio frequency
- TX power
- Antenna gain
- Antenna radiation pattern
- Elevation and azimuth angles of the antenna
- Height of the antenna center point relative to neighboring land
- Acceptable criterion for maximum permissible exposure
- Highlands and lowlands in the vicinity of antenna
- Height of the location of interest

6.6.7 Safety Standard

6.6.7.1 Introduction

To determine safe, controlled, and hazardous zones around a radar station, it is a crucial requirement to specify necessary criterion. Institute of Electric and Electronic Engineering (IEEE) has presented document C95.1-1999 for this purpose which may be used for further clarification. An outstanding issue for a high-power transmitter used in a satellite ground station or fixed radar site is its equivalent radiated power, ERP, which includes TX power and related antenna gain.

This standard specifies the maximum permissible exposure, (MPE), and its relevant conditions. According to its definition, the peak value of radiated power density related to radiowaves, which is not harmful to human health, is called maximum permissible exposure.

Noting the relation of electric field intensity (E) and magnetic field intensity (H) with power density, this definition can be generalized for electric field intensity and magnetic field intensity. This section will provide tables and charts related to safety criterion using the above-mentioned standard.

6.6.7.2 Criterion for Controlled Zone

Maximum permissible exposure of electromagnetic waves with a power density S, electric field intensity E, and magnetic field intensity H, in the frequency range of 3 KHz to 300 GHz for a specified duration of time, is presented in the IEEE/C95.1-1999. In the case of continuous wave transmission, the value of peak power density is equal to MPE obtained from the table; therefore,

$$S_{\text{peak}} = \text{MPE}, \quad \text{CW} \tag{6.123}$$

Table 6.5 Maximum permissible exposure in controlled zone[a]

Electromagnetic fields				
Frequency limit (MHz)	Electric field intensity E (V/m)	Magnetic field intensity H (A/m)	Electromagnetic power density S (mW/Cm^2)	Average time (min) $\|E^2\|, \|H^2\|$, S
0.003~0.1	614	163	100	6
0.01~3	614	$16.3/f$	$100/f^2$	6
0.003~0.1	$1{,}842/f$	$16.3/f$	$900/f^2$	6
30~100	61.4	$16.3/f$	$1/f^2$	6
100~300	6.4	0.163	1	6
300~3,000	–	–	$f/300$	6
3,000~15,000	–	–	10	6
15,000~3,000	–	–	10	$616{,}000/f^{1/2}$

[a]IEEE standard C95.1-1999 titled *Safety Levels with Respect to Human Exposure to Radio Frequency Electromagnetic Fields, 30 KHz to 300 GHz* reprinted with permission from IEEE, 3 Park Avenue, New York, NY 10016-5997 USA, copyright 1999, by IEEE

However, for pulsed transmitted waves, the peak value of power density is equal to

$$S_{\text{peak}} = (\text{MPE}) \cdot T_{\text{avg}}/(5T_{\text{d}}) \tag{6.124}$$

where:

MPE: Maximum permissible exposure of power density according to Table 6.5 in terms of mW/cm^2

T_{avg}: Average time in terms of seconds

T_{d}: Pulse duration time in terms of seconds

Noting that in pulsed radar systems, the following relation exists between duty cycle and the mentioned times:

$$d_c = T_{\text{d}}/T_{\text{avg}} \tag{6.125}$$

Thus, relation (6.124) concerning pulsed radars converts into

$$S_{\text{peak}} = \text{MPE}/(5d_c) \tag{6.126}$$

The MPE chart for controlled zone for permanent stay in the radar antenna main lobe of less than 6 h is presented in Fig. 6.36.

6.6.7.3 Criterion for Safe Zone

Maximum permissible exposure to electromagnetic waves including power, electric and magnetic densities in frequency range of 3 KHz up to 300 GHz is presented in the IEEE/C95.1-1999 (Table 6.6). Steady presence of humans in areas with less than MPE is not harmful and can be regarded as safe zone.

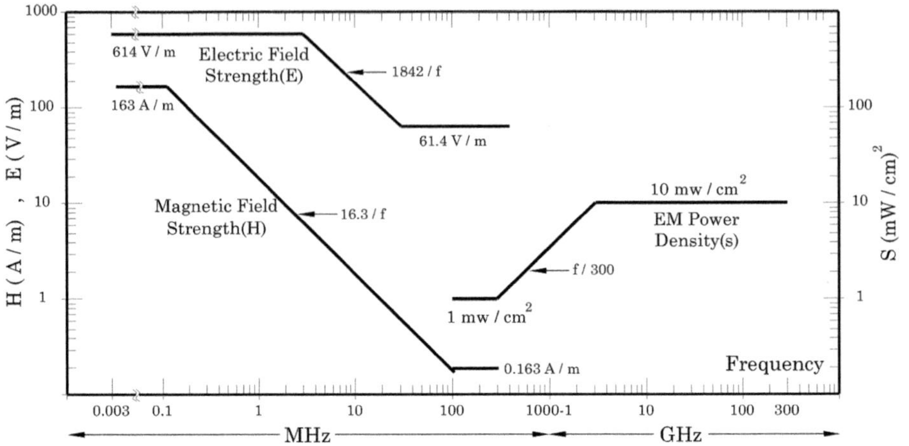

Fig. 6.36 MPE chart in controlled zone (IEEE standard C95.1-1999 titled *Safety Levels with Respect to Human Exposure to Radio Frequency Electromagnetic Fields, 30 KHz to 300 GHz* reprinted with permission from IEEE, 3 Park Avenue, New York, NY 10016-5997 USA, copyright 1999, by IEEE)

Table 6.6 Maximum permissible exposure in safe zone[a]

Electromagnetic fields					
Frequency limit (MHz)	Electric field intensity E (V/m)	Magnetic field intensity H (A/m)	Electromagnetic power density S (mW/Cm^2)	Average time (min) $\|H^2\|$	$\|E^2\|$, S
0.1∼0.003	614	163	100	6	6
1.34∼0.1	614	$16.3/f$	$100/f^2$	6	6
3∼1.34	$823.8/f$	$16.3/f$	$180/f^2$	$f^2/0.3$	6
30∼3	$823.8/f$	$16.3/f$	$180/f^2$	30	6
100∼30	27.5	$158.3/f^{1.668}$	$0.2/f^{3.336}$	30	$0.0636/f^{1.337}$
300∼100	27.5	0.0729	0.2	30	30
3,000∼300	–	–	$f/1{,}500$	30	–
3,000∼15,000	–	–	$f/1{,}500$	$90{,}000/f$	–
300,000∼150,000	–	–	10	$616{,}000/f^{1.2}$	–

[a]IEEE standard C95.1-1999 titled *Safety Levels with Respect to Human Exposure to Radio Frequency Electromagnetic Fields, 30 KHz to 300 GHz* reprinted with permission from IEEE, 3 Park Avenue, New York, NY 10016-5997 USA, copyright 1999, by IEEE

In such a position, the relations (6.121)–(6.124) are also valid, and particularly peak power density can be calculated for pulsed radars using relation (6.124). The maximum permissible exposure chart for safe zones is shown in Fig. 6.37.

Example 6.14. For a pulsed radar operating in C-band with the following specifications:

$$f = 5.6\,\text{GHz}\,,\quad P_t = 250\,\text{kW}\,,\quad G_t = 43\,\text{dB}_i\,,\quad d_c = 0.12\,\%$$

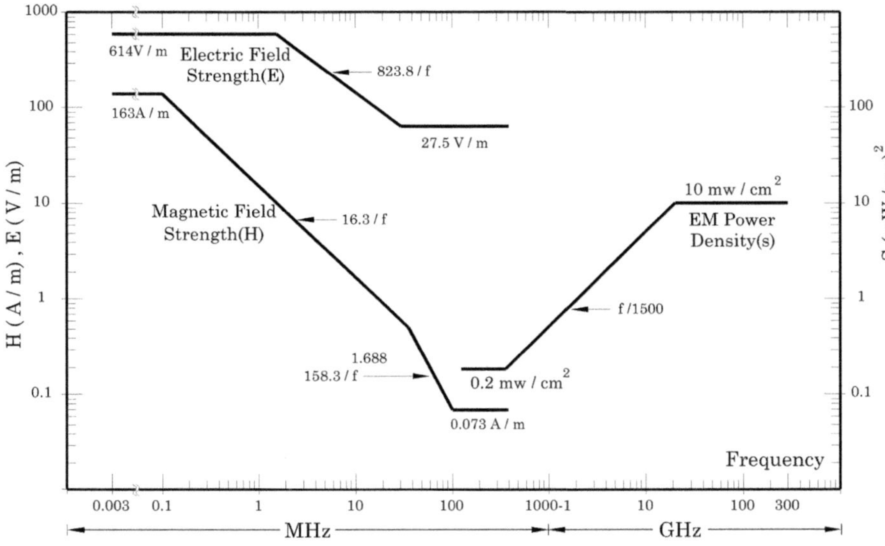

Fig. 6.37 MPE chart in safe zone. IEEE standard C95.1-1999 titled "Safety Levels with Respect to Human Exposure to Radio Frequency Electromagnetic Fields, 30 KHz to 300 GHz" reprinted with permission from IEEE, 3 Park Avenue, New York, NY 10016-5997 USA, copyright 1999, by IEEE

1. Calculate maximum permissible exposure density in the controlled and safe zones.
2. Calculate radius of the hazardous and controlled areas.
3. Plot safety diagram of the proximity of radar antenna.

Solution. 1.

$$\mathrm{MPE}(c) \xrightarrow[3{,}000<f<15{,}000]{\mathrm{IEEE}/C95.1-1{,}999} MPE(c) = 10\,\mathrm{mW/cm^2} = 100\,\mathrm{W/m^2}$$

$$S_{\mathrm{peak}(c)} = 100 \times \frac{1}{5 \times 0.12\%} = 16{,}666\,\mathrm{W/m^2} = 16.7\,\mathrm{kW/m^2}$$

$$\mathrm{MPE}(s) \xrightarrow[3{,}000<f<15{,}000]{\text{Table 6.5}} \mathrm{MPE}(s) = f/1{,}500\,\mathrm{mW/Cm^2} = f/150\,\mathrm{W/m^2}$$

$$S_{\mathrm{peak}(s)} = \frac{f}{150} \times \frac{1}{5d_c} = \frac{5{,}600}{150} \times \frac{1}{5 \times 0.12\%} = 6{,}222\,\mathrm{W/m^2}$$

$$S_{\mathrm{peak}(s)} = 6.2\,\mathrm{KW/m^2}$$

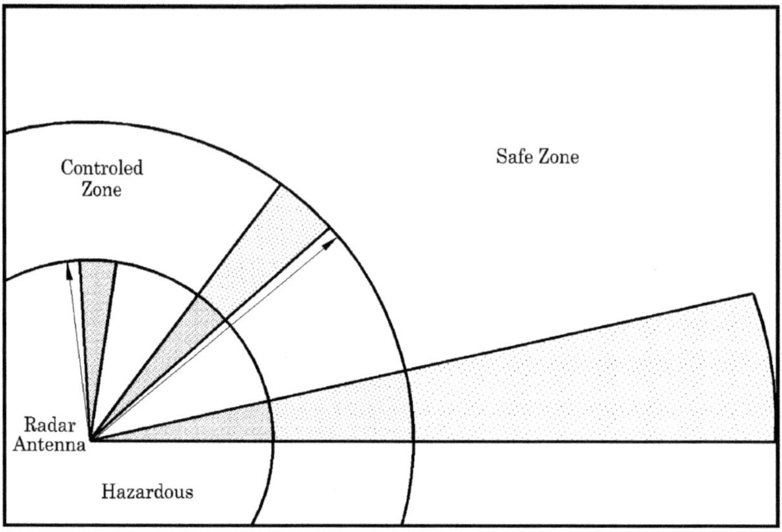

Fig. 6.38 Radar safety chart

2. The radius of hazardous area is equal to the radius of the minimum controlled distance calculated as follows:

$$G_t = 43\ \text{dB}_i \implies g_t = 20{,}000$$

$$\text{EIRP} = P_t \cdot g_t = 250 \times 20{,}000 = 5 \times 10^6\,\text{kW}$$

$$S_{\text{peak}(c)} = P_t \cdot g_t/4\pi R^2 = \text{ERP}/4\pi R^2$$

$$\implies R_h = \sqrt{\frac{5 \times 10^6}{4\pi \times 16.7}} \implies R_h = 154.4\,\text{m}$$

Radius of the controlled area, equivalent to minimum safe distance denoted by R_c, is calculated by the following relation:

$$R_c = \sqrt{\frac{\text{ERP}}{\pi S_{\text{peak}(s)}}} \implies R_c = 253.4\,\text{m}$$

3. The safety diagram around the intended antenna is presented in Fig. 6.38. ■

6.7 Exercises

Questions

1. Define radar and list its applications.
2. What factors make fixed and mobile radars different?
3. Explain types of radar network structure complete with simple schematics.
4. Determine frequency bands used in radars and prepare a report to describe how the waves propagate in each band and what are the major factors affecting this process.
5. Express the average and maximum value of the power in each type of CW and pulsed radars and write pertaining relations.
6. Define the radar cross section and mention its application in the related calculations.
7. Describe the waves scattering when they encounter flat surfaces and determine its effect on the reception of radar waves.
8. Refer to Figs. 6.8 and 6.10 and determine the effects of geometrical shape of the plates on their radiation pattern.
9. Explain the radar equation in the simplified case and extend it to the complex cases.
10. Specify reception power and signal-to-noise ratio in a simple situation for monostatic radars and for a condition where the route of radar waves is reflective and noisy.
11. Determine the place of free space loss in radar equations and define its relation with frequency and distance.
12. Noting Fig. 6.12, evaluate the medium of radar wave propagation and summarize the main losses under real conditions in the form of a list
13. Considering the graphs of Fig. 6.13 and neglecting the losses below 0.1 dB/km, then tabulate frequency bands at which the effects of gases, rain, and water vapor are significant.
14. Explain the effects of troposphere and ionosphere layers on the radar waves.
15. Define Doppler effect, state the reason for its occurrence, and express its pertaining relation.
16. Define clutter in radar links and specify its sources and types.
17. Define the radar cross section of surface and volume clutters and their relations.
18. Define the radar horizon and express its equation. Describe how the K-factor affects the magnitude of this horizon.
19. Define path gain factor in propagation of radar waves and determine relationship between the received power and this factor. Plot the normalized graph of the received power against the power on direct path in terms of radar antenna height.
20. Specify the main noise components in radar links and express the RX noise figure and thermal noise and their pertaining relations.

21. Determine the application of the radar coverage diagrams and their different types. Also state affecting factors on these diagrams.
22. Referring to the diagrams provided in the book, define the distance, angle, AGL, and MSL diagrams complete with their applications.
23. Give a list of main issues in site selection of radar stations.
24. Referring to the safety chart of the radar emissions as per Fig. 6.34, define each one of the hazardous, safe, and controlled zones and their pertinent criterion.
25. List radar parameters which may be applied by radar system engineer to increase its range in the rain clutter. Which parameter has the most impact on radar detection capability, explain why.
26. Specify limiting factors for the maximum radar range by giving related equations and providing a brief discussion about each of them.

Problems

1. A radar TX is connected to a 50 ohms matched antenna. The equation of output signal in metric system is given below:

$$S(t) = 2{,}000\ \sin(5 \times 10^{p} \pi t + 30^{\circ})$$

   (a) Calculate the maximum output power of the antenna.
   (b) Also calculate the average power in CW and pulsed radars assuming that $T_R = 1$ ms and $\tau = 50\,\mu$s.

2. If the gain of antenna used in the previous problem is 43 dB$_i$, then calculate:

   (a) Equivalent transmission power in the direction of its main lobe
   (b) Equivalent transmission power in the direction of the side lobe if its gain is 23 dB less than antenna gain in its principal direction

3. Assume a rectangular plate having dimensions of $1 \times 2\,\text{m}^2$. If this plate is rotated by 30° relative to the direction of the main radar wave, what is the value of its radar cross section.
4. Passive reflectors are extensively used as plain metallic plates in the LOS radio communications to alter the wave direction and/or to pass the obstructions in 2–15-GHz frequency bands. In a communication link working in a 7.5-GHz band, a flat rectangular reflector is used having dimensions of $3 \times 5\,\text{m}^2$.

   (a) Calculate the radar cross section of this passive reflector.
   (b) Obtain the maximum gain of the reflector in terms of dB.
   (c) Assuming the angle between main incident and reflected waves equal to 100°, then calculate its radar cross section and gain under these conditions.

5. A radar is installed on a naval vessel at the height of AMSL = 20 m. TX power is 10 kW and antenna gain is 45 dB$_i$ in the X-band. To cover an aircraft at 5-km

altitude, calculate the maximum distance of the aircraft in order to be detected by the radar. Assume radar frequency $f = 10\,\text{GHz}$ and $\sigma = 10\,\text{m}^2$ for the aircraft.

6. Received noise power in a radar system is equal to 100 pW and the least signal-to-noise ratio required is 8 dB, calculate:

   (a) The signal power required for detection
   (b) The maximum range if the radar system of problem 5 is used for detection of targets having a radar cross section of $8\,\text{m}^2$

7. Conduct the calculations of problem 5 in logarithmic system and finally compare the results.
8. A radar system is tracking a target with the following specifications and having a radar cross section of $1\,\text{m}^2$:

$$P_t = 100\,\text{kW}\ ,\ \ G_R = 37\,\text{dB}_i\ ,\ \ f = 2.5\,\text{GHz}$$

$$R = 20\,\text{km}\ ,\ \ T_s = 600°\text{K}\ ,\ \ L_t = 5\,\text{dB}$$

   (a) Calculate the signal-to-noise ratio.
   (b) If a minimum SNR = 10 dB is required for a proper detection, then calculate the maximum target distance.

9. For a radar system in the K-band, there is a heavy raining of 10-mm/h intensity, and the target distance from radar is 20 km. Calculate the additional losses from gases and rain. (You may use the graph of Fig. 6.13.)
10. An aircraft is flying at a fixed altitude of 6 km from the ground level. The aircraft radar identifies the target at a 20 km distance. Assuming that the velocity of the plane is 900 km/h and its radar frequency is 10 GHz, calculate:

   (a) Doppler frequency for fixed target and the value of received waves' frequency
   (b) Doppler frequency when the target is moving away from the plane with a velocity of 100 km/h in the direction of 30°, related to the plane direction

11. Calculate radar cross section of surface clutter for a wave having the following specifications:

$$R = 20\,\text{km}\ ,\ \ \Phi = \theta = 1°\ ,\ \ \psi = 5°\ ,\ \ f = 3\,\text{GHz}$$

   (a) Solve this problem for sea surface.
   (b) Solve this problem for cultivated land.

12. A radar system with 100-kW output power is using an antenna with 45-dB$_i$ gain in 5.6-GHz frequency. If its RX noise figure is 4.5 dB and all miscellaneous losses are 25 dB, ambient temperature is 27°C and $B_n = 5$ KHz, find:

   (a) The signal-to-noise ratio for a high-altitude target with radar cross section of $3\,\text{m}^2$ when the direction of moving target makes an angle of 60° with the wave path
   (b) The maximum radar range to achieve SNR = 11 dB

13. Repeat Example 6.8 for $\alpha_k = 3$ and $\theta = 22.5°$.
14. (a) Noting Fig. 6.29 for a minimum elevation angle of 1°, find minimum and maximum radar range.
    (b) Using graph of Fig. 6.32, determine the least antenna elevation angle to cover distances of up to 100 km in the horizontal direction of 180–200°.
15. Using AGL and MSL diagrams indicated in Figs. 6.33 and 6.34 of this chapter, determine:
    (a) Heights from the ground surface covered in the azimuth angles of 90 and 180°.
    (b) Specify the covered areas with a height more than 2,000 m above the sea level.
    (c) Is it possible to find the antenna elevation angle for the above results.
16. Solve the Example 6.14 of the book for the following assumptions:

$$f = 2.5\,\text{GHz}\,,\ P_t = 750\,\text{kW}\,,\ G_t = 37\,\text{dB}_i\,,\ d_c = 0.15\,\%$$

17. The boresight direction (main axis) of the antenna in previous problem is adjusted for elevation angle of 0°. Assuming the antenna is installed on a tower at 30 m high, then calculate the minimum and maximum hazardous, safe, and controlled distances on the ground level surrounding the radar tower.
18. Repeat problem 17 for conditions where antenna elevation angle being adjustable between -2 to $+10°$ and antenna horizontally rotating 360° at heights of 1.8 and 3 m from the ground level, then plot its safety chart.
19. Maximum range of an S-band radar for detecting a 1-m^2 target is 250 km, find:
    (a) Range of detection of a single bird with RCS = 0.01 m^2
    (b) Range of detection of a small fighter with RCS = 2 m^2
20. Calculate the attenuation of uniform rainfall distributed throughout the radar coverage and the radar cross section of its antenna pointing with 1.5° by 3° beamwidth and a 2 μs pulse width at the following condition:
    (a) $r = 4$ mm/h, $f = 10$ GHz, $R = 20$ km
    (b) $r = 4$ mm/h, $f = 35$ GHz, $R = 10$ km
    (c) $r = 10$ mm/h, $f = 3$ GHz, $R = 30$ km
21. Determine the range at which the radar echoes equal RX thermal noise for the following system:

$$P_t = 100\,\text{kW},\ f = 10\,\text{GHz},\ \tau = 0.1\,\mu\text{s}$$

$$\text{Azimuth HBW} = 2°\,,\ \text{Elevation HBW} = 10°\,,\ N = 4\,\text{dB}$$

where the radar is located at a height of 1 km over a flat sea surface with $\sigma° = -40$dB at 3° grazing angle.

22. A radar system is used for detection of flying targets around an airport. Technical specifications of the radar equipment are:

$$P_t = 100\,\text{kW}\,,\ f = 5.6\,\text{GHz}\,,\ \tau = 1.5\,\mu\text{s}\,,\ G_t = 44\,\text{dB}_i$$

(a) Find maximum range of radar when there is clear sky condition for detection of targets with $2\,\text{m}^2$.
(b) Calculate reduction percent of the radar range in reflective routes where the path gain magnitude is $|F| = 0.7$.
(c) Repeat the case for situation when there is typical surface clutter (assume typical values for θ , Φ, E_l, σ_0, and A_c).
(d) Repeat the case for a typical rain clutter.

Chapter 7
Short-Range Radio Communications

7.1 Introduction

Short-range radio communications generally refer to communications over a distance of up to 1 km. Currently, this type of information exchange has found very extensive applications, and as a result, it is necessary to evaluate short-range radiowaves propagation and study how it can be properly modeled.

The nature and applications of this type of communications is categorized as part of mobile radio systems. Propagation of mobile radiowaves for long distances has been extensively assessed in the other chapter, and pertaining relations are presented on the basis of different models. This chapter is devoted to the study and evaluation of radiowaves propagation for short-range distances which are basically divided into the following two types:

- Short-range outdoor radio communications
- Indoor radio communications

Different positions of short-range radio communications are depicted in Fig. 7.1 for which radio section of ITU has presented the following recommendations:

- ITU-R, P.1411 for short-range radio communications in outdoor areas for frequency range of 300 MHz to 100 GHz.
- ITU-R, P.1238 for radio communications in indoor areas for frequency range of 900 MHz to 100 GHz.

Considering the far field criterion of the electromagnetic waves and noting that the wavelength for UHF waves is in the range of 0.1–1 m, the waves in this situation shall be evaluated under far field conditions. This chapter is not concerned with indoor radio communications pertaining to MF, HF, and VHF waves which are basically dedicated to long-distance communications systems especially public broadcasting.

A. Ghasemi et al., *Propagation Engineering in Radio Links Design*,
DOI 10.1007/978-1-4614-5314-7_7,

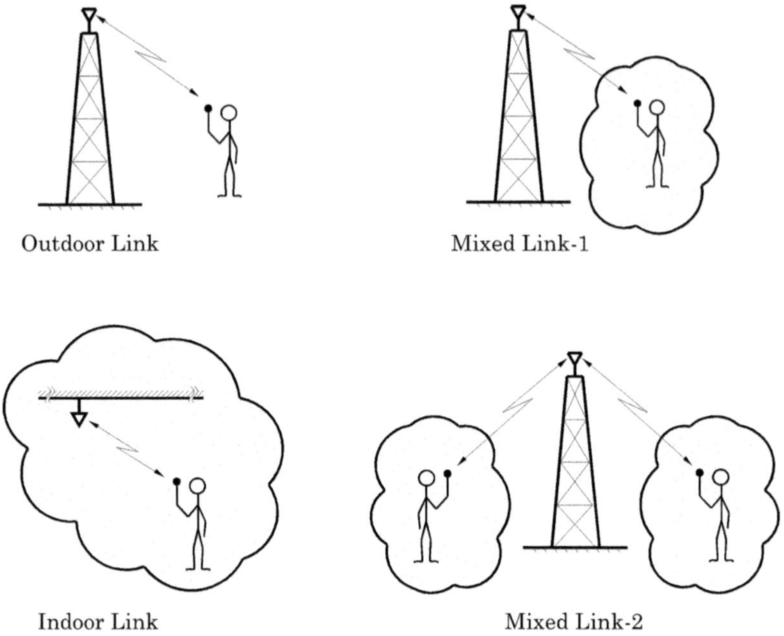

Fig. 7.1 Main types of short-range communications

7.2 General Considerations

7.2.1 Applications

Among many and ever expanding applications of short-range radio communications, the following examples are of more interest:

- Radio local area network, RLAN, and/or wireless local area network, WLAN
- Wireless local loop, WLL
- Wireless private branch exchange, WPBX
- Electronic games
- Telemetry and remote sensing systems in industrial plants
- Remote control devices
- Access systems and communications with backbone networks
- Video transfer systems over short distances
- Personal communications such as cordless telephone sets
- Radio services in mobile, personal, and in flight environments

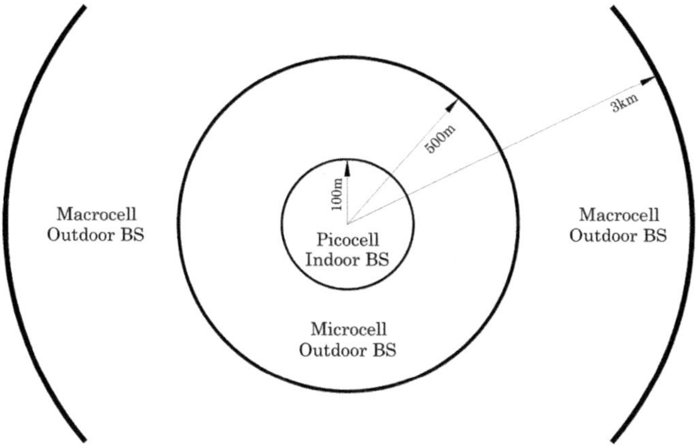

Fig. 7.2 Types of radio cells in short-range communications

7.2.2 *Frequency Bands*

All UHF, SHF, EHF, and even infrared and laser bands are used for short-range radio communications. For this purpose, frequency spectrum within the specified bands is not allocated entirely for short-range radio links, but specific slots are used for short-range communications as indicated in the Article 5 of radio regulations.

The key point concerning the intended frequency bands is the frequency reuse policy as a planning basis for this type of communications due to the following reasons:

- Low power of the transmitter output and, as a result, their limited propagation range
- Very short periods of usage in most cases
- No need for frequency diversity techniques
- Low bandwidth due to low bit rate requirements

7.2.3 *Types of Radio Cells*

The radio cells in short-range communications, as depicted in Fig. 7.2, are categorized into three categories:

- *Small Macro-cells.*
  Small macro-cells have a radius of 500 m up to 3 km, and related BS (base station) equipments are generally installed outdoors and propagate inside open areas. BS antennas are installed on the top of simple towers or on the rooftops.

- *Micro-cell.*
  Micro-cells have a radius of 100–500 m, and the location of BS antenna is generally outdoors and positioned at a lower height than the heights of buildings.
- *Pico-cell.*
  Pico-cells have a radius not more than 100 m, and their BS antennas are installed indoor.

7.2.4 Main Features

The main and basic features of short-range communications which make it unique compared to other mobile radio systems are given below:

- Using radio units with low TX power.
- Simplicity of BS sites and no need for sophisticated and expensive antennas.
- Heavy and high radio towers are not required.
- Standby systems are not required in most cases.
- Low electric power consumption.
- Utilization of CT, DCT, DECT, TDMA, FDMA, CDMA, and SCDMA technologies.

7.3 Quality Aspects of Short-Range Radio Links

Quality of radiowave propagation on short distances can be characterized by the following issues.

7.3.1 Propagation Impairments

The main phenomena of radiowaves propagation over short-range radio links include, but not limited to, the following items:

- *Reflection of the waves* from surfaces with large dimensions relative to their wavelengths. Such surfaces may be the floor, the ceiling, and/or the walls of the rooms also including the ground surface and rooftops.
- *Diffraction* of the waves caused by the edge of obstructions along radiowaves.
- *Scattering of the waves* caused by the waves encountering surfaces and bodies with dimensions less than the radiowaves wavelength such as road side traffic signs, electric pylons, and tree bushes or foliage.
- *Temporal and spatial variation* of mobile terminals resulting in the multipath, scattering, and shadowing mechanisms.

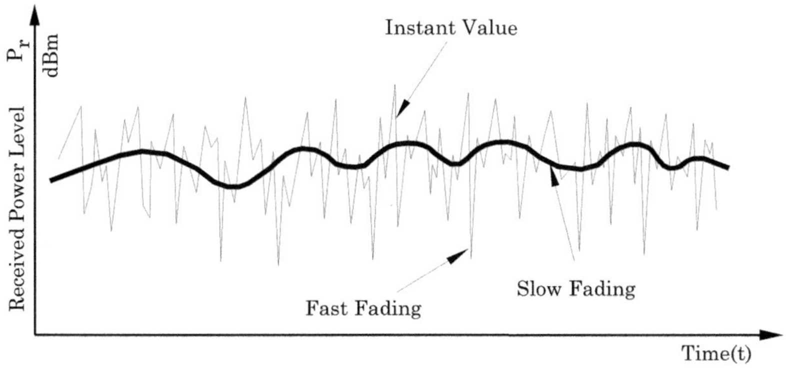

Fig. 7.3 Fading in received signal level

- *Effects of motions* made by persons and obstacles resulting in the fluctuation of the received signal level according to Fig. 7.3. As the amount of speed increases, the changes will be more rapid. For instance, in cellular systems, the losses will reach up to 30 dB along with delay spread of 3 μs.
- *Polarization mismatch* due to random alignment of mobile terminal.
- *Air ducting* or channeling of electromagnetic energy, especially in corridors at high frequencies.
- *Atmospheric phenomena* such as rain, snow, hail, rain shower, water vapor, cloud, mist, dust, and soil. Noting that radio channels may be impaired severely on the outdoor links, these phenomena do not exist for the indoor links, and consequently, their effects can be disregarded.
- *Large-scale fading* due to obstacles and objects on radio links being established in their shadows causing considerable losses in the received signal.
- *Small-scale fading* caused by multipath effects creating abrupt changes in the received signal level in small areas and/or different time periods. This phenomenon is caused by changes emanating from Doppler effect and frequency modulation. This effect is generated by moving objects even if the position of mobile terminal is fixed.
- *Delay spread* due to the reception of radiowaves by the RX antenna from different routes as shown in Fig. 7.4. Obviously, its value depends on the structure of medium and RX position.

  These types of delays are capable of interfering with the radiowaves resulting in the changes of data rate. It is defined with the following relation:

$$R < \frac{1}{2\tau_d} \tag{7.1}$$

where R is the rate of data transfer in terms of b/s and τ_d is the delay spread in terms of second.

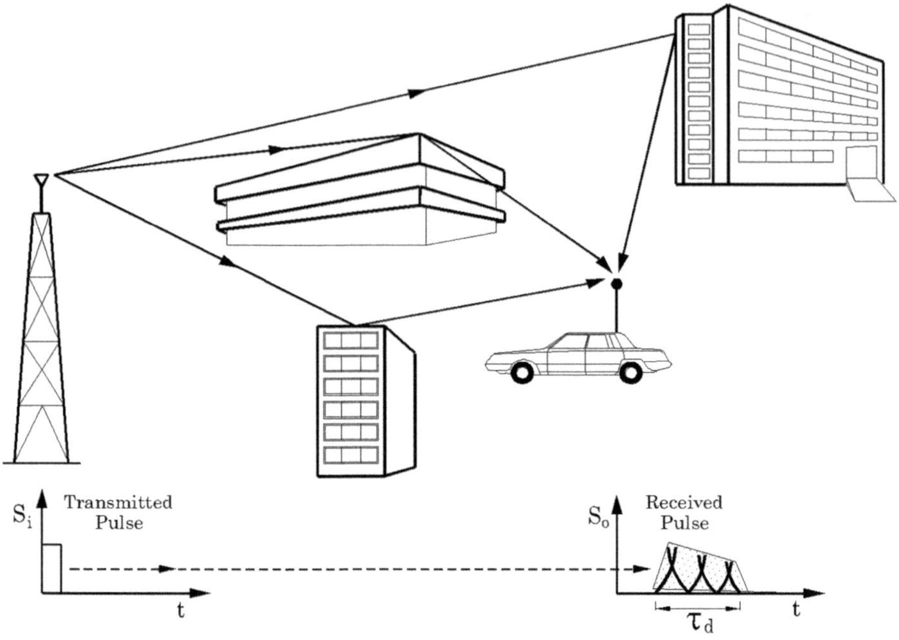

Fig. 7.4 Multipath reception and delay spread with distance

– Adverse effects on the radio channels are listed below:

  - Multipath dispersion
  - Delay spread
  - Frequency selective fading
  - Rayleigh fading

7.3.2 Outdoor Radiowave Propagation

7.3.2.1 Losses

Losses in radiowaves propagation in the outdoor environment can be divided into four categories as indicated below:

- Free-space losses
- Open mediums losses
- Losses in the Suburbs
- Losses in the urban areas

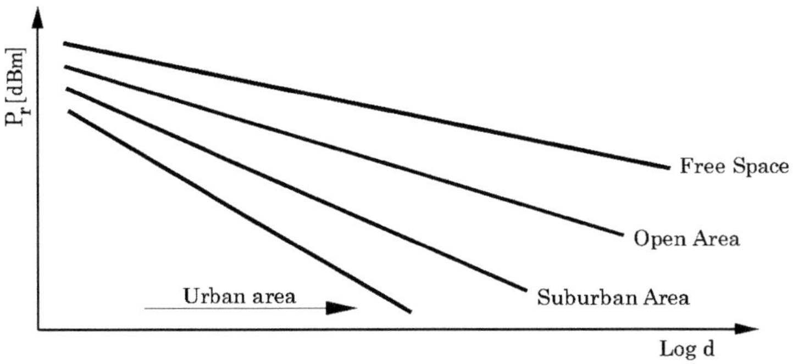

Fig. 7.5 Relative attenuation in outdoor media

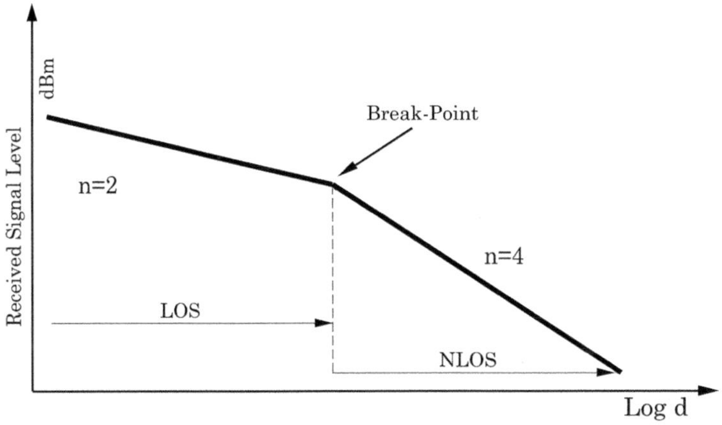

Fig. 7.6 Received power variations vs. distance

7.3.2.2 Signal Level

The received signal level in typical environments and under the same conditions concerning transmitter power, frequency of the radio channel, and distance is relatively illustrated in Fig. 7.5.

The received signal level can be simply defined for a single fixed frequency by

$$P_r = K\, d^{-n} \tag{7.2}$$

In the above equation, P_r is the relative received power, K is a constant, d is the distance in terms of meters, and n is the distance power loss exponent which is $n = 2$ for the open areas; otherwise, it will be more than 2 up to 4 $(2 < n \leq 4)$ depending on the medium conditions. Figure 7.6 displays the graph of the received power at the RX site versus the distance. As it may be noted, on the regions where there is LOS radio communications, the value of n is 2, whereas, for NLOS conditions when the scattering phenomenon prevails, then the value of n is 4.

Table 7.1 Comparison of main characteristics short-range vs. typical mobile radio communications

Description	Short-range communications	Mobile radio communications
Cell radius	Up to 1 km	Tens of km(s)
TX power	0.1–1 W	1–25 W
Fading distribution	Rician	Rayleigh
Delay spread	10–100 ns	0.1–10 μs

7.3.2.3 Main Characteristics

Main characteristics of short-range radio communications in comparison with ordinary mobile radio communications systems are indicated in the Table 7.1.

Example 7.1. In an outdoor communication link at 1,800 MHz, calculate the path losses when $d = 500$ m for the following two positions:

1. Line of sight, LOS
2. Non-line of sight, NLOS

Solution. 1.

$$L_{b1} = 32.4 + 20 \log\left(\frac{f.d}{1{,}000}\right) = 91.5\,\text{dB}$$

2.

$$L_{b2} = 32.4 + 40 \log\left(\frac{f.d}{1{,}000}\right) = 150.6\,\text{dB}$$

■

7.3.3 *Indoor Radiowaves Propagation*

7.3.3.1 Propagation Medium

In this situation, the propagated radiowaves within indoor areas are attenuated much faster. Under such conditions, usually the operating radius is less than 100 m, and there are obstacles such as the walls, ceilings, and/or human traffic in the path of radiowaves propagation. One of the distinct features of indoor media is lack of atmospheric phenomena such as rainfall, snow, hail, and rain showers.

7.3.3.2 Radiowave Path Loss

The losses of radiowaves within the indoor media for a frequency of 900 MHz can be simply stated by the following relation (suggested in some references):

$$L_b = 30 + N \log d + K_1 F + K_2 W \tag{7.3}$$

In the above relation, each of the components and their pertaining units are indicated below:

L_b: Path loss in terms of dB
N : Distance power loss coefficient
d : Distance between TX and RX units in m
K_1: Number of floors between TX and RX units
K_2: Number of the walls between TX and RX units
F : Loss of each floor in terms of dB
W : Loss of each wall in terms of dB

It must be explained that the distance power loss coefficient denoted by N lies within the limits of 20–60 with prevailing values of 25–40.

7.3.3.3 Other Factors

The other effective factors increasing radiowave loss within indoor areas are as follows:

- Human movements resulting in formation of new routes for radiowaves causing more losses up to 10 dB.
- Metallic plates and/or partitions which create the radiowaves delay spread of 30–60 ns.
- Walls with losses of about 10–15 dB.
- Ceilings and/or floors with losses of around 12–27 dB.
- Received power at the RX location depending on the composition and structure of the said location, construction materials, their density, and related hardwares.
- Radiowaves crossing through the window glasses suffer attenuation about 6 dB less than the walls.

Example 7.2. In an indoor radio link at a distance of $d = 80$ m, the 900-MHz band is used. Calculate the path losses for the following cases based on (7.3):

1. LOS radio link
2. NLOS radio link at a floor with $N = 45$ and a wall in between with 12 dB losses
3. The link between the third and fifth floors including an average of 20 dB per each floor and $N = 38$

Solution. 1.

$$L_{b1} = 30 + 20 \log 80 = 68 \text{ dB}$$

2.

$$L_{b2} = 30 + 45 \log 80 + 1 \times 12 = 127.6 \text{ dB}$$

3.

$$L_{b3} = 30 + 38 \log 80 + 2 \times 20 = 142.3 \text{ dB}$$ ■

7.4 ITU-R Method for Short-Range Outdoor Radio Communications

7.4.1 Introduction

Based on ever-increasing demand for short-range outdoor radio communications covering small distances and noting the following facts:

1. Numerous services provided in mobile communications for limited distances such as remote control services, telemetry systems, etc.
2. Extensive requirements for establishing RLAN and/or WLAN services
3. Possibility of using low-power transmitters with tens of mw and reducing their frequency pollution
4. Low usage time per each user
5. Possibility of using wireless local loop, WLL capabilities
6. Using such systems for distribution networks of WLL as a quick and suitable alternative for the existing cable networks and their future expansions

International Telecommunication Union (ITU) in 1999 presented the first draft of its recommendation No. ITU-R, P.1411 and revised it several times later on. Noting the complexity and versatility of this topic, its modifications still continue. This section is treated based on the subjects and method of this recommendation and shall be pursued accordingly.

7.4.2 Propagation Medium

Radiowave propagation on short routes being used in urban and suburban media are basically influenced by the buildings and trees. The role of buildings and different structures is greater and more distinct. The mobile terminals are mostly used by pedestrian users of urban areas and at the presence of vehicles.

In this position, typical environments are categorized in four classes as summarized in Table 7.2. Recognizing that there is a wide variety of environments within each category, it is not intended to model every possible case but to the given models that are more common and representative.

Table 7.2 Propagation impairments for different physical operating environments (Ref.: ITU-R, 1411-4, Tables 7.1 and 7.2)

Environment	Propagation impairments	Velocity of vehicular users
Urban areas with high-rise buildings	• Urban canyon, characterized by streets lined with tall buildings of several floors each • Building height makes significant contributions from propagation over rooftops unlikely • Rows of tall buildings provide the possibility of long path delays • Large numbers of moving vehicles in the area act as reflectors adding Doppler shift to the reflected waves	50 km/h
Urban/suburban with low-rise buildings	• Typified by wide streets • Building heights are generally less than three stories making diffraction over rooftop likely • Reflections and shadowing from moving vehicles can sometimes occur • Primary effects are long delays and small Doppler shifts	50~100 km/h
Residential areas	• Single and double storey dwellings • Roads are generally two lanes wide with cars parked along sides • Heavy to light foliage possible • Motor traffic usually light	40 km/h
Rural areas	• Small houses surrounded by large gardens • Influence of terrain height (topography) • Heavy to light foliage possible • Motor traffic sometimes high	80~100 km/h

7.4.3 Geometry of Radiowave Path

7.4.3.1 Main Positions

Figure 7.7 illustrates four main situations of the radiowave path geometry as described below. It must be mentioned that BS_a is a fixed station with an antenna installed at a height higher than the rooftops of the buildings and BS_l is a fixed station with an antenna installed at a height lower than the rooftops of the buildings:

- First position relates to a direct (line of sight, LOS) path between BS_a and MS_1 terminal
- Second position relates to a direct (line of sight, LOS) path between BS_l and MS_2 terminal

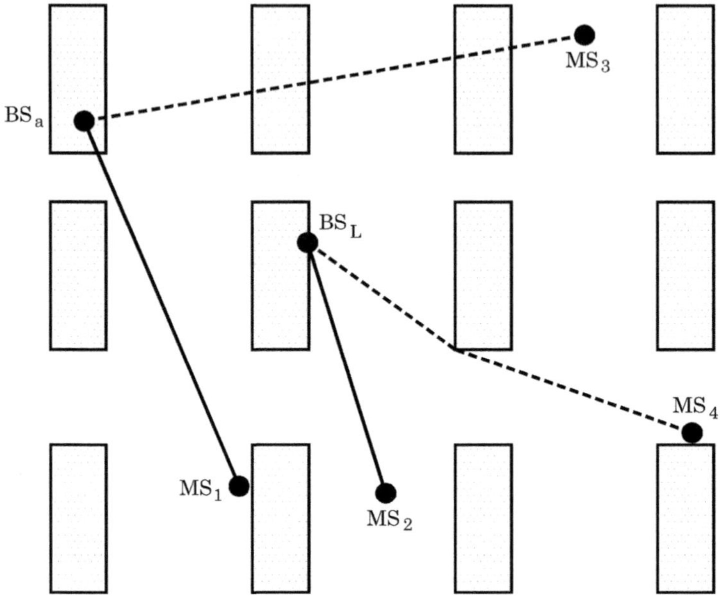

Fig. 7.7 Typical propagation situations in urban areas (Ref.: ITU-R, P.1411-4)

- Third position relates to an indirect (non-line of sight, NLOS) path between BS_a and MS_3 terminal
- Fourth position relates to an indirect (non-line of sight, NLOS) path between BS_l and MS_4

7.4.3.2 Line-of-Sight Paths

Radiowave propagation paths of line of sight, as indicated in Fig. 7.7, include the paths of $BS_a - MS_1$ and $BS_l - MS_2$. Noting that the radio wavelength (λ) is much less than the distance between the TX and RX(d), the electromagnetic waves shall be considered under far field conditions for which transmission loss must be calculated according to the methods of LOS communications.

7.4.3.3 Non-line-of-Sight Paths

Indirect paths are categorized in two non-line of sight, NLOS types named NLOS1 and NLOS2. The former relates to the paths over the rooftops, and the latter relates to the paths through the roads. The following descriptions provide some details about each one of these situations.

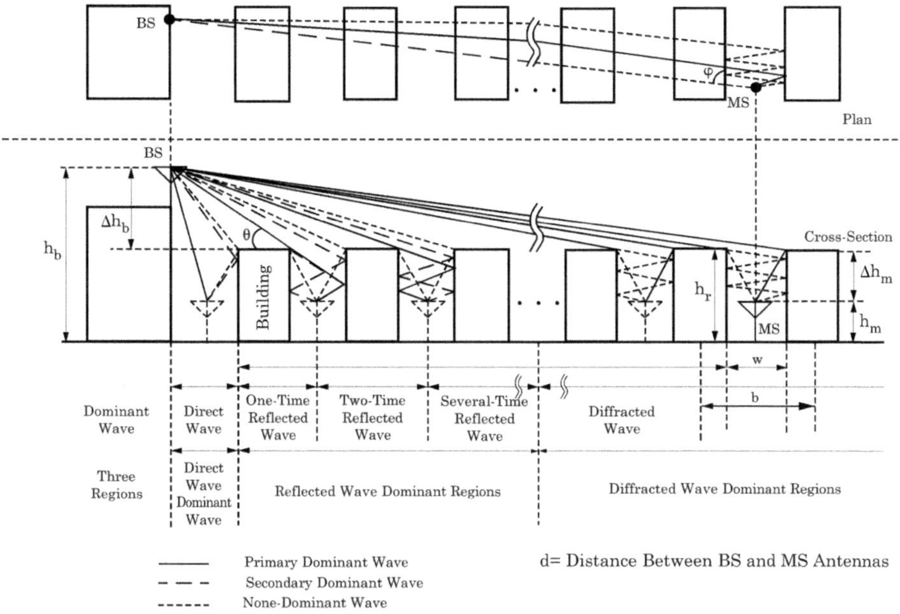

Fig. 7.8 Definition of parameters in NLOS links, model 1 (Ref.: ITU-R, P.1411-4)

- *Model 1*

  Radiowave propagation in this case of NLOS condition is illustrated in Fig. 7.8 for which the parameters and related units are given below:

  h_r : Average height of buildings in meters
  W : Street width in terms of meters
  b : Average distance between the buildings in meters
  φ : Angle between street axis and direct path in degrees
  h_b : BS antenna height in meters
  h_m : MS antenna height in meters
  l : Length of the path covered by buildings in meters
  d : Distance between BS and MS in meters

  The NLOS1 case frequently occurs in residential/rural environments for all types of radio cells and is predominant for small macro-cells in urban/suburban low-rise environments. The parameters h_r, b, and l can be derived from building data along the line between the antennas. However, the determination of w and φ requires a two-dimensional analysis of the area around the mobile. Note that l is not necessarily normal to the building orientation.

- *Model 2*

  Observing Fig. 7.9, NLOS2 type of radiowaves propagation generally includes micro-cells on the routes of the roads which is equivalent to the route from BS to MS. Parameters and units related to this model are stated below:

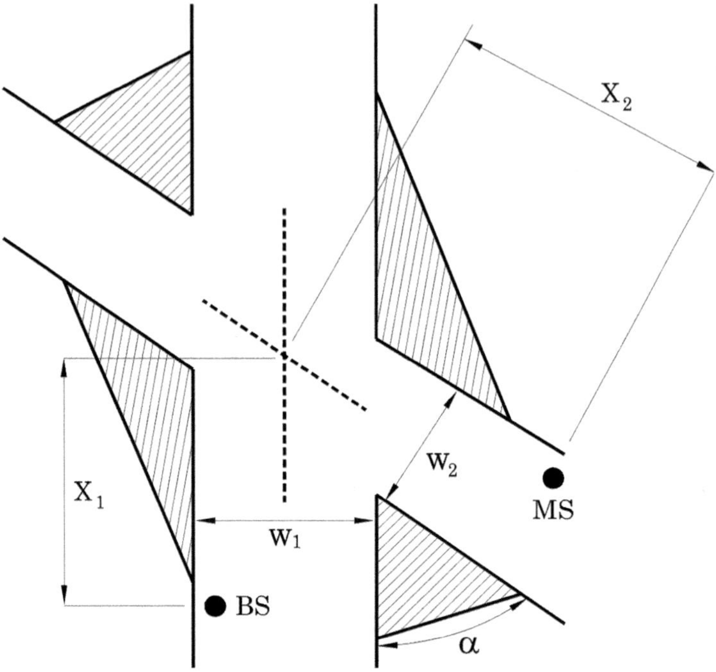

Fig. 7.9 Definition of parameters for NLOS links, model 2 (Ref.: ITU-R, P.1411-4)

W_1: Street width at the BS position in meters
W_2: Street width at the MS position in meters
X_1: Distance of BS to the crossing point of the streets in meters
X_2: Distance of MS to the crossing point of the streets in meters
α : Angle between axes of streets directions in radians

The NLOS2 model is dominant model in urban environments with tall buildings and occurs frequently in micro-cells and pico-cells in urban low-height environments. It shall be essential to specify all parameters of this model through 2-dimensional analysis of the surrounding environment of the mobile terminal.

7.4.3.4 Data Requirements

For exclusive calculations of each communications case (based on the stated models and/or other ones), the following data are required as a minimum:

- Building structures
- Absolute and relative heights of the buildings
- Trees, green lands, and vegetation patterns

The location data must have an accuracy of the order 1–2 m, and if 50 m contours are used for high and low lands of the related region, noting the application (dense/moderate urban, suburban, etc.), a suitable coefficient may be dedicated. These data along with vector data of the roads are used to extract street orientation angles.

7.4.4 Radio Path Loss Calculations

For typical scenarios in urban areas, some closed-form algorithms can be applied. These propagation models can be used both for general and site-specific calculations. The corresponding propagation situations are defined in Sect. 7.4.3. Suitable type of the model depends also on the frequency range. Different models have to be applied for UHF propagation and for millimetric-wave propagation. In the UHF frequency range, LOS and NLOS situations are considered. In millimetric-wave propagation, LOS is considered only. Additional attenuation by oxygen and hydrometeors has to be considered in the latter frequency range.

7.4.4.1 LOS Model Inside the Roads

- *UHF Waves Propagation*
  In the UHF frequency range, the basic transmission loss, as defined in the first chapter, can be characterized by two bounds and a single breakpoint. These bounds are denoted by $L_{\mathrm{LOS-l}}$ and $L_{\mathrm{LOS-u}}$ as lower and upper extremes:

$$L_{\mathrm{LOS-l}} \leq L_{\mathrm{LOS}} \leq L_{\mathrm{LOS-u}} \tag{7.4}$$

Approximate values for L_{LOS} extremes where $d \leq R_{\mathrm{bp}}$ are

$$L_{\mathrm{LOS-l}} = L_{\mathrm{bp}} + 20\ \log(d/R_{\mathrm{bp}}) \tag{7.5}$$

$$L_{\mathrm{LOS-u}} = L_{\mathrm{bp}} + 20 + 25\ \log(d/R_{\mathrm{bp}}) \tag{7.6}$$

Also approximate values for L_{LOS} extremes where $d > R_{\mathrm{bp}}$ are

$$L_{\mathrm{LOS-l}} = L_{\mathrm{bp}} + 40\ \log(d/R_{\mathrm{bp}}) \tag{7.7}$$

$$L_{\mathrm{LOS-u}} = L_{\mathrm{bp}} + 20 + 40\ \log(d/R_{\mathrm{bp}}) \tag{7.8}$$

Figure 7.10 presents variations of lower and upper extremes of the relative basic transmission loss against distance.

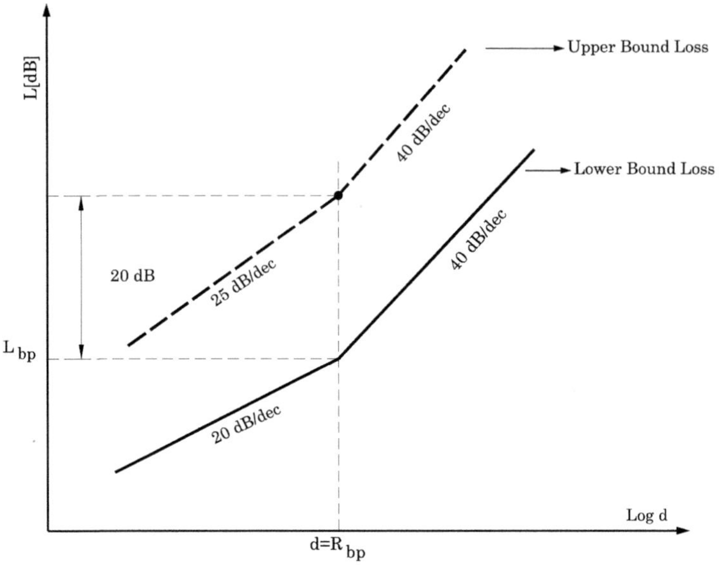

Fig. 7.10 Variations of basic transmission loss vs. distance

As it can be observed from the graph, the slope of variations of the basic transmission loss has a step variation on the breakpoint. The approximate value of this distance for UHF band is equal to

$$R_{\text{bp}} \approx \frac{4 \cdot h_b . h_m}{\lambda} \tag{7.9}$$

Also the value of the basic transmission loss at the break-point, L_{bp} can be calculated from the following relation:

$$L_{\text{bp}} = \left| 20 \log \left(\frac{\lambda^2}{8\pi \cdot h_b . h_m} \right) \right| \tag{7.10}$$

Example 7.3. Line-of-sight communications is assumed for mobile radio links with antenna heights of $h_b = 12$ m and $h_m = 1.8$ m at 900 MHz frequency, find:

1. Distance of the breakpoint
2. Lower and upper extremes of the basic transmission loss at a distance of 800 m

Solution. 1.

$$f = 900\,\text{MHz} \implies \lambda = \frac{C}{f} = \frac{1}{3}\,\text{m}$$

$$R_{\text{bp}} = \frac{4 \times 12 \times 1.8}{\frac{1}{3}} \implies R_{\text{bp}} = 259.2\,\text{m}$$

Table 7.3 Effective height of road, h_s (heavy traffic) (Ref.: ITU-R, P.1411-4)

Frequency (GHz)	h_b (m)	h_s(m) $h_m = 2.7$	$h_m = 1.6$
3.35	4	1.3	**
	8	1.6	**
8.45	4	1.6	**
	8	1.6	**
15.75	4	1.4	**
	8	*	**

*There is no breakpoint
**Breakpoint is beyond 1 km

Table 7.4 Effective height of road, h_s (light traffic) (Ref.: ITU-R, P.1411-4)

Frequency (GHz)	h_b (m)	h_s(m) $h_m = 2.7$	$h_m = 1.6$
3.35	4	0.59	0.23
	8	*	*
8.45	4	**	0.43
	8	**	*
15.75	4	**	0.74
	8	**	*

*Not measured
**Breakpoint is beyond 1 km

2. Since $d > R_{\mathrm{bp}}$ thus:

$$L_{\mathrm{bp}} = \left| 20 \log \left(\frac{1}{9 \times 8\pi \times 12 \times 1.8} \right) \right| = 73.78 \text{ dB}$$

$$L_{\mathrm{LOS-l}} = 73.77 + 40 \log \left(\frac{800}{259.2} \right) = 93.35 \text{ dB}$$

$$L_{\mathrm{LOS-u}} = 73.77 + 20 + 40 \log \left(\frac{800}{259.2} \right) = 113.35 \text{ dB}$$ ■

- *SHF Waves Propagation Up to 15 GHz*
  In the SHF band for the waves propagation on a small distance up to about 1 km, the road traffic and its effective height play a major role altering the breakpoint. In this situation, the approximate value of break-point distance is calculated from the following equation:

$$R_{\mathrm{bp}} = 4 \times \frac{(h_b - h_s)(h_m - h_s)}{\lambda} \tag{7.11}$$

where h_s is the effective road height due to such objects like vehicles on the road and pedestrians near the roadway and the rest of parameters are according to the previous definitions. The values of h_s are provided in the Tables 7.3 and 7.4 for two situations of heavy and light traffic conditions, respectively, in different frequencies based on the daytime and nighttime measurements.

Heavy traffic corresponds to 10–20% of the roadway covered with vehicles and 0.2–1% of the footpath occupied by pedestrians. Light traffic was 0.1–0.5% of the roadway and less than 0.001% of the footpath occupied. The roadway was 27 m wide, including 6 m wide footpaths on either side.

When $h_m > h_s$, the approximate value of upper and lower limits of basic transmission loss for the SHF frequency band can be calculated using (7.4)–(7.8) with L_{bp} given by

$$L_{\mathrm{bp}} = \left| 20 \log \left[\frac{\lambda^2}{8\pi(h_b - h_s)(h_m - h_s)} \right] \right| \tag{7.12}$$

Also when $h_m \leq h_s$, no breakpoint exists. The area near the $BS(d < R_s)$ has a basic propagation loss similar to that of the UHF range, but the area distant from the BS has propagation characteristics in which the attenuation coefficient is cubed. Therefore, the approximate lower and upper limits for $d \geq R_s$ are given by

$$L_{\mathrm{LOS-l}} = L_s + 30 \log\left(\frac{d}{R_s}\right), \quad d \geq R_s \tag{7.13}$$

$$L_{\mathrm{LOS-u}} = L_s + 20 + 30 \log\left(\frac{d}{R_s}\right), \quad d \geq R_s \tag{7.14}$$

The value of basic propagation loss is equivalent to the minimum losses value at $d = R_s$, and it is equal to

$$L_s = \left| 20 \log\left(\frac{\lambda}{2\pi R_s}\right) \right| \tag{7.15}$$

R_s in the above equations has been experimentally determined to be 20 m.

Example 7.4. In a short-range outdoor radio communications link at 3.35 GHz considering $h_b = 8$ m and $h_m = 2.7$ m, calculate upper and lower limits of transmission loss at a distance of 400 m from the end terminals.

Solution.

$$(h_b = 8\,\mathrm{m},\ h_m = 2.7\,\mathrm{m}) \xrightarrow{\text{Table 7.3}} h_s = 1.6\,\mathrm{m}$$

$$\lambda = C/f \implies \lambda = 0.0896\,\mathrm{m}$$

$$R_{\mathrm{bp}} = 4 \times (8 - 1.6)(2.7 - 1.6)/0.0896 \implies R_{\mathrm{bp}} = 314.3\,\mathrm{m}$$

$$L_{\mathrm{bp}} = \left| 20 \log \left[\frac{\lambda^2}{\pi \cdot \lambda(h_b - h_s)(h_m - h_s)} \right] \right| = 86.86\,\mathrm{dB}$$

$$L_{\mathrm{LOS-l}} = L_{\mathrm{bp}} + 40 \log(d/R_{\mathrm{bp}}) = 86.86 + 4.2 \approx 91\,\mathrm{dB}$$

$$L_{\mathrm{LOS-u}} = L_{\mathrm{LOS-l}} + 20 = 111\ \mathrm{dB}$$

■

- *Millimetric-Waves Propagation*
  At frequencies above 10 GHz, the break-point distance according to (7.9) is much more than the maximum cell radius (500 m). This means that the rule of fourth power is not valid in this frequency band, and therefore, the rate of power decay with the distance nearly follows the free-space loss and is about 2.2.
  In this frequency band, it is essential to consider losses pertaining to the existing atmospheric gases and rainfall. To calculate the mentioned losses, the following recommendations may be used:
  - ITU-R, P.676 : Attenuation by atmospheric gases
  - ITU-R, P.837 : Characteristics of precipitation for propagation modeling
  - ITU-R, P.838 : Specific attenuation model for rain, to be used in prediction methods

7.4.4.2 NLOS Models

NLOS signals can arrive at the BS or MS based on mechanisms such as diffraction, multipath, or a combination of them. This section develops models that relate to diffraction:

- *Propagation for urban area*
  Models are defined for the two situations described in Sect. 7.4.3.3. The models are valid for:

  h_b: 4–50 m
  h_m: 1–3 m
  f : 800–5,000 MHz
  : 2–16 GHz for $h_b < h_r$ and $w_2 < 10$ m (or sidewalk)
  d : 20–5,000 m

- *Propagation for suburban area*
  A model is defined for the situation of $h_b > h_r$ described in Sect. 7.4.3.3. The model is valid for:

  h_r : Any height in terms of meter
  Δh_b: 1–100 m
  Δh_m: 4–10 (less than h_r) m
  h_b : $h_r + \Delta h_b$ in m
  h_m : $h_r - \Delta h_m$ in m
  f : 0.8–20 GHz
  w : 10–25 m
  d : 10–5,000 m

- *Millimetric-Waves Propagation*
  Millimetric-waves coverage is considered only for LOS situations because of the large diffraction losses experienced when obstacles cause the propagation path to become NLOS. For these situations, multipath reflections and scattering will be the most likely signal propagation method.

7.4.4.3 NLOS Propagation Over Rooftops for Urban Area

The multiscreen diffraction model given below is valid if the rooftops are all about the same height. Assuming the rooftop heights differ only by an amount less than the first Fresnel-zone radius over a path of length l (see Fig. 7.8), the rooftop height to use in the model is the average rooftop height. If the rooftop heights vary by much more than the first Fresnel-zone radius, a preferred method is to use the highest buildings along the path in a knife-edge diffraction calculation to replace the multiscreen model.

In the model for transmission loss in the NLOS1 (see Fig. 7.8) for rooftops of similar height, the loss between isotropic antennas is expressed as the sum of free-space loss, L_{bf}; the diffraction loss from rooftop to street, L_{rts}; and the reduction due to multiple screen diffraction past rows of buildings, L_{msd}.

In this model, L_{bf} and L_{rts} are independent of the BS antenna height, while L_{msd} is dependent on whether the base station antenna is at, below, or above building heights:

$$L_{\mathrm{NLOS1}} = \begin{cases} L_{\mathrm{bf}} + L_{\mathrm{rts}} + L_{\mathrm{msd}} & \text{for } L_{\mathrm{rts}} + L_{\mathrm{msd}} > 0 \\ L_{\mathrm{bf}} & \text{for } L_{\mathrm{rts}} + L_{\mathrm{msd}} \leq 0 \end{cases} \tag{7.16}$$

The free-space loss is given by

$$L_{\mathrm{bf}} = 32.4 + 20 \log 10(d/1{,}000) + 20 \log 10(f) \tag{7.17}$$

where:

d: Path length (m)
f: Frequency (MHz)

The term L_{rts} describes the coupling of the wave propagating along the multiple screen path into the street where the mobile station is located. It takes into account the width of the street and its orientation:

$$L_{\mathrm{rts}} = -8.2 - 10 \log(w) + 10 \log(f) + 20 \log(\Delta h_m) + L_{\mathrm{ori}} \tag{7.18}$$

$$L_{\mathrm{ori}} = \begin{cases} -10 + 0.354\,\varphi & \text{for } 0^\circ \leq \varphi < 35^\circ \\ 2.5 + 0.075(\varphi - 35) & \text{for } 35^\circ \leq \varphi < 55^\circ \\ 4.0 - 0.114(\varphi - 55) & \text{for } 55^\circ \leq \varphi \leq 90^\circ \end{cases} \tag{7.19}$$

where

$$\Delta h_m = h_r - h_m \tag{7.20}$$

L_{ori} is the street orientation correction factor, which takes into account the effect of rooftop diffraction into streets that are not perpendicular to the direction of propagation (see Fig. 7.8).

The multiple screen diffraction loss from the BS due to propagation past rows of buildings depends on the BS antenna height relative to the building heights and on the incidence angle. A criterion for grazing incidence is the *settled field distance*, d_s:

$$d_s = \frac{\lambda d^2}{\Delta h_b^2} \tag{7.21}$$

where (see Fig. 7.8)

$$\Delta h_b = h_b - h_r \tag{7.22}$$

For the calculation of L_{msd}, d_s is compared to the distance l over which the buildings extend. The calculation for L_{msd} uses the following procedure to remove any discontinuity between the different models used when the length of buildings is greater or less than the *settled field distance*.

The overall multiple screen diffraction model loss is given by

$$L_{msd} = \begin{cases} -\tanh\left[\frac{\log(d)-\log(d_{bp})}{\chi}\right] & \\ \times\left[L1_{msd}(d) - L_{mid}\right] + L_{mid} & \text{for } l > d_s \text{ and } dh_{bp} > 0 \\ \tanh\left[\frac{\log(d)-\log(d_{bp})}{\chi}\right] & \\ \times\left[L2_{msd}(d) - L_{mid}\right] + L_{mid} & \text{for } l \le d_s \text{ and } dh_{bp} > 0 \\ L2_{msd}(d) & \text{for } dh_{bp} = 0 \\ L1_{msd}(d) - \tanh\left[\frac{\log(d)-\log(d_{bp})}{\xi}\right] & \\ \times\left(L_{upp} - L_{mid}\right) - L_{upp} + L_{mid} & \text{for } l > d_s \text{ and } dh_{bp} < 0 \\ L2_{msd}(d) + \tanh\left[\frac{\log(d)-\log(d_{bp})}{\xi}\right] & \\ \times\left[L_{mid} - L_{low}\right] + L_{mid} - L_{low} & \text{for } l \le d_s \text{ and } dh_{bp} < 0 \end{cases} \tag{7.23}$$

where

$$dh_{bp} = L_{upp} - L_{low} \tag{7.24}$$

$$\xi = (L_{upp} - L_{low}) \cdot \upsilon \tag{7.25}$$

$$L_{mid} = \frac{(L_{upp} + L_{low})}{2} \tag{7.26}$$

$$L_{upp} = L1_{msd}(d_{bp}) \tag{7.27}$$

$$L_{low} = L2_{msd}(d_{bp}) \tag{7.28}$$

and

$$d_{\text{bp}} = |\Delta h_b| \sqrt{\frac{l}{\lambda}}$$
$$\upsilon = [0.0417]$$
$$\chi = [0.1] \tag{7.29}$$

where the individual model losses, $L1_{\text{msd}}(d)$ and $L2_{\text{msd}}(d)$, are defined as follows:

Calculation of $L1_{\text{msd}}$ *for* $l > d_{\text{s}}$

(Note: this calculation becomes more accurate when $l \gg d_{\text{s}}$.)

$$L1_{\text{msd}}(d) = L_{\text{bsh}} + k_a + k_d \log(d/1{,}000) + k_f \log(f) - 9 \log(b) \tag{7.30}$$

where L_{bsh} is a loss term that depends on the BS height:

$$L_{\text{bsh}} = \begin{cases} -18 \log(1 + \Delta h_b) & \text{for } h_b > h_r \\ 0 & \text{for } h_b \le h_r \end{cases} \tag{7.31}$$

Other factors are

$$k_a = \begin{cases} 71.4 & \text{for } h_b > h_r \text{ and } f > 2{,}000\,\text{MHz} \\ 73 - 0.8\Delta h_b & \text{for } h_b \le h_r, f > 2{,}000\,\text{MHz and } d \ge 500\,\text{m} \\ 73 - 1.6\Delta h_b\, d/1{,}000 & \text{for } h_b \le h_r, f > 2{,}000\,\text{MHz and } d < 500\,\text{m} \\ 54 & \text{for } h_b > h_r \text{ and } f \le 2{,}000\,\text{MHz} \\ 54 - 0.8\Delta h_b & \text{for } h_b \le h_r, f \le 2{,}000\,\text{MHz and } d \ge 500\,\text{m} \\ 54 - 1.6\Delta h_b\, d/1{,}000 & \text{for } h_b \le h_r, f \le 2{,}000\,\text{MHz and } d < 500\,\text{m} \end{cases} \tag{7.32}$$

$$k_d = \begin{cases} 18 & \text{for } h_b > h_r \\ 18 - 15\frac{\Delta h_b}{h_r} & \text{for } h_b \le h_r \end{cases} \tag{7.33}$$

$$k_f = \begin{cases} -8 & \text{for } f > 2{,}000\,\text{MHz} \\ -4 + 0.7(f/925 - 1) & \text{for medium sized city and suburban centers with medium tree density and } f \le 2{,}000\,\text{MHz} \\ -4 + 1.5(f/925 - 1) & \text{for metropolita centers and } f \le 2{,}000\,\text{MHz} \end{cases} \tag{7.34}$$

Calculation of $L2_{\text{msd}}$ for $l < d_s$

In this case, a further distinction has to be made according to the relative heights of the BS and the rooftops:

$$L2_{\text{msd}}(d) = -20\log(Q_M) \tag{7.35}$$

where

$$Q_M = \begin{cases} 2.35\left(\frac{\Delta h_b}{d}\sqrt{\frac{b}{\lambda}}\right)^{0.9} & , h_b > h_r + \delta h_u \\ \frac{b}{d} & , h_b \le h_r + \delta h_u \text{ and } h_b \ge h_r + \delta h_l \\ \frac{b}{2\pi d}\sqrt{\frac{\lambda}{\rho}}\left(\frac{1}{\theta} - \frac{1}{2\pi+\theta}\right) & , h_b < h_r + \delta h_l \end{cases} \tag{7.36}$$

and

$$\theta = \arctan\left(\frac{\Delta h_b}{b}\right) \tag{7.37}$$

$$\rho = \sqrt{\Delta h_b^2 + b^2} \tag{7.38}$$

$$\delta h_u = 10^{-\log\left(\sqrt{\frac{b}{\lambda}}\right) - \frac{\log(d)}{9} + \frac{10}{9}\log\left(\frac{b}{2.35}\right)} \tag{7.39}$$

and

$$\delta h_l = \frac{0.00023b^2 - 0.1827b - 9.4978}{[\log(f)]^{2.938}} + 0.000781b + 0.06923 \tag{7.40}$$

7.4.4.4 NLOS Propagation Over Rooftops for Suburban Area

A propagation model for the NLOS1 based on geometrical optics theory is shown in Fig. 7.8. This figure indicates that the composition of the arriving waves at the MS changes according to the BS-MS distance. A direct wave can arrive at the MS only when the BS-MS distance is very short.

The several-time (one-, two-, or three-time) reflected waves, which have a relatively strong level, can arrive at the MS when the BS-MS separation is relatively short. When the BS-MS separation is long, the several-time reflected waves cannot arrive, and only many-time reflected waves, which have weak level beside that of diffracted waves from building roofs, arrive at the MS.

Based on these propagation mechanisms, the loss due to the distance between isotropic antennas can be divided into three regions in terms of the dominant arrival waves at the MS. These are direct wave, reflected wave, and diffracted wave dominant regions. The loss in each region is expressed as follows based on GO theory:

$$L_{\mathrm{NLOS1}} = \begin{cases} 20\log\left(\frac{4\pi d}{\lambda}\right) & \text{for } d < d_0 \\ & \text{(Direct wave domniant region)} \\ L_{0n} & \text{for } d_0 < d \le d_n \\ & \text{(Reflected wave dominant region)} \\ 32.1\log\left(\frac{d}{d_n}\right) + L_{d_n} & \text{for } d > d_n \\ & \text{(Diffracted wave domniant region)} \\ & n = \begin{cases} 2\ (0.8\text{ GHz} \le f < 5\text{ GHz}) \\ 3\ (5\text{ GHz} \le f < 20\text{ GHz}) \end{cases} \end{cases} \tag{7.41}$$

where

$$L_{0n} = \begin{cases} \text{when } d_k < d \le d_{k+1} & (k = 0,\ldots,n-1) \\ & n = \begin{cases} 2\ (0.8\text{ GHz} \le f < 5\text{ GHz}) \\ 3\ (5\text{ GHz} \le f < 20\text{ GHz}) \end{cases} \\ L_{d_k} + \frac{L_{d_{k+1}} - L_{d_k}}{d_{k+1} - d_k} \cdot (d - d_k) \end{cases} \tag{7.42}$$

$$d_k = \frac{1}{\sin\varphi} \cdot \sqrt{B_k^2 + (h_b - h_m)^2} \tag{7.43}$$

$$L_{d_k} = 20\log\left\{\frac{4\pi d_{kp}}{0.4^k.\ \lambda}\right\} \tag{7.44}$$

$$d_{kp} = \frac{1}{\sin\varphi_k} \cdot \sqrt{A_k^2 + (h_b - h_m)^2} \tag{7.45}$$

$$A_k = \frac{w.(h_b - h_m)(2k+1)}{2(h_r - h_m)} \tag{7.46}$$

$$B_k = \frac{w.(h_b - h_m)(2k+1)}{2(h_r - h_m)} - k \cdot w \tag{7.47}$$

$$\varphi_k = \tan^{-1}\left(\frac{B_k}{A_k} \cdot \tan\varphi\right) \tag{7.48}$$

7.4.4.5 NLOS Propagation in Street Canyons, $800 \le f \le 2{,}000$ MHz

For NLOS2 situations where both antennas are below rooftop level, diffracted and reflected waves at the corners of the street crossings have to be considered (see Fig. 7.9):

$$L_{\mathrm{NLOS2}} = -10\log\left(10^{-L_r/10} + 10^{-L_d/10}\right) \quad \text{dB} \tag{7.49}$$

where:

L_r: reflection path loss defined by

$$L_r = 20\log(x_1 + x_2) + x_1 x_2 \frac{f(\alpha)}{w_1 w_2} + 20\log\left(\frac{4\pi}{\lambda}\right) \quad \text{dB} \tag{7.50}$$

where

$$f(\alpha) = \frac{3.86}{\alpha^{3.5}} \text{ dB}, \quad 0.6 < \alpha[\text{rad}] < \pi \tag{7.51}$$

L_d: diffraction path loss defined by

$$L_d = 10\log[x_1 x_2 (x_1 + x_2)] + 2D_a - 0.1 \times \left(90 - \alpha\frac{180}{\pi}\right) + 20\log\left(\frac{4\pi}{\lambda}\right) \quad \text{dB} \tag{7.52}$$

$$D_a = \left(\frac{40}{2\pi}\right)\left[\arctan\left(\frac{x_2}{w_2}\right) + \arctan\left(\frac{x_1}{w_1}\right) - \frac{\pi}{2}\right] \quad \text{dB} \tag{7.53}$$

7.4.4.6 NLOS Propagation in Street Canyons, $2 \le f \le 16$ GHz

The propagation model for the NLOS2 situations as described in Sect. 7.4.3.3 with the corner angle $\alpha = \pi/2$ rad is derived based on measurements at a frequency range from 2 to 16 GHz, where $h_b < h_r$ and w_2 is up to 10 m (or sidewalk). The path loss characteristics can be divided into two parts: the corner loss region and the NLOS region. The corner loss region extends for d_{corner} from the point which is 1 m down the edge of the LOS street into the NLOS street. The corner loss (L_{corner}) is expressed as the additional attenuation over the distance d_{corner}. The NLOS region lies beyond the corner loss region, where a coefficient parameter (β) applies. This is illustrated by the typical curve shown in Fig. 7.11. Using x_1, x_2, and w_1, as shown in Fig. 7.11, the overall path loss (L_{NLOS2}) beyond the corner region ($x_2 > w_1/2 + 1$) is found using

$$L_{\text{NLOS2}} = L_{\text{LOS}} + L_c + L_{\text{att}} \tag{7.54}$$

$$L_c = \begin{cases} \frac{L_{\text{corner}}}{1 - \log(1 + d_{\text{corner}})}[1 - \log(x_2 - w_1/2)] & x_2 \le w_1/2 + 1 + d_{\text{corner}} \\ d_{\text{corner}} & x_2 > w_1/2 + 1 + d_{\text{corner}} \end{cases} \tag{7.55}$$

$$L_{\text{att}} = \begin{cases} 10\beta\log\left(\frac{x_1 + x_2}{x_1 + w_1/2 + d_{\text{corner}}}\right) & x_2 > w_1/2 + 1 + d_{\text{corner}} \\ 0 & x_2 \le w_1/2 + 1 + d_{\text{corner}} \end{cases} \tag{7.56}$$

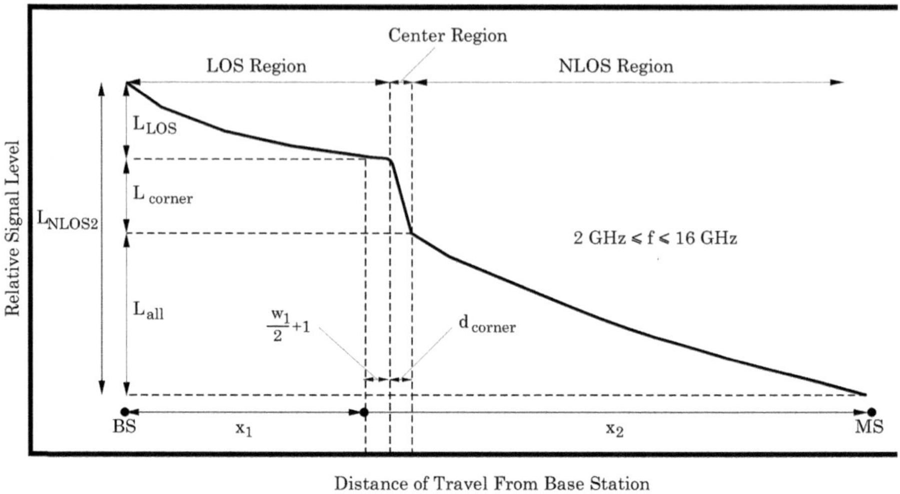

Fig. 7.11 Propagation concept along street Canyons with low base station height (Ref.: ITU-R, P.1411-4)

where L_{LOS} is the path loss in the LOS street for x_1 (>20 m), as calculated in Sect. 7.4.4.1. In (7.55), L_{corner} is given as 20 dB in an urban environment and 30 dB in a residential environment. In (7.56), β is given by 6 and d_{corner} is 30 m in both environments.

In a residential environment, the path loss does not increase monotonically with distance, and thus, the coefficient parameter may be lower than the value in an urban environment due to the presence of alleys and gaps between the houses.

With a high base station antenna in the small macro-cell, the effects of diffraction over rooftops are more significant. Consequently, the propagation characteristics do not depend on the corner loss.

7.4.4.7 Propagation Between Terminals Located Below Rooftop Height at UHF

The model described below is intended for calculating the basic transmission loss between two terminals of low height in urban environments. It includes both LOS and NLOS regions and models the rapid decrease in signal level noted at the corner between the LOS and NLOS regions.

The model includes the statistics of location variability in the LOS and NLOS regions and provides a statistical model for the corner distance between the LOS and NLOS regions. Figure 7.12 illustrates the LOS, NLOS and corner regions, and the statistical variability predicted by the model.

This model is recommended for propagation between low-height terminals where both terminal antenna heights are near street level well below rooftop height. It is

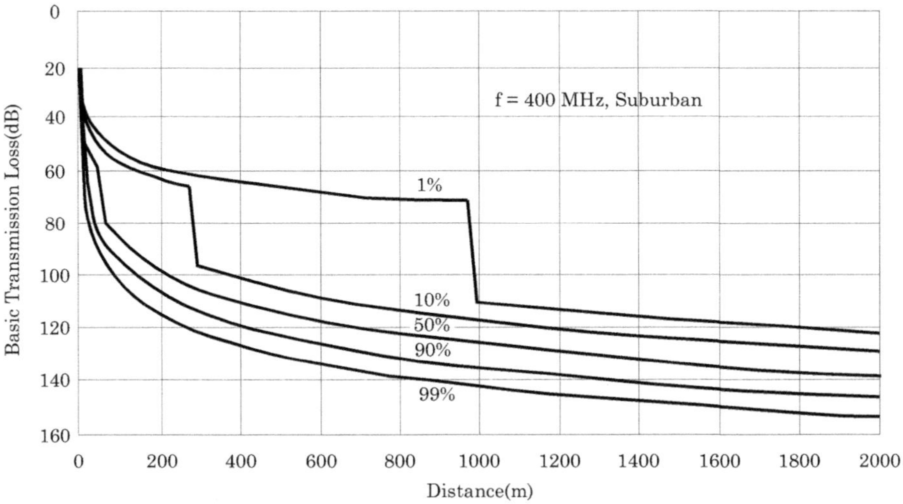

Fig. 7.12 Basic transmission loss not exceeded for 1, 10, 50, 90 and 99% of locations (Ref.: ITU-R, P.1411-4)

reciprocal with respect to transmitter and receiver and is valid for frequencies in the range 300–3,000 MHz. The model is based on measurements made in the UHF band with antenna heights between 1.9 and 3.0 m above ground and transmitter-receiver distances up to 3,000 m.

The parameters required are the frequency f (MHz) and the distance between the terminals d (m), then step-by-step calculation procedure is specified below:

Step 1: Calculate the median value of the line-of-sight loss:

$$L_{\mathrm{LOS}}^{\mathrm{median}}(d) = 32.45 + 20\,\log f + 20\,\log(d/1{,}000) \qquad (7.57)$$

Step 2: For the required location percentage, p (%), calculate the LOS location correction with $\sigma = 7\,\mathrm{dB}$:

$$\Delta L_{\mathrm{LOS}}(p) = 1.564\,\sigma(\sqrt{-21n(1-p/100)} - 1.1774) \qquad (7.58)$$

Alternatively, values of the LOS correction for $p = 1$, 10, 50, 90 and 99% are given in Table 7.5.

Step 3: Add the LOS location correction to the median value of LOS loss:

$$L_{\mathrm{LOS}}(d,p) = L_{\mathrm{LOS}}^{\mathrm{median}}(d) + \Delta L_{\mathrm{LOS}}(p) \qquad (7.59)$$

Step 4: Calculate the median value of the NLOS loss:

$$L_{\mathrm{NLOS}}^{\mathrm{median}}(d) = 9.5 + 45\,\log\,f + 40\,\log(d/1{,}000) + L_{\mathrm{urban}} \qquad (7.60)$$

L_{urban} depends on the urban category and is 0 dB for suburban, 6.8 dB for urban, and 12.3 dB for dense urban/high rise.

Table 7.5 LOS and NLOS location variability corrections (Ref.: ITU-R, P.1411-4)

p (%)	ΔL_{LOS} (dB)	ΔL_{NLOS} (dB)	d_{LOS} (m)
1	−11.3	−16.3	976
10	−7.9	−9.0	276
50	0.0	0.0	44
90	10.6	9.0	16
99	20.3	16.3	10

Step 5: For the required location percentage, $p(\%)$, add the NLOS location correction:

$$\Delta L_{\text{NLOS}}(p) = \sigma N^{-1}(p/100) \quad \text{with } \sigma = 7\text{dB} \tag{7.61}$$

$N^{-1}(.)$ is the inverse normal cumulative distribution function. An approximation to this function, good for p between 1 and 99%, is given by the location variability function $Q_i(x)$ of recommendation ITU-R P.1546. Alternatively, values of the NLOS location variability correction for typical values of p including 1, 10, 50, 90 and 99% are given in Table 7.5.

Step 6: Add the NLOS location correction to the median value of NLOS loss:

$$L_{\text{NLOS}}(d,p) = L_{\text{NLOS}}^{\text{median}}(d) + \Delta L_{\text{NLOS}}(p) \tag{7.62}$$

Step 7: For the required location percentage, p (%), calculate the distance d_{LOS} for which the LOS fraction F_{LOS} equals p:

$$\begin{aligned} d_{\text{LOS}}(p) &= 212[\log(p/100)]^2 - 64\log(p/100) \quad \text{if } p < 45 \\ d_{\text{LOS}}(p) &= 79.2 - 70(p/100) \quad \text{otherwise} \end{aligned} \tag{7.63}$$

Values of d_{LOS} for p = 1, 10, 50, 90 and 99% are given in Table 7.5. This model has not been tested for $p < 0.1\%$. The statistics were obtained from two cities in the United Kingdom and may be different in other countries. Alternatively, if the corner distance is known in a particular case, set $d_{\text{LOS}}(p)$ to this distance.

Step 8: The path loss at the distance d is then given as:

(a) : If $d < d_{\text{LOS}}$, then $L(d,p) = L_{\text{LOS}}(d,p)$
(b) : If $d > d_{\text{LOS}} + w$, then $L(d,p) = L_{\text{NLOS}}(d,p)$
(c) : Otherwise, linearly interpolate between the values $L_{\text{LOS}}(d_{\text{LOS}},p)$ and $L_{\text{NLOS}}(d_{\text{LOS}}+w,p)$:

$$\begin{aligned} L_{\text{LOS}} &= L_{\text{LOS}}(d_{\text{LOS}},p) \\ L_{\text{NLOS}} &= L_{\text{NLOS}}(d_{\text{LOS}}+w,\ p) \\ L(d,p) &= L_{\text{LOS}} + (L_{\text{NLOS}} - L_{\text{LOS}})(d - d_{\text{LOS}})/w \end{aligned}$$

The width w is introduced to provide a transition region between the LOS and NLOS regions. This transition region is seen in the data and typically has a width of $w = 20\,\text{m}$.

7.4.4.8 Default Parameters for General Calculations

If the data on the structure of buildings and roads are unknown (site-general situations), the following default values are recommended:

$$\begin{aligned} h_r &= 3 \times (\text{number of floors}) + \text{roof height (m)} \\ \text{roof height} &= 3\text{ m for pitched roofs} \\ &= 0\text{ m for flat roofs} \\ w &= b/2 \\ b &= 20 - 50\text{ m} \\ \varphi &= 90^\circ \end{aligned}$$

7.4.4.9 Influence of Vegetation

The effects of propagation through vegetation (primarily trees) are important for outdoor short-path predictions. Two major propagation mechanisms as shown in Fig. 7.13 can be identified:

- Propagation through (not around or over) trees
- Propagation over trees

The first mechanism predominates for geometries in which both antennas are below the tree tops and the distance through the trees is small, while the latter predominates for geometries in which one antenna is elevated above the tree tops.

The attenuation is strongly affected by multipath scattering initiated by diffraction of the signal energy both over and through the tree structures. For propagation through trees, the specific attenuation in vegetation should be taken into account applying suitable procedure.

In situations where the propagation is over trees, diffraction is the major propagation mode over the edges of the trees closest to the low antenna. This propagation mode can be modeled most simply by using an ideal knife-edge diffraction model. Though the knife-edge model may underestimate the field strength because it neglects multiple scattering by tree tops, a mechanism that may be modeled by radiative transfer theory.

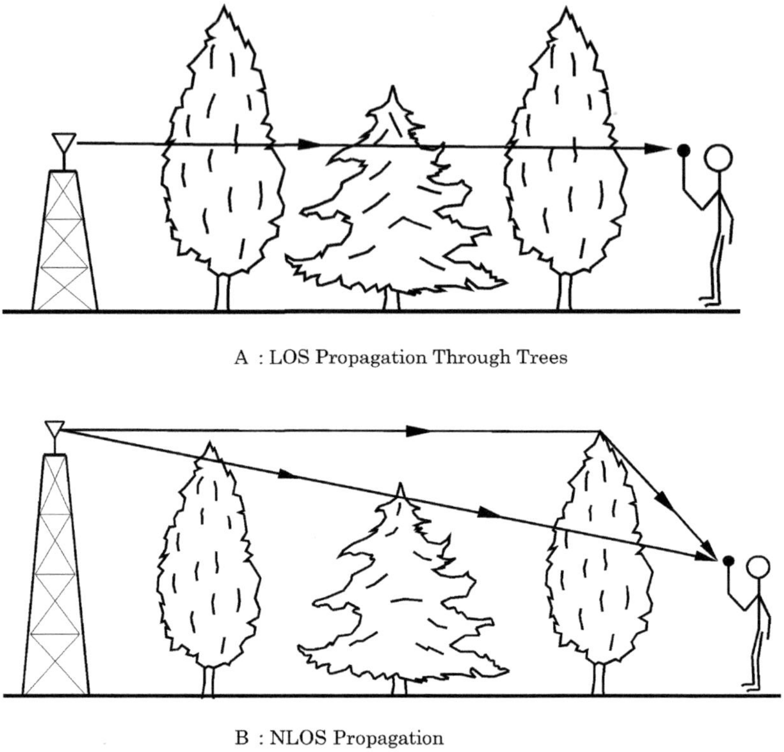

Fig. 7.13 Propagation mechanisms in woodlands

7.4.5 *Building Entry Loss*

Building entry loss is the excess loss due to the presence of a building wall (including windows and other features). It is defined as the difference between the signal levels outside and inside the building at the same height. Account must also be taken of the incident angle. (When the path length is less than about 10 m, the difference in free-space loss due to the change in path length for the two measurements should be taken into account in determining the building entry loss. For antenna locations close to the wall, it may also be necessary to consider near field effects.)

Additional losses will occur for penetration within the building as explained in the next Sect. 7.5. It is believed that, typically, the dominant propagation mode is one in which signals enter a building approximately horizontally through the wall surface (including windows) and that for a building of uniform construction, the building entry loss is independent of height.

Table 7.6 Office building entry loss (Ref.: ITU-R, P.1411-4)

Frequency (GHz)	Mean loss	Standard deviation
5.2	12 dB	5 dB

Table 7.7 Loss due to stone block wall at various incident angles (Ref.: ITU-R, P.1411-4)

Incident angle (degrees)	0	15	30	45	60	75
Loss due to wall (dB)	28	32	32	38	45	50
Standard deviation (dB)	4	3	3	5	6	5

Building entry loss should be considered when evaluating the radio coverage from an outdoor system to an indoor terminal. It is also important for considering interference problems between outdoor and indoor systems.

The experimental results shown in Table 7.6 were obtained at 5.2 GHz through an external office building wall made of brick and concrete with glass windows. The wall thickness was 60 cm, and the window-to-wall ratio was about 2:1.

Table 7.7 shows the results of measurements at 5.2 GHz through an external wall made of stone blocks at incident angles from 0–75°. The wall was 400 mm thick with two layers of 100-mm thick blocks and loose fill between. Particularly at larger incident angles, the loss due to the wall was extremely sensitive to the position of the receiver as evidenced by the large standard deviation.

7.4.6 Multipath Models

7.4.6.1 Street Canyon Environment

Characteristics of multipath delay spread for the LOS case in an urban high-rise environment for dense urban micro-cells and pico-cells have been developed based on measured data at frequencies from 2.5 to 15.75 GHz at distances from 50 to 400 m. The r.m.s. delay spread S at distance of d follows a normal distribution with the mean value given by

$$a_S = C_a d^{\gamma_a} \quad ns \tag{7.64}$$

and the standard deviation given by

$$\sigma_S = C_\sigma d^{\gamma_\sigma} \quad ns \tag{7.65}$$

where C_a, γ_a, C_σ, and γ_σ depend on the antenna height and propagation environment. Table 7.8 lists some typical values of the coefficients for distances of 50–400 m based on measurements made in urban and residential areas.

Table 7.8 Typical coefficients of r.m.s. delay spread (Ref.: ITU-R, P.1411-4)

Measurement conditions				a_s		σ_s	
Area	f (GHz)	h_b (m)	h_m (m)	C_a	γ_a	C_σ	γ_σ
Urban	2.5	6.0	3.0	55	0.27	12	0.32
	3.35–15.75	4.0	2.7	23	0.26	5.5	0.35
			1.6	10	0.51	6.1	0.39
Residential	3.35–8.45		0.5				
	3.35	4.0	2.7	2.1	0.53	0.54	0.77
	3.35–15.75		1.6	5.9	0.32	2.0	0.48

From the measured data at 2.5 GHz, the average shape of the delay profile was found to be

$$P(t) = P_0 + 50(e^{-t/\tau} - 1) \quad \text{dB} \tag{7.66}$$

where:

P_0: Peak power (dB)
τ: Decay factor

and t is in ns.

From the measured data, for an r.m.s. delay spread S, τ can be estimated as

$$\tau = 4\,S + 266 \text{ ns} \tag{7.67}$$

A linear relationship between τ and S is only valid for the LOS case.

From the same measurement set, the instantaneous properties of the delay profile have also been characterized. The energy arriving in the first 40 ns has a Rician distribution with a K-factor of about 6–9 dB, while the energy arriving later has a Rayleigh or Rician distribution with a K-factor of up to about 3 dB.

7.4.6.2 Over Rooftop Propagation

Characteristics of multipath delay spread for both LOS and NLOS cases in an urban high-rise environment for micro-cells have been developed based on measured data at 1,920–1,980 MHz and 2,110–2,170 MHz using omnidirectional antennas. The median r.m.s. delay spread S in this environment is given by

$$S_u = \exp(A \cdot L + B) \quad \text{ns} \tag{7.68}$$

where $A = 0.038$, $B = 2.3$, and L is path loss (dB).

From the same measurement set, values of r.m.s. delay spread at different frequency bands (190 MHz apart) were compared at each location. More than 10 % of locations showed larger than 300 ns differences in r.m.s. delay spread with 25 dB threshold and larger than 2 μs differences in delay interval using 15 dB threshold.

Table 7.9 Typical r.m.s. delay spread values[a] (Ref.: ITU-R, P.1411-4)

Measurement conditions					r.m.s. delay spread (ns)	
		Antenna height				
Area	Frequency	$h_{BS}(m)$	$h_r(m)$	Range (m)	50%	95%
Suburban	5.2	20	2.8	100–1,000	189	577

[a]Threshold value of 30 dB was used for r.m.s. delay spread calculation

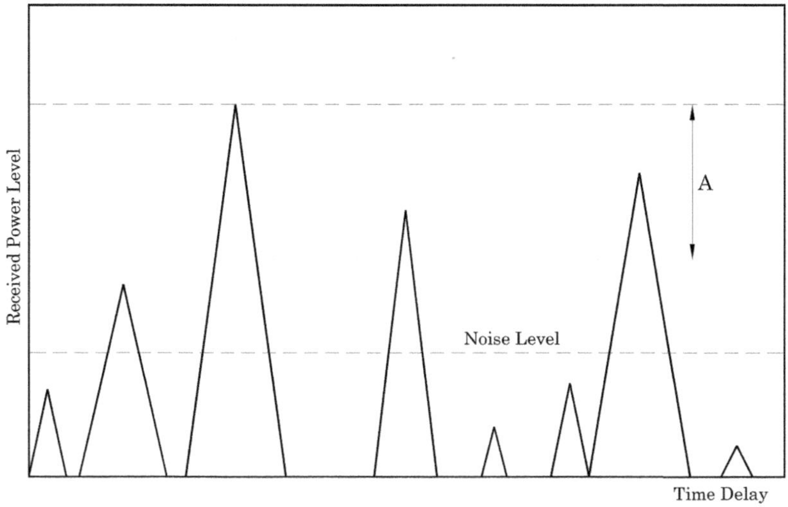

Fig. 7.14 Determination of the number of peaks (Ref.: ITU-R, P.1411-4)

The distributions of the multipath delay characteristics for the 5.2-GHz band in a suburban environment with a BS antenna height of 20 m and MS antenna height of 2.8 m were derived from measurements. Table 7.9 lists the measured r.m.s. delay spread for the 5.2-GHz band for cases where the cumulative probability is 50 and 95%.

7.4.7 Number of Signal Components

For the design of high data rate systems with multipath separation and synthesis techniques, it is important to estimate the number of signal components (i.e., a dominant component plus multipath components) arriving at the receiver. The number of signal components can be represented from the delay profile as the number of peaks whose amplitudes are within A dB of the highest peak and above the noise floor, as shown in Fig. 7.14.

Table 7.10 Maximum number of signal components (low BS antenna in an urban area) (Ref.: ITU-R, P.1411-4)

Frequency (GHz)	Antenna height (m)		Range (m)	Maximum number of signal components					
				$A = 3$ dB		$A = 5$ dB		$A = 10$ dB	
	h_b	h_m		80%	95%	80%	95%	80%	95%
3.35	4	1.6	0–200	2	3	2	4	5	6
			0–1,000	2	3	2	4	5	9
8.45	4	1.6	0–200	1	3	2	3	4	6
			0–1,000	1	2	2	4	4	8
15.75	4	1.6	0–200	1	3	2	3	4	5
			0–1,000	2	3	2	4	6	10

Table 7.11 Maximum number of signal components (high BS antenna in an urban area) (Ref.: ITU-R, P.1411-4)

Frequency (GHz)	Antenna height (m)		Range (m)	Maximum number of signal components					
				$A = 3$ dB		$A = 5$ dB		$A = 10$ dB	
	h_b	h_m		80%	95%	80%	95%	80%	95%
3.35	55	2.7	150–590	2	2	2	3	3	13
8.45	55	2.7	150–590	2	2	2	3	3	12

Table 7.12 Maximum number of signal components (low BS antenna in a residential area) (Ref.: ITU-R, P.1411-4)

Frequency (GHz)	Antenna height (m)		Range (m)	Maximum number of signal components					
				$A = 3$ dB		$A = 5$ dB		$A = 10$ dB	
	h_b	h_m		80%	95%	80%	95%	80%	95%
3.35	4	2.7	0–480	2	2	2	2	2	3

Tables 7.10–7.12 show the results of measurements for three different scenarios (a low BS antenna in an urban area, a high BS antenna in an urban area, a low BS antenna in a residential area). The temporal resolution in the measurements was 20 ns. Table 7.13 shows the results of measurements for a high BS antenna in a suburban environment. The temporal resolution for this measurement was 50 ns. These tables list the maximum number of signal components which have been observed at 80 and 95% of locations in each measurement section.

7.4.8 *Polarization Characteristics*

Cross-polarization discrimination (XPD) differs between LOS and NLOS areas in an SHF micro-cellular environment. Measurements indicate a median XPD of 13 dB

Table 7.13 Maximum number of signal components (high BS antenna in a suburban area) (Ref.: ITU-R, P.1411-4)

Frequency (GHz)	Antenna height (m)		Range (m)	Maximum number of signal components					
				$A = 3$ dB		$A = 5$ dB		$A = 10$ dB	
	h_b	h_m		80 %	95%	80%	95%	80%	95%
3.67	40	2.7	0–5,000	1	2	1	3	3	5

for LOS paths and 8 dB for NLOS paths and a standard deviation of 3 dB for LOS paths and 2 dB for NLOS paths at SHF. These median values are compatible with the UHF values for open and urban areas, respectively.

7.4.9 *Characteristics of Direction of Arrival*

The r.m.s. angular spread in the azimuthal direction in a micro-cell or pico-cell environment in an urban area was obtained from the measurement made at a frequency of 8.45 GHz. The receiving base station had a parabolic antenna with a half-power beamwidth of 4°. The antenna heights of the transmitting mobile station and the receiving base station were 2.7 m and 4.4 m, respectively.

In the LOS situation, the r.m.s. angular spread has an average value of 30° (standard deviation of 11°). In the NLOS situation, the r.m.s. angular spread has an average value of 41° (standard deviation of 18°).

7.4.10 *Fading Characteristics*

The fading depth, which is defined as the difference between the 50% value and the 1% value in the cumulative probability of received signal levels, is expressed as a function of the product ($2\,\Delta f \Delta L_{\max}$ MHz·m) of the received bandwidth in Fig. 7.15. $\Delta L_{\max}$ is the maximum difference in propagation path lengths between components whose level is larger than the threshold, which is 20 dB lower than the highest level of the indirect waves as shown in Fig. 7.16. In this figure, a in decibels is the power ratio of the direct to the sum of indirect waves, and $a = -\infty$ dB represents a non-line-of-sight situation.

When $2\Delta f \Delta L_{\max}$ is less than 10 MHz·m, the received signal levels in line-of-sight and non-line-of-sight situations follow Rayleigh and Nakagami-Rice distributions, corresponding to a narrow-band fading region. When it is larger than 10 MHz·m, it corresponds to a wideband fading region, where the fading depth becomes smaller and the received signal level follows neither Rayleigh nor Nakagami-Rice distributions.

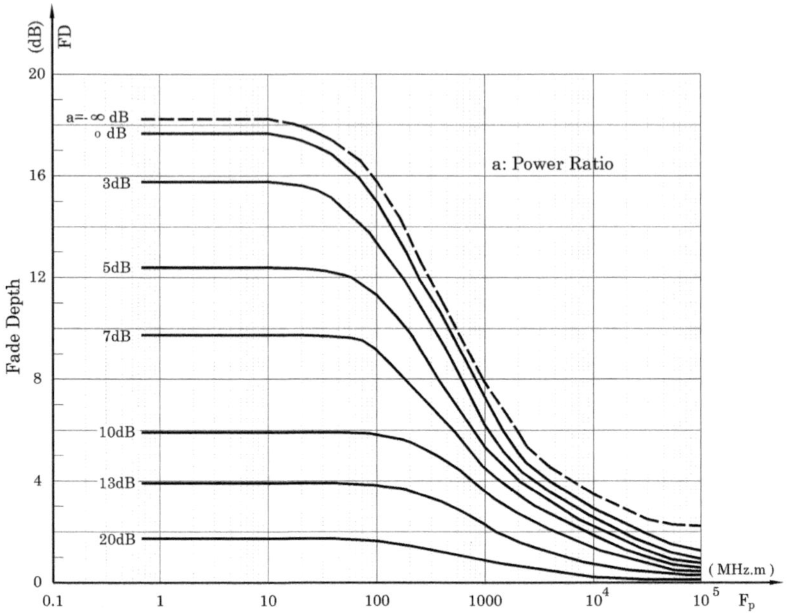

Fig. 7.15 Fading depth vs. $2\Delta f \cdot \Delta L_{max}$(Ref.: ITU-R, P.1411-4)

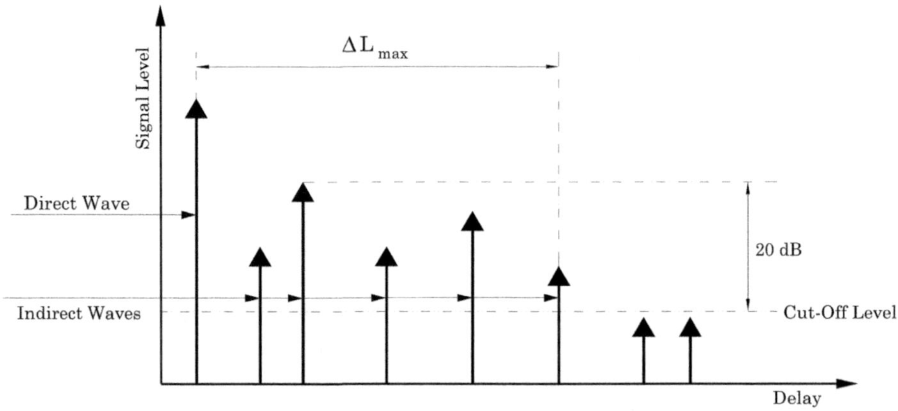

Fig. 7.16 Model for calculating ΔL_{max}(Ref.: ITU-R, P.1411-4)

7.5 ITU-R Method for Indoor Radio Communications

7.5.1 Introduction

Indoor wireless communications are classified virtually as short-range radio links (operating within less than 1 km) having many applications including, but not limited to, the following fields:

- Personal communications within surrounded areas
- Radio local area network (RLANs)
- Wireless private branch exchanges, WPBXs
- Remote control units for building electrical and mechanical utilities

Also, noting the following issues:

- Determination of necessary standards for RLAN communications to conform with wireless or wired systems
- Very low-power consumption of the equipment related to indoor radio systems in mobile services and personal communications
- Difference between indoor and outdoor radiowaves propagation environments

In 1997, ITU-R prepared a method for calculations and predictions of losses due to the radiowaves propagation in indoor areas. This method presented initially by recommendations No.ITU-R, P.1238 for 900 MHz to 100 GHz frequency band which has been revised four times up to now. This section briefly outlines indoor radio communications on the basis of the above-mentioned recommendation.

7.5.2 Propagation Medium

In the case of radiowaves propagation in covered areas, the following points can be noted:

- Indoor communications have differences with those of outdoor communications including radius of coverage and lack of atmospheric phenomena such as cloud, snow, and rain.
- The limits of coverage area in indoor communications can be specified with an acceptable accuracy based on building and structure dimensions.
- Buildings dimensions and their composition of materials have a major influence on the loss of radiowaves propagating between TX and RX.
- Considering high traffic volume and limitation on the allocated frequency band, the frequency reuse policy will increase the radio interference probability in the medium.
- Because of using high frequencies in indoor areas, small displacements create obstructions resulting in great losses in the received signal level.

- Complexity of effective factors on the waves propagation within covered areas requires accurate planning. To do this, it is necessary to specify proper details of wave path including the structure of the medium, its material composition, en route furniture, walls, and floors.
- For initial and preliminary planning which is limited only to specifying the number of fixed stations and suitable coverage, it may be possible to meet this objective through overall data.

7.5.3 Propagation Impairments

7.5.3.1 Main Factors

Main factors affecting the radiowaves propagation in the indoor areas are:

- Free-space losses related to the radiowave propagation
- Reflection from, and diffraction around, objects (including walls and floors) within the rooms
- Transmission loss through walls, floors, and other obstacles
- Channeling of energy, especially in corridors at high frequency
- Motion of persons and objects in the room, including possibly one or both ends of the radio link

7.5.3.2 Effective Factors

Phenomena increasing the adverse effects of the medium on the radiowaves propagation are:

- Additional loss due to obstacles and transmission through building materials, and possible mitigation of free space by channeling
- Temporal and spatial variation of path loss
- Multipath effects from reflected and diffracted components of the wave
- Polarization mismatch due to random alignment of mobile terminal

7.5.3.3 Main Features

Main features of indoor wireless communications can be characterized as follows:

- High/medium/low data rate
- Coverage area of each base station (e.g., room, floor, building)
- Different kinds of terminals including mobile/portable/fixed
- Real-time/non-real-time/quasi-real-time operations

Table 7.14 Typical services and propagation impairments

Type of service	Characteristics	Propagation impairments
Wireless local area network (WLAN)	High data rate, single or multiple rooms, portable, non-real time, point to multipoint or each point to each point	• Path loss-temporal and spatial distribution • Multipath delay • Ratio of desired-to-undesired mode strength
Wireless private branch exchange (WPBX)	Medium data rate, multiple rooms, single floor or multiple floors, real time, mobile, point to multipoint	• Path loss-temporal and spatial distribution
Indoor paging	Low data rate, multiple floors, non-real time, mobile, point to multipoint	• Path loss-temporal and spatial distribution
Indoor wireless video	High data rate, multiple rooms, real time, mobile or portable, point to point	• Path loss-temporal and spatial distribution • Multipath delay

- Network topology (e.g., point to point, point to multipoint, and each point to each point)
- Different network architectures like point to point and point to multipoint

7.5.3.4 Types of Services

Types of the main services provided by indoor communications, namely, audio, data transfer with different speeds, image transfer, and other video services including their characteristics and propagation impairments, are specified in the Table 7.14.

7.5.4 Path Loss Models

Two models are considered for calculation of radiowave path loss for which it is assumed that both the base station and mobile/portable terminal are located inside the same building:

- General model
- Site-specific model

7.5.4.1 General Models

The models described in this section are considered to be in general and typical location as they require little path or site information. The indoor radio path loss is characterized by both an average path loss and its associated shadow fading statistics.

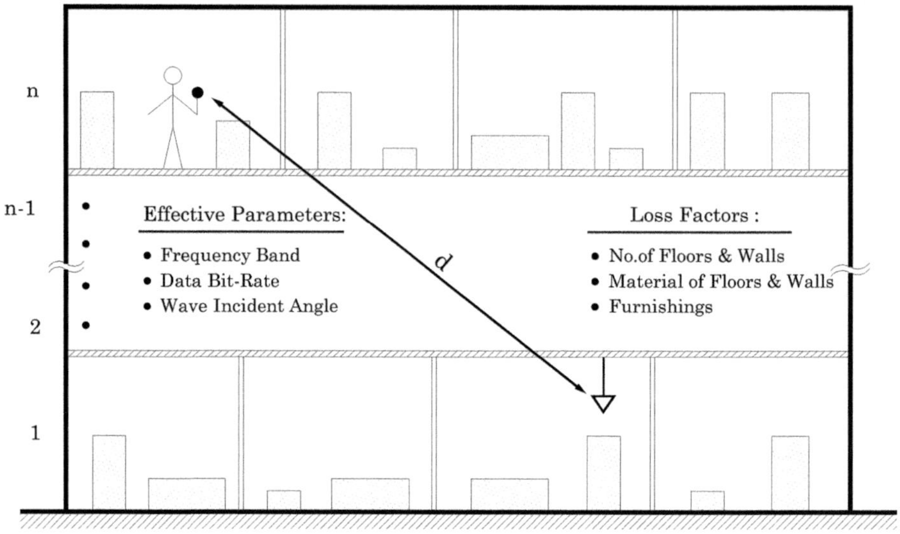

Fig. 7.17 Main factors in radiowaves propagation in indoor environment

Several indoor path loss models account for the attenuation of the signal through multiple walls and/or multiple floors. The model described in this section accounts for the loss through multiple floors to allow for such characteristics as frequency reuse between floors.

The distance power loss coefficients given below include an implicit allowance for transmission through walls, over and through obstacles, and for other loss mechanisms likely to be encountered within the same floor of a building. Site-specific models would have the option of explicitly accounting for the loss due to each wall instead of including it in the distance model.

Total transmission loss, L_t, for the basic model as shown in Fig. 7.17 has the following form:

$$L_t = 20 \log f + N \log d + L_f(n) - 28 \qquad (7.69)$$

where:

N : Distance power loss coefficient
f : Frequency (MHz)
d : Separation distance (m) between the base station and portable terminal (where $d > 1$ m)
L_f: Floor penetration loss factor (dB)
n : Number of floors between base station and portable terminal ($n \geq 1$)

Typical parameters, based on various measurement results, are given in Tables 7.15 and 7.16. Additional general guidelines are given at the end of the section.

Table 7.15 Power loss coefficient, N, for indoor transmission loss calculation (Ref.: ITU-R, P.1411-4)

Frequency	Residential	Office	Commercial
900 MHz	–	33	20
1.2–1.3 GHz	–	32	22
1.8–2 GHz	28	30	22
4 GHz	–	28	22
5.2 GHz	–	31	–
60 GHz[a]	–	22	17
70 GHz[a]	–	22	–

[a]In Table 7.15, measured values at 60 and 70 GHz propagation are within a single room or space and do not include any allowance for transmission through walls. Gaseous absorption around 60 GHz is also significant for distance greater than about 100 m which may influence frequency reuse distances

Table 7.16 Floor penetration loss factor, L_f, in dB (Ref.: ITU-R, P.1411-4)

Frequency	Residential	Office	Commercial
900 MHz	–	9(1 floor) 19(2 floors) 24(3 floors)	–
1.8–2 GHz (n floors)	$4n$	$15 + 4(n - 1)$	$6+3(n - 1)$
5.2 GHz	–	16(1 floor)	–

For the various frequency bands where the power loss coefficient is not stated for residential buildings, the value given for office buildings could be used.

It should be noted that there may be a limit on the isolation expected through multiple floors. The signal may find other external paths to complete the link with less total loss than that due to the penetration loss through many floors.

When the external paths are excluded, measurements at 5.2 GHz have shown that at normal incidence, the mean additional loss due to a typical reinforced concrete floor with a suspended false ceiling is 20 dB, with a standard deviation of 1.5 dB. Lighting fixtures increased the mean loss to 30 dB, with a standard deviation of 3 dB, and air ducts under the floor increased the mean loss to 36 dB, with a standard deviation of 5 dB. These values, instead of L_f, should be used in site-specific models such as ray-tracing.

Example 7.5. Indoor communications are required in an administrative environment at 1,900 MHz between the floors 2 and 6. For a distance of 80 m between TX and RX units, find:

1. The distance power loss coefficient of the path and floor penetration loss factor
2. The path loss on the basis of ITU-R model

Table 7.17 Standard deviation in shadow fading for indoor transmission (Ref.: ITU-R, P.1411-4)

Frequency (GHz)	Residential	Office	Commercial
1.8–2	$\sigma = 8\,\text{dB}$	$\sigma = 10\,\text{dB}$	$\sigma = 10\,\text{dB}$
5.2 GHz	–	$\sigma = 12\,\text{dB}$	–

Solution. 1.

$$(\text{office area, } f = 1.9\ \text{GHz}) \xrightarrow{\text{Table (7.15)}} N = 30$$

$$(n = 6 - 2 = 4) \xrightarrow{\text{Table (7.16)}} L_f = 15 + 4(4 - 1) = 27\ \text{dB}$$

2.

$$L_t = 20\ \log\ 1{,}900 + 30\ \log\ 80 + 27 - 28 = 121.67\ \text{dB}$$ ■

The indoor shadow fading statistics are log-normal, and standard deviation values (dB) are given in Table 7.17.

Although available measurements have been under various conditions which make direct comparisons difficult and only selective frequency bands have been reported upon, a few general conclusions can be drawn, especially for the 900–2,000-MHz band:

- Paths with a line-of-sight (LOS) component are dominated by free-space loss and have a distance power loss coefficient of around 20.
- Large open rooms also have a distance power loss coefficient of around 20; this may be due to a strong LOS component in most areas of the room. Examples include rooms located in large retail stores, sports arenas, open-plan factories, and open-plan offices.
- Corridors exhibit path loss less than that of free space, with a typical distance power coefficient of around 18. Grocery stores with their long, linear aisles exhibit the corridor loss characteristic.
- Propagation around obstacles and through walls adds considerably to the loss which can increase the power distance coefficient to about 40 for a typical environment. Examples include paths between rooms in closed-plan office buildings.
- For long unobstructed paths, the first Fresnel-zone breakpoint may occur for which the distance power loss coefficient may change from about 20 to about 40.
- The decrease in the path loss coefficient with increasing frequency for an office environment (Table 7.15) is not always observed or easily explained. On the one hand, with increasing frequency, loss through obstacles (e.g., walls, furniture) increases, and diffracted signals contribute less to the received power; on the other hand, the Fresnel zone is less obstructed at higher frequencies leading to lower loss. The actual path loss is dependent on these opposing mechanisms.

7.5.4.2 Site-Specific Models

For estimating the path loss or field strength, site-specific models are also useful. Models for indoor field strength prediction based on the uniform theory of diffraction (UTD) and ray-tracing techniques are available. Detailed information of building structure is necessary for the calculation of the indoor field strength. These models combine empirical elements with the theoretical electromagnetic approach of UTD.

The method takes into account direct, single-diffracted, and single-reflected rays and can be extended to multiple diffraction or multiple reflection as well as to combinations of diffracted and reflected rays. By including reflected and diffracted rays, the path loss predication accuracy is significantly improved.

7.5.5 Delay Spread Models

7.5.5.1 Multipath

The mobile/portable radio propagation channel varies in time, frequency, and with spatial displacement. Even in the static case, where the transmitter and receiver are fixed, the channel can be dynamic since scatterers and reflectors are likely to be in motion. The term multipath arises from the fact that through reflection, diffraction, and scattering, radiowaves can travel from a transmitter to a receiver by many paths.

There is a time delay associated with each of these paths that is proportional to the path length. A very rough estimate of the maximum delay time to be expected in a given environment may be obtained simply from the dimensions of the room and from the fact that the time (ns) for a radio pulse to travel distance d (m) is approximately 3.3 d. These delayed signals, each with an associated amplitude, form a linear filter with time-varying characteristics.

7.5.5.2 Impulse Response

The goal of channel modeling is to provide accurate mathematical representations of radio propagation to be used in radio link and system simulations for the system deployment modeling. Since the radio channel is linear, it is fully described by its impulse response. Once the impulse response is known, one can determine the response of the radio channel to any input. This is the basis of link performance simulation.

The impulse response is usually represented as a power density function of excess delay, relative to the first detectable signal. This function is often referred to as a power delay profile.

The channel impulse response varies with the position of the receiver and may also vary with time. Therefore, it is usually measured and reported as an average of profiles measured over one wavelength to reduce noise effects or over several wavelengths to determine a spatial average. It is important to define clearly which average is meant and how the averaging was performed.

The recommended averaging procedure is to form a statistical model for each impulse response estimate (power delay profile) and then to locate the times before and after the average delay T_D beyond which the power density does not exceed specific values (-10, -15, -20, -25, -30 dB) with respect to the peak power density. The median, and if desired the 90th percentile, of the distributions of these times forms the model.

7.5.5.3 r.m.s Delay Spread

Power delay profiles are often characterized by one or more parameters, as mentioned above. These parameters should be computed from profiles averaged over an area having the dimensions of several wavelengths. (The parameter r.m.s delay spread is sometimes found from individual profiles, and the resulting values averaged, but, in general, the result is not the same as that found from an averaged profile.) A noise exclusion threshold, or acceptance criterion, for example, 30 dB below the peak of the profile, should be reported along with the resulting delay spread which depends on this threshold.

Although the r.m.s delay spread is very widely used, it is not always a sufficient characterization of the delay profile. In multipath environments where the delay spread exceeds the symbol duration, the bit error ratio for phase shift keying modulation depends not on the r.m.s delay spread but rather on the received power ratio of the desired wave to the undesired wave. This is particularly pronounced for high-symbol-rate systems but is also true even at low symbol rates when there is a strong dominant signal among the multipath components (Rician fading).

However, if an exponentially decaying profile can be assumed, it is sufficient to express the r.m.s delay spread instead of the power delay profile. In this case, the impulse response can be reconstructed approximately as

$$h(t) = \begin{cases} \mathrm{e}^{-t/S} & \text{for } 0 \le t \le t_{\max} \\ 0 & \text{otherwise} \\ t_{\max} >> S \end{cases} \tag{7.70}$$

where:

S : r.m.s delay spread
$t_{\max}$: Maximum delay

The advantage in using the r.m.s delay spread as the model output parameter is that the model can be expressed simply in the form of a table. Typical delay spread parameters, estimated from averaged delay profiles, for three indoor environments

Table 7.18 r.m.s delay spread parameters (Ref.: ITU-R, P.1411-4)

Frequency	Environment	A(ns)	B(ns)	C(ns)
1,900 MHz	Indoor residential	20	70	150
1,900 MHz	Indoor office	35	100	460
1,900 MHz	Indoor commercial	55	150	500
5.2 GHz	Indoor office	45	75	150

are given in Table 7.18. These values are based on measurements at 1,900 MHz and 5.2 GHz using omnidirectional antennas. (There is little evidence of a strong frequency dependence in these parameters when omnidirectional antennas are used.)

In Table 7.18, column B represents median values that occur frequently and column A represents lower, but not extreme, values that also occur frequently, while column C represents extremely high delay values that occur only rarely. The values given in the table represent the largest room sizes likely to be encountered in each environment.

Within a given building, the delay spread tends to increase as the distance between antennas increases and hence to increase as path loss increases. With greater distances between antennas, it is more likely that the path will be obstructed and that the received signal will consist entirely of scattered paths.

The r.m.s delay spread S is roughly in proportion to the area of the floor space, F_S, and is given by the following equation:

$$\log S = 1.1 + 0.23 \log F_S \tag{7.71}$$

where the units of F_S and S are m^2 and ns, respectively.

This equation is based on measurements in the 2-GHz band for several room types such as office, lobby, corridor, and gymnasium. The maximum floor space for the measurements was 1,000 m^2. The median value of the estimation error is -1.6 ns and the standard deviation is 24.3 ns.

When the delay spread S is represented in dB, the standard deviation of S is in the range of about 0.7–1.2 dB.

7.5.6 *Effect of Polarization and Antenna Radiation Pattern*

In an indoor environment, there is not only a direct path but also reflected and diffracted paths between the transmitter and receiver. The reflection characteristics of a building material depend on polarization, incidence angle, and the material's complex permittivity, as represented by Fresnel's reflection formula. The angles of arrival of multipath components are distributed depending on the antenna beamwidths, building structures, and siting of transmitter and receiver. Therefore, polarization and the effective antenna radiation pattern can significantly affect indoor propagation characteristics.

Table 7.19 Example of antenna directivity dependence of static r.m.s. delay spread (Ref.: ITU-R, P.1411-4)

Frequency (GHz)	TX antenna	RX antenna beamwidth (degrees)	Static r.m.s. delay spread (90th percentile) (ns)	Room size (m)
60	Omnidirectional	Omnidirectional	17	13.5 × 7.8
		60	16	Empty office
		10	5	room
		5	1	

7.5.6.1 LOS Paths

It is widely accepted that, in line-of-sight (LOS) channels, directional antennas reduce r.m.s delay spread as compared to omnidirectional antennas and that circular polarization (CP) reduces it compared to linear polarization (LP). Therefore, in this case, a directional CP antenna offers an effective means of reducing the delay spread.

The prime mechanism of the polarization dependence can be attributed to the fact that, when the CP signal is incident on a reflecting surface at an incidence angle smaller than the Brewster angle, the handedness of the reflected CP signal is reversed. The reversal of the CP signal at each reflection means that multipath components arriving after one reflection are orthogonally polarized to the LOS component; this eliminates a significant proportion of the multipath interference.

This effect is independent of frequency as predicted theoretically and supported by indoor propagation experiments in the frequency range of 1.3–60 GHz and applies equally indoors and outdoors.

Since all existing building materials have Brewster angles greater than 45°, multipath due to single reflections (i.e., the main surface of multipath components) is effectively suppressed in most room environments irrespective of the interior structure and materials in the room. The possible exceptions are environments where very large incident angles dominate the multipath, such as in a long hallway. The variation in r.m.s delay spread on a moving link is also reduced when CP antennas are used.

Since multipath propagation components have an angle-of-arrival distribution, those components outside the antenna beamwidth are spatially filtered out by the use of directional antenna resulting in reduction of delay spread. Indoor propagation measurement and ray-tracing simulations performed at 60 GHz, with an omnidirectional transmitting antenna and four different types of receiving antennas (omnidirectional, wide-beam, standard horn, and narrow-beam antennas) directed toward the transmitting antenna, revealed that the suppression of delayed components is more effective with narrower beamwidths.

Table 7.19 shows an example of the antenna directivity dependence of static r.m.s delay spread not exceeded at the 90th percentile obtained from ray-tracing simulations at 60 GHz for an empty office. It may be noted that a reduction in r.m.s delay spread may not necessarily always be desirable as it can mean increased dynamic ranges for fading of wideband signals as a result of missing inherent frequency diversity. In addition, it may be noted that some transmission schemes take advantage of multipath effects.

7.5.6.2 NLOS Paths

When the direct path is obstructed, the polarization and antenna directivity dependence of delay spread may be more complicated than those in the LOS path. There are few experimental results relating to the obstructed case. However, an experimental result obtained at 2.4 GHz suggests that the polarization and antenna directivity dependence of delay spread in the obstructed path is significantly different from that in the LOS path. For instance, an omnidirectional horizontally polarized antenna at the transmitter and a directional CP receiving antenna gave the smallest r.m.s. delay spread and lowest maximum excess delay in the obstructed path.

7.5.6.3 Orientation of Mobile Terminal

In the portable radio environment, propagation is generally dominated by reflection and scattering of the signal. Energy is often scattered from the transmitted polarization into the orthogonal polarizations. Under these conditions, cross-polarization coupling increases the probability of adequate received levels of randomly oriented portable radios. Measurement of cross-polarization coupling carried out at 816 MHz showed a high degree of coupling.

7.5.7 Effect of Transmitter and Receiver Siting

There are few experimental and theoretical investigations regarding the effects of transmitter and receiver sites on indoor propagation characteristics. In general, however, it may be suggested that the base station should be placed as high as possible near the room ceiling to attain LOS paths as far as possible.

In the case of handheld terminals, the user terminal position will of course be dependent on the user's motion rather than any system design constraints. However, for non-handheld terminals, it is suggested that the antenna height be sufficient to ensure LOS to the base station whenever possible. The choice of station siting is also very relevant to system configuration aspects such as spatial diversity arrangements and zone configuration.

Table 7.20 Complex permittivity of interior construction materials (Ref.: ITU-R, P.1411-4)

	1 GHz	57.5 GHz	70 GHz	78.5 GHz	95.9 GHz
Concrete	7-j0.85	6.5-j0.43	–	–	6.2-j0.34
Lightweight concrete	2-j0.5	–	–	–	–
Floorboard (synthetic resin)	–	3.91-j0.33	–	3.64-j0.37	3.16-j0.39
Plaster board	–	2.25-j0.03	2.43-j0.04	2.37-j0.1	2.25-j0.06
Ceiling board (rock wool)	1.2-j0.01	1.59-j0.01	–	1.56-j0.02	1.56-j0.04
Glass	6.76-j0.09	6.76-j0.16	6.76-j0.17	6.76-j0.18	6.76-j0.19
Fiberglass	1.2-j0.1	–	–	–	–

7.5.8 Effect of Building Materials, Furnishing, and Furniture

Indoor propagation characteristics are affected by reflection from and transmission through the building materials. The reflection and transmission characteristics of those materials depend on their complex permittivity. Site-specific propagation prediction models may need information on the complex permittivity of building materials and on building structures as basic input data.

The complex permittivity of typical building materials, experimentally obtained at 1, 57.5, 78.5, and 95.9 GHz, is tabulated in Table 7.20. These permittivities indicate significant difference from one material to another while showing little frequency dependence in the range 60–100 GHz, except for floorboard which varied by 10%.

7.5.9 Effect of Movement of Objects

The movement of persons and objects within the room causes temporal variations of the indoor propagation characteristics. This variation, however, is very slow compared to the data rate likely to be used and can therefore be treated as virtually a time-invariant random variable. Apart from people in the vicinity of the antennas or in the direct path, the movement of persons in offices and other locations in and around the building has a negligible effect on the propagation characteristics.

Measurements performed when both of the link terminals are fixed indicate that fading is bursty (statistics are very nonstationary) and is caused either by the perturbation of multipath signals in areas surrounding a given link or by shadowing due to people passing through the link. The results of some experiments are given below:

- Measurements at 1.7 GHz indicate that a person moving into the path of a LOS signal causes a 6–8 dB drop in received power level, and the K-value of the Nakagami-Rice distribution is considerably reduced. In the case of NLOS conditions, people moving near the antennas did not have any significant effects on the channel.

- In the case of a handheld terminal, the proximity of the user's head and body affects the received signal level. At 900 MHz with a dipole antenna, measurements show that received signal strength decreased by 4–7 dB when the terminal was held at the waist and 1–2 dB when the terminal was held against the head of the user, in comparison to received signal strength when the antenna was several wavelengths away from the body.
- When the antenna height is lower than about 1 m, for example, in the case of a typical desktop or laptop computer application, the LOS path may be shadowed by people moving in the vicinity of the user terminal. For such data applications, both the depth and the duration of fades are of interest.
- Measurements at 37 GHz in an indoor office lobby environment have shown that fades of 10–15 dB were often observed. The duration of these fades due to body shadowing, with people moving continuously in a random manner through the LOS, follows a log-normal distribution, with the mean and standard deviation dependent on fade depth. For these measurements, at a fade depth of 10 dB, the mean duration was 0.11 s and the standard deviation was 0.47 s. At a fade depth of 15 dB, the mean duration was 0.05 s and the standard deviation was 0.15 s.
- Measurements at 70 GHz have revealed that the mean fade durations due to body shadowing were 0.52 s, 0.25 s, and 0.09 s for the fade depth of 10 dB, 20 dB, and 30 dB, respectively, in which the mean walking speed of persons was estimated at 0.74 m/s with random directions and human body thickness was assumed to be 0.3 m.
- Measurements indicate that the mean number occurrence of body shadowing in an hour caused by human movement in an office environment is given by

$$\bar{N} = 260 \times D_p \tag{7.72}$$

where $D_p (0.05 \leq D_p \leq 0.08)$ is the number of persons per square meter in the room. Then, the total fade duration per hour is given by

$$T = \bar{T}_S \times \bar{N} \tag{7.73}$$

where $\bar{T}_S$ is mean fade duration.

The number of occurrences of body shadowing in an hour at the passage in an exhibition hall was 180–280, where D_p was 0.09–0.13.

- The distance dependency of path loss in an underground mall is affected by human body shadowing. The path loss in an underground mall is estimated by the following equation with the parameters given in Table 7.21:

$$L(x) = -10 \cdot \alpha(1.4 - \log f - \log x) + \delta \cdot x + C, \text{ dB} \tag{7.74}$$

Table 7.21 Path loss parameters for a typical underground mall (Ref.: ITU-R, P.1411-4)

	LOS			NLOS		
	α	δ (m^{-1})	C (dB)	α	δ (m^{-1})	C (dB)
Off-hour	2.0	0	5	3.4	0	−20
Rush hour	2.0	0.065	5	3.4	0.065	−20

where:

f: Frequency (MHz)
x: Distance (m)

Parameters for the NLOS case are verified in the 5-GHz band and those of the LOS case are applicable to the frequency range of 2–20 GHz. The range of distance x is 10–200 m.

The environment of the underground mall is a ladder type mall that consists of straight corridors with glass or concrete walls. The main corridor is 6 m wide, 3 m high, and 190 m long. The typical human body is considered to be 170 cm tall and 45 cm wide at the shoulders. The passage densities are approximately 0.008 persons/m^2 and 0.1 persons/m^2 for a quiet period (early morning, off-hour) and a crowded period (lunchtime or rush hour), respectively.

Example 7.6. In a typical underground mall, calculate upper and lower attenuation limits of radiowaves propagating a distance of 100 m at 5 GHz.

Solution. The upper limit of attenuation corresponds to the NLOS links in rush hours for which $\alpha = 3.4$, $\delta = 0.065\,\mathrm{m}^{-1}$, and $C = -20$ dB, then using (7.74) yields

$$L_\mathrm{u} = -10\alpha(1.4 - \log f - \log x) + \delta x + C$$

$$L_\mathrm{u} = 132.7 \text{ dB}$$

The lower limit of attenuation corresponds to the LOS links in off-hours for which $\alpha = 2$, $\delta = 0$, and $C = 5$ dB, and then applying (7.74) results in

$$L_\mathrm{l} = 91 \text{ dB}$$ ■

7.6 RF Leaky Cable

One of semi-guided media for radiowaves propagation is RF leaky cable which provides radio communications inside surrounded locations like tunnels and underground mines where communications are crucial for safety and emergency cases. To meet the requirements, slotted transmission lines known as RF leaky cables are used widely. As shown in Fig. 7.18, this type of RF cables is a special form of heliax cables including a number of elliptical slots that are uniformly arranged on the cable.

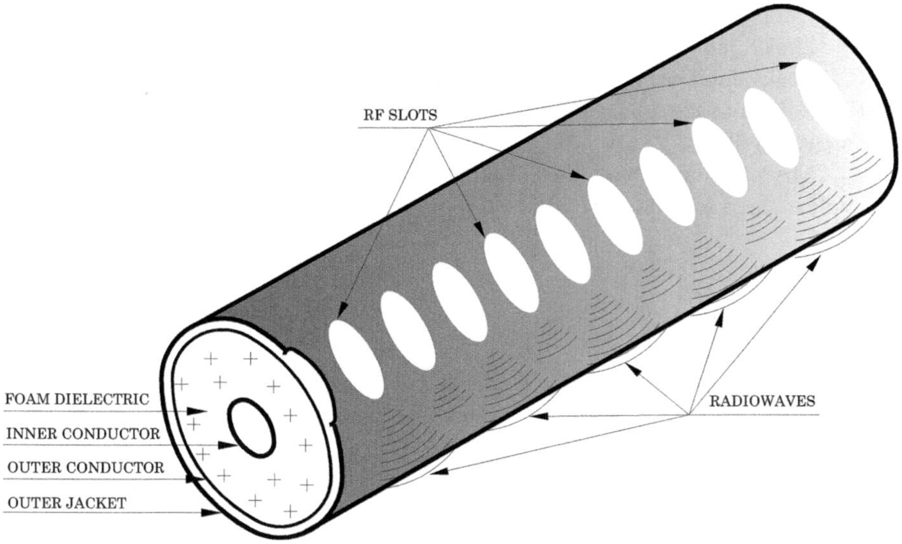

Fig. 7.18 RF leaky cable structure

When a cable is properly fed by VHF/UHF amplified radiowave signals, they will propagate inside the cable and will leak from its slots continuously which is detectable within few meters by proper receivers. Among a variety of limiting factors for radiowave propagation inside and outside of leaky cable, the following items are of major importance:

- Attenuation of RF leaky cable
- Coupling loss between air and RF leaky cable
- Free-space loss
- Diffraction loss
- Multipath fading

The last three items are related to the propagation of radiowaves in open area suffering some attenuations due to abnormal conditions specified in other chapters. The first two items are related to the RF leaky cable which will be explained briefly.

RF cable attenuation is usually in the order of several dB/100 m depending on its structure, material, dimension, length, and operating frequency. Also, the coupling loss in the air interface is considerable and usually in the range of 60–70 dB.

A typical formula for design calculations of mobile network using leaky cable is as follows:

$$\mathrm{RSL} = P_\mathrm{t} + G_\mathrm{M} - \alpha_\mathrm{L} - C_\mathrm{L} - \mathrm{FSL} - L_\mathrm{m} \tag{7.75}$$

- RSL: Received signal level in dB_m
- P_t : Input signal level of leaky cable in dB_m

- G_M : Mobile antenna gain in dB_i
- α_L : Leaky cable longitudinal loss in dB
- C_L : Leaky cable air interface coupling loss in dB
- FSL : Free-space loss of radiowaves between leaky cable and mobile radio antenna in dB
- L_m : Miscellaneous losses such as jumper cable, RF connectors, splitter, diplexer, duplexer, and RF coupler in dB

7.7 Exercises

Questions

1. Short-range radio communications refer to which mode of radio links and what are its main types?
2. Evaluate frequency bands for short-range communications and prepare a report presenting the results of the investigation. Millimeter and micrometer radiowaves include in which part of the frequency spectrum.
3. Name the main applications of short-range indoor and outdoor radio communications. Can you mention any new application not stated in the book?
4. State the types of radio cells for short-range radio communications and indicate main aspects of each of them.
5. List the basic principles in the short-range radiowaves propagation.
6. Define delay spread and express its effects on the radiowaves reception and specify under which conditions this phenomenon has more effect.
7. Define the breakpoint in short-range radiowaves propagation and evaluate the graph shown in Fig. 7.6 on this basis.
8. For short-range communications, specify the classification of outdoor radiowave propagation media and briefly define the characteristics of each one.
9. Define LOS and NLOS waves and their types in the outdoor radio links indicating the main parameters of each type.
10. Assess and comment about the graph given in Fig. 7.10, and particularly write down the answers of the following cases:
    - The upper and lower losses
    - Maximum difference between the upper and lower losses at the same distance (prior, after, and on the breakpoint)
    - Rate of change in the path loss at different locations
11. Evaluate the main differences of radiowaves propagation in UHF and SHF bands for indoor and outdoor environments.
12. Express the effects of heavy and light traffic on the road over the radiowaves path loss for short-range radio links.
13. Explain the L_{bf}, L_{rts}, and L_{msd} in the radiowaves propagation over the buildings.

14. Determine major effects of vegetation and woodlands on the radiowaves propagation. Which modes of propagation may be used for this situation?
15. What effects do the radiowaves encounter when they collide with the trees, and what are the main propagation approaches under such conditions?
16. Evaluate the limits of the losses created by the radiowaves entering a building at 5-GHz band and also the effect of incidence angle in this respect.
17. Describe the calculation procedures set by ITU-R for short-range indoor/outdoor radio communications. Specify the important parameters in each one and particularly state their role in the high-speed data transfer.
18. Consider Tables 7.9–7.12 and specify how the number of received components affects the analysis of the received signal quality.
19. Explain the effects of the following phenomena in short-range communications for outdoor/indoor radiowaves.
    - Radiowaves polarization
    - Fading of the received signal
20. List the characteristics of radiowaves propagation medium in covered areas.
21. Outline the affecting factors in the indoor waves propagation.
22. Specify types of main services that may be provided by the indoor radio links.
23. Explain path losses related to the general model in the indoor radio communications, and define effective components and their pertinent units. Also, specify the limits of the power coefficient variations of the distance.
24. Which type of antenna and polarization is suitable for indoor radio communications.
25. Explain the effect of the following issues in indoor radio communications:
    - TX and RX antennas positions
    - Material of the objects in indoor areas
26. Discuss the effects of multipath mechanism on indoor radio links.

Problems

1. Using path loss model of indoor radiowaves according to (7.3), calculate the transmission basic loss of radiowaves propagation at 900 MHz within the following environments:
   (a) : Sports center $d = 100\,\text{m}$ and $N = 35$
   (b) : Administrative building $d = 60\,\text{m}$, $N = 42$, $K_1 = K_2 = 3$, $F = 12\,\text{dB}$, and $W = 18\,\text{dB}$
2. In the previous problem, assume 500 mW and $5\,\text{dB}_i$ for the TX output power and antenna gain, respectively, and then find the power level at the RX antenna location

3. We generalize (7.3) for UHF band by adding a frequency-dependent term 20 log $f/900$ (f in MHz). Solve problem 1 for frequency of $f = 2{,}400\,\text{MHz}$. Assume that other losses are the same.
4. In an outdoor radio communications, the heights of antennas are $h_b = 15\,\text{m}$ and $h_m = 2.7\,\text{m}$, assuming $f = 1{,}500\,\text{MHz}$ calculate:

   (a) : Distance of the path loss breakpoint
   (b) : The maximum losses for distances of $d_1 = 300\,\text{m}$ and $d_2 = 800\,\text{m}$

5. For $h_b = 8\,\text{m}$, $h_m = 2.7\,\text{m}$, and $f = 3.35\,\text{MHz}$, calculate:

   (a) : The effective height of road
   (b) : Distance of the path loss breakpoint

6. For the previous problem, calculate the path loss upper and lower bounds at distances of $d_1 = 150\,\text{m}$ and $d_2 = 450\,\text{m}$.
7. Parameters of a radio link according to the Fig. 7.8 are:

$$l = 600\,\text{m}\,, \quad W = 30\,\text{m}\,, \quad b = 60\,\text{m},$$

$$h_r = 21\,\text{m}\,, \quad h_m = 2\,\text{m}\,, \quad h_b = 30\,\text{m}$$

   Calculate value of d_s, L_{bf}, $(L_{\text{msd}} + L_{\text{rts}})$, and L_{NLOS}.
8. In urban and residential media, use the Table 7.7 for a radio link operating at $f = 5.2\,\text{GHz}$ and path length equal to 300 m and find:

   (a) : Standard deviation for delay time
   (b) : Power decay relative to its maximum value with $\tau = 2\,\mu\text{s}$ and $S = 200\,\text{ns}$

9. On the second floor of an administrative environment and frequency band of 5.2 GHz, a radio link is established at a maximum distance of 100 m. If a 15 dB additional loss is considered for objects and the walls, then find:

   (a) : The distance power loss coefficient of the path and loss factor for the floor penetration.
   (b) : Total transmission loss based on the ITU-R model.

10. In an underground shopping center, calculate decay in the radiowave propagation over a distance of 120 m at 2.5 GHz.

   (a) : Normal, off-hours, and LOS propagation
   (b) : For rush hours and NLOS propagation

Chapter 8
Noise in Radiowave Propagation

8.1 Introduction

Based on the great influence of noise on the radio communications and its role in the detection and quality of the main signal whether analog or digital, this chapter is devoted to the assessment of this topic. In view of the noise regarded as an impairment factor in communications, it requires evaluation and recognition from different perspectives specifically including the following points:

- Main sources of radio noises
- Noise nature and aspects
- Prediction of system performance in noisy media

Previous chapters provided limited information about the noise which is dependent on radiowave propagation in different frequency bands and services. Initially, we will provide information particularly about the types of noises and sources generating them, the interrelation and basic principles, and radio noises, and finally, it provides a brief look at some specific issues concerning the noise affecting telecommunication systems such as satellite and mobile radio communications. It must be reminded that the radio section of ITU has elaborated on this issue with its recommendation No. ITU-R, P. 372 in 1951 and revised the same several times, and the latest updated version has been published in 2007.

8.2 Noise in Radio Communications

8.2.1 Definition of the Noise and Its Source

Generally, the noise includes unwanted signals and waves which may be attached to the main signal in different stages of the main signal being generated, transmitted,

A. Ghasemi et al., *Propagation Engineering in Radio Links Design*,
DOI 10.1007/978-1-4614-5314-7_8,

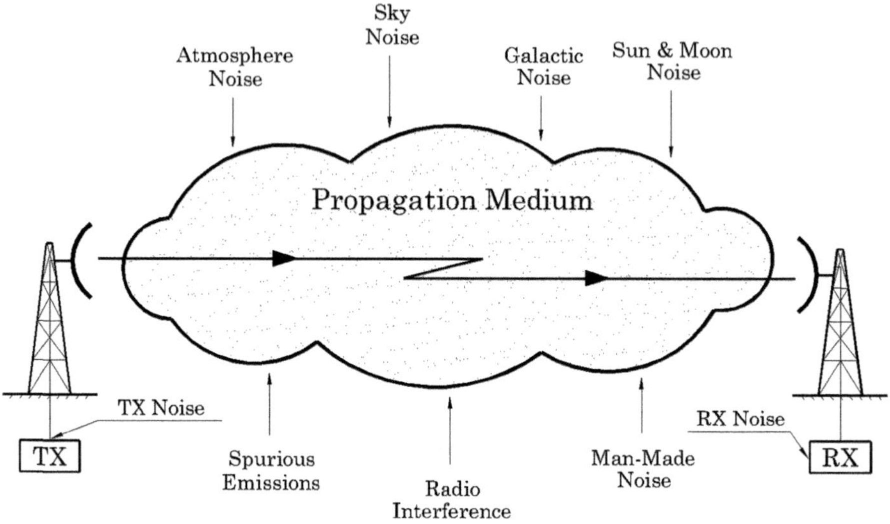

Fig. 8.1 Major sources of radio noise

processed, and amplified. The added noises will cause interference and disruption of the main signal in the last stage of detection and its quality. Based on the ITU-R recommendation No. V.573, radio noise is defined as follows:

A time-varying electromagnetic phenomenon having components in the radio-frequency range, apparently not conveying information and which may be super imposed on, or combined with, a wanted signal.

Although on the basis of what was stated, radio interference and spurious emissions are regarded as a part of external noises, these cases are beyond the scope of this chapter and can be evaluated separately. Given the above definition and as illustrated in Fig. 8.1, the main types of the noise can be classified as follows:

- Natural noises generated by the Earth and its atmosphere, the Sun, radio stars, and the galaxies as a whole
- Artifact or man-made noises related to electrical appliances, engines, electromotors, unwanted signals, etc.
- Urban noises
- Noises related to industrial plants and activities
- Noise generated by TX and RX equipments and other radio devices (as thermal noise)
- Temporal noise such as noise from thunder storms and rain

Noise generated from all equipments and devices used for radiowave transmission and reception entitled as internal noise while remaining noises are categorized as external noise. The external noises related to the radiowaves propagation (also called the radio noises) are evaluated in the following sections.

8.2.2 Main Effects

Radio noises have significant impacts on the following issues in the radio communications field:

- Classification of the frequency spectrum for allocation to various telecom services by the international, regional, and national institutions and authorities
- Frequency band selection in radio system design
- Limitations exerted on radio systems performance and the received signal quality
- The antenna effective noise figure, that is the equivalent antenna noise temperature which is a planning parameter to specify the performance of telecom systems

8.3 Theoretical Principles

This section will review some of the theoretical principles related to telecom systems including the following outstanding points:

- Thermal noise power
- Noise factor and figure
- Noise temperature
- Equivalent antenna noise temperature

8.3.1 Thermal Noise Power

According to Fig. 8.2 in a resistor, R with absolute temperature degree, t in terms of degrees Kelvin, a voltage with a magnitude of V_s will be generated at the output terminal due to the random motions of electrons. According to Planck's rule, power produced by this noise denoted as P_R can be expressed by the following equation:

$$P_R = 4h.f.R.B_n/(e^{hf/kt} - 1) \tag{8.1}$$

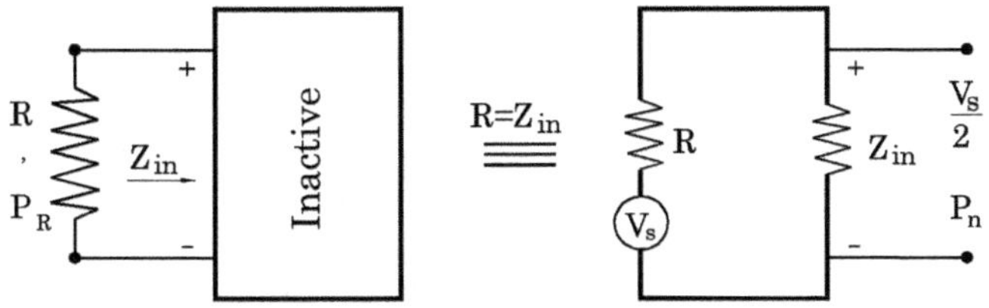

Fig. 8.2 Noise power in a resistor and connected passive network

The components and their related units in the above equation are given below:

h : Planck's constant equal to 6.63×10^{-34}
k : Boltzmann's constant equal to 1.38×10^{-23}
R : Resistance in ohms
t : Temperature in degrees Kelvin
f : Frequency in hertz
B_n : Noise bandwidth in hertz

Noting that $hf/kt \ll 1$ thus using Taylor's expansion gives the following result:

$$e^{hf/kt} \approx 1 + hf/kt \tag{8.2}$$

The received power is equal to

$$P_R = 4k.t.R.B_n \tag{8.3}$$

and the equivalent voltage of the noise source which generates power P_R is equal to

$$V_s = 2\sqrt{k.t.R.B_n} \tag{8.4}$$

Noting that in relation (8.3), the noise power density is independent of frequency f, and the spectrum of the generated noise is of white noise type. As indicated in Fig. 8.2, in case of connecting this resistor to a matched passive network ($R = Z_{in}$), then the maximum noise power P_n is transferred to this network which is equal to

$$P_n = k.t.B_n \tag{8.5}$$

8.3.2 Noise Factor and Figure

Noise effects on the reception of a signal in the telecommunication systems result in signal impairment. Therefore, signal-to-noise ratio denoted as SNR is a significant parameter in telecommunication such as radio systems and is defined by the following expression:

$$SNR = (\text{power of the main signal})/(\text{power of noise}) = \mathrm{S/N} \tag{8.6}$$

To facilitate the noise calculations in a professional manner, it is a common practice to express this ratio in the logarithmic form as indicated below:

$$SNR[\mathrm{dB}] = 10\ \log S - 10\ \log N = S[\mathrm{dB}_m] - N[\mathrm{dB}_m] \tag{8.7}$$

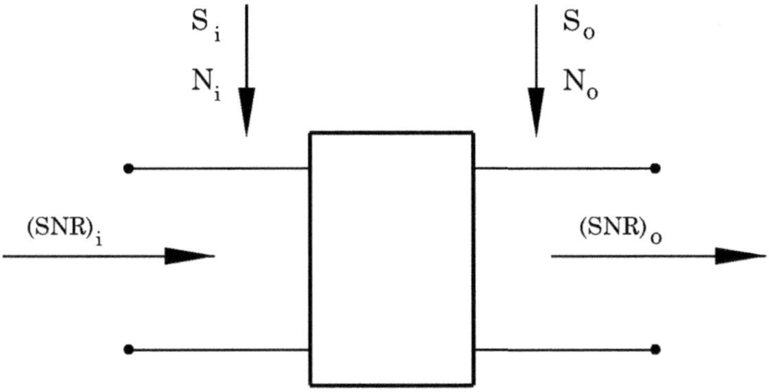

Fig. 8.3 Simple concept to define noise factor and noise figure

8.3.2.1 Standard Conditions

Noting the Fig. 8.3 and the above discussion, noise factor (of an amplifier/attenuator) denoted as f at a conventional temperature ($t_0 = T_0 = 290°\text{K}$) and in a perfectly matched system is defined as

$$\mathtt{f} = (S_i/N_i) : (S_0/N_0) = (SNR)_i/(SNR)_0 \tag{8.8}$$

Noise figure denoted by F is equal to the value of noise factor in logarithmic scale and expressed as

$$F = 10 \log \mathtt{f} = (SNR)_i[\text{dB}] - (SNR)_0[\text{dB}] \tag{8.9}$$

Assuming an ideal system without any noise being added, then the value of f is one, and the value of F is zero. In a real situation, each system adds some noise to the input resulting in higher noise factor/figure. If the amplification coefficient of the system is assumed to be A, then

$$S_0 = A \cdot S_i \tag{8.10}$$

$$N_0 = AN_i + P_{nA} = A.k.t_0.B_\text{n} + P_{nA} \tag{8.11}$$

In the latter relation, P_{nA} is the internal noise power of the amplifier. Combining it with relation (8.8) yields

$$\mathtt{f} = 1 + P_{nA}/A.k.t_0.B_\text{n} \tag{8.12}$$

And subsequently the additional noise power of the amplifier is simplified to

$$P_{nA} = A(\mathtt{f} - 1)k.t_0.B_\text{n} \tag{8.13}$$

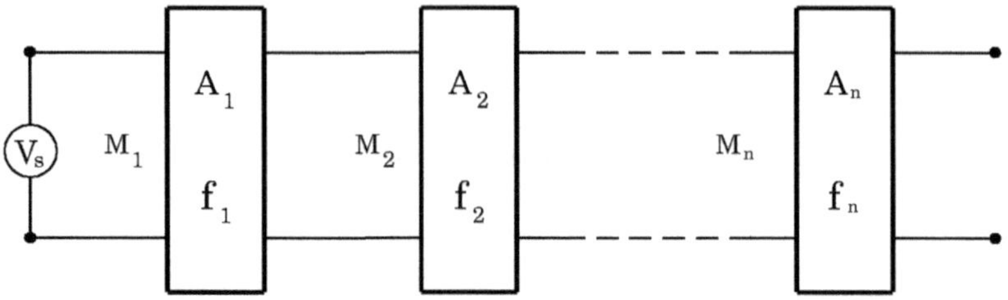

Fig. 8.4 Noise factor for cascaded amplifiers

8.3.2.2 Nonstandard Conditions

The above-mentioned relations in the previous section were achieved under standard conditions, in other words:

- Temperature equal to the reference value ($t_0 = 290$ K)
- Perfect matching of noise source to the passive network ($M = 1$)

In case one or both of the above conditions do not hold, we have a nonstandard condition, and the system noise factor denoted as $\mathtt{f}'$ will be as indicated below:

$$\mathtt{f}' = t/t_0 + (\mathtt{f} - 1)/M \tag{8.14}$$

$$F' = 10 \log \mathtt{f}' \tag{8.15}$$

$\mathtt{f}'$ and F' are the noise factor and figure, respectively, under nonstandard conditions.

8.3.2.3 Noise Factor in a Cascaded System

As shown in Fig. 8.4, when the system consists of a cascaded section of several amplifiers with a specified noise factor, in that position, the total noise factor of the said system on the output and under standard conditions can be expressed by the following relation:

$$\mathtt{f} = \mathtt{f}_1 + \frac{\mathtt{f}_2 - 1}{A_1} + \frac{\mathtt{f}_3 - 1}{A_1 \cdot A_2} + \cdots + \frac{\mathtt{f}_n - 1}{A} \tag{8.16}$$

$$A = \prod_{i=1}^{n-1} A_i \tag{8.17}$$

If according to Fig. 8.5 the system constitutes of several attenuators each having a specified noise factor and attenuation coefficient, then in such circumstance the total system noise on the output under standard conditions can be defined by the following relation:

$$\mathtt{f} = \mathtt{f}_1 + l_1(\mathtt{f}_2 - 1) + l_1 \cdot l_2(\mathtt{f}_3 - 1) + \cdots + l(\mathtt{f}_n - 1) \tag{8.18}$$

$$l = \prod_{i=1}^{n-1} l_i \tag{8.19}$$

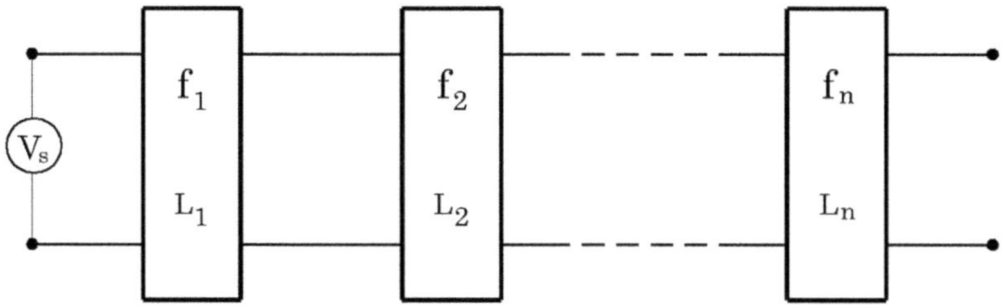

Fig. 8.5 Noise factor for cascaded attenuators

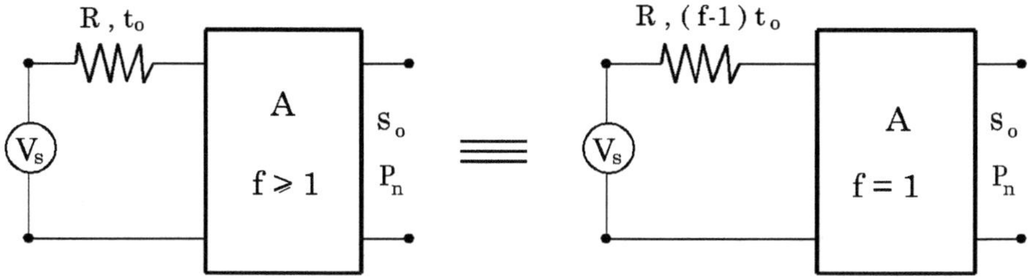

Fig. 8.6 Equivalent circuit of a noisy amplifier

It is also possible to derive the following general equation through combining of the relations (8.16) and (8.18):

$$\mathtt{f} = \mathtt{f}_1 + \frac{l_1(\mathtt{f}_2 - 1)}{A_1} + \frac{l_1 \cdot l_2(\mathtt{f}_3 - 1)}{A_1 \cdot A_2} + \cdots + \frac{l(\mathtt{f}_n - 1)}{A} \tag{8.20}$$

8.3.3 Equivalent Noise Temperature

By a precise analysis of relation (8.13), it is revealed that given the value of amplifier noise power with a noise factor of f, as shown in Fig. 8.6, it is possible to replace it with a combination of a resistor with noise power at a temperature of $(\mathtt{f} - 1)t_0$ and an ideal noiseless amplifier. This effective temperature is denoted by T_{eq} and called the amplifier noise temperature.

$$T_{eq} = (\mathtt{f} - 1)t_0 \tag{8.21}$$

If the source of noise temperature is assumed to be T degrees, then the equivalent system noise temperature will be given as:

$$T_s = T + [(f - 1)/M] \cdot t_0 \tag{8.22}$$

Under perfect matching condition, the latter relation can be simplified as

$$M = 1 \implies T_s = T + (f - 1)t_0 \tag{8.23}$$

8.3.4 Antenna Equivalent Noise Temperature

In radio communications, the external noises are received along with the main radiowaves through the antenna. There is certain supplementary information in section (8.4) about radio noises and their sources and values completed with some expressions and related diagrams and curves. Basically, the merit of a reception system in satellite communications is expressed by the RX antenna G/T; hence there is also some additional information about this topic in section (8.5).

Generally, RX antenna noises coming from cosmic and galaxy bodies such as the Sun and radio stars are received with an extensive and continuous spectrum in the form of white noise. In addition to radio noises, the antenna collects noises emanating from further absorption and emission of electromagnetic energy by these bodies including sky particles.

The total external noises which the antenna absorbs can be given by an equivalent noise temperature in terms of degrees Kelvin denoted as T_A. The equivalent noise power with this temperature is generated through a resistor equal to the antenna radiation resistance and is same as total generated noises. The antenna equivalent noise temperature is a function of frequency and direction of the antenna which varies considerably with the amount of the antenna received noise in terms of time, day, and its particular position. In fact, T_A is an average value which may be calculated using the following integral expression:

$$T_A = \frac{1}{4\pi} \int_0^{\pi} \int_0^{2\pi} g \cdot v \cdot t \cdot \sin\theta \, d\theta \cdot d\varphi \tag{8.24}$$

All three parameters of g, v, and t, as defined below, are functions of θ and φ expressed as

g : RX antenna gain
v : Electromagnetic absorption coefficient
t : Temperature at the intended point

It must be reminded that in addition to thermal noises, the antenna also absorbs noises related to lightning, man-made noises generated by electrical machinery, and other active elements with non-thermal nature. In diagrams (8.11)–(8.13) which will be presented in future section, the amount of external noise temperature including their related noise figures is presented in the frequency range of 0.1 Hz–100 GHz.

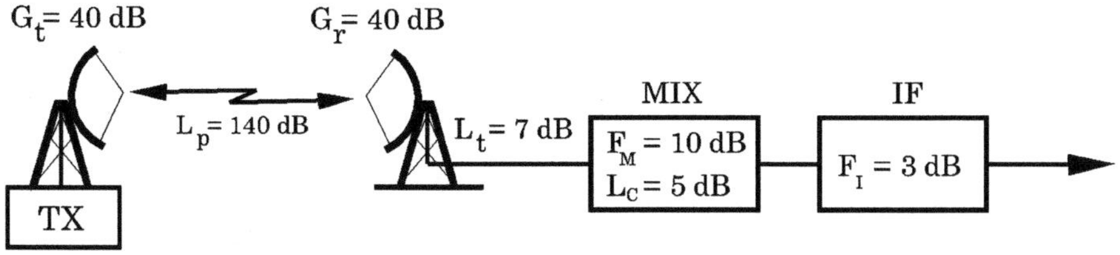

Fig. 8.7 Microwave radio link (Example 8.1)

Example 8.1. A microwave link is employed to exchange signals with a bandwidth of $B_n = 22$ MHz. The RX antenna noise temperature is 120°K. The values of other parameters are depicted in Fig. 8.7.

1. Calculate the noise power at the output of IF amplifier.
2. Assuming the net losses from TX output to RX input equal to 140 dB, find the minimum TX output power by accounting for a 30 dB safety margin and SNR = 23 dB.

Solution. 1.

$$L_t = 7\text{ dB} \implies l_t = 5$$

$$F_M = 10\text{ dB} \implies f_M = 10$$

$$L_c = 5\text{ dB} \implies l_c = 3$$

$$F_I = 3\text{ dB} \implies f_I = 2$$

$$T_s = T_A + (f_t - 1)T_0 + (f_M - 1)l_t T_0 + (f_{IF} - 1)l_t l_c T_0$$

$$T_s = 120 + 4T_0 + 9 \times 5T_0 + 1 \times 5 \times 3T_0$$

$$\implies T_s = 18,680\text{ K}$$

$$P_n = k \cdot T_s \cdot B_n \implies P_n = 5.671 \times 10^{-12} = 5.671\text{ pW}$$

2.

$$P_n[\text{dB}_m] = 10\log(5.671 \times 10^{-9}) = -82.46\text{ dB}_m$$

$$S[\text{dB}_m] = P_n[\text{dB}_m] + SNR[\text{dB}] + FM$$

$$= -29.46\text{ dB}_m$$

$$P_t[\text{dB}_m] = S[\text{dB}_m] + L_p[\text{dB}] - G_t - G_r$$

$$= -29.46 + 140 - 40 - 40 = 30.54\text{ dB}_m$$

■

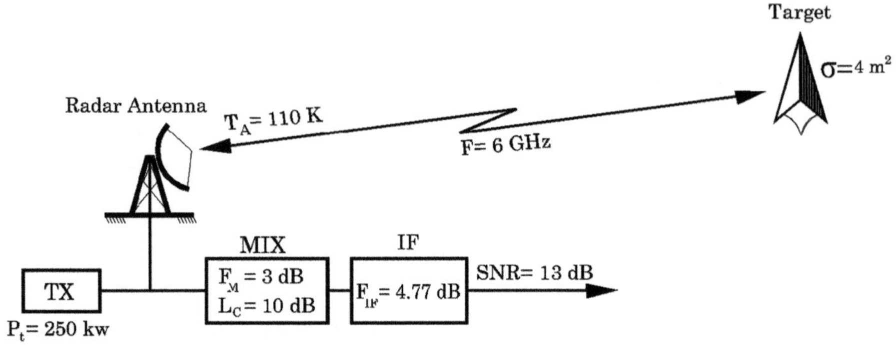

Fig. 8.8 Radar system (Example 8.2)

Example 8.2. The main aspects of a radar link are given in Fig. 8.8. Assuming that the transmission line losses between the antenna and TX/RX are ignored and the signal bandwidth is 1 MHz, find:

1. The system noise temperature.
2. The external noise figures and total noise figure.
3. The maximum radar range.

Solution. 1.

$$\begin{aligned}
F_M &= 3 \text{ dB} \implies \mathtt{f}_M = 2 \\
F_{IF} &= 4.77 \text{ dB} \implies \mathtt{f}_{IF} = 3 \\
L_c &= 10 \text{ dB} \implies l_c = 10 \\
T_s &= T_A + (\mathtt{f}_M - 1)T_0 + l_c(\mathtt{f}_{IF} - 1)T_0 \\
&= 6,100°\text{K}
\end{aligned}$$

2.

$$\begin{aligned}
F_A &= 10 \log(k.T_A.B_n) = -148.2 \text{ dB} \\
F_s &= 10 \log(k.T_s.B_n) = -130.7 \text{ dB}
\end{aligned}$$

3. Values of the related components in non-logarithmic scale (numerical scale) are:

$$\begin{aligned}
G &= 37 \text{ dB}_i \implies g = 5,000 \\
f &= 6 \text{ GHz} \implies \lambda = 0.05 \text{ m} \\
(SNR)_0[\text{dB}] &= 13 \implies P_r/P_n = 20 \\
P_n &= k.T_s.B_n = 8.418 \times 10^{-14} \text{ W} \\
P_r &= 20\, P_n \implies P_r = 1.684 \times 10^{-12} \text{ W} = 1.684 \text{ pW}
\end{aligned}$$

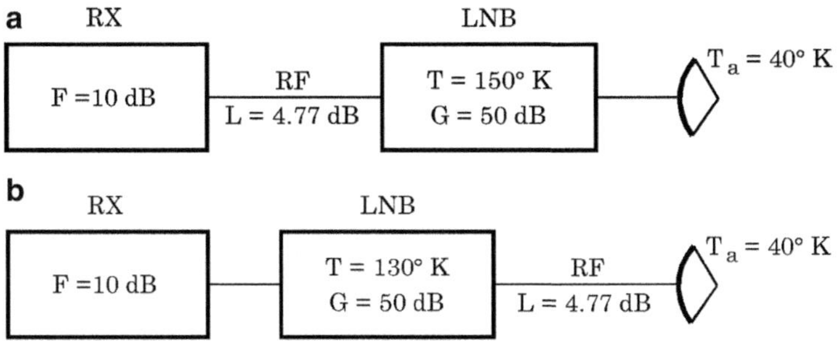

Fig. 8.9 (**a**) Outdoor type LNB in Satellite TVRO (**b**) Indoor type LNB in Satellite TVRO

Besides, the received power of the monostatic radar (according to relation specified in Chap. 6) can be expressed by

$$P_r = \frac{P_t \cdot g^2 \cdot \lambda^2 \cdot \sigma}{(4\pi)^3 . R^4}$$

$$R_{max} \approx 6.5 \times 10^4 \text{ m} \approx 65 \text{ km}$$ ■

Example 8.3. To receive the signals from a TV broadcasting satellite by a single LNB (low-noise block) with the following specifications:

- LNB gain: 50 dB
- RX noise figure: 10 dB
- LNB noise temperature: 150°K
- Antenna noise temperature: 40°K

And assuming the transmission line losses at the reception frequency equal to 4.77 dB, for two cases depicted in Fig. 8.9, find the equivalent noise temperature, and describe the reason for mounting LNB on the receiving antenna.

Case 1: LNB mounted outdoor on the receiving antenna.
Case 2: LNB is moved to an indoor location through the transmission line where the equivalent noise temperature is reduced to 130°K.

Solution. Since F, G, and L are in decibels, in order to be able to use the related relations, first their non-logarithmic (numerical) values must be calculated:

$$f = 10, \quad l = 3, \quad g = 10^5$$

Case 1: Noting diagram (8.9a) and utilizing formulas (8.16) and (8.18) yields:

$$T_s = 40 + 150 + \frac{(3-10) \times 290}{10^5} + \frac{3 \times (10-1) \times 290}{10^5} \approx 190°\text{K}$$

Case 2: Noting diagram (8.9b) in this position, the value of equivalent noise temperature is:

$$T_{s2} = 40 + (3-1)\times 290 + 3\times 130 + \frac{3\times(10-1)\times 290}{10^5} \approx 1{,}010^\circ\mathrm{K}$$

The results reveal that in case 2, the amount of equivalent noise temperature is larger than case 1. Therefore, it is a common practice to install LNB of satellite TVRO terminal very close to the antenna. ■

8.4 Radio Noises

The radio section of ITU has prepared and presented comprehensive and useful information and graphs for evaluation of noise in long-distance communications through the recommendation No. ITU-R, P.372 titled *Radio Noises*. It should be noted that some of the legends and notations in the relations of this section are different from those used in this book; therefore special attention must be paid in using and comparing them with the relations of other sections.

8.4.1 Source of Radio Noises

External radio noises are caused by the following factors:

- Radiations related to the electrical discharge resulting from lightning in the Earth's atmosphere
- Unintentional radiations related to electrical machinery, electrical and electronics devices, electric power transmission lines, and/or internal combustion engine ignition generally called man-made noises
- Emissions from atmospheric gases and hydrometeors
- Ground or other obstructions within the antenna beam
- Radiation from celestial radio sources

Radiowaves interference and spurious emissions related to TX and RX systems are not considered as external radio noises and should be studied and evaluated separately.

8.4.2 Noise Intensity Specifications and Their Interrelations

Noise factor denoted by f for a receiving system consists of several noise sources at the RX terminal. In this situation, all internal and external noises must be accounted for. The only proper reference point to specify the overall operating noise factor of the entire system, is the lossless RX antenna input (this reference point does not exist physically, and it is only a hypothetical reference point).

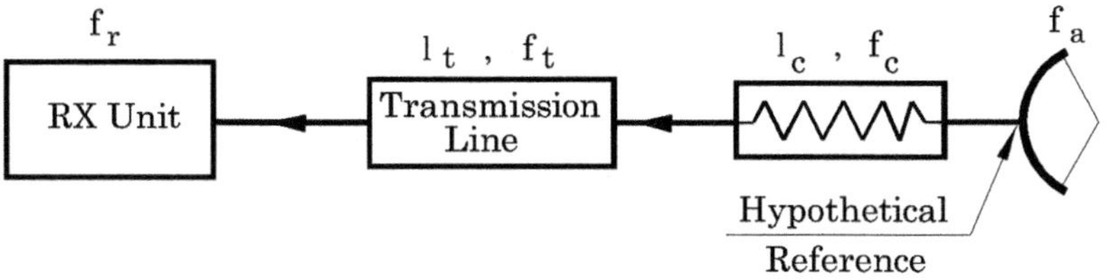

Fig. 8.10 Noise factor in reception system

For receivers without spurious emissions and observing Fig. 8.10, it will be possible to express the noise factor by the following relation:

$$\mathtt{f} = \mathtt{f}_a + (\mathtt{f}_c - 1) + l_c(\mathtt{f}_t - 1) + l_c \cdot l_t(\mathtt{f}_r - 1) \tag{8.25}$$

where supplementary details of the above relation are:

$\mathtt{f}_a$: External noise factor as defined below:

$$\mathtt{f}_a = P_n/k \cdot t_0 \cdot b \tag{8.26}$$

The following relation exists between noise figure F_a and noise factor $\mathtt{f}_a$:

$$F_a[\mathrm{dB}] = 10 \log \mathtt{f}_a \tag{8.27}$$

P_n : Noise power produced by the equivalent lossless antenna
k : Boltzmann's constant equal to 1.38×10^{-23} J/K
t_0 : Reference temperature equal to 290°K
b : Noise bandwidth of RX system in terms of Hz
l_c : Loss factor of the antenna circuit (equivalent to the ratio of its input power to output power)
l_t : Loss factor of the transmission line (equivalent to the ratio of its input to output power)
$\mathtt{f}_r$: Rx noise factor, keeping in mind that the following relation exists between the RX noise figure denoted as F_r and noise factor denoted as $\mathtt{f}_r$:

$$F_r[\mathrm{dB}] = 10 \log \mathtt{f}_r \tag{8.28}$$

$\mathtt{f}_c$ is the noise factor associated with the antenna circuit given by the following equation:

$$\mathtt{f}_c = 1 + (l_c - 1) \cdot (t_c/t_0) \tag{8.29}$$

$\mathtt{f}_t$ is the noise factor associated with the losses of the transmission line given by the following relation:

$$\mathtt{f}_t = 1 + (l_t - 1) \cdot (t_t/t_0) \tag{8.30}$$

t_c : Antenna actual temperature and the surrounding land in terms of degrees Kelvin
t_t : Actual temperature of the transmission line in terms of degrees Kelvin

When t_c and t_t are equal to the reference temperature of the medium (i.e., $t_c = t_t = t_0$), the relation (8.25) can be simplified to

$$\mathtt{f} = (\mathtt{f}_a - 1) + \mathtt{f}_c \cdot \mathtt{f}_t \cdot \mathtt{f}_r \tag{8.31}$$

Equation (8.26) can be expressed in the logarithmic scale as:

$$P_\mathrm{n}[\mathrm{dB}] = F_a + B - 204 \tag{8.32}$$

In this equation, the components are defined as follows:

$$P_\mathrm{n}[\mathrm{dB}_w] = 10 \log P_\mathrm{n} \tag{8.33}$$

$$B = 10 \log b \tag{8.34}$$

$$10 \log kt_0 = 10 \log(1.38 \times 10^{-23} \times 290) = -204 \tag{8.35}$$

For a short vertical monopole antenna ($h << \lambda$) over a perfect ground plane, the effective value (rms) of the vertical component of the field intensity is equal to:

$$E_n[\mathrm{dB}_{\mu V/m}] = F_a + 20 \log f + B - 95.5 \tag{8.36}$$

In relation (8.36), each of the variables and their corresponding units are

E_n : Field intensity with the bandwidth b, in terms of $\mathrm{dB}_{\mu V/m}$
f : Center frequency in terms of MHz

Similarly for a half-wave dipole antenna in free space, the effective value of the vertical component of the field intensity can be derived as:

$$E_n[\mathrm{dB}_{\mu V/m}] = F_a + 20 \log f + B - 99 \tag{8.37}$$

As usual, the external noise factor in the form of noise temperature denoted as t_a is expressed by:

$$\mathtt{f}_a = t_\mathrm{a}/t_0 \tag{8.38}$$

t_a, which sometimes is indicated by T_a, is called the effective antenna noise temperature due to the external noises. Given the relations F_a and E_n according to Eqs. (8.36) and (8.37), it is possible to obtain the value of E_n. The above noise power is essential for specifying the signal-to-noise ratio, but in most instances, it is not sufficient. As an example, to determine the system performance, it requires proper and additional explanations about the shape of the received random noise waves. Since in this section the concerning noises have a uniform phase spread,

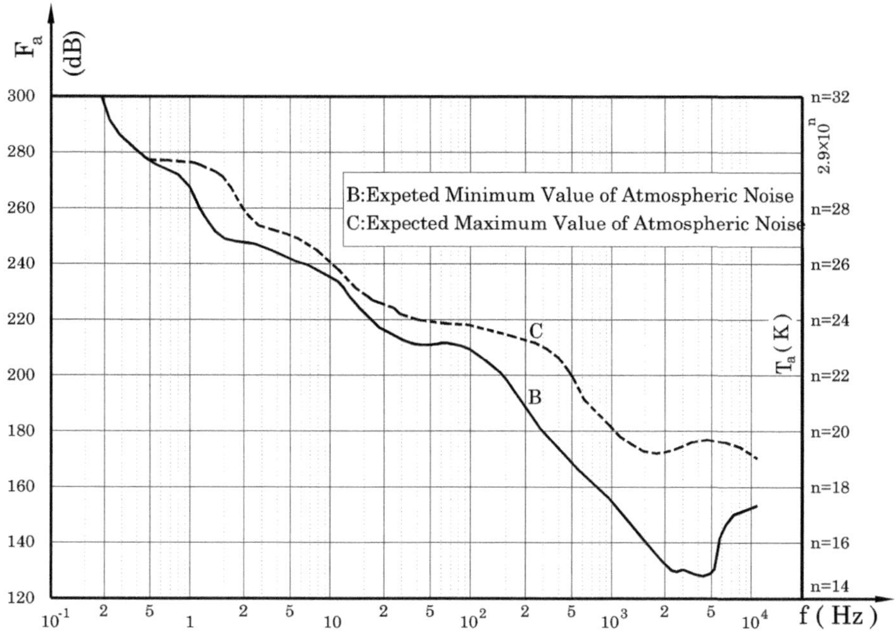

Fig. 8.11 F_a variations in frequency range of 0.1 Hz–10 KHz (Ref.: ITU-R, P.372-9)

their overlapping shall be also determined. To process impulse noises at higher frequencies (that is greater than 1 GHz), the amount of F_a is very small, and only impulses with high amplitudes possess noise greater than RX noise.

8.4.3 Noise Level in Terms of Frequency

In order to determine the noise level and its variations relative to the frequency, three figures are introduced for providing the values of noise figures denoted as F_a in the frequency range from 0.1 Hz to 100 GHz along with the noise levels of interest.

Figure 8.11 covers the frequency range 0.1 Hz–10 KHz. The solid curve is the minimum expected hourly median values of F_a based on measurements (taking into account the entire Earth's surface, all seasons, and times of day), and the dashed curve gives the maximum expected values. Note that in this frequency range, there is very little seasonal, diurnal, or geographic variation. The larger variability in the 100–10,000 Hz range is due to the variability of the Earth-ionosphere waveguide cutoff.

Figure 8.12 covers the frequency range 10^4–10^8 Hz, that is, 10 KHz–100 MHz for various categories of noise. The minimum expected noise is shown by the solid curves. For atmospheric noise, the minimum values of the hourly medians expected

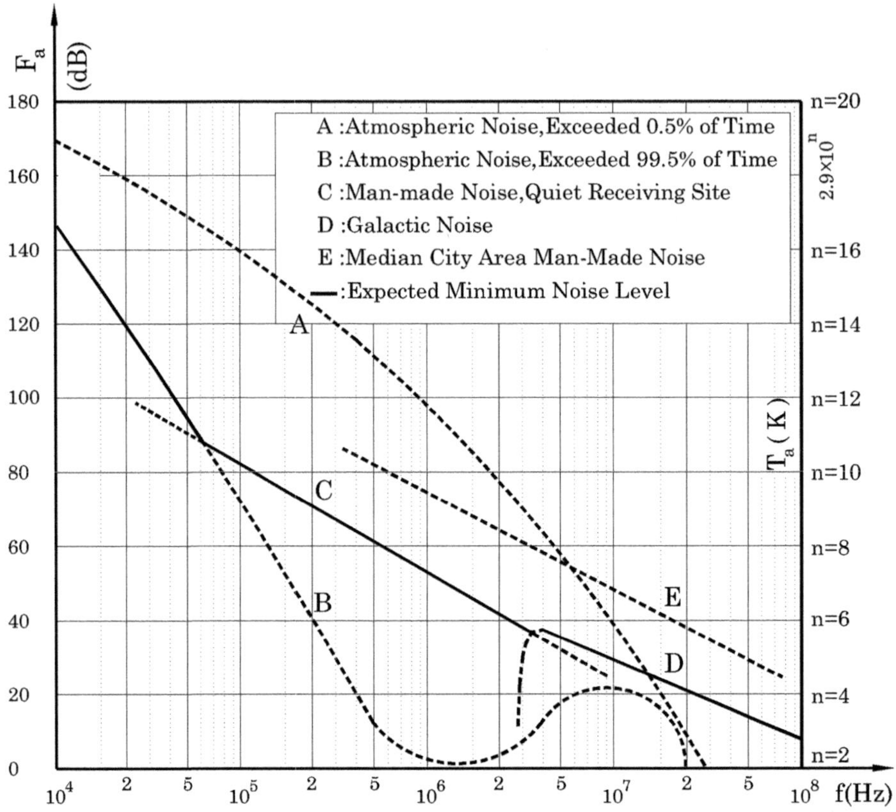

Fig. 8.12 F_a variations in frequency range of 10 KHz–100 MHz (Ref.: ITU-R, P.372-9)

are taken to be those values exceeded 99.5% of the hours, and the maximum values are those exceeded 0.5% of the hours. For the atmospheric noise curves, all times of day, seasons, and the entire Earth's surface have been taken into account.

Figure 8.13 covers the frequency range 10^8–10^{11} Hz, that is, 100 MHz–100 GHz. Again, the minimum noise is given by solid curves, while some other noises of interest are given by dashed curves.

The majority of the results shown in the three figures are for omnidirectional antennas (except as noted on the figures). For directional antennas, however, studies have indicated that at HF, for example, for atmospheric noise from lightning for very narrow beam antennas, there can be as much as 10 dB variation (5 dB above to 5 dB below the average F_a value shown) depending on antenna pointing direction, frequency, and geographical location.

For galactic noises, the average value (over the entire sky) is given by the solid curve labelled galactic noise (Figs. 8.12 and 8.13). Measurements indicate a ± 2 dB variation about this curve, neglecting ionospheric shielding. The minimum galactic noise (narrow beam antenna towards galactic pole) is 3 dB below the solid galactic

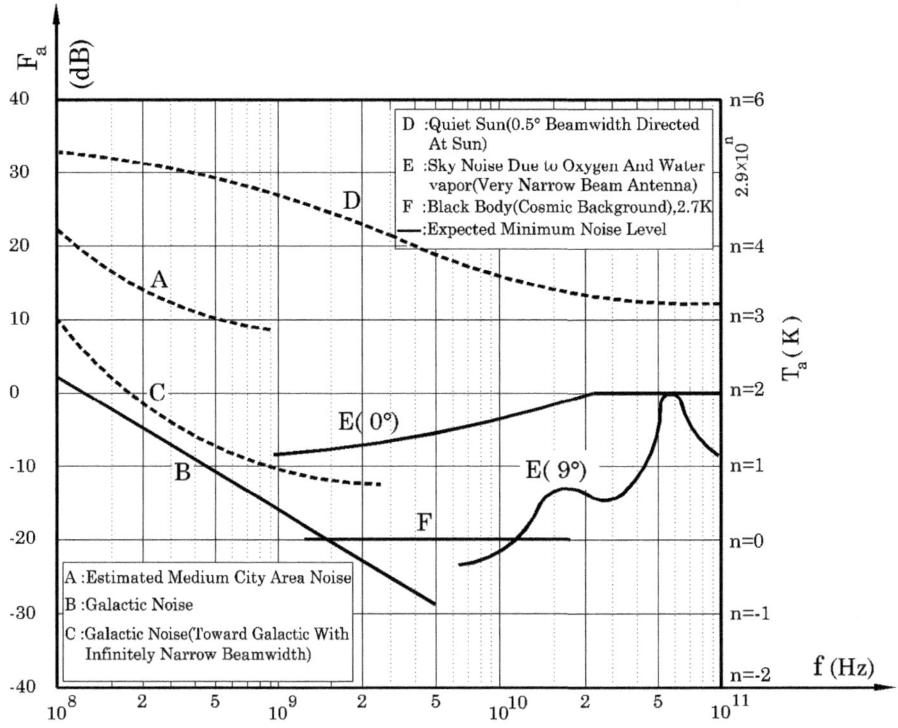

Fig. 8.13 F_a variations in frequency range of 100 MHz–100 GHz (Ref.: ITU-R, P.372-9)

noise curve shown on Fig. 8.13. The maximum galactic noise for narrow beam antennas is shown via a dashed curve in Fig. 8.13.

Example 8.4. 1. Specify the radio noise values and their equivalent noise temperature for frequencies of 4 and 30 GHz using Fig. 8.13.
2. Calculate the noise factor and noise figure at the antenna input for a 4 GHz frequency.
3. Obtain the antenna input noise power for a 5 MHz bandwidth.
4. Calculate the effective noise temperature at the RX input assuming 5 dB for transmission line loss.

Solution. 1. To specify radio noise figures and also equivalent noise temperatures, considering the operating frequencies and utilizing Fig. 8.13, it can be concluded that

$$f_1 = 4\text{ GHz} \Longrightarrow \textit{Galactic Noise} \xrightarrow{\textit{graph B}} F_a = -27\text{ dB},\ T_a = 2^\circ\text{K}$$

$$\textit{Cosmic Noise} \xrightarrow{\textit{graph F}} F_a = -20\text{ dB},\ T_a = 2.9^\circ\text{K}$$

$$Sky\ Noise \xrightarrow{graph\ E(0)} F_a = -4 \text{ dB},\ T_a = 1.74 \times 10^{2^\circ}\text{K}$$

$$Sun\ Noise \xrightarrow{graph\ D} F_a = +20 \text{ dB},\ T_a = 2.9 \times 10^{4^\circ}\text{K}$$

$$f_2 = 30 \text{ GHz} \Longrightarrow Galactic\ Noise\ is\ Negligible$$

$$Cosmic\ Noise \xrightarrow{graph\ F} F_a = -20 \text{ dB},\ T_a = 2.9^\circ\text{K}$$

$$Sky\ Noise \xrightarrow{graph\ E(90)} F_a = -15 \text{ dB},\ T_a = 14.5^\circ\text{K}$$

$$Sky\ Noise \xrightarrow{graph\ E(0)} F_a = 0 \text{ dB},\ T_a = 2.9 \times 10^{2^\circ}\text{K}$$

$$Sun\ Noise \xrightarrow{graph\ D} F_a = +12 \text{ dB},\ T_a = 3.48 \times 10^{3^\circ}\text{K}$$

2. If the antenna axis is directed toward the Sun, then:

$$f_1 = 4 \text{ GHz},\quad T_t \approx 2.9 \times 10^{4^\circ}\text{K}$$

$$\mathtt{f}_a = \frac{T_t}{T_0} = \frac{2.9 \times 10^4}{290} = 100$$

$$F_a = 10 \log \mathtt{f}_a = 20 \text{ dB}$$

If the antenna axis is not directed toward the Sun and ignoring the Sun noises, it gives the following result:

$$T_t' = 1.74 \times 10^2 + 2.9 + 2 = 179^\circ\text{K}$$

$$\mathtt{f}_a' = \frac{T_t'}{T_0} = 0.62$$

$$F_a' = 10 \log \mathtt{f}_a' = -2.08 \text{ dB}$$

3. Given the (8.32) and at a position directed toward the Sun,

$$P_\text{n}[\text{dB}_W] = F_a + B - 204 = -117 \text{ dB}_W$$

$$P_\text{n} = \text{Antilog}(-11.7) \approx 2 \text{ pW}$$

for second case:

$$P_n' = -139.08 \text{ dB}_W \quad\Longrightarrow\quad P_n' = 1.26 \times 10^{-14} \text{ W}$$

4.

$$\mathtt{f}_c = 1,\quad l_t[\text{dB}] = L_t = 5 \text{ dB} \Longrightarrow l_t = 7$$

$$\mathtt{f}_t = 1 + (l_t - 1) = 7$$

$$\mathtt{f} = \mathtt{f}_a - 1 + \mathtt{f}_c \cdot \mathtt{f}_t = 100 - 1 + 1 \times 7 = 106$$

$$\mathtt{f} = \frac{T_r}{T_0} \implies T_r = 30,740^\circ\mathrm{K}$$

$$\mathtt{f}' = \mathtt{f}'_a - 1 + \mathtt{f}_c \cdot \mathtt{f}_t = 0.62 - 1 + 7 = 6.62$$

$$\mathtt{f}' = \frac{T'_r}{T_0} \implies T'_r = 1,919.8^\circ\mathrm{K}$$

■

8.4.4 Noise from Atmospheric Gases and the Earth's Surface

Noise related to individual sources such as the Sun, gases persisting in the atmosphere, and the Earth's surface are usually defined with brightness noise temperature denoted as t_b. The effective antenna temperature denoted as t_a in fact is the convolution of the antenna radiation pattern and the sky and the Earth's brightness temperature. For antennas whose radiation pattern constitute a single noise source, the value of antenna temperature and the brightness temperature are identical (e.g., the curves of C, D, and E in Fig. 8.13).

Figure 8.14 illustrates the atmosphere brightness temperature for terrestrial RX(s) in the 1–340 GHz frequency range. In these curves, the share of cosmic noise has a magnitude of 2.7°K which has been considered along with other additional terrestrial sources. Also, Fig. 8.15 illustrates the same specifications on an expanded horizontal scale (i.e., frequency band of 1–60 GHz.)

The mentioned schematics are calculated for seven different values of the elevation angle and an average atmospheric conditions (with a water vapor density of 7.5 g/m^3, temperature of 288°K and 2 km height for water vapor). The US standard atmosphere of 1976 is considered for a dry atmosphere as a reference.

In Earth-space telecommunications, if the signal attenuation from spacecraft TX is known, then there will be a proper estimation of the brightness temperature for the frequency range of 2–30 GHz in the said direction which can be derived from the following formula:

$$t_\mathrm{b} = t_\mathrm{e}(1 - \mathrm{e}^{-d}) + 2.7 \qquad (8.39)$$

where:

d : Optical depth = attenuation (dB/4.343)
t_e : Effective temperature, usually taken to be 275°K

In the above relation, there are certain conclusions that can be indicated with a precision of 0.1 dB in the frequency range less than 30 GHz. In higher frequencies, the composition of components related to radiowaves propagation occurs, and the magnitude of brightness temperature will be very high. The brightness temperature

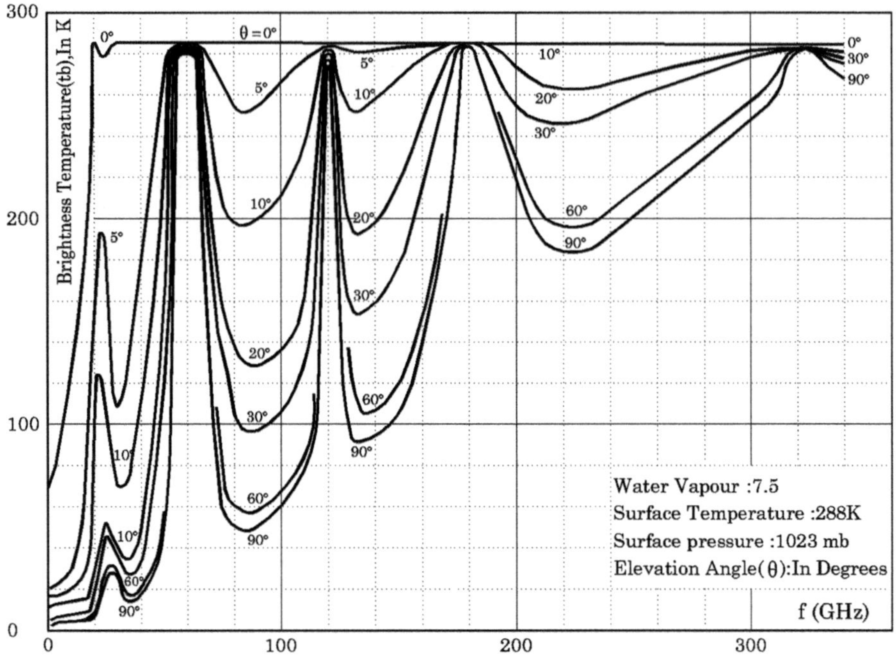

Fig. 8.14 Brightness temperature of clear air (Ref.: ITU-R, P.372-9)

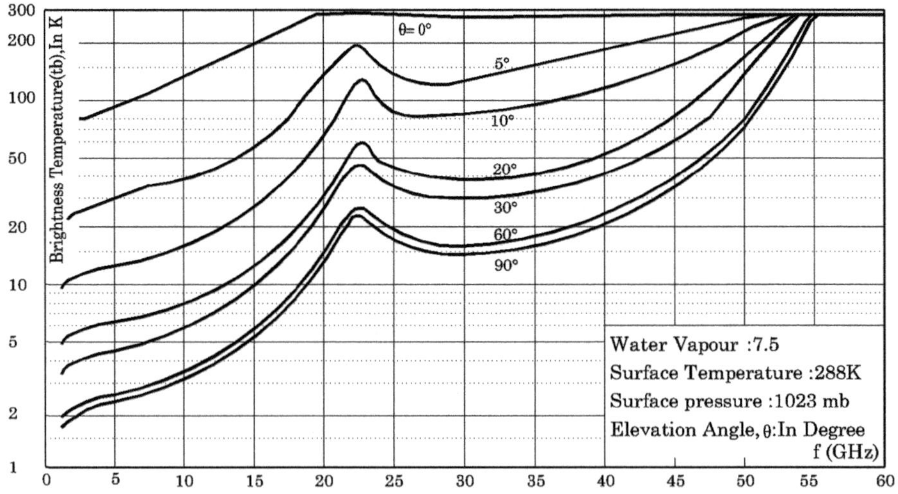

Fig. 8.15 Brightness temperature of clear air (expanded frequency axis) (Ref.: ITU-R, P.372-9)

of the Earth's surface can be approximated by the following simple relation:

$$T = \varepsilon \, T_{\text{surf}} + \rho \, T_{\text{atm}} \tag{8.40}$$

In the above relation, each one of the components and their corresponding units are given as:

ε : Effective emissivity of the Earth's surface
ρ : Effective reflection coefficient
T_{surf} : Physical temperature of the Earth's surface in degrees Kelvin (°K)
T_{atm} : Brightness temperature of the sky

Up to about 100 GHz and specially below 10 GHz, the reflection coefficient ρ is generally high, and emissivity ε is low. Figure 8.16a illustrates the emissivity and brightness temperature of a smooth water surface for horizontal and vertical polarizations and two incident angles. It must be reminded that for frequencies greater than 5 GHz, fresh and salted water are not distinguishable.

Figure 8.16b shows the nadir brightness temperature of the sea surface at three different frequencies as a function of physical temperature for a salinity of 36 parts per thousand. The rise in brightness temperature of the sea surface in terms of the sea surface with wind velocity is shown in Fig. 8.16c and d which also provides a useful tool for storm detection.

The emissivities (and hence the brightness temperatures) of land surfaces are higher than those of water surfaces due to the lower dielectric constant of land. In Fig. 8.17a, the brightness temperature of a smooth field for different moisture contents is shown, while in Fig. 8.17b, the brightness temperature for different types of roughness is presented. The curves are given for vertical, horizontal, and circular polarizations. For higher moisture content, the brightness temperature decreases, while for higher roughness, the brightness temperature increases.

Figure 8.18 shows calculations of brightness temperature as seen from geostationary orbit by a satellite using an Earth-coverage beam (Earth fills the main beam between 3 dB points). As the satellite moves around its orbit, one can see the effect of the African land mass (hot) at 30°E longitude and of the Pacific Ocean (cold) at 180–150°W longitude. Brightness temperature increases with increasing frequency, largely due to gaseous absorption. Curves are for US standard atmosphere with 7.5 g/m^3 water vapour and 50% cloud cover. The Earth-coverage antenna pattern is given by $G(\varphi) = -3(\varphi/8.715)^2$ dB for $0 \leq \varphi \leq 8.715$ where φ is the angle off antenna boresight.

8.4.5 *Man-Made Noise*

Median values of man-made noise power related to some radiowaves propagation media are indicated by Fig. 8.19 for lossless and grounded short vertical monopole type antenna. As it can be observed from this figure in all positions, the results are in accordance with linear variations of median value of F_{am} with a frequency f in terms of MHz of the form:

$$F_{am} = c - d \log f \tag{8.41}$$

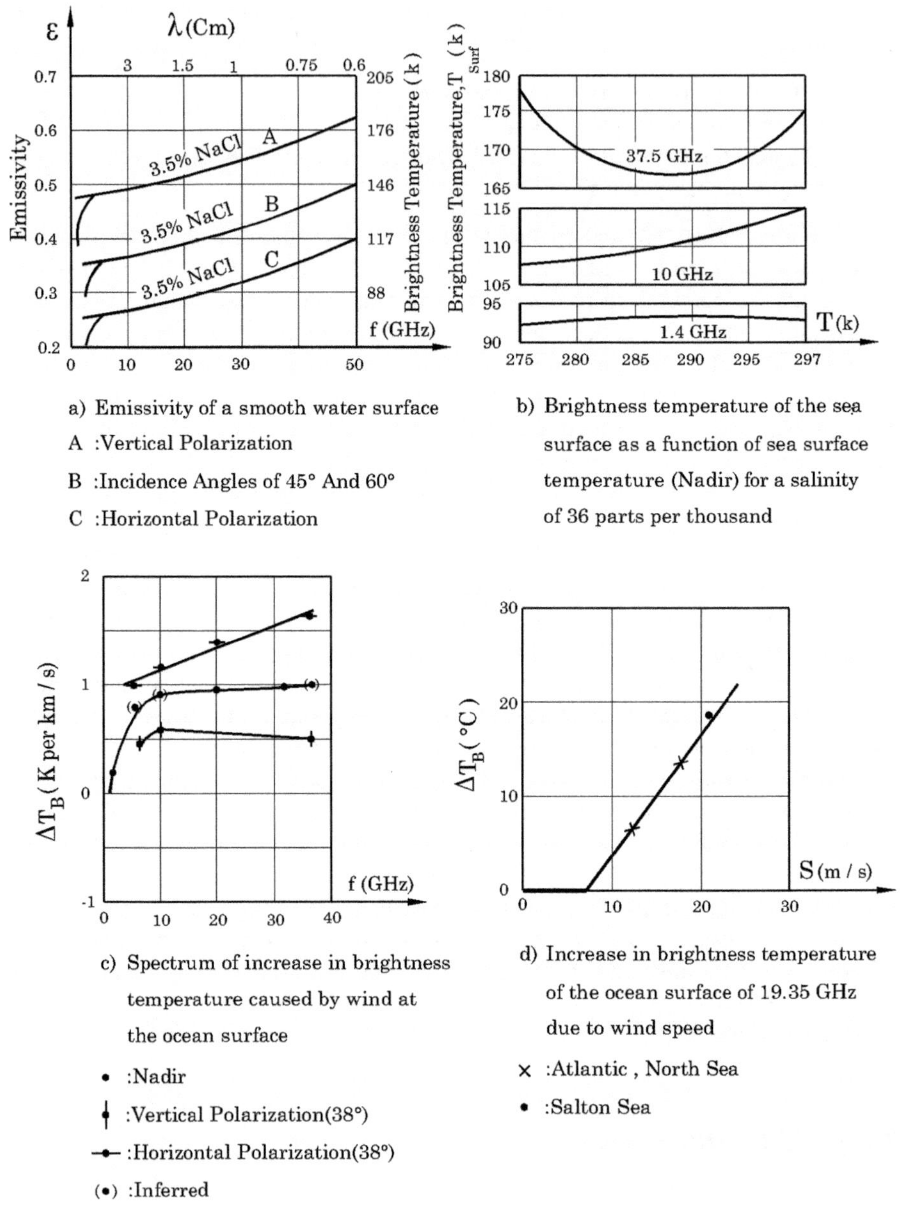

a) Emissivity of a smooth water surface

A :Vertical Polarization

B :Incidence Angles of 45° And 60°

C :Horizontal Polarization

b) Brightness temperature of the sea surface as a function of sea surface temperature (Nadir) for a salinity of 36 parts per thousand

c) Spectrum of increase in brightness temperature caused by wind at the ocean surface

• :Nadir

┼ :Vertical Polarization(38°)

–•– :Horizontal Polarization(38°)

(•) :Inferred

d) Increase in brightness temperature of the ocean surface of 19.35 GHz due to wind speed

× :Atlantic , North Sea

• :Salton Sea

Fig. 8.16 Emissivity and brightness temperature variations of sea surface (Ref.: ITU-R, P.372-9)

In this relation, the *c* and *d* constants for different media can be selected according to the Table 8.1.

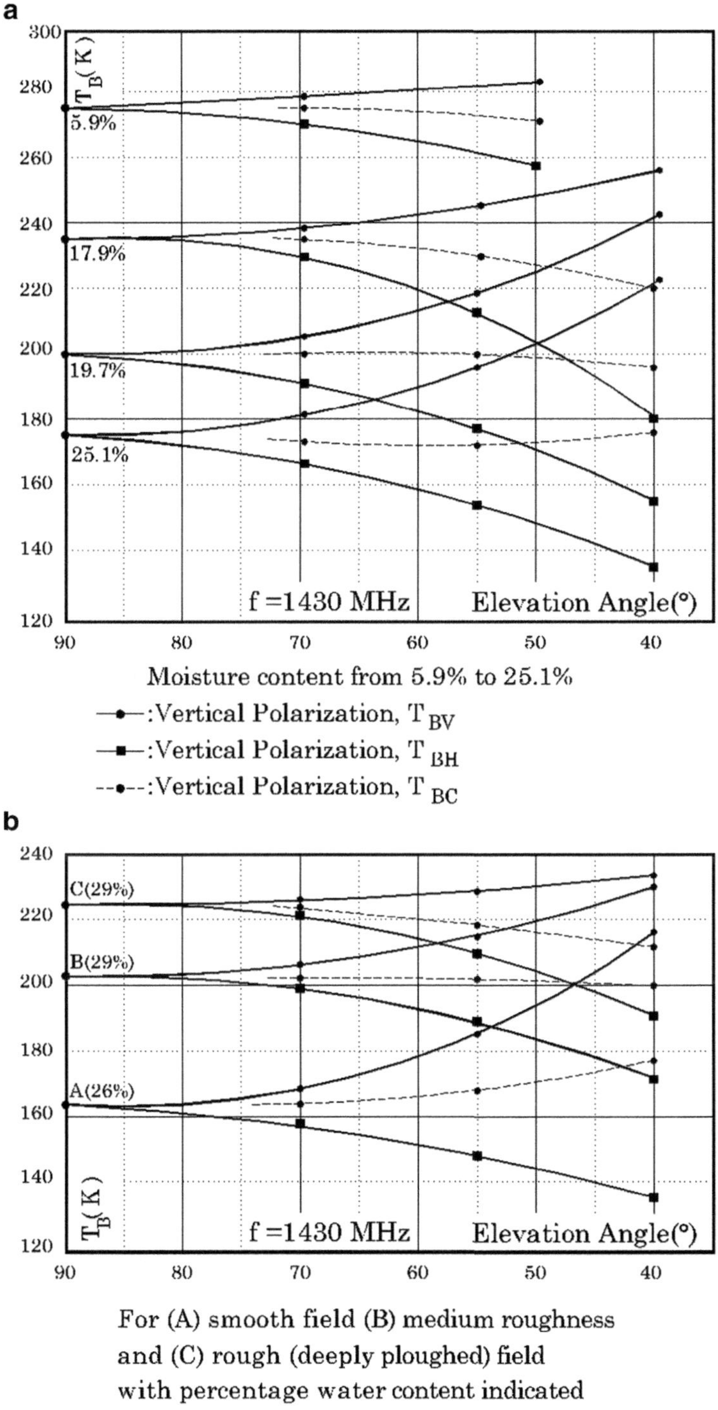

Fig. 8.17 Brightness temperature of ground versus elevation angle (Ref.: ITU-R, P.372-9)

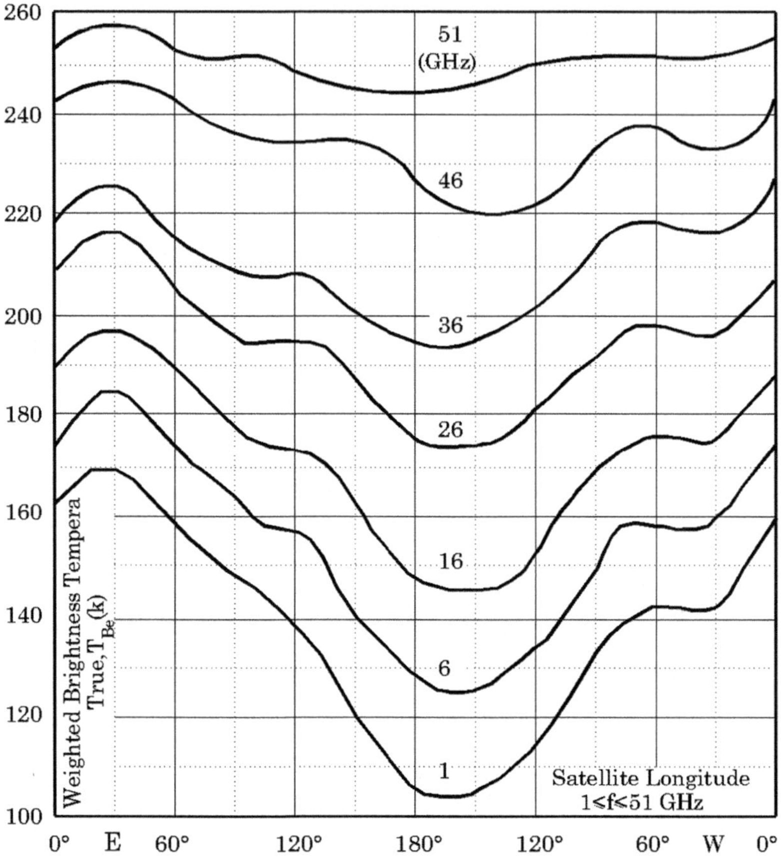

Fig. 8.18 Weighted brightness temperature of earth versus longitude (Ref.: ITU-R, P.372-9)

Table 8.1 Values of constants *c* and *d*

Environmental category	*c*	*d*
Business (curve A)	76.8	27.7
Residential (curve B)	72.5	27.7
Rural (curve C)	67.2	27.7
Quiet rural (curve D)	53.6	28.6
Galactic noise (curve E)	52.0	23.0

For the business, residential, and rural environments, the decimal noise power deviation (relative to the time) is provided with D_u and D_l symbols in the Table 8.2. The last column of this table also contains the deviation values related to the location. These deviations are uncorrelated, and log-normal distributions are suitable for both halves. The values related to Table 8.2 were measured in the 1970s and may somewhat differ with the values of later decades because of man-made noise generation.

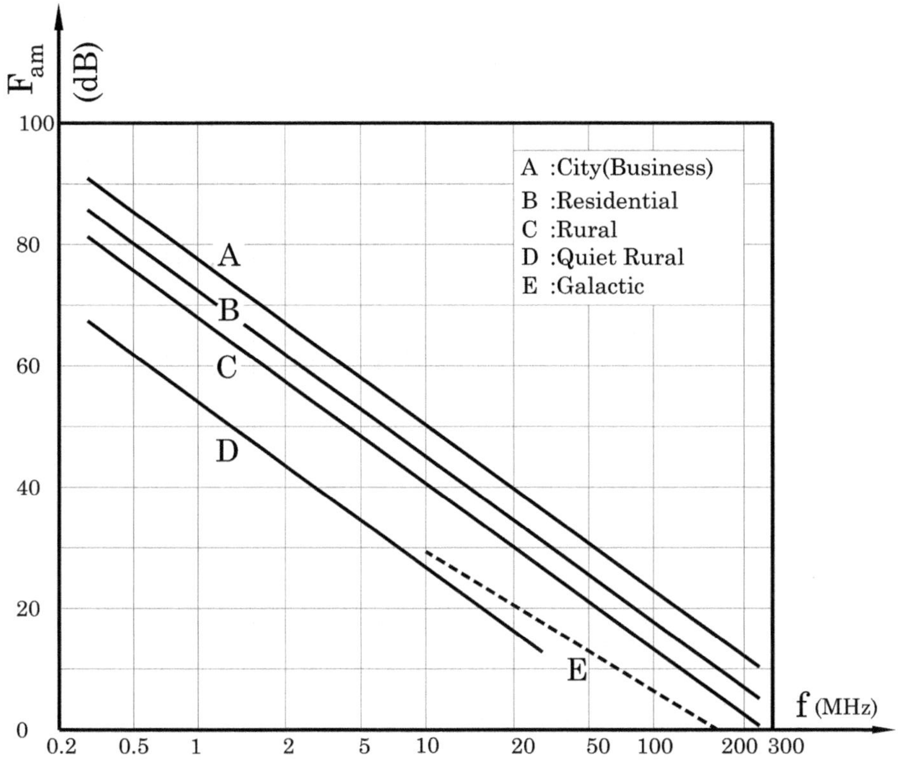

Fig. 8.19 Median values of man-made noise power (Ref.: ITU-R, P.372-9)

Table 8.2 Values of deviations of man-made noise (Ref.: ITU-R, P.372-9)

Category	Decile	Variation with time (dB)	Variation with location (dB)
Business	Upper	11.0	8.4
	Lower	6.7	8.4
Residential	Upper	10.6	5.8
	Lower	5.3	5.8
Rural	Upper	9.2	6.8
	Lower	4.6	6.8

Data analysis concerning the available measurements for commercial environments in the 200–900 MHz frequency range indicates linear variations of the man-made noises relative to the logarithm of frequency with a more gradual slope as expressed below:

$$F_{am} = 44.3 - 12.3 \log f, \quad 200 < f < 900 \text{ MHz} \tag{8.42}$$

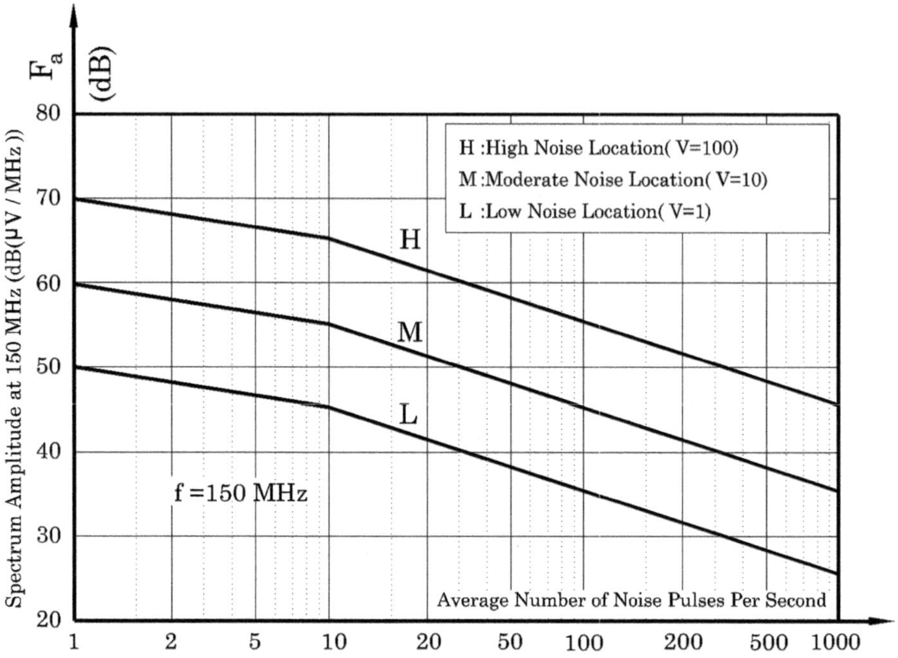

Fig. 8.20 Sample of F_a variations versus NAD (Ref.: ITU-R, P.372-9)

A basic component of man-made noise in VHF frequency band is related to the ignition impulses from car engines. In this situation, the noise contribution is in the form of impulsive noise amplitude distribution or NAD. Figure 8.20 illustrates a sample of noise amplitude distribution on 150 MHz frequency for three types of car engine density. The NAD value for other frequencies can be obtained from the following equation:

$$A = C + 10 \log V - 28 \log f \tag{8.43}$$

The components and pertinent units are described below:

C : A constant equal to 106 $\mathrm{dB}_{\mu V/MHz}$
V : Traffic density in terms of the number of car engines per one square kilometer
f : Frequency in terms of MHz

Example 8.5. 1. Find the median value of the man-made noise figure for frequencies of 450 and 900 MHz in business environments.

2. Calculate its variation range assuming the log-normal distribution and for 90% time.
3. Specify the ratio of the man-made noise to the galactic noise at the 450 MHz frequency.

Solution. 1. Using Table 8.1 $\Longrightarrow$ $c = 76.8$ and $d = 27.7$

$$f = 450\text{ MHz} \Longrightarrow F_{am} = 76.8 - 27.7 \log 450 \Longrightarrow F_{am} = 3.3\text{ dB}$$

$$f = 900\text{ MHz} \Longrightarrow F_{am} = -5\text{ dB}$$

2. Business area $\xrightarrow{\textit{Table 5.2}}$ $\sigma_t = 11$ dB

$$FM = q_0(90\%) \times \sigma_t = 1.282 \times 11 = 14.1\text{ dB}$$

3. For calculation of the galactic noise figure, using (8.44) yields:

$$f = 450\text{ MHz} \Longrightarrow F'_{am} = 52 - 23 \log 450 = -9\text{ dB}$$

$$R = \frac{f_a}{f'_a} = Antilog\left(\frac{3.3+9}{10}\right) = 17$$

■

8.4.6 Brightness Temperature from Extra-Terrestrial Sources

It is essential for radio communications at frequencies below 2 GHz to consider the noise from the Sun and galaxy (the Milky Way) as a broad belt containing strong radio emissions. The median value of galactic noise regardless of the ionosphere layer can be specified by the following equation:

$$F_{am} = 52 - 23 \log f \tag{8.44}$$

In this relation, F_{am} is in terms of decibels and f in terms of MHz. For frequencies higher than 2 GHz, considering the Sun and some powerful sources such as Cassiopeia A, Cygnus A, Cygnus X, and Crab Nebula will be sufficient, since the cosmic background contributes only 2.7 K and the Milk Way acts as a narrow strip (with magnified intensity). The range of brightness temperatures which are routine for extraterrestrial noise sources is provided in the frequency range 0.1–100 GHz as indicated in Fig. 8.21.

In recommendation No. ITU-R, P.372, some graphs are presented as equi-level countours for the total radio noise temperature at 408 MHz frequency with 5° angular resolution. There are samples of these graphs provided in Fig. 8.22. For more information, refer to the latest version of the said recommendation. The background cosmic noise radiation varies with the frequency. If the value of brightness temperature for frequency f_0 is equal to $t_b(f_0)$, then to calculate the brightness temperature at frequency f_i, the following equation may be used:

$$t_b(f_i)[K] = t_b(f_0) \times (f_i/f_0)^{-2.75} + 2.7^\circ\text{K} \tag{8.45}$$

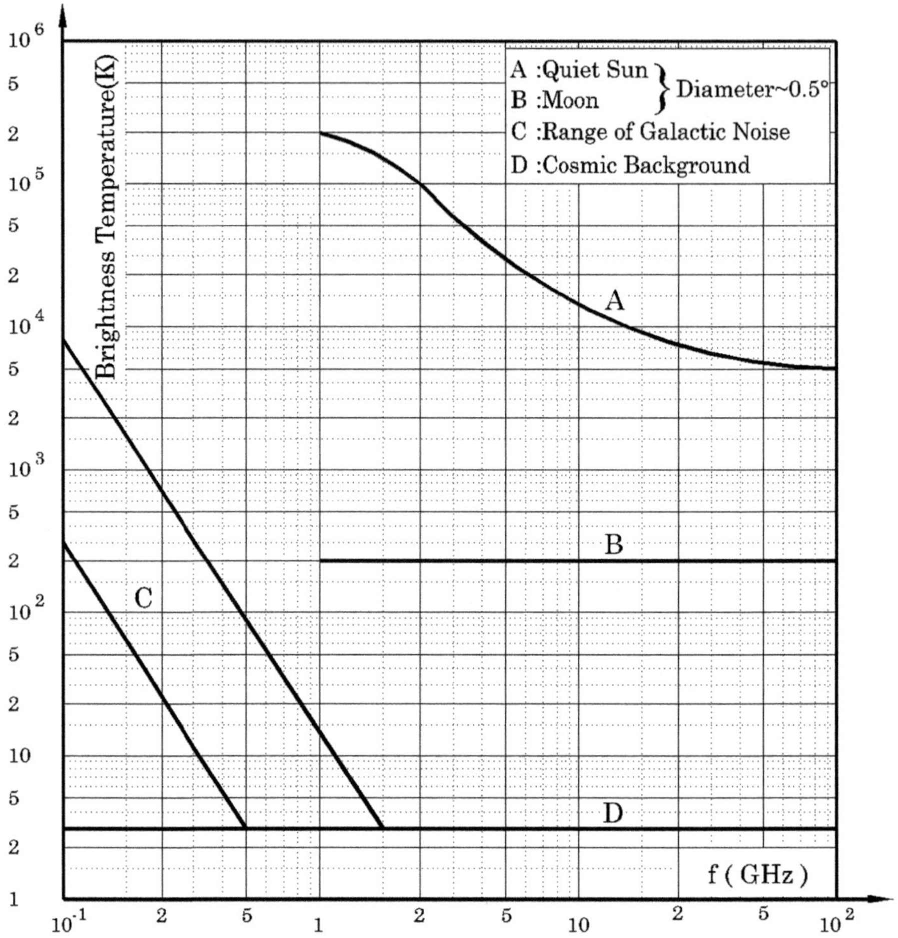

Fig. 8.21 Extraterrestrial noise sources (Ref.: ITU-R, P.372-9)

Example 8.6. Assuming the brightness temperature of cosmic noise equal to 200°K, calculate its value at 1 GHz.

Solution.

$$f_0 = 408 \text{ MHz}, \quad t_b(f_0) = 200°\text{K}, \quad f_i = 1 \text{ GHz}$$

$$\Longrightarrow t_b(1 \text{ GHz}) = 19.7°\text{K}$$ ■

For more accuracy in using the above equation, it requires to consider the exponent variations in relation (8.45) over frequency and its position in the sky. The intensity variations with frequency for point sources depend on its physical conditions. In satellite communications, which employ a synchronized orbit with

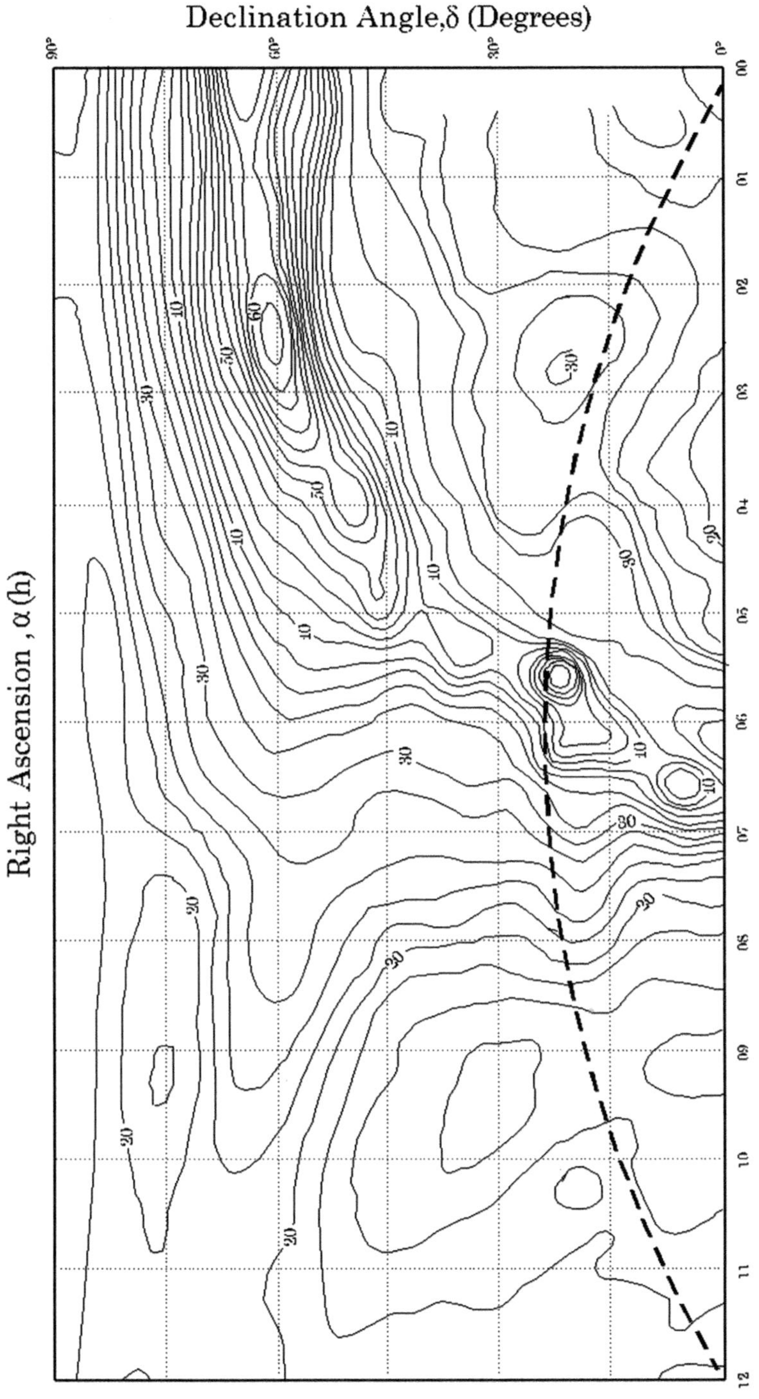

Fig. 8.22 Radio sky temperature at 408 MHz (Ref.: ITU-R, P.372-9)

Earth and at the height of around 36,000 km above the Earth surface, as mentioned in Chap. 3, a limited portion of the sky with a maximum solid angle of 17.3° is considered for this purpose.

The Sun is a strong variable noise source with a temperature of about $10^{6°}$K at 50–200 MHz and at least $10^{4°}$K at 10 GHz under quiet conditions. During turbulences in the Sun, these values increase drastically. The Moon brightness temperature at frequencies higher than 1 GHz to a great extent is independent of the frequency, and it is about 140°K for the crescent Moon and 280°K for full Moon brightness.

8.4.7 Atmospheric Noise Due to Lightning

The ITU-R has prepared some charts showing the expected median values of background atmospheric radio noise, F_{am} (dB) above kT_0b, at 1 MHz for each season, 4-h time block in local time. These charts consist of three groups of graphs as follows:

- The relative value of the Earth's atmospheric radio noise, F_{am} [dB] (relative to kT_0b at 1 MHz) for which a sample is given by Fig. 8.23
- Radio noise variations in terms of frequency for which a sample is given by Fig. 8.24
- Noise variation specifications for which a sample is given by Fig. 8.25

In preparing these charts, the reference antenna is a short vertical monopole positioned on a fully conductive plate. Field intensity of the incident ray can be derived according to details of Sect. 8.4.2. The conclusions obtained from the measurements reveals that the atmospheric noises are less than the amount of man-made and galactic noises. These charts should be used with caution since they present only estimates of what atmospheric noise levels would be recorded if the other types of noise are ignored. For more details and also definition of each parameter, refer to the 9th revision of recommendation ITU-R, P.372 published in 2005.

Atmospheric noises generated by lightning are generally non-Gaussian in character, and its probability density function has a significant role to specify the digital system performance. Normally, the amplitude probability distribution, APD, of this noise is described in terms of voltage deviation, V_d which is the ratio of r.m.s voltage to the average noise envelope overlapping voltage.

APD curves related to different values of V_d are presented in Fig. 8.26. The value of effective r.m.s voltage, A_{rms}, is used as the basis of calculation. The measured values of V_d fluctuates around the median value, V_{dm} by σV_d.

APD graphs can be used for a wide range of bandwidths. The values of V_d provided in Fig. 8.26 are intended for a 200-Hz bandwidth, and Fig. 8.27 shows how it is transformed to the corresponding V_d for a bandwidth other than 200 Hz.

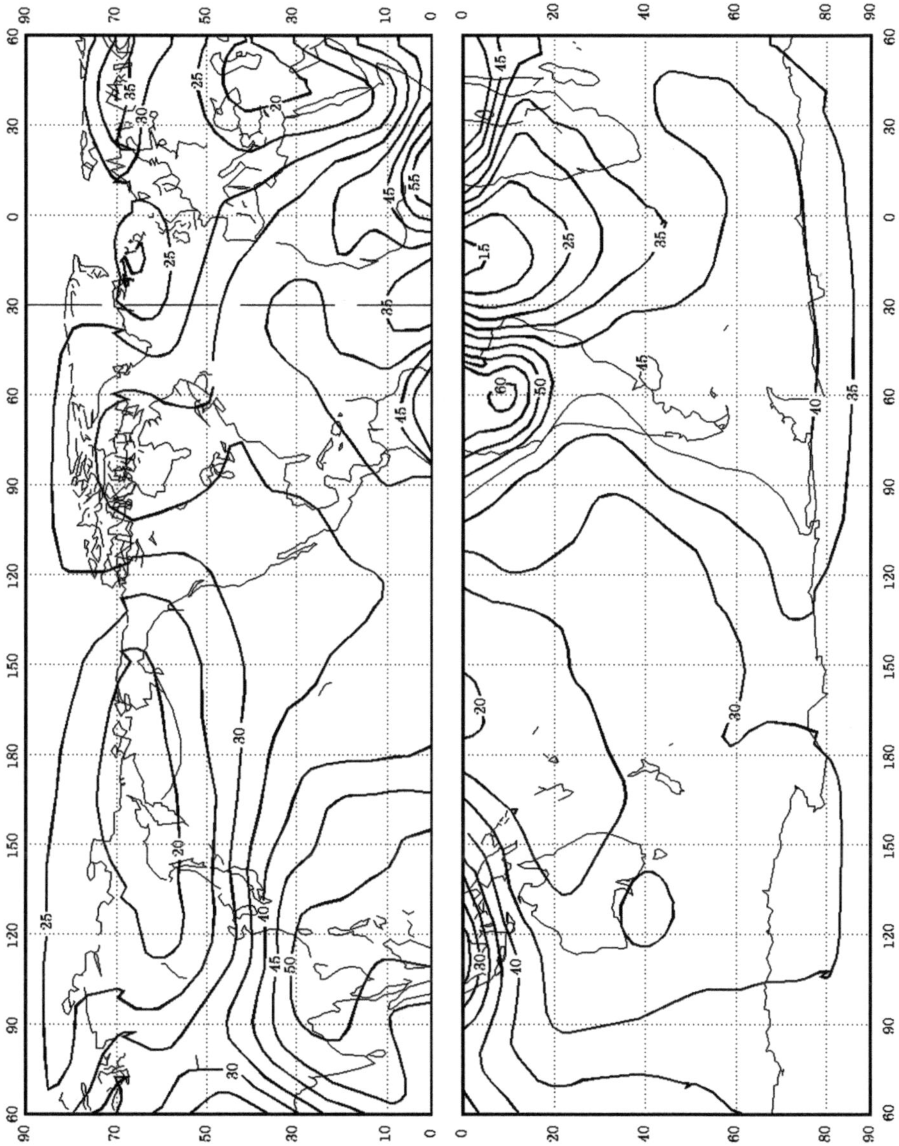

Fig. 8.23 Expected values of atmospheric radio noise, F_{am} (dB above kT_0b at 1 MHz), Winter Local Time 00.80∼12.00 (Ref.: ITU-R, P.372-9)

8.4.8 Noise From Several Sources

There are occasions where more than one type of noise needs to be considered because two or more types are of comparable size. This can be true at any frequency,

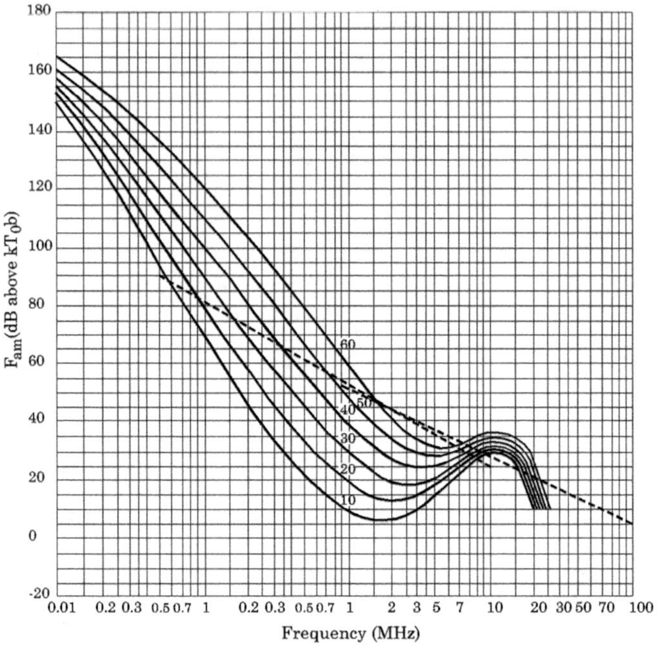

Fig. 8.24 Variation of radio noise with frequency (Ref.: ITU-R, P.372-9)

Fig. 8.25 Data on noise variability and character (Ref.: ITU-R, P.372-9)

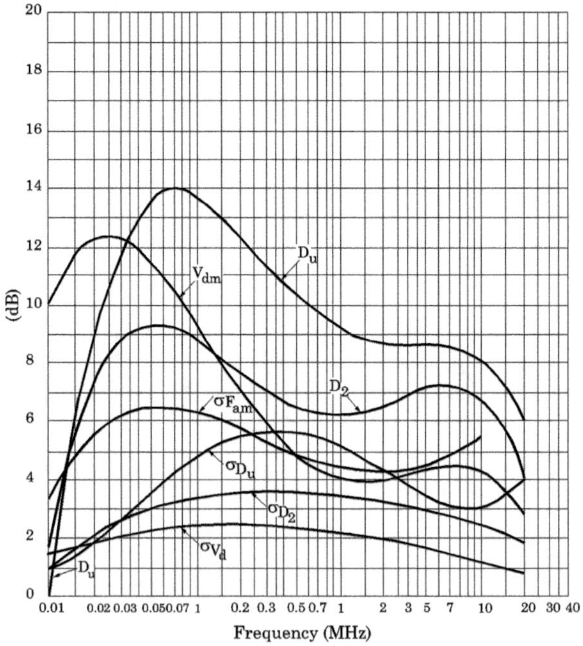

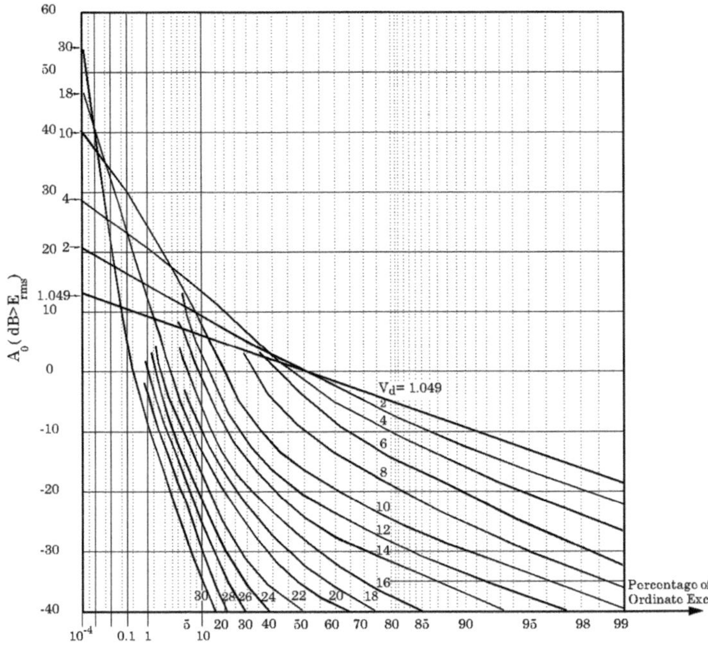

Fig. 8.26 Amplitude probability distribution for atmospheric radio noise (Ref.: ITU-R, P.372-9)

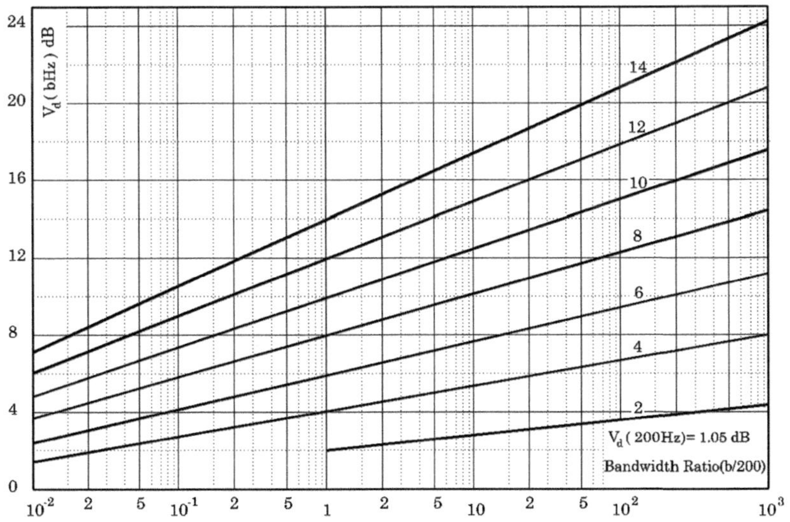

Fig. 8.27 Translation of a 200-Hz bandwidth V_d to other bandwidths, b(Ref.: ITU-R, P.372-9)

in general, but occurs most often at HF band where atmospheric, man-made, and galactic noises can be of comparable size. For example, the indicated median values for the specified kinds of noises in the Fig. 8.12 should be combined for further radio link analysis working at 10 MHz. The values given are median F_a values, F_{am}. The f_a values have distributions about the median f_a. As noted earlier, these distributions are log-normal distributions on each side of the median.

An appropriate method for obtaining the median value and distribution for the sum of two or more noise processes has been developed in which the resultant noise is also assumed to be log-normally distributed. In this method, the resultant median noise power is given by the sum of the median noise powers of the individual noise processes. The standard deviation of the resultant noise is obtained by summing noise powers determined one standard deviation above the median power for each of the noise processes involved and then subtracting the resultant median noise power from that result.

8.5 Noise in Satellite Communications

8.5.1 Introduction

One of the main subjects in satellite communications is the received noise power by the RX antenna. Normally, the noise in satellite systems and the related calculations are expressed in terms of equivalent noise temperature. The equivalent noise temperature of an antenna has effective role in the reception merit of the RX system, that is, G/T. Also, this factor is a major parameter in power budget calculations of satellite links.

8.5.2 External Noise Sources in Satellite Communications

The sources of incoming noise to the antenna from the surrounding medium can be grouped as:

- Galactic noise
- Sun noise
- Sky noise
- Earth's noise
- Man-made noise

The total amount of different types of noise as stated above is dependent on the frequency and shown in the graphs of Fig. 8.13 on satellite communications frequency band.

8.5.3 Galactic Noise

The galactic noise in deep space is due to the spread of clouds of ionized gases such as hydrogen which its great quantity lies toward the center of galaxy. This type of noise (sometimes called cosmic noise) as indicated in Fig. 8.13, is more effective at low frequencies, but at frequencies above 1 GHz (the desirable frequency for satellite communications), they are inherently trivial and can be disregarded. To obtain an approximate value of the galactic noise temperature, the following formula can be used:

$$T_g[^\circ\mathrm{K}] = T_0 \cdot \lambda_0^2 \tag{8.46}$$

where $T_0 = 290^\circ\mathrm{K}$ and λ_0 is the wavelength of interest in meters.

8.5.4 Solar Noise

The main radiation source in the Solar System is the Sun, and its most effective position occurs at the time when the directions of Sun rays and antenna axis are identical. The Sun radiation covers a vast frequency band stretching from HF to the light frequency. The radiation intensity includes the following types:

- Quiet Sun (inactive Sun)
- Active Sun

Generally in both cases, the noise decreases with the increase of frequency. At frequencies between 0.1 and 1.5 GHz, the Sun noise temperature is in the range of 10^4–10^6 °K. To obtain an approximate value of the Sun noise temperature, the following formula may be used:

$$T_{\mathrm{Sun}} = 2250\, T_0 \cdot \lambda_0 \tag{8.47}$$

where $T_0 = 290^\circ\mathrm{K}$ and λ_0 is the wavelength of interest in meters.

The share of the Sun noise in the equivalent noise temperature of antenna depends on the antenna gain in the direction of the Sun and its beamwidth. For example, the terrestrial parabolic antenna with a diameter of 30 m has a 0.18° beamwidth at 4 GHz frequency which in the quiet condition of the Sun and collinear Sun beam with the antenna axis, the share of Sun noise is around 4,000°K. This amount of noise is so massive that the proper reception of main signals by the antenna would be simply impossible. Normally, the antenna receives the Sun noise on the side edges, in such a way that if this occurs in suitable conditions, the antenna noise temperature may decrease to 4°K.

Example 8.7. Find noise temperature at $f = 1$ GHz and check the results with the values given in the Fig. 8.13:

1. Galactic noise
2. Sun noise

Solution. 1.

$$f = 1\ \text{GHz} \implies \lambda_0 = 0.3\ \text{m}$$

$$T_g = 290 \times (0.3)^2 = 26.1°\text{K}$$

Using Fig. 8.13 for f = 1 GHz yields:

$$T_g = 2.9 \times 10^{0.8} = 18.3°\text{K}$$

2.

$$T_{\text{Sun}} = 2250 \times 290 \times 0.3 \implies T_{\text{Sun}} = 195{,}750°\text{K}$$

Using Fig. 8.13 for f = 1 GHz yields:

$$T_{\text{Sun}} = 2.9 \times 10^{4.8} = 182{,}978°\text{K}$$ ■

8.5.5 *Sky Noise*

8.5.5.1 General Concepts

Sky noise is a combination of noises due to the troposphere (atmosphere) and ionosphere and also the noises related to the clouds and rain. At frequencies above 1 GHz, ionospheric noise can be disregarded. However, the atmospheric noise does exist at higher frequencies since the gases in this layer and also the existing water vapor absorb the energy of electromagnetic waves, which results in emissions called the sky noise.

Since the thermal noise of the RX system on the uplink and downlink is directly influenced by the sky noise, specifically on the paths from the ground toward the space, they are of great significance for RX systems having a low noise figure. The effective sky noise temperature (T_{sky}) on a specific direction is given by the following relation in degrees Kelvin:

$$T_{\text{sky}} = \int_0^{\infty} T(s) \cdot A(s) e^{-\int_0^s A(s') \cdot ds'} . ds \tag{8.48}$$

where $T(s)$ is the temperature of medium, $A(s)$ is the absorption coefficient of the medium per each kilometer, and s is the path length in terms of kilometer from the

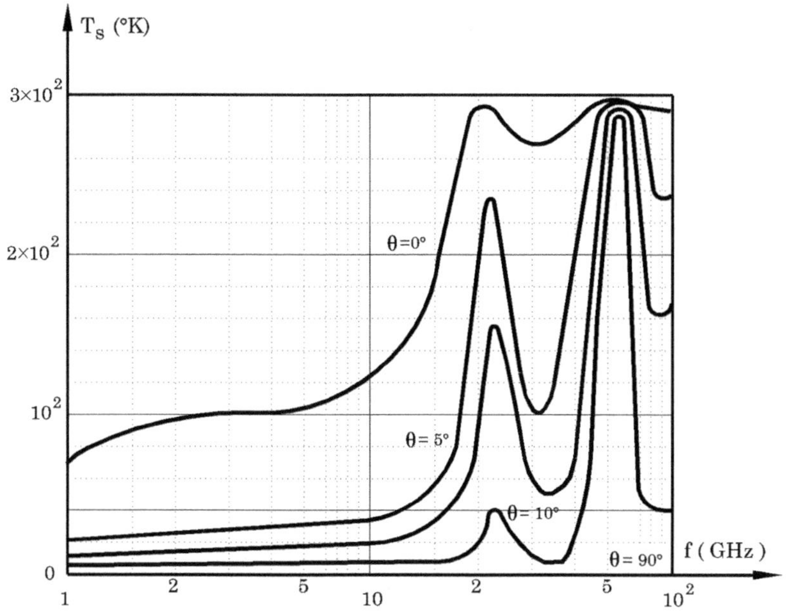

Fig. 8.28 Sky noise temperature due to atmosphere

antenna. If we replace $T(s)$ with the mean noise temperature T_m, relation (8.48) is simplified into:

$$T_{\text{sky}} = T_m \left(1 - \frac{1}{L}\right) \tag{8.49}$$

where L is the loss factor caused by the absorbing medium. If it is assumed that atmosphere is grouped horizontally, then T_{sky} in terms of elevation angle, θ can be expressed by:

$$T_{\text{sky}}(\theta) = T_m(1 + L_0 - \text{cosec}\ \theta), \quad \theta > 15^\circ \tag{8.50}$$

L_0 is the loss factor in the direction of Zenith ($\theta = 90^\circ$).

8.5.5.2 Radiation from Atmosphere Gases

The values of sky noises generated in the atmosphere are shown in Fig. 8.28. The curves are plotted against surface temperature of 20°C, pressure of 1 atmosphere on the sea surface, and 10 g/m^3 of water vapor existing in the air (equivalent to 58% humidity). It must be noted that under such conditions, these graphs have peaks around 22 and 60 GHz for which the maximum atmospheric attenuation occurs under the same conditions. The noise generated by the atmosphere gases can have a great impact on a RX system with a very low noise characteristic of less than 300°K.

8.5.5.3 Radiation from the Clouds and Rain-Drops

The noise existing in the sky occurs because of the absorption by clouds and rain and may be defined by the following relation:

$$T_{\text{sky}} = T_m(1 - 10^{-A/10}) \tag{8.51}$$

where, A is the path attenuation in terms of decibel. T_m is derived approximately from measuring satellite signal with radiometer which is equal to 270°. An increase of thermal noise emanating from sky noise for 1 dB rain attenuation is equal to 56°K and for 5 dB rain attenuation is equal to 180°K.

This type of noise determines directly the signal attenuation related to rain and finally affects the carrier-to-noise ratio (C/N). As an example, consider a system which has a thermal noise equal to 290°K (with noise figure of 3 dB). If the rain attenuation for the system is 5 dB, then the noise temperature of the system will be 470°K, and as a result, the difference of C/N ratio compared with similar value for clear sky will be 6.2 dB ($1.2 + 5$).

8.5.6 Ground Noise

The ground noise generally emanates from the two following sources:

- Emission from the black body of the Earth
- Reflection of noise from the Earth

The average Earth temperature is 270°K with a wide spread frequency spectrum. The terrestrial stations, through side corners and satellite antenna, collect the Earth's emissions. In both of the above cases, the Earth plays a major role in the antenna noise temperature of the ground stations which will be around 30% of the total noise value. In the case of satellite antenna, the noise share of the Earth will depend on the orientation of satellite antenna toward a part or the whole Earth's surface.

8.5.7 Man-Made Noise

The main sources of man-made noise are from power lines, electric equipment, combustion engines of cars, and other types of noises related to urban and industrial activities. This sort of impairment causes a rise in thermal noise of RX system. The amount of the man-made noise in most cases influences the site selection for satellite ground stations specially for those including large antennas (say

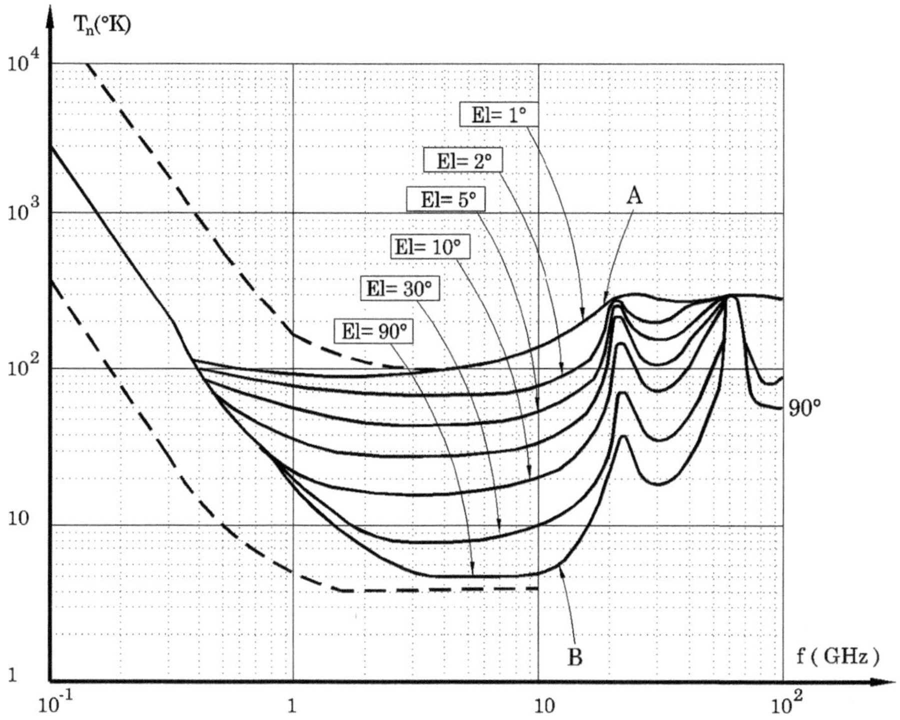

Fig. 8.29 Sky noise versus frequency and antenna elevation angle

10∼30 m diameter). The main satellite ground stations are selected outside urban and industrial centers to keep the antenna noise temperature low. Where it is inevitable to place the antenna close to a city, this type of man-made noise shall be kept low to the extent possible by using antenna with a narrow beamwidth orientated toward the satellite. Normally, the antenna of large terrestrial stations is placed at location which is well away from man-made noise and/or radio-interfering stations.

8.5.8 Antenna Thermal Noise

Figure 8.29 indicates thermal noise of an ideal antenna (lossless and without side corners toward the Earth) installed on the Earth's surface. These graphs have been calculated for desirable atmospheric conditions. The curve shown with dashed line at the bottom is sketched with the minimum cosmic noise, without the noise from the Sun and elevation angle of 90°. The curve given with dashed line at the top relates to a position of maximum cosmic and solar noises with a value one hundred times the value of the noise from the Sun in quiet condition and elevation angle of 0°.

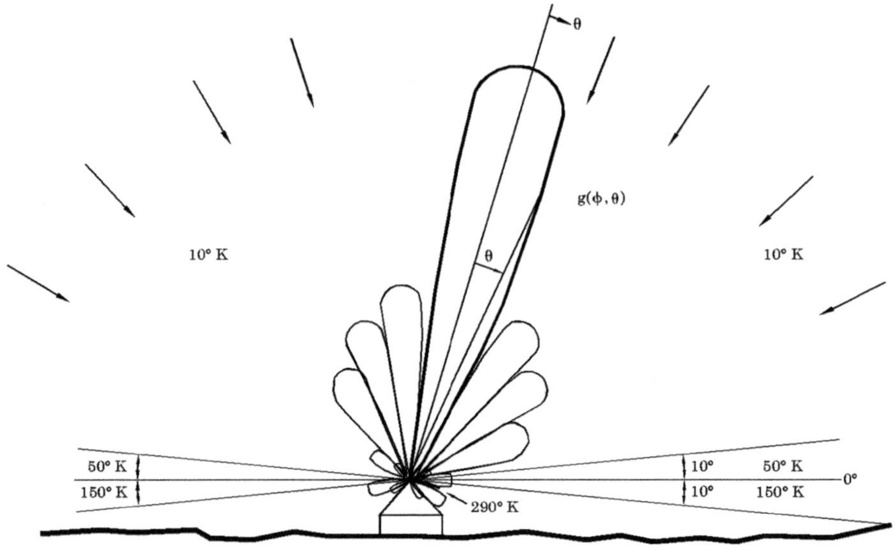

Fig. 8.30 Sky and ground equivalent noise temperature versus satellite antenna elevation angle

For presentation of a sample of approximate thermal noise of the antenna, practically and in most cases, for a 4 GHz frequency as indicated in Fig. 8.30, the temperature of T_a will be around the following values:

- 290°K for an angle of more than 10° under horizon
- 150°K for an angle between zero and 10° under horizon
- 50°K for an angle between zero and 10° above horizon
- 10°K for an angle more than 10° above horizon

In general, by accounting for the sum of different noises, the range of equivalent noise temperature for satellite antennas is 60°K for C band and 80°K for Ku band. The mentioned temperature values are measured at the input feed point of the antenna. As stated in Example 8.3, to calculate the noise temperature of RX system, the attenuation effect and RX noise temperature must be also added to that value.

8.5.9 *Equivalent Noise Temperature*

8.5.9.1 Introduction

In the LOS communications, a long distance between TX and RX results in large free-space losses; hence the received signal level, RSL at the RX location, is low. Particularly, this case in satellite communications is critical and limited to few picowatts due to the increased distance of several hundreds of kilometers in LEO satellites and several thousands of kilometers in GEO satellites. However, the low received signal level is not a problem individually, and weak signals can be detected by satellite low-noise amplifiers (LNA), but in some occasions, the added radio noises from various sources are the limiting factor for proper detection of the main signal.

The undesired noises can only be tolerated when the main signal level relative to the noise level is higher with a certain magnitude. This factor is called the signal-to-noise ratio and denoted by SNR. The low value of SNR cannot be improved with high amplification of the RX system, and the final output will be noisy. In the next sections, the effects and main points related to the noise and equivalent temperature in satellite links will be discussed in more detail.

8.5.9.2 Noise Power Density

The main cause of electrical noise is random thermal motion of electrons which occur in various active and passive parts of the radio link as described below:

- Sky and terrestrial noises received by the antenna.
- Active and passive parts of the RX which generally is determined by the noise figure.
- Noises made by attenuating parts.

In addition to the above noises, it may be produced by receiver intermodulation because of the existence of non-linear elements which must be discussed separately. Normally for evaluation of noise effects, its power and density (denoted by P_n and N_0, respectively) are significant parameters. The thermal noise power relation is given by

$$P_n = k \cdot T_n \cdot B_n \tag{8.52}$$

In the above relation, each parameter and related units are as defined below:

P_n : Thermal noise power in watts (W).
K : Boltzmann constant in joule per degree Kelvin (J/K) and its value is 1.38×10^{-23}.
T_n : Equivalent noise temperature in degree Kelvin (K).
B_n : Equivalent noise bandwidth in Hertz (Hz).

The equivalent noise bandwidth is always more than half power bandwidth (-3 dB) of amplitude-frequency characteristics, and normally it is considered 12% more than this value, that is,

$$B_n \approx 1.12 \times B(-3\ \text{dB}) \tag{8.53}$$

The main characteristics of the thermal noise is its flat frequency spectrum and thus can be defined as "*noise power per unit bandwidth*" denoted by N_0 or ρ, and it is known as the noise power density given below:

$$N_0 = \frac{P_\mathrm{n}}{B_\mathrm{n}} = kT_\mathrm{n} \tag{8.54}$$

Example 8.8. A single antenna with equivalent noise temperature of 40°K is connected to the input of a RX unit with an equivalent noise temperature of 90°K. Find the noise density and noise power for a 30 MHz bandwidth.

Solution. Using relation (8.54), we have

$$N_0 = (40 + 90) \times 1.38 \times 10^{-23} = 1.8 \times 10^{-21}\ \mathrm{W/Hz}$$

To calculate the noise power in the specified bandwidth by using relation (8.53), we have:

$$P_\mathrm{n} = 1.8 \times 10^{-21} \times 1.12 \times 30 \times 10^6 = 6.048 \times 10^{-14}\ \mathrm{W}$$

$$P_\mathrm{n} = 0.06\ \mathrm{pW}$$

■

8.5.9.3 Equivalent Noise Temperature

The noise temperature is defined using relations (8.52) and (8.54) as:

$$T_\mathrm{n} = \frac{P_\mathrm{n}}{kB_\mathrm{n}} = \frac{N_0}{k} \tag{8.55}$$

The equivalent noise temperature is directly related to the physical temperature but generally is not equal to it. The equivalent noise temperature of the troposphere is very important in satellite communications, especially at frequencies higher than 10 GHz. In the two following cases, the equivalent noise temperature has a steep increase and causes disruption in satellite communications:

1. At frequencies around 22 GHz and 60 GHz, because of special composition of the atmosphere
2. Small elevation angles of ground station antenna, less than 5° in C band and below 10° in Ku band

8.5.9.4 Antenna Noise

A satellite antenna receives an impressive amount of noise along with the main signal which is radiated either from space or terrestrial stations. Although their origin is the same, they defer in value. The antenna noises in a general classification consist of the following types:

- Sky noise
- Noise from lossy components

Sky noise refers to radiowave emissions in the world which is generated by warm bodies. Such emissions have a bandwidth exceeding the microwave band (1–10 GHz). Figure 8.29 provides the equivalent sky noise temperature which can be received by a terrestrial station antenna on two extremes A and B in logarithmic scales. Graph A is related to an antenna with a small elevation angle, and graph B is related to an antenna with an elevation angle of around 90°.

The reason for the noise temperature being high in graph A is the reception of thermal radiation of the Earth which occurs for communications in C band at elevation angles below 5° and in Ku band for elevation angles below 10°. The graphs A and B indicate that in frequencies less than 1 GHz, the sky noise temperature rises drastically. Within the interval of 1–10 GHz for a given elevation angle, its variation is very limited, and within the interval 10–100 GHz, it has particular variations.

There are two distinct peaks at 22 GHz and 60 GHz. The first peak is caused by the resonance of water vapor molecules, and the second peak occurs due to resonance of oxygen molecules. As mentioned earlier, the graph of Fig. 8.29 relates to the equivalent noise temperature on the terrestrial antennas.

The satellite antennas which are positioned toward the Earth, receive a great amount of the Earth's thermal noise, and hence, their equivalent noise temperatures without accounting for antenna losses are around 290°K, whereas for large ground station, antennas including antenna losses are within the following limits:

- About 60°K for C band
- About 80°K for Ku band

It should be noted that any type of energy absorption within the radiowave path is considered as a source of thermal noise. In fact, there is a direct relationship between the path losses and equivalent noise temperature. For instance, rain impairs the radiowave propagation in two ways, including direct attenuation and by generating additional noise. It must be also noted that the losses occurring in the reception system (including antenna, connectors, and feeder) add some noise to the received signal and finally result in higher noise temperature.

8.5.9.5 Total Noise Temperature

Noting Fig. 8.9 which shows a typical satellite reception system, to calculate the total noise temperature of a RX system, the following formula can be used:

$$T_s = T_a + T_{\mathrm{eq}} + \frac{(l-1)T_0}{g_1} + \frac{l(\mathtt{f}-1)T_0}{g_1} \tag{8.56}$$

In the above relation, each of the parameters is defined below:

T_s : Equivalent noise temperature of the system
T_a : Equivalent noise temperature of the RX antenna
g_1 : Gain of low-noise amplifier
T_{eq} : Equivalent noise temperature of low-noise amplifier
l : Transmission line losses
$\mathtt{f}$: RX noise factor
T_0 : Reference temperature (290°K)

It must be reminded that the above relation is in numeric (nonlogarithmic) system, and because of the common usage of decibel units, it is essential to have proper conversion in dealing with the above formula.

8.5.9.6 Thermal Noise of Rain

Rain is the major cause of fading in the satellite communications which is intensified at higher frequencies. The rain attenuation has negative impact in the following two ways:

- Radiowaves scattering
- Absorption of electromagnetic energy and consequential effects in the increase of equivalent sky noise temperature

Therefore, the rain, apart from its imposed attenuation, is a major factor in the increase of RX system noise temperature which results in a drop in the received carrier-to-noise ratio, C/N_0. The adverse effects in the uplink caused by rain due to high amount of received noise on satellite RX unit are relatively low and negligible.

The reason for receiving a great amount of noise by satellite antenna is its direction toward the Earth. It is important in the uplink to stabilize the received signal level within a specific limit. This can be accomplished through a control system of the carrier power at the ground station to compensate additional losses caused by the rain.

Normally, some fade margin should be considered to compensate the rain attenuation in satellite links. This margin is proportionate with the frequency being used and the intensity of raining. Obviously, at the times of heavy raining, special care must be considered.

The effect of rain on the downlink is more severe. Specially this case is more significant for the equivalent noise temperature relative to clear sky noise. If A_R [dB] is assumed to indicate attenuation due to rain absorption, then the power reduction coefficient in the non-logarithmic form is $A_R = 10^{\frac{A_R}{10}}$, and the effective rain noise temperature T_R is expressed as:

$$T_R = T_a\left(1 - \frac{1}{A_R}\right) \quad (8.57)$$

In the above formula, T_{R} is the rain apparent temperature which is a function a number of factors including physical temperature and radiowaves scattering effects by rain drops. The rain apparent temperature is around 270–290°K. Applying additional rain equivalent noise temperature, the total sky noise will be:

$$T_{\text{sky}} = T_{\text{CS}} + T_{\text{R}} \tag{8.58}$$

In the above relation, T_{CS} is identical with T_{S} as calculated in relation (8.56). In the case where rain attenuation emanates exclusively from electromagnetic energy absorption which normally occurs for L and C frequency bands during mild raining with an intensity of 1 mm/h, the following relation holds true between the ratio of final carrier to noise ratio $[C/N]_R$ and the ratio of carrier-to-clear sky noise $[C/N]_{\text{CS}}$:

$$[C/N]_R^{-1} = [C/N]_{\text{CS}}^{-1} \times \left[A_{\text{R}} + (A_{\text{R}} - 1)\frac{T_a}{T_{\text{S,CS}}}\right] \tag{8.59}$$

In the above relation, $T_{\text{S, CS}}$ and T_a are the system noise temperature for clear sky and apparent rain noise temperature, respectively. It must be noted that in heavy raining particularly for high frequencies such as Ku and K_a bands, radiowaves scattering will be significant. Under these conditions, the total attenuation caused by the absorption and scattering of the radiowaves must be taken into account for calculation of carrier power decrease, and absorption attenuation must be used for calculation of noise temperature increase.

8.6 Noise in Mobile Communications

8.6.1 Introduction

The quality of mobile radio communications is influenced by the noises generated from different sources having various characteristics. The following are major types of noises in radio mobile communications:

- RX unit noise with Gaussian distribution.
- Atmosphere noise which rapidly drops as frequency is increased and can be disregarded for UHF and higher frequency bands.
- Galactic noises which are less than background noises on VHF/UHF frequency bands.
- Solar noises which relate to the Sun activities and the angle between the direction of the sun-shine and antenna main axis.
- Man-made noises generated by the operation of electrical machinery and vehicular ignition motors. This type of noise, in contrast with thermal noise, has an impulsive nature and plays a key role in the mobile communications.

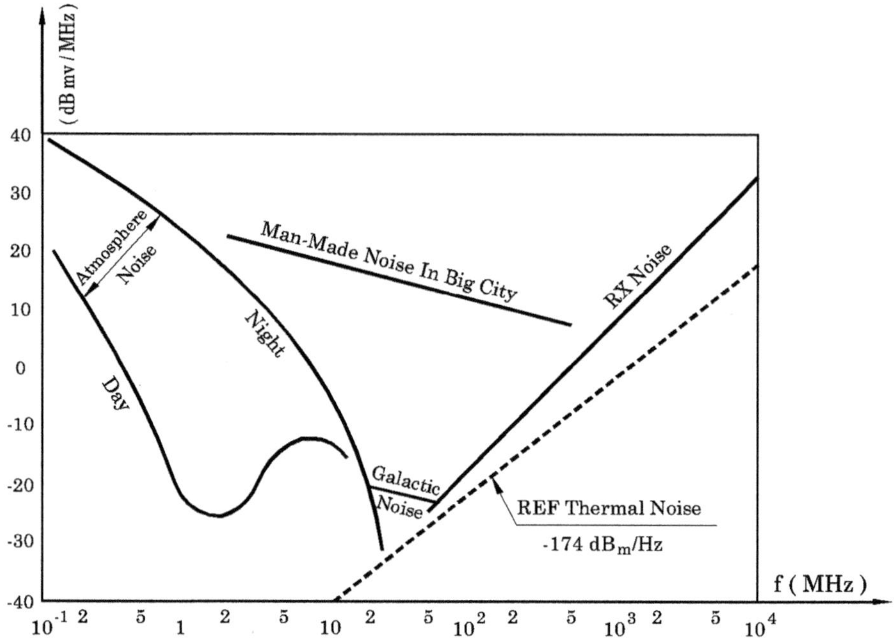

Fig. 8.31 Noise level in radio communications versus frequency

Figure 8.31 indicates the value of different kinds of noise in an audio channel having a bandwidth of 6 KHz and at a frequency range of up to 10 GHz. As it can be observed, the man-made noise is dominant on VHF band and lower portion of UHF band where thermal noises play a significant role on upper portion of UHF band.

In practical situations, the value of man-made noise has time and location variations. It is therefore essential to apply the median noise value along with the standard deviation obtained from various methods.

8.6.2 Natural Noises

Natural noises in the mobile radio communications can be classified into the following categories:

- Atmospheric noise
- Galactic noise
- Solar noise

Figure 8.32 illustrates the amount of natural noise compared to basic thermal noise. The basic noise is assumed to be equal to 10 $\log(kT_oB)$ where k is the

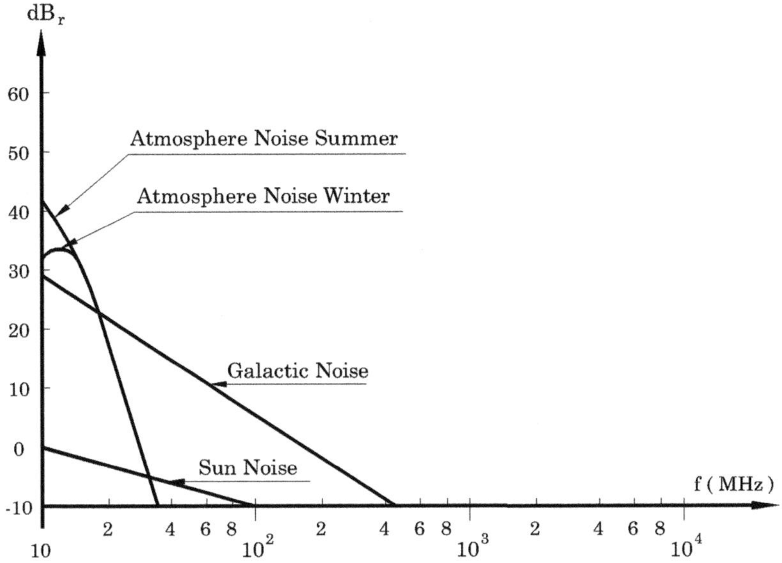

Fig. 8.32 Relative natural noise level versus frequency

Boltzmann constant, B is the effective RX bandwidth, and T_o is the reference noise temperature and equivalent to 290°K, as follows:

$$N_{\mathrm{th}} = 10\ \log(kT_0B) = -174\ \mathrm{dB}_m/\mathrm{Hz} \tag{8.60}$$

As it can be observed, concerning natural noises in VHF and UHF frequency bands and noting their origin and also operating frequency, the following points are outstanding and must be taken into account for the mobile radio link design:

- Natural noise level falls with frequency increase.
- Basically, the natural noises in VHF and UHF bands are negligible, and only galactic noise in VHF band has undesired effects.

8.6.3 Man-Made Noises

This kind of noise in the mobile communications is generally generated by unintentional sources such as combustion engines of cars, electric power lines, and operations of industrial equipment and mobile radio units as well. Man-made noise has a great role in mobile communications, and this type of noise can be divided into the four following categories:

- Urban noise (commercial/industrial)
- Suburb and residential noise

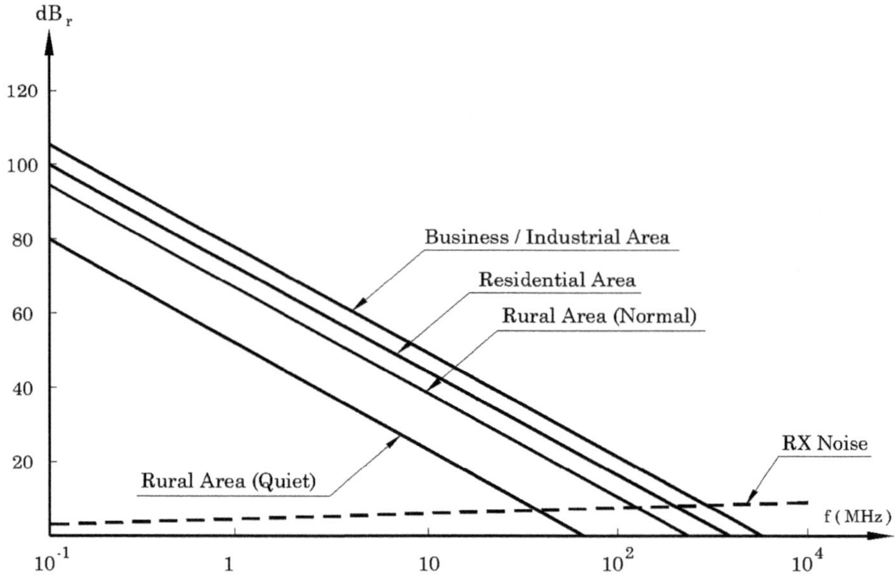

Fig. 8.33 Man-made noise level in various regions

- Rural area noise
- Internal noise of mobile radio units

The relative value of these noises is plotted in Fig. 8.33 in terms of frequency. It is also noted in this case that urban and suburban noises (industrial, commercial, residential, and rural areas) decrease with the frequency increase, and it is only the noise related to RX unit which as the frequency rises, the noise increases as well. Figure 8.34 also shows the amount of standard deviation of different regions based on the measurements.

8.6.4 *Traffic Noise*

Because of wide spread usage of mobile radio services in vehicles and the presence of significant amount of noises on the roads, highways, main roads, and sideways, there are certain studies conducted about this case, and the pertinent results are provided in the form of graphs of the mean traffic noise values. As it is shown in the curve of Fig. 8.35, this noise is in fact a function of vehicle density and also frequency.

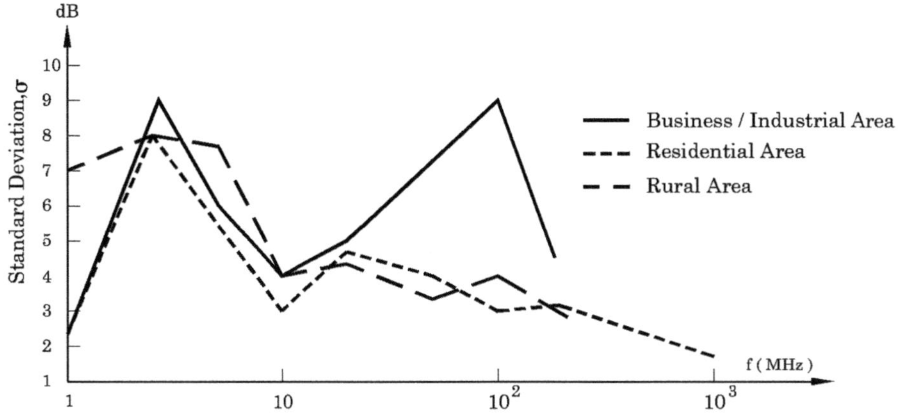

Fig. 8.34 Standard deviation of man-made noise level

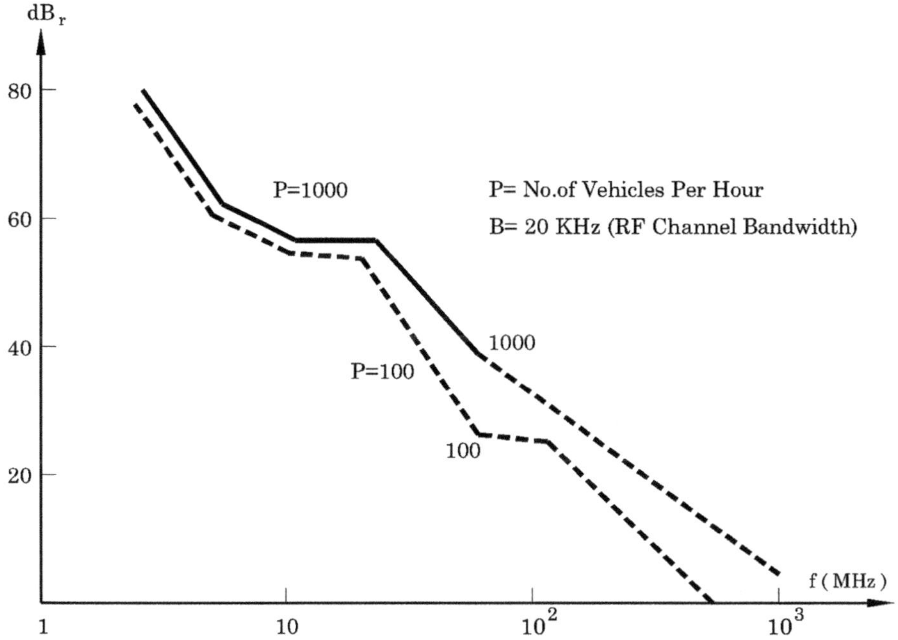

Fig. 8.35 Average traffic noise

8.7 Exercises

Questions

1. What are the main noise types and present their categorization?
2. Evaluate the main effects of the noise in telecommunication systems and explain it briefly.
3. Define the noise factor and noise figure and describe their differences.
4. Express the generated noise power in a resistor and determine units of its main components.
5. Prove the relations of (8.16) and (8.18) concerning the noise factor of hybrid systems.
6. Express the equation of noise temperature in a hybrid system and explain the affecting factors.
7. Prove the equation of (8.32) for noise power.
8. Explain how the equations of (8.36) and (8.37) are derived.
9. Given the curves of (8.13), determine the noise factor, noise figure, and noise temperature in the frequency range of 1–10 GHz, related to the following sources:
    - Solar noise in quiet condition
    - Sky noise
    - Background cosmic noise
10. Noting the Figs. 8.14 and 8.15, determine at what frequencies the noise peaks occur and explain the reasons briefly.
11. Specify the median values of man-made noise figure at 1, 10, 100 MHz for the following areas using Fig. 8.19.
    - Business areas
    - Residential areas
    - Rural areas
12. Define types of external noise sources in satellite communications and explain them briefly. Discuss the case of received noise via antenna according to Fig. 8.30.
13. Define the noise power and density and determine the pertinent relation including the unit of each parameter.
14. Explain affecting factors on the satellite antenna noise temperature. What are their normal values on C and Ku bands?
15. Explain why the first stage of the RX amplifiers must be installed in the vicinity of satellite antenna. Prove your answer by an example.
16. Specify the main kinds of noises emanating from terrestrial sources.
17. Describe the thermal noise of rain in satellite communications and express its relation with the equivalent noise temperature.

18. Given Fig. 8.31, evaluate contribution of different noises in mobile radio communications within the 100–2,000 MHz frequency range.
19. Noting the graphs of Fig. 8.33, evaluate the man-made noises and their range of variations in mobile radio communications.
20. Specify affecting parameters in the city traffic noise in mobile radio communications by evaluating Fig. 8.35.

Problems

1. Signal-to-noise ratios at the input and output of a system are 15 and 12 decibels, respectively, find:

   (a) The noise figure and noise factor of the system.
   (b) The input noise power in case of receiving the signal power level equal to 15 pW.

2. Calculate the noise density produced in a resistor at 27°C and calculate its power for 1 MHz bandwidth.
3. Assuming 5 dB and 20 dB as the noise figure and amplification coefficient for an amplifier, calculate,

   (a) Noise factor and its equivalent noise temperature.
   (b) Total noise factor and its equivalent noise temperature when 3 units of these amplifiers are interconnected in series by certain connectors each having a 3-dB attenuation in a perfect matching condition.

4. Solve Example 8.1 with the following parameters:

$$g_t = g_r = 500,\ \ L_p = 138\ \text{dB},\ \ L_t = 5\ \text{dB},\ \ F_M = 7\ \text{dB}$$

$$L_c = 3\ \text{dB},\ \ F_I = 5\ \text{dB},\ \ SNR = 20\ \text{dB}$$

5. Solve Example 8.2 with the following parameters:

$$P_t = 200\ \text{kW},\ \ T_A = 140^\circ\text{K},\ \ L_c = 7\ \text{dB},\ \ F_M = 5\ \text{dB}$$

$$F_{IF} = 3\ \text{dB},\ \ SNR = 10\ \text{dB},\ \ f = 10\ \text{GHz},\ \ \sigma = 5\ \text{m}^2$$

6. Repeat Example 8.3 by setting LNB gain to 47 dB and its noise figure to 8 dB where the equivalent noise temperature of the antenna is assumed 100°K.
7. If the equivalent noise temperature of an antenna in standard conditions is equal to 120°K,

   (a) Obtain the antenna noise factor at 47°C.
   (b) Assume the above antenna including coupling loss of L_c = 3 dB and connected to a receiver with a noise figure of 3 dB via transmission line

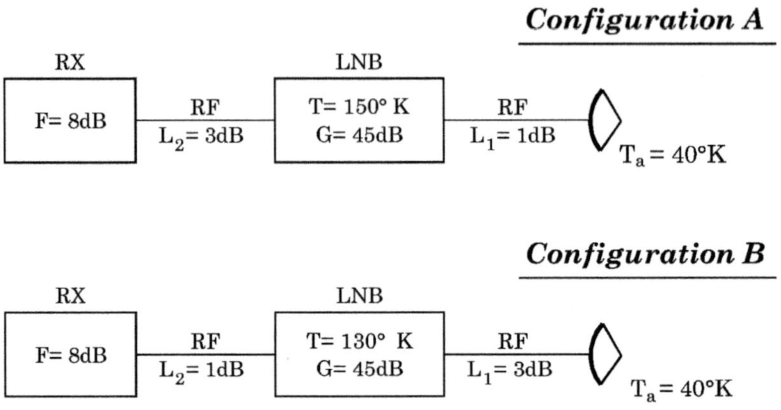

Fig. 8.36 Typical configuration for Problem 11

with a loss of $L_t = 4.77$ dB, then calculate the total RX system noise factor and its equivalent noise temperature for standard conditions.

(c) Obtain the above values for the medium temperature of 47°C.

(d) Determine the input noise power for both cases.

8. (a) Obtain the radio noise values and their equivalent noise temperature for frequencies of 10 MHz, 200 MHz, 5 GHz, and 20 GHz in standard conditions.
   (b) Determine the input noise power of an antenna when the bandwidth of the input signal is equal to 1 MHz.
9. For a mobile radio system in 150 MHz find the following:
   (a) The median value of man-made noise figure for suburban and rural environments
   (b) Location and time standard deviation
   (c) Safety margin for 95% and 90% location and time coverages, respectively.
10. (a) Specify the brightness temperature of the Moon, Sun, and galaxy for 10–20 GHz frequency range using the Fig. 8.21.
    (b) Assuming the galactic brightness noise temperature equal to 50°K at 500 MHz, calculate its value for 800 MHz frequency and compare the results with the graph (8.21).
11. Given Fig. 8.36, calculate the equivalent noise temperature for each of the configurations A and B, and specify the difference:
12. In a Ku band satellite terminal, an antenna with $\phi = 1.8$ m is used. The equivalent noise temperature of the system is 240°K when connected to a TX with 8 W output power.
    (a) Calculate the G/T and EIRP of the system.
    (b) Recalculate the G/T and EIRP when the same antenna is used at C band and total noise temperature of the system is decreased to 170°K.
    (c) Assuming $A_R = 1.5$ dB and $T_S = 420$°K, how much the safety margin will decrease for rainy condition with $T_A = 280$°K.

Chapter 9
Improvement Techniques in Radiowave Reception

9.1 Fading

Fading is a phenomenon specific to radiowaves when they are propagating within a time-varying medium. This phenomenon affects the received signal level changing with time even where TX power and antenna gain remain constant. Most of radio links including terrestrial line of sight, mobile radio, troposcattering, and satellite links in the troposphere and ionosphere experience signal fluctuation around their median values.

Fading has a very strong influence on the service quality and system performance of a radio network. Therefore, it is essential to have good knowledge about it and the ways to counteract this adverse mechanism. Major sources of the fading include the following:

- Atmospheric variations
- Multipath reception
- Abnormal conditions in the radiowaves propagation channel
- Tropospheric precipitation such as rain showers, snow, and hail

In evaluation of the received signal level variations with time, as shown in Fig. 9.1, the following two types of fluctuations with different nature may be identified:

- Fast fading, lasting from a fraction of a second to few minutes
- Slow fading, lasting much longer than the first type, that is, with periods of few hours or more

This section is followed by presenting some examples of either fast or slow fadings dedicated to the terrestrial line-of-sight radio links.

1. The received signal level is usually associated with variations of atmosphere refraction index. In the sub-refractive conditions, by increasing the air refractive

A. Ghasemi et al., *Propagation Engineering in Radio Links Design*,
DOI 10.1007/978-1-4614-5314-7_9,

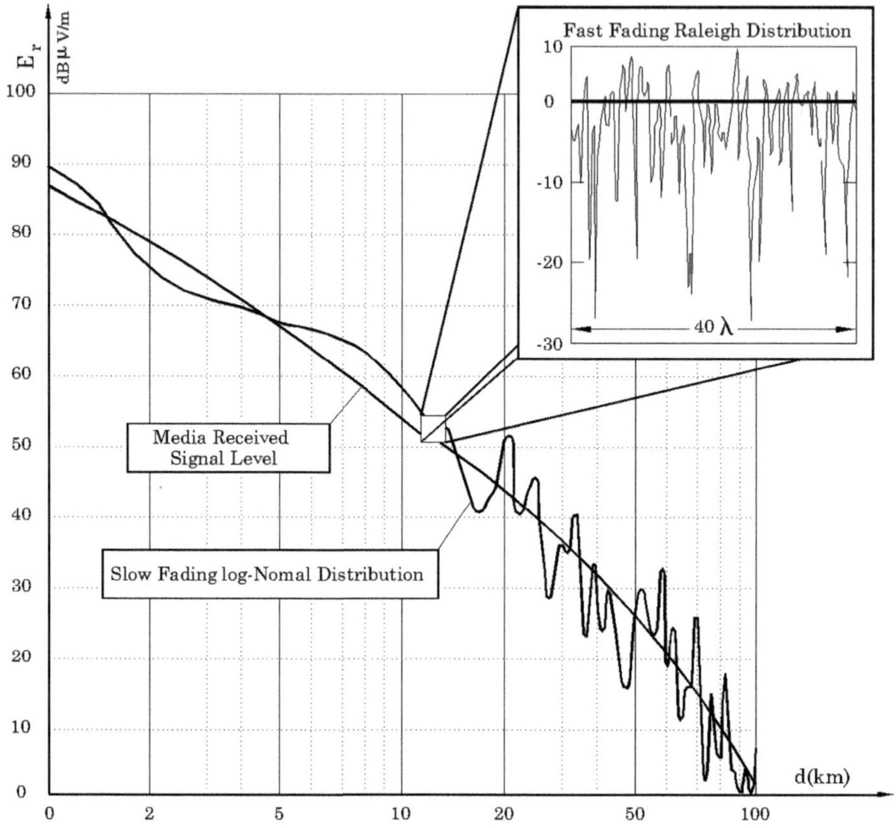

Fig. 9.1 Received signal level in a typical radio link

index, K-factor and Earth effective radius will be reduced. Also, this condition will increase the terrain heights which eventually may obstruct the route of the radiowaves, resulting in high attenuation or total blockage.

2. Radio links with enough clearance are subject to more reflections for high K-factor values. On the other hand, in the obstructed links, super refraction condition where $K_e > 1.33$ may reduce the path attenuation significantly and diminish the obstacles temporarily up to few hours. This situation does not reduce the main radio link performance but it will be a potential source of interference to other RF channels working in the same frequency.
3. Changes of arrival angle by a fraction of a degree in vertical or horizontal planes decreases the antenna effective gain.
4. Many experiments indicate frequent reflections in the atmospheric layers near the ground that leads to the formation of air ducts. This mechanism occurs with greater probability when there is fog, mist, or humidity over valleys, swamps, lakes, and rivers. Formation of air ducts is more frequent where the atmosphere is well stratified and there is no wind at dawn and early morning.

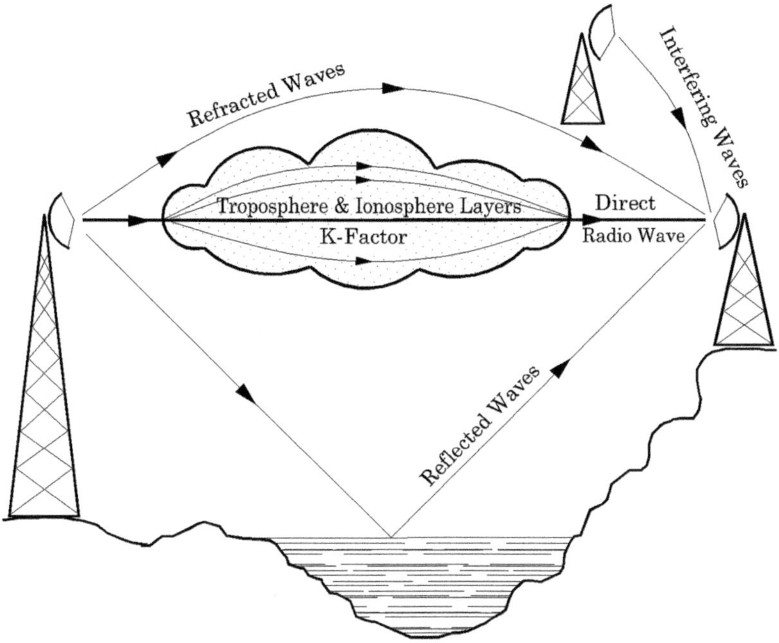

Fig. 9.2 Multipath reception in a typical radio links

5. In addition to ground reflections, sometimes there is fast fading due to random changes in the atmospheric refraction index which will produce multiple paths between TX and RX antennas.
6. Radio links over smooth terrains or seas are subject to the multipath effects. In this situation, the median received signal level is multiplied by the path gain factor varying from 0 to 2 ($0 < |F| < 2$). This type of multipath leads to deep fading for very small values of $|F|$ and up-fading when $|F| > 1$.
7. Multipath fading, as indicated in Fig. 9.2, has great impact on the system performance and in some occasions results in total outage of radio links. In addition to adverse effects in the signal-to-noise ratio (SNR), it will increase the bit error rate (BER). Multipath fading is a frequency-sensitive mechanism that introduces server signal distortion in high-capacity digital networks with a big bandwidth.
8. In radio links where the orthogonal polarizations are employed, multipath situation may be the main cause for cross-polarization interference.
9. Also, the line-of-sight links may experience fast changes in the amplitude of the received signal level due to scintillation which does not depend on the path length and RF channel frequency. This phenomenon is effective in the frequency range from 10 to 40 GHz and results in the amplitude attenuation of few decibels. However, in comparison with the normal fade margin for a radio link, about 20–40 dB, this kind of fading is not of concern in the design stage.

10. Radio links at frequencies above 10 GHz and path lengths more than 10 km are subject to considerable attenuation during heavy rainfall when its intensity is high (50 mm/h or more). This type of fading may be a dominant factor in the design stage for locations with heavy rainfalls.

9.2 Improvement in Radiowaves Reception

9.2.1 Fade Margin

Fading in radio communications is a stochastic process, some which on occasions results in a sharp reduction of the received signal level for a small period of time and may happen frequently. When the magnitude of fading exceeds a predetermined value, it is called deep fading, which leads to the total outage of the radio link.

In the design of a radio network, some additional power is considered for the received signal as "*fade margin*" to combat the fading phenomenon. Indeed, the fade margin is the difference between the standard received signal level (when there is no fading) and the minimum receiver sensitivity or the RX threshold level. The fade margin values vary greatly for different systems based on radio link nature, operating frequency, and RF channel bandwidth. This may be up to 40 dB or more for some links. Sometimes increasing the fade margin is not useful, concerning the following issues:

- Frequency selective fading for high-capacity digital radio links (i.e., bit rates more than 100 Mb/s).
- Frequency pollution and producing undesired interference resulting in non-efficient usage of radio spectrum.

However, using improvement and coding techniques are very common in radio link design. In these techniques, radiowaves are received through diverse RF channels (mostly 2 RF channels) enabling the receiver to obtain an acceptable SNR (or signal to interference ratio).

The diversity reception includes those techniques considered mainly in the domain of space, frequency or angle, and their combinations. The application of diversity techniques is useful in reducing the adverse effects of fading on system performance making an efficient utilization of the radio spectrum.

9.2.2 Improvement Techniques

The most popular improvement techniques in the reception of radiowaves include the following methods:

- System diversity
- Route diversity

- Site diversity
- Station height diversity
- Antenna space diversity
- RF channel frequency diversity
- RF channel cross band diversity
- Hybrid diversity
- Radiowave polarization diversity
- Radiowave arrival angle diversity
- Field component diversity
- Time diversity

ITU-R has performed a number of studies in respect of selecting improvement techniques for reception of radiowaves and operating frequencies based on new technologies.

In the next sections, more details are provided for the above techniques including their applications in some of main radio systems such as fixed terrestrial radio networks, mobile radio systems, satellite communications, and trans-horizon troposcattering links.

9.2.3 Combination of Received Signals

To extract the final input signal by diversity reception of the radiowaves, there are different methods to combine the received signals. The major methods include the following cases:

- Selection combiner, which compares the received signal levels and selects the strongest one.
- Equal gain combiner, which compares the received signal levels and simply adds them together.
- Maximal gain combiner (sometimes called optimal addition combiner) provides linear combination of synchronized basic signals maximizing the output SNR.

Figure 9.3 illustrates a sample of improvement magnitude in the Rayleigh fading depth by using of 1, 2, 3, and 4 channels when their combination for selection is based on the best signal quality.

Figure 9.4 displays a typical relative comparison for 2-channel reception when different types of combiners are employed. This figure is prepared based on two uncorrelated channels ($\rho = 0$).

In case of some correlation between different reception channels, then the improvement factor will decrease when the correlation coefficient increases between them. Subject to complete dependence of reception channels, that is, $\rho = 1$, then all channels will act as single channel and there will be no improvement. Figure 9.5 indicates fading depth for two reception channels improvement for some typical values of correlation coefficient.

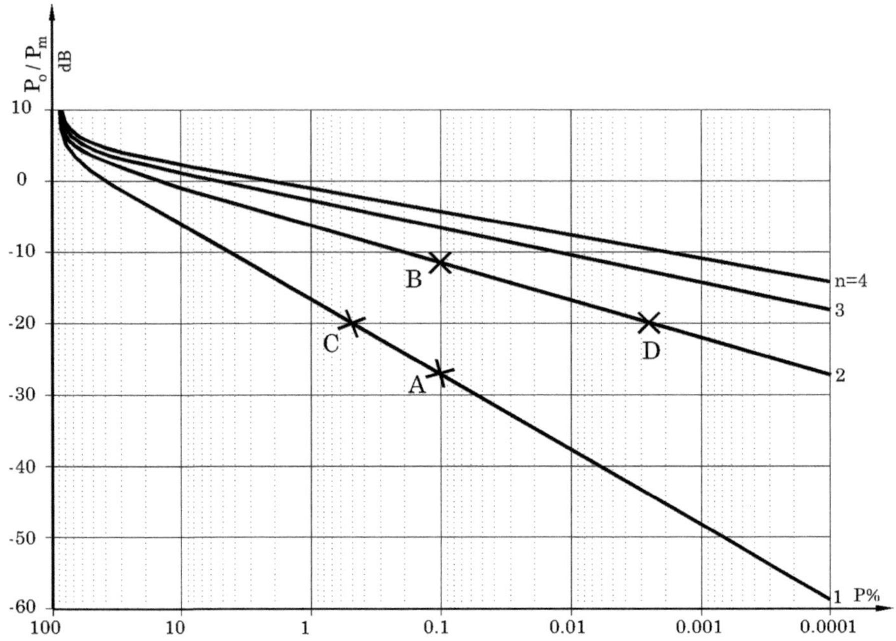

Fig. 9.3 Rayleigh fading single and multiple reception channels

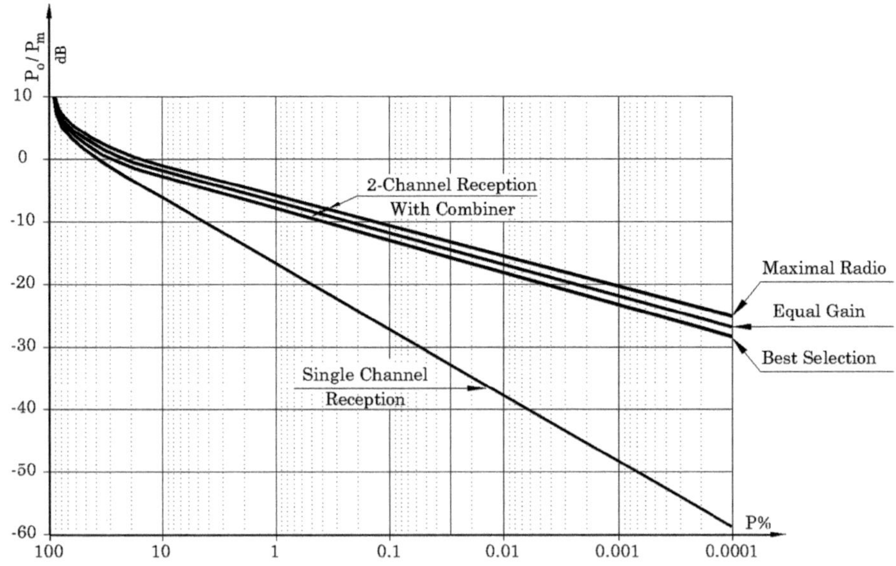

Fig. 9.4 Typical fading for 2 uncorrelated reception channels for different combiner types

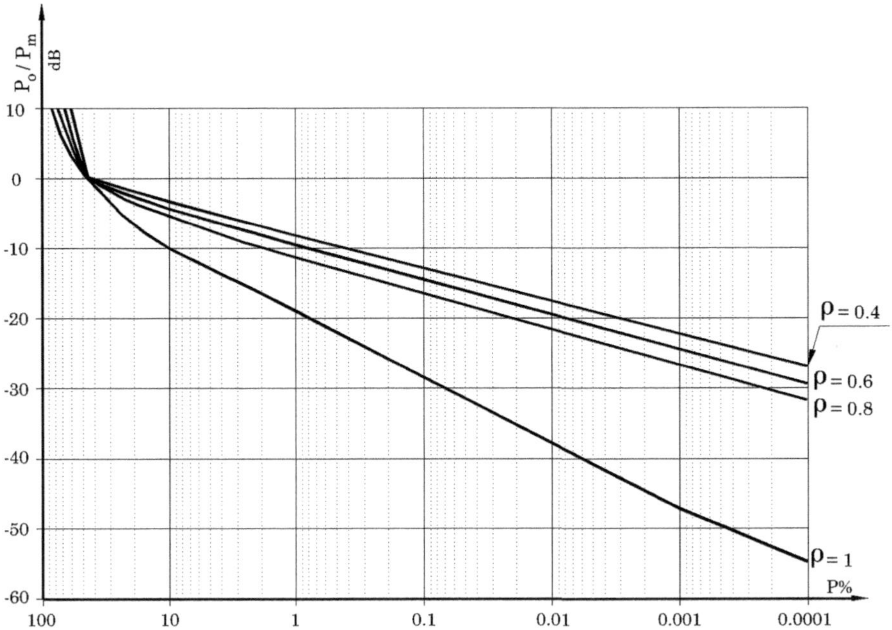

Fig. 9.5 Rayleigh fading of 2 correlated reception channels

9.2.4 Improvement Parameter

Considering Fig. 9.3, now we can introduce an improvement parameter to express quality or strength of improvement in diversity reception of radiowaves. This parameter may be defined in a number of ways among which the following types are most common:

- *Improvement Gain* denoted by IG in decibels is defined for each fixed value of time percentage of fading probability as the difference in dB between fading depth with and without diversity reception. As an example given in Fig. 9.3, vertical line for $P = 0.1\%$ intersects single and two channels of reception curves at points A and B, respectively, where the received signal level is less than its median value by

$$RSL(A) = -28 \text{ dB}$$

$$RSL(B) = -11.5 \text{ dB}$$

$$IG = RSL(B) - RSL(A) = +16.5 \text{ dB}$$

- *Improvement Factor* denoted by IF is the ratio of two percentages of time for a fixed level of fading depth with and without diversity receptions. As an example using Fig. 9.3, horizontal line for $P_0/P_m = -20$ intersects single and two channels reception at points C and D, respectively, where percentages of time probability are:

$$P_C(P_0/P_m = -20\text{ dB},\ n = 1) = 0.5\%$$
$$P_D(P_0/P_m = -20\text{ dB},\ n = 2) = 0.003\%$$
$$IF = (0.5)/(0.003) \approx 170$$

- Sometimes the improvement parameter is expressed by the ratio of received power with and without diversity reception:

$$IF = \frac{P_d}{P_r} = \frac{|E_d|^2}{|E_r|^2}$$
$$IG[\text{dB}] = P_d[\text{dB}_m] - P_r[\text{dB}_m]$$
$$IG = 20\ \log|E_d| - 20\ \log|E_r|$$

9.3 Improvement Techniques in Mobile Radiowaves Reception

9.3.1 Overall Concepts and Definitions

This section is devoted to providing details of improvement techniques of radiowaves reception in mobile communications. However, these techniques may be very diverse in theory, but practically only the effective ones with acceptable performance and having less costs are evaluated and used mainly in the terrestrial radio mobile communications in UHF band. For details concerning improvement of mobile satellite communications networks, you can refer to the pertinent section.

In general, improvement of amplitude and quality of radiowaves reception at the RX unit requires using special methods. Based on these concepts, the radiowaves should be processed through combiner circuit prior to arriving at the RX input.

The combination and selection method of the final signal for applying to the RX input requires special electronics circuits, which at some occasions their function as a logic circuit is exclusively intended for the selection of the best incoming signal and, in some instances, includes complicated circuits to combine them for obtaining a robust and powerful signal.

As indicated in Fig. 9.6, the probability of intense simultaneous fading on different routes is very low, and suitable improvement techniques are capable of upgrading the communication to a great extent and establish the link on a desirable level.

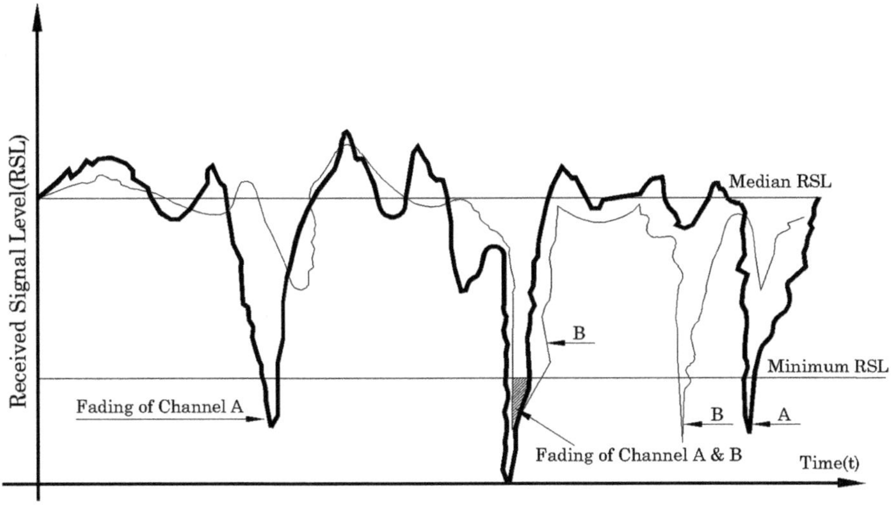

Fig. 9.6 Fading of RF channels

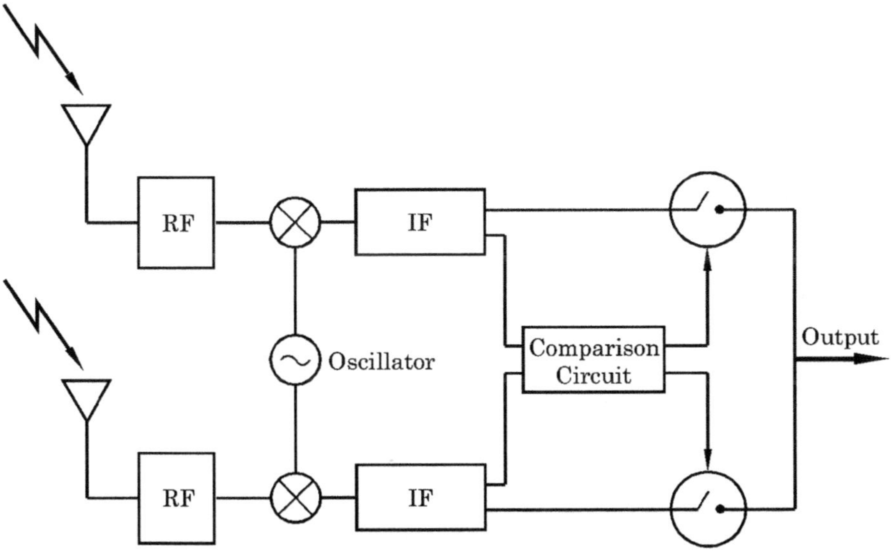

Fig. 9.7 Simple circuit for selection combiner

Figure 9.7 illustrates a simple circuit for selection of the best received signal through comparison of two or more incoming signals. In such a situation, the best signal is transferred to the RX input using proper electronics circuits. Also, Fig. 9.8 presents a simple circuit for the phase conversion of the received signal, and their summation is indicated at the output.

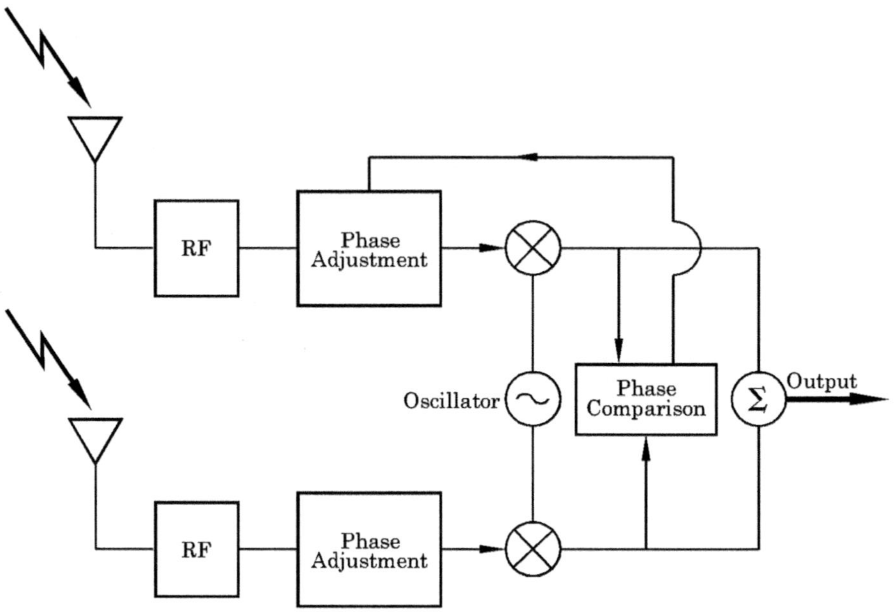

Fig. 9.8 Simple circuit for addition combiner

Generally, the improvement factor designated by I is defined for each of selected methods by:

$$I = \frac{P_{\mathrm{d}}}{P_{\mathrm{r}}} = \frac{|E_d|^2}{|E_r|^2} \tag{9.1}$$

where P_{d} and E_d represent the RX input power and field strength levels, respectively, using combiner circuit for diversity reception, while P_{r} and E_r are the same quantities without using any type of improvement techniques. Equation (9.1) can be expressed in logarithmic units as:

$$I[\mathrm{dB}] = 10(\log P_{\mathrm{d}} - \log P_{\mathrm{r}}) = 20 \log |E_d| - 20 \log |E_r| \tag{9.2}$$

Example 9.1. 1. Determine the improvement gain when a selection combiner is used for two independent channels related to the graphs of Fig. 9.4 for time percentages of less than 0.01%.

2. In case of two channels having a correlation with $\rho = 0.6$, find the reduction value in the improvement gain.

Solution. 1.

$$(one\ channel,\ \rho = 0.01\%) \xrightarrow{\text{Fig. 9.4}} P_0/P_m = -37\ \text{dB}$$

$$(two\ channel,\ \rho = 0.01\%) \xrightarrow{\text{Fig. 9.4}} P_0/P_m = -18.5\ \text{dB}$$

$$IG = -18.5 - (-37) = +18.5\ \text{dB}$$

2.

$$(\rho = 0.6,\ 2 - channel,\ P = 0.01\%) \xrightarrow{\text{Fig. 9.5}} P_0/P_m = -20\ \text{dB}$$

$$\Delta(IG) = (IG) - (IG)' = 1.5\ \text{dB}$$

■

The above definition for I signifies its instantaneous value. The overall improvement factor may be expressed by the ratio of deep fading occurrence probabilities without and with an improvement technique for diversity reception of radiowaves.

Improvement techniques, which are common in the mobile radio systems, are not exactly identical methods for other systems. For example, in the LOS radio links at both ends, similar diversity techniques are used in the individual (SD, FD, ...) or combined (SD+FD) forms, while these approaches are not usually used in mobile radio systems. Also, in many cases, diversity techniques used in the BTS and mobile radio units are quite different. The main diversity methods in the mobile radio networks are

- BTS location diversity
- Antenna space diversity
- RF channel frequency diversity
- Radiowaves polarization diversity
- Field component diversity
- Time diversity

9.3.2 BTS Location Diversity

Noting the nature of mobile communications and movements of handheld and vehicular radio units as illustrated in Fig. 9.9, there are locations, due to fixed or temporary obstacles; the radio link is severely attenuated and results in outage of the link between the mobile terminal and fixed station.

These kinds of obstacles are capable of producing long- or short-term fadings which should be compensated through the proper selection of stations or providing additional fixed stations at suitable locations to complete coverage area to improve service quality.

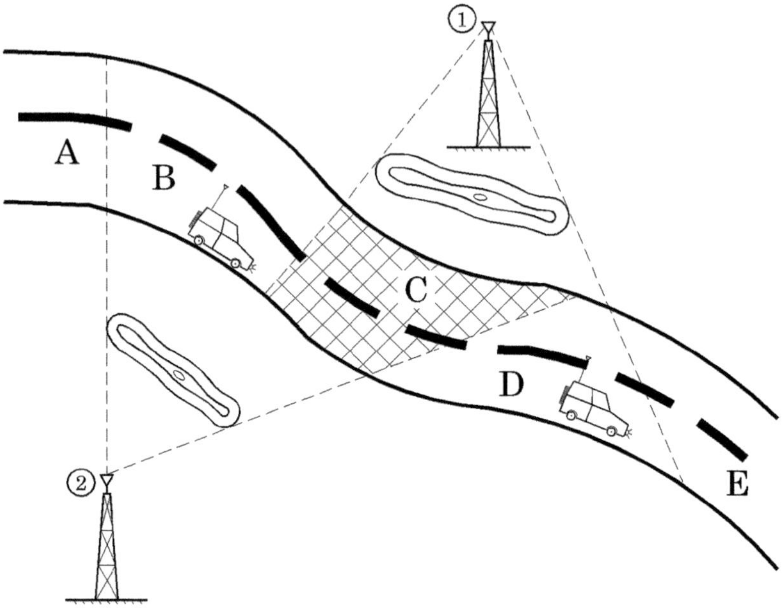

Fig. 9.9 Illustration of site diversity concept for mobile base stations

In practice, for large networks due to the necessity of using a great number of base stations for traffic handling and area coverage, many places are covered with more than just one fixed station. Whereas with the BTS location diversity, the main objective is to improve the quality of the received radiowaves in the mobile terminals, and increasing the traffic capacity is not an initial necessity but is taken as a secondary objective.

As depicted in Fig. 9.9, the BTS location diversity has the following two main roles in alleviating the adverse effects:

- Compensating long-term fadings caused by natural obstacles such as low- and highlands or mountainous areas.
- Compensating short-term fadings such as moving obstacles situated on the path of radiowaves.

Figure 9.9 illustrates how long-term fading of radiowaves due to the obstacles of each area is determined. As it can be observed:

- Areas marked by A and E are covered by both BTS-1 and 2.
- Areas (A + B) are covered by BTS-1, and areas (D + E) are covered by BTS-2.
- Area marked by C is uncovered zone.

In zone C, in which the radiowaves from both base stations are subject to fading, the received signals are not detectable. In addition to what was discussed in the above concerning the long-duration fading, the versatility of the location of

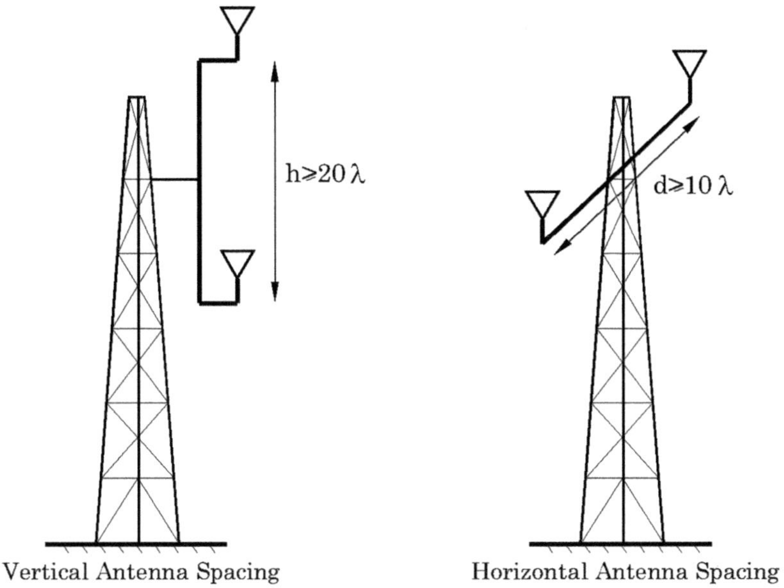

Fig. 9.10 BTS antenna space diversity

base stations is also utilized to counteract short-term fading. Suppose a vehicle as indicated in Fig. 9.9 is overtaking a high-speed truck on the left-hand side, in these circumstances if this action takes place in the zone-D causes complete outage of the link but if this overtaking takes place in the zone-B, its link can be maintained through station No.1.

In this technique, the criterion for proper BTS selection to establish the link with mobile terminal noting its position of location based on the received signal power relative to the mobile terminal through each one of the BTS(s) and comparison will be made via a logic circuit.

9.3.3 Antenna Space Diversity

9.3.3.1 General

At the receiving site, by using two antennas at different positions, it is possible to achieve improvement in the reception of radiowaves. This method is called antenna space diversity which may be classified in the following basic types as shown in Fig. 9.10:

- Antenna horizontal space diversity
- Antenna vertical space diversity

Due to the special conditions of land mobile communications, such as the location of stations or nature of movements and based on the experimental test results, the ITU-R suggests a distance not less than 10λ for horizontal antenna spacing and not less than 20λ for vertical antenna spacing in order to provide an effective improvement factor. The essential condition to improve the reception of mobile radiowaves by antenna space diversity is receiving them on fairly independent RF channels ($\rho < 0.2$).

This method is practical and cost-effective only when the distance between the antennas is reasonable. Noting that antenna spacing compared to the distance between the TX and the RX stations is negligible, hence to counteract long-term fading resulting from natural terrain topography is not suitable rather a technique to prevent short-term fading caused by atmospheric adverse effects or moving obstacles.

Example 9.2. For RF channel frequencies of $f_1 = 400\,\text{MHz}$ and $f_2 = 1{,}800\,\text{MHz}$, calculate the minimum distance between the antennas to obtain a suitable correlation factor enabling them to act as a space diversity technique.

Solution.

$$f_1 = 400\ \text{MHz} \quad \Longrightarrow \quad \lambda_1 = 0.75\ \text{m}$$

The minimum distance between the antennas, which is obtained in horizontal position (i.e., the angle of incoming waves is equal to zero), is given by

$$d_1 = 10\lambda_1 \quad \Longrightarrow \quad d_1 = 7.5\ \text{m}$$

Also, for frequency f_2, the following distance should be considered:

$$f_2 = 1{,}800\ \text{MHz} \quad \Longrightarrow \quad \lambda_2 = 0.167\ \text{m}$$

$$d_2 = 10\lambda_2 \quad \Longrightarrow \quad d_2 = 1.67\ \text{m}$$

■

9.3.3.2 Mobile Terminal

In the land mobile communications, the correlation factor of the received radiowaves noting the velocity and time is determined from the following relation in which the Bessel function J_0 is of zero order, β is the propagation constant, V is the mobile terminal velocity, and t is time.

$$\rho = J_0^2(\beta V t) \tag{9.3}$$

For using the above relation to specify the minimum proper distance and receiving uncorrelated signals, Vt is replaced by d converting the above relation into the following form:

$$\rho = J_0^2(\beta d) = J_0^2(2\pi d/\lambda) \tag{9.4}$$

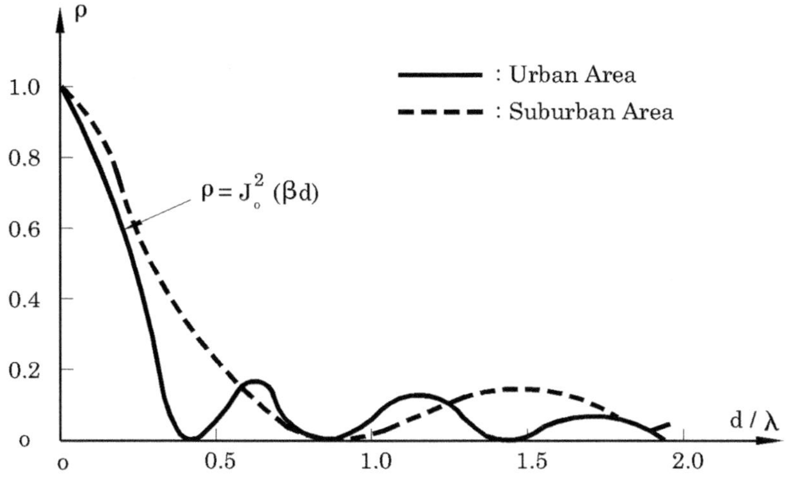

Fig. 9.11 Variation of correlation factor with normalized distance

The normalized sketch of correlation factor in d/λ is illustrated with Fig. 9.11 for uniform angular distribution of the received radiowaves at urban environments. As it can be observed, the first zero of this function is obtained for $d = 0.4\lambda$. In suburban environments, due to non-uniform angular distribution of radiowave determined by the dashed line, the first zero of the ρ function is obtained at a distance of $d = 0.8\lambda$ which is greater compared with the one obtained for the uniform distribution.

Based on the above discussions, it is concluded that for the reception of uncorrelated radiowaves in urban environments at a frequency of 850 MHz, the minimum distance of the antennas to receive uncorrelated radiowaves must be at least 14 cm, but the same value for suburban areas must be at least 28 cm. Considering that one of the main practical criteria for handheld terminals is their compactness and low weight, hence utilization of antenna space diversity for mobile terminals is not recommended.

9.3.3.3 Base Station

In a base station, using antenna space diversity technique for the reception of radiowaves is a normal practice and is relatively popular which is capable of effectively improving the received signal level up to maximum of 6 to 9 dB. The factors affecting this mechanism as given in Fig. 9.12 are the following:

- Height of antennas, h
- Distance between antennas, d
- Radio channel frequency, f
- Direction of waves, α

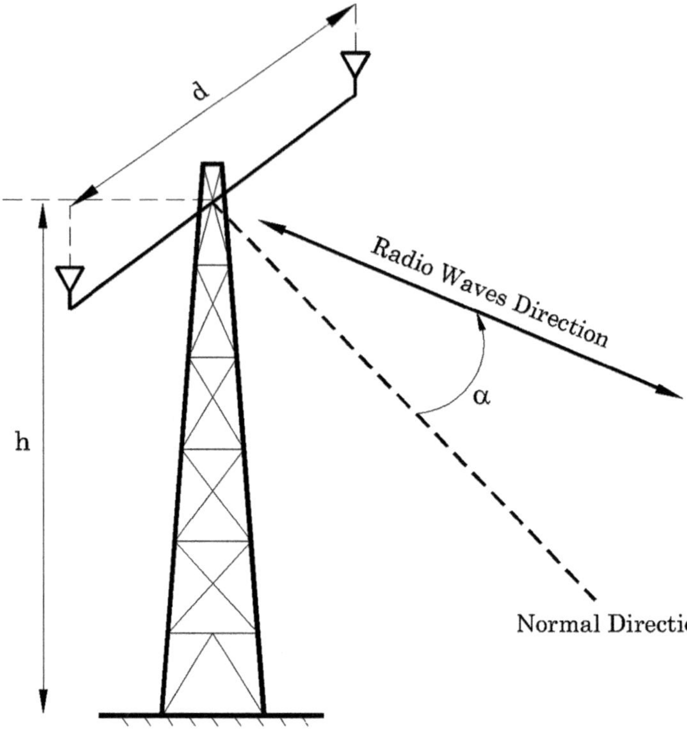

Fig. 9.12 Geometrical parameters of antenna space diversity

Correlation factor has a direct relation with the height (h) and angle (α) and an inverse relation with the distance (d) and frequency (f). The antenna space diversity is characterized with space diversity parameter denoted as η_d and defined below:

$$\eta_d = \frac{h}{d} \tag{9.5}$$

Normally, available charts indicating variations of ρ_d versus η_d for different values of α are used for planning of mobile radio systems. Figure 9.13 is a typical variations of ρ_d in terms of η_d for reference frequency of $f_0 = 850\,\text{MHz}$.

For RF channel frequencies other than f_0, it is possible to replace "d" with its modified value "d'" in the relation (9.5):

$$d' = d \times \frac{f_0}{f} \tag{9.6}$$

Example 9.3. To achieve $\rho_d = 0.6$ with antennas at a height of 30 m, find:

1. The minimum distance of antennas on 850 MHz frequency
2. The minimum distance of antennas on 425 MHZ frequency

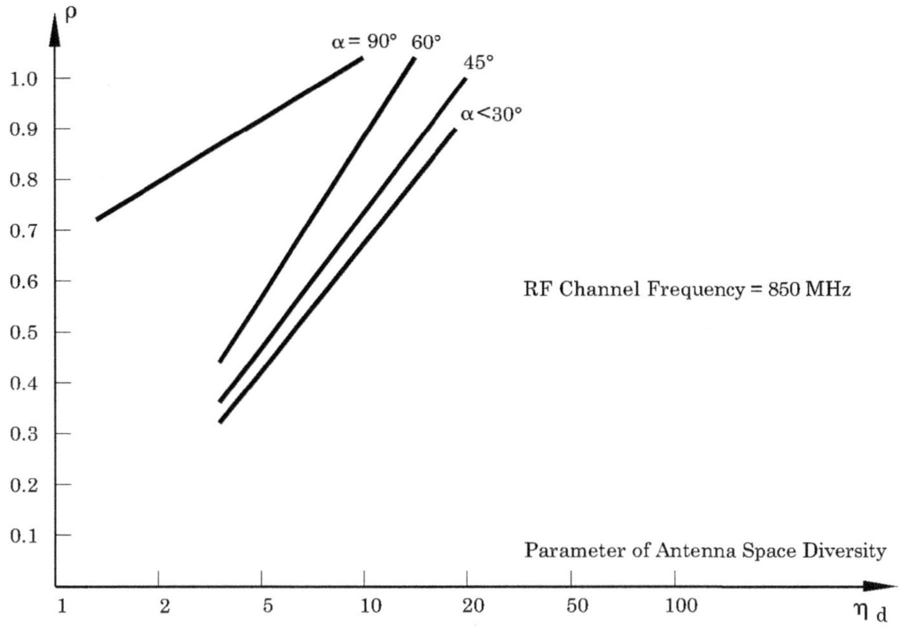

Fig. 9.13 Variation of correlation factor in antenna space diversity

Solution. 1. The minimum distance will be obtained for $\alpha = 0$ noting the graph of Fig. 9.13:

$$(\rho_d = 0.6\,, \quad \alpha = 0) \implies \eta_d \approx 9$$

$$\eta_d = \frac{h}{d} \implies d = \frac{30}{9} = 3.3\ \text{m}$$

2. By changing the frequency to $f' = 425\,\text{MHz}$, using relation (9.15) yields:

$$d' = 3.3 \times \frac{850}{425} \implies d' = 6.6\ \text{m}$$

■

9.3.4 *Frequency Diversity*

9.3.4.1 Definition

Another commonly used method to improve radiowaves reception is the transmission of the link traffic over two different radio channels with frequencies of f_1 and f_2. This method is called frequency diversity for which the amount of improvement factor will be a function of frequency separation of the radio channels denoted by Δf. In the fixed radio links, because of a large capacity of each radio channel, generally the frequency diversity is allocated in the form of (N+1), that is, one radio channel supporting N main radio channels.

In today's mobile radio systems, because of the limitations of the number of radio channels, normally several radio channels are used to increase the traffic handling capacity. With a proper selection of frequency separation between them in fact in addition to increasing the traffic handling capacity, frequency diversity is also achieved.

9.3.4.2 Correlation of RF Channels

The correlation between two radio channels in a mobile radio link with a frequency separation of Δf is according to the following relation:

$$\rho_f = \frac{1}{\sqrt{1 + (2\pi \cdot \Delta f \cdot \Delta)^2}} \tag{9.7}$$

By satisfying the condition of $\Delta f \cdot \Delta > 0.5$, then ρ_f can be simply derived as:

$$\rho_f \approx \frac{1}{2\pi \cdot \Delta f \cdot \Delta} \tag{9.8}$$

In the above relation, Δf is the frequency separation between the RF channels in terms of MHz, and Δ is the value of standard deviation of time delays of the radiowaves relative to the direct route between the TX and the RX units designated as T in terms of microseconds. The value of Δ is calculated by the following relation:

$$\Delta = \sqrt{< T^2 > - < T >^2} \tag{9.9}$$

Example 9.4. The value of standard deviation of time delays in a mobile radio link using frequency diversity is 0.3 μs; find:

1. The value of frequency separation for $\rho_f = 0.3$.
2. The minimum value of Δf for $\rho = 0.1$ and an improvement factor of 4.

Solution. 1.

$$\rho_f = 0.3 \implies 0.3 = \frac{1}{2\pi \cdot \Delta f \cdot 0.3} \implies \Delta f = 1.8 \text{ MHz}$$

2.

$$\rho'_f = 0.1 \implies \Delta f = 5.3 \text{ MHz}$$

■

9.3.5 Polarization Diversity

In a radio link, radiowaves are transmitted in the form of electromagnetic waves with a specific polarization. In the related RX unit, some radiowaves on orthogonal polarization are detected by using a suitable combiner unit. This process is called polarization diversity which may include an improvement up to 3 dB in the receiver.

Although, the gain of polarization diversity is low, but sometimes it is an optimum choice due to flexibility and low cost.

In general, the mobile antennas are vertical with linear polarization (horizontal, vertical, or tilted). By using antenna with right or left circular polarization, these waves can be received with two identical components in the form of orthogonal circular polarization which their combination provide a gain up to 3 dB.

In case of not using antennas with circular polarization, this method is only effective when the received orthogonal signals have a difference of less than 2 dB, otherwise, since their combination do not provide a significant improvement and stronger component will be selected as the main signal.

9.3.6 Field Component Diversity

A planar wave emitted from the antenna of a base station or mobile terminal having vertical polarization consists of E_z, H_x, and H_y components. Reception of these components is possible through special circuits called field component diversity. The equation related to each of these components is described below:

$$E_z = \hat{a} \cdot \exp[-j\beta Vt \cdot \cos(\varphi - \alpha)] \tag{9.10}$$

$$H_x = \hat{a} \cdot \sin\varphi \cdot \exp[-j\beta Vt \cdot \cos(\varphi - \alpha)] \tag{9.11}$$

$$H_y = -\hat{a} \cdot \cos\varphi \cdot \exp[-j\beta Vt \cdot \cos(\varphi - \alpha)] \tag{9.12}$$

The above components at any instance at the time of reception are uncorrelated, and the following power relation exists between them continuously:

$$|E_z^2| = |H_x^2| + |H_y^2| \tag{9.13}$$

Due to the difference of radiation patterns of circular and dipole antennas, the three above components are uncorrelated and can produce an improvement in the reception of mobile radiowaves functioning as field component diversity. This function is practical only through the use of suitable electronics circuits in one of the following arrangements:

$$V_1 = E_z + H_x + H_y \tag{9.14}$$

$$V_2 = |E_z| + |H_x| + |H_y| \tag{9.15}$$

$$V_3 = |E_z|^2 + |H_x|^2 + |H_y|^2 \tag{9.16}$$

9.3.7 Time Diversity

The time diversity is basically intended for the reception of digital signals in the channels which are affected by fading. In this approach, similar data are transmitted with a specified time intervals. This time interval is equal to the inverse of channel fading cycle time:

$$f_b = 2f_m \tag{9.17}$$

In the mobile radio communications, the time interval can be expressed by:

$$\tau_s \geq \frac{1}{2f_m} = \frac{1}{2(V/\lambda)} \tag{9.18}$$

Example 9.5. For a vehicle with a velocity of 80 km/h, calculate the time of repeating messages with time diversity. Assume the radio channel frequency is 800 MHz.

Solution.

$$V = 80\ \mathrm{km/h},\quad f = 800\ \mathrm{MHz}$$

$$\lambda = C/f = 0.375\ \mathrm{m}$$

$$V = \frac{80 \times 1{,}000}{3{,}600} = 22.22\ \mathrm{m/s}$$

$$\tau_s = \frac{0.375}{2 \times 22.22} = 0.008\ \mathrm{s} \implies \tau = 8\ \mathrm{ms}$$

■

9.4 Improvement Techniques for Reception of LOS Waves

9.4.1 Introduction and General Concepts

To decrease outage time of propagation, as a countermeasure, diversity techniques may be employed. The basic objective of using any form of diversities is to avoid intense attenuation of the received signal in the RX unit. Normally, in analog microwaves radio systems, the flat fading has a key role in the signal attenuation, whereas in digital radio systems, the selective fading has a great influence and is a dominant factor particularly in the case of using high-capacity systems. As an example, in a specified route for a system with 34 Mb/s capacity, the effects of selective fading are ten times more effective than flat fading, and the same ratio for 155 Mb/s capacity can reach to 200 and higher.

The amount of improvement of each type of diversity is called improvement factor and indicated as I. The improvement factor is expressed as

$$I = \frac{P_0}{P_0(D)} \tag{9.19}$$

In the relation (9.19), each of the parameters is stated below:

I : Improvement factor
P_0 : Percentage of outage time for unprotected link
$P_0(D)$: Percentage of outage time for protected system using diversity reception techniques

The performance of improvement techniques can only be effective if the outage and distortions resulting from combining and/or selecting signals be kept minimal for which the selection of a proper and correct system is very essential and forthcoming.

In planning a digital radio network in the first stage, the acceptable amount of overall outage time of propagation must be specified and then proceed by dividing it for various hops of the network. Upon fixing of the amount of outage of each path, the proper improvement techniques must be used to achieve an acceptable value.

The slow effects of flat fading (non-frequency selective) due to beam spreading, and fast frequency selective fading due to multipath propagation must be taken into account in the LOS radio link design. There are a number of techniques available to alleviate these effects, most of which act simultaneously. The same techniques often alleviate the reductions in the cross-polarization discrimination as well. They can be categorized as techniques that do not require any kind of diversity reception or transmission techniques.

Since it is desirable to avoid diversity reception to the extent possible, techniques without additional devices are preferred because of cost impacts. These strategies and techniques are also effective for diversity systems and should be employed when convenient even though they may be less necessary.

9.4.2 *Improvements Without Diversity Reception*

To reduce the negative impact of multipath fading without using any kind of diversity, there are several techniques which are based on one or more of the following strategies:

- Reducing the occurrence of significant flat fading due to atmospheric mechanisms such as beam spreading, antenna decoupling, and atmospheric multipath.

- Reducing the occurrence of significant Earth surface reflections
- Reducing the relative delay time of the surface reflections with respect to the atmospheric radiowaves.

The most common practices to improvement techniques without using diversity reception or transmission are listed below:

- Increase of radiowaves path inclination
- Reduction of effects of surface reflections
- Shielding of the reflection point (or zone)
- Moving of reflection zone to more poor reflecting surface
- Optimum selection of antenna heights
- Optimum choice of antenna polarization
- Use of antenna discrimination
- Reduction of radiowaves path clearance

9.4.3 Techniques for Diversity Reception

The main improvement techniques to combat the propagation adverse effects and reduction of outage time normally used are:

- Antenna space diversity, SD
- Frequency diversity, FD
- Cross band diversity, CBD
- Combined space and frequency diversity, (SD + FD)
- Hybrid diversity, HD
- Angle diversity, AD

In addition to the above-mentioned techniques, there are other popular methods some of which are listed below:

- Adaptive equalizer, AE
- Hot standby, HS
- Space diversity in hot standby configuration, (SD + HS)
- Space diversity complete with adaptive equalizer, (SD + AE)
- Frequency diversity complete with adaptive equalizer, (FD + AE)

In a more generalized position, there are certain techniques which can be also utilized in large networks such as:

- Radio routes diversity
- System diversity

Due to the significant effects of the improvement techniques in digital radio links and decreasing the propagation outage time, research works along with experimental tests have been conducted by the experts of different countries especially regarding

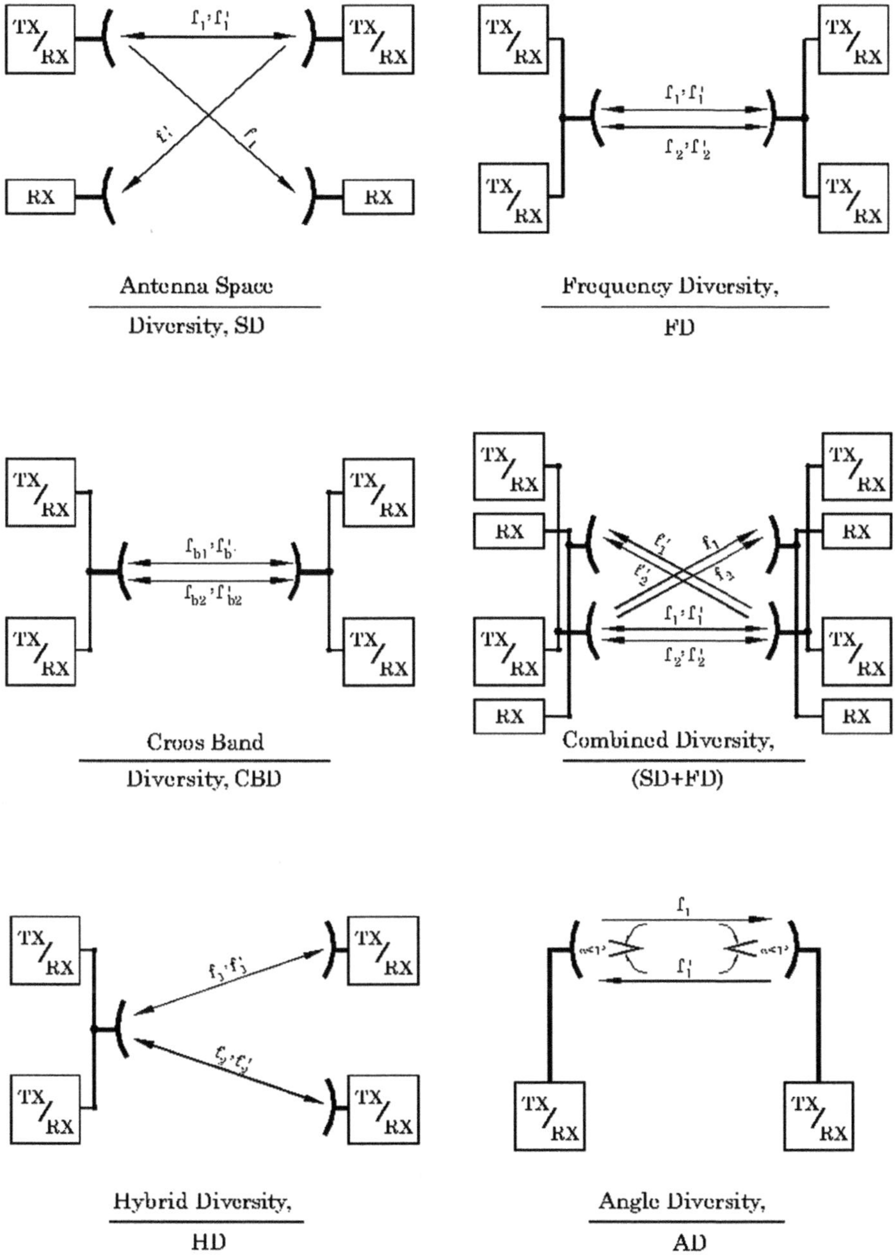

Fig. 9.14 Main types of diversity techniques in LOS microwave links

SD, FD, and AE. In Fig. 9.14, the schematic diagrams of the main variety of improvement techniques are presented for which some details are given in the next sections. The selection of these methods depends on various factors such as

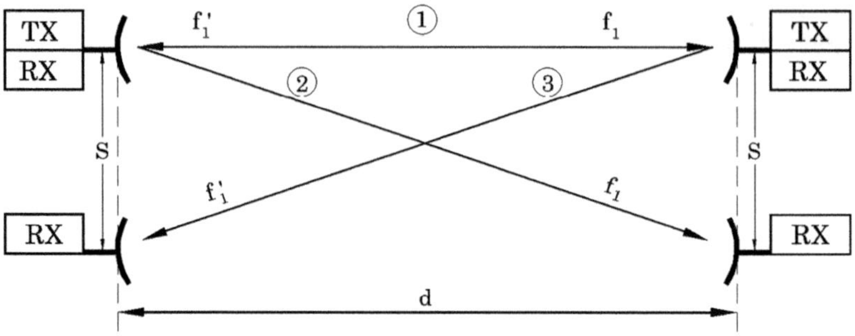

Fig. 9.15 Simple concepts of antenna space diversity in LOS microwave links

congestion of frequencies in the operating zone and limitations exerted by regulatory authorities, the hop distance, atmospheric conditions, improvement requirements, and the system capacity.

9.4.4 Antenna Vertical Space Diversity

One of the most common methods to improve the reception quality of the LOS radio links due to multipath mechanism is antenna vertical space diversity. As indicated in Fig. 9.15, in each of the stations of both ends of the radio hop, one antenna and one RX unit are added to protect the radio link availability against propagation adverse effects.

In space diversity reception, the signal reaches the receiver through more than one transmit/receive antenna path. On loss path, space diversity is usually, in implemented using two antennas at the receiver with a large enough vertical separation to provide two signals in which the impairments due to multipath fading are sufficiently decor related, where impairments are signal distortion and loss of signal power.

The major advantages of antenna vertical space diversity are:

- High improvement factor in radiowaves reception
- Good performance in multipath condition due to anti-reflective effects
- Saving RF channels
- More equipment reliability because of additional receivers

The combination of antenna vertical space diversity and adaptive equalizer results in synergetic effect, that is, overall improvement will be more than the simple product of their individual improvement factors. As indicated in Fig. 9.15, the reason for cancelation of multipath effects in radio hops using antenna space diversity is based on the reception of radiowaves through different routes. These waves are detected by main and space diversity antennas in different phases; thus, the output of the combiner is always improved compared to each output.

The improvement factor related to antenna space diversity is possible within the range of 10–200 where the vertical distance of the antennas is 150–200 times of the carrier wavelength. This method is a suitable choice for the following cases:

- Locations where the frequency congestion does not permit the use of additional RF channels
- Highly reflective path over sea or low-altitude smooth terrain
- High-capacity radio links with long path length

Selecting height of antenna centerlines in both ends of a LOS radio link is a key point to satisfy the radio path clearance criteria; otherwise, the system will not work properly.

For the microwave radio systems with low and medium capacity, it is possible to use the following empirical formula devised by Vigants:

$$I_{\mathrm{SD}} = 100 \times \frac{(S/9)^2 \cdot (f/4)}{d/40} \times \upsilon \times (10^{-4}/L^2) \tag{9.20}$$

where

I_{SD}: Improvement factor
f : Frequency (GHz)
S : Antennas vertical spacing (m)
υ : Ratio of gains of space antenna to antenna the main antenna
L : Safety margin
d : Path length (km)

Noting that $FM = -20 \log L$ and considering the difference of the main and space antennas gain is $V = -20\,\mathrm{Log}\upsilon$ and by applying equivalent switching system efficiency of η, then formula (9.29) is converted into

$$I_{\mathrm{SD}} = \frac{1.2 \times 10^{-3} \times f \times S^2 \times 10^{(FM-V)/10}}{d} \times \eta \tag{9.21}$$

In this formula, the parameters are identical with those of formula (9.29), and for new parameters we may state:

FM : Fade margin (safety margin) in dB
V : Difference of the main and standby antennas gain in dB
η : Switching system efficiency

Validity of (9.20) and (9.21) is limited to the variation range of the main parameters given below:

- Frequency in terms of GHz $2 \leq f \leq 11$
- Distance in terms of km $24 \leq d \leq 70$
- Distance of main and space antennas in m $5 \leq S \leq 15$
- Fade margin in terms of dB $30 \leq FM \leq 50$

In low- and medium-capacity systems for which the above equations are valid, fade margin is taken as flat fade margin (FFM). For digital systems with high or medium capacities where the magnitude of selective fading is considerably immense, empirical work has been performed by different experts and also by ITU-R in recommendation No.P.530 which are available for further references.

Example 9.6. A digital radio link of medium capacity (10∼100 Mb/s) at 7.5 GHz frequency band employs main antennas with 3 m diameter. The path length is 50 km, and fade margin is considered to be 40 dB. To combat undesired selective fading effects, it is intended to use space antennas with 2.4 m diameter at a distance of $S = 200\,\lambda$ from the main antennas. Assuming a switching efficiency of 75% and 2 dB difference in the antenna (3 m and 2.4 m) gains, calculate the improvement factor for this radio link.

Solution. Noting the stated assumptions, the parameters of the link are:

$$f = 7.5\ \text{GHz},\ S = 200\,\lambda,\ FM = 40\ \text{dB},\ V = 2\ \text{dB},\ \eta = 0.75$$

$$\lambda = C/f \implies \lambda = 4\ \text{cm} \implies S = 8\ \text{m}$$

Equation relation (9.21) yields:

$$I_{\text{SD}} = (1.2 \times 10^{-3} \times 7.5 \times 8^2 \times 10^{(40-2)/10} \times 0.75)/(50)$$

$$\implies I_{\text{SD}} = 54.5$$

■

9.4.5 *Frequency Diversity of Radio Channels*

Frequency diversity of RF channels is another improvement technique to compensate outage of radio links and is considered by the planners and widely used for locations where there is no frequency congestion. In the frequency diversity techniques, the same information is transmitted over more than one radio channel.

This technique in addition to improving propagation outage, due to using additional transmitter and receiver at each station, provides more equipment availability leading to lower outage because of the equipment failure. Figure 9.16 shows an (1 + 1) frequency diversity radio link with RF channel frequency spacing equal to Δf.

The main features of this technique are:

- Cost-effective method
- Average improvement factor
- Increased of equipment availability (TX and RX units)

In case of using several main RF channels (N) to convey required traffic, by adding one additional RF channel, the system is upgraded to frequency diversity

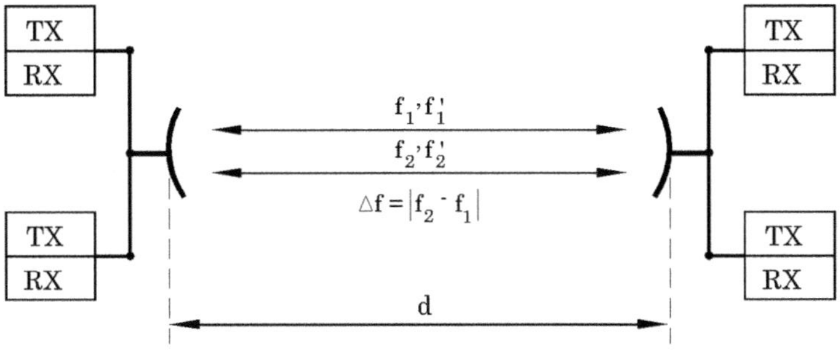

Fig. 9.16 Simple concepts of frequency diversity in LOS microwave links

with (N + 1) configuration. By allocating priority to the most important RF channel, a good improvement is available for the protected channel.

In order to calculate the improvement factor for frequency diversity of RF channels, it is possible to use the following empirical formula for analog and digital systems having low or medium capacity:

$$I_{\text{FD}} = \frac{80}{d} \cdot \frac{\Delta f}{f^2} \cdot 10^{FM/10} \tag{9.22}$$

where:

I_{FD}: Improvement factor of frequency diversity
d: Path length in km
f: Main RF frequency channel in GHz
FM: Safety margin (fade margin) in dB
Δf : Frequency difference of main and standby RF channels in GHz

For systems having low and medium capacity, the fade margin value must be set equal to FFM, and for systems with high capacity, the value of FM must be set to the average of FFM plus NFM (net fade margin), that is, FM = (FFM + NFM)/2. In (N + 1) FD systems, the amount of improvement factor can be calculated using the following equation:

$$I_{FD(N+1)} = I_{FD(1+1)}/K\,, \quad 1 < K < 2 \tag{9.23}$$

In (9.22), f should be replaced by f_{eq} when several f are designated to RF channels.

As it may be concluded from (9.22), the amount of I_{FD} depends on frequency spacing of RF channels denoted by Δf.

Frequency spacing should be according to the network frequency planning based on the RF channel arrangement presented by F-series of ITU-R recommendations. This means that frequency spacing of RF channels is not arbitrary and depends on frequency band, channel bandwidth, traffic capacity per channel, and observing the competent regulations/recommendations.

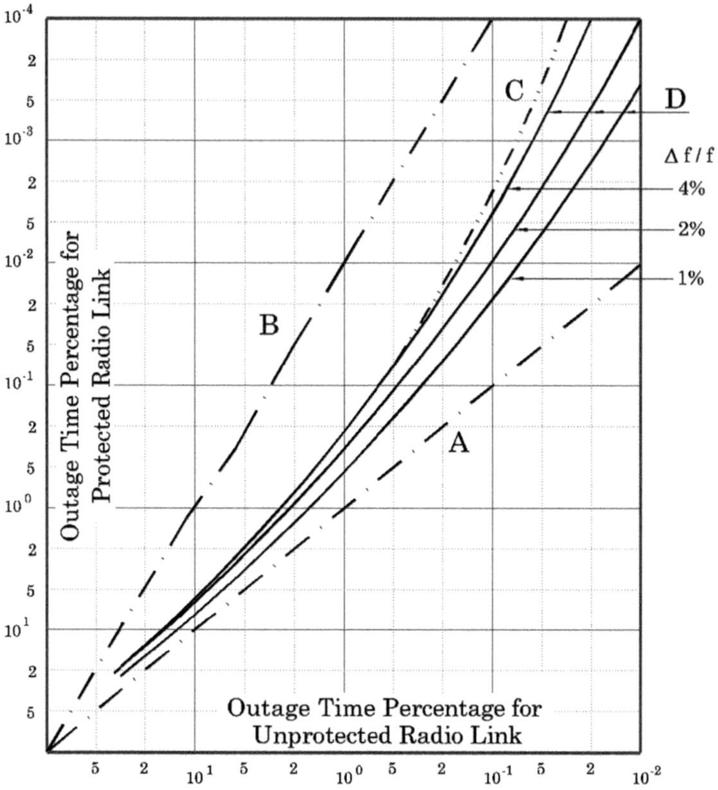

Graph A **:** Outage time percent limit for unprotected radio link

Graph B **:** Outage time percent limit for full protected radio link

Graph C**:** Outage time percentage for protected radio link with antenna space diversity (S ⩾ 150λ)

Graph D **:** Outage time percentage for protected radio link with frequency diversity in terms of Δf/f

Fig. 9.17 Percentage of outage time of radio link protected link versus unprotected link

In countries or places where the fixed radio networks are not used extensively, utilization of frequency diversity is a common practice, whereas in advanced countries, due to immense use of this type of networks and limitations in the frequency bands, a great effort has made other protection systems instead of frequency diversity.

Figure 9.17 provides typical graphs based on numerous experiment results implemented by different countries which is reflected in ITU-R report. By using this figure, it is possible to find the improvement factor of a LOS radio link by applying the following steps:

Step 1: Calculate the outage time percentage for the unprotected link of interest using a suitable procedure.
Step 2: Determine the outage time percentage protected link based on the used diversity technique.
Step 3: Calculate improvement factor using (9.19).

ITU-R has introduced a complicated technique but at the same time a more precise way of calculating the effects of frequency diversity in the recommendation ITU-R, P.530.

Example 9.7. Frequency diversity in (2 + 1) configuration is used to improve outage time of a radio link with 30-km path length working at 13 GHz including 40 dB as fade margin. Frequency planning is based on the following relations, and RF channels 1, 3, and 5 are selected as main and protection channels, all without any mutual correlation, calculate the improvement factor of frequency diversity for each of them.

$$f_n = f_0 - 259 + 28n, \quad f_0 = 13{,}000\ \text{MHz}$$
$$f'_n = f_0 + 7 + 28n, \quad n = 1, 2, \ldots, 8$$

Solution.

$$\Delta f = |f_3 - f_1| = 56\ \text{MHz}$$
$$(\Delta f)' = |f_3 - f_5| = 56\ \text{MHz}$$

Noting relation (9.22) yields:

$$I_{\text{FD}} = (I_{\text{FD}})' = \frac{80}{30} \times \frac{0.056}{(13)^2} \times 10^4 = 8.84$$ ■

9.4.6 *Cross Band Diversity*

This technique for the protection of main radio channel(s) is based on using one radio channel on a different frequency band. As an example, the bands of 6 GHz and 11 GHz or the bands of 2 and 7 GHz may be used for this purpose. Noting that the possibility of severe fading to occur simultaneously in two different bands is much less than one frequency band (this technique has an improvement factor of around 50–150). According to the Fig. 9.18, auxiliary TX and RX units are used in this technique. The basic advantages of this method are:

- High improvement factor
- Increasing of equipment availability using more radio units
- Prevent system being severely faded on high-frequency bands due to intense raining

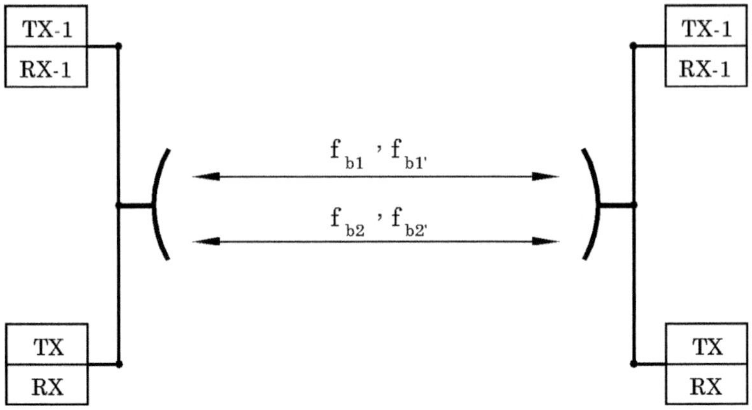

Fig. 9.18 Simple concept of cross band diversity in LOS microwave link

A unique feature of the cross band diversity (CBD), is its ability to combat against deep fading due to the intense rain. In this situation, neither antenna space diversity nor frequency diversity can be an effective countermeasure. As an example, using 6-GHz and 11-GHz bands for CBD method will provide protection against intense rain. When signal attenuation in 11-GHz band is high, its value in 6-GHz band is quite low. It should be noted that during the rainfall, the atmosphere is well mixed with very low probability of deep fading.

9.4.7 Combined Diversity

Combined antenna space and RF channel frequency diversity technique is employed to protect high-capacity radio links against propagations and adverse effects. This method is used where high improvement factor is required.

Figure 9.19 illustrates a simple concept of a radio link using (SD + FD) diversity in a (3 + 1) configuration. Since the (SD + FD) method is expensive, therefore, it may be used only on long routes, high capacities, and harsh climatic conditions where occurrence probability of deep fading is high. The amount of improvement factor of this configuration is approximately equal to the product of improvement factors related to SD diversity and FD diversity, that is,

$$I_{(\mathrm{SD+FD})} \approx I_{\mathrm{SD}} \cdot I_{\mathrm{FD}} \tag{9.24}$$

This method includes advantages of both frequency diversity and space diversity. The clearance criterion, set for antenna space diversity, should be satisfied for the combined diversity as well.

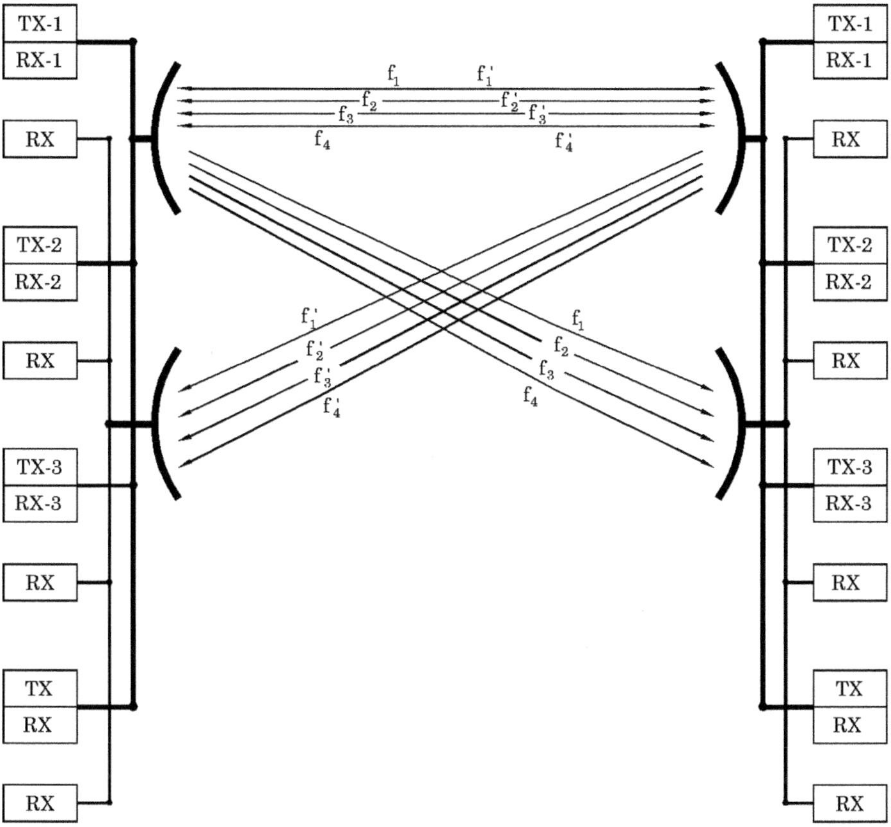

Fig. 9.19 Combined frequency and space diversity (FD + SD) in (3 + 1) radio link configuration

Example 9.8. In the case of using frequency diversity to enhance the improvement factor of the example (9.6) with frequency spacing of $\Delta f = 56\,\text{MHz}$, find the overall improvement factor value for the specified (SD + FD) configuration.

Solution.

$$\Delta f = 56\ \text{MHz},\quad f = 7.5\ \text{GHz}\ ,\quad d = 50\ \text{km}$$

$$I_{\text{FD}} = \frac{80}{50} \times \frac{0.056}{(7.5)^2} \times 10^4 \times 0.75 \approx 12$$

$$I_{(\text{SD+FD})} = I_{\text{SD}} \cdot I_{\text{FD}} = 54.5 \times 12 = 654$$

■

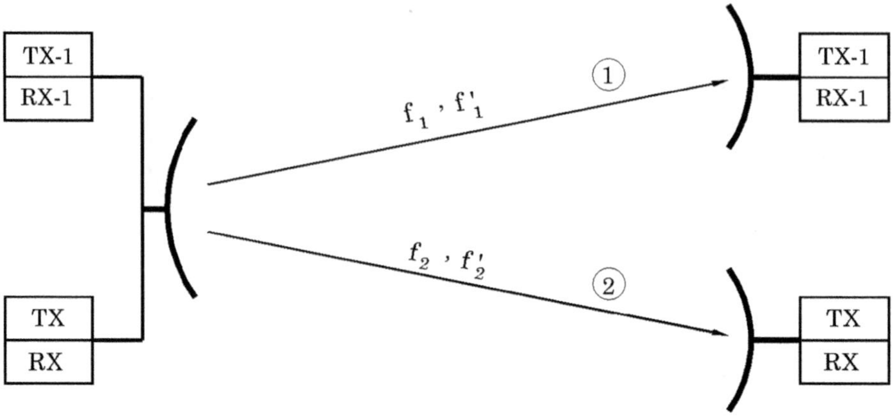

Fig. 9.20 Simple concept of hybrid diversity in LOS microwave link

9.4.8 Hybrid Diversity

Figure 9.20 illustrates the concept of hybrid diversity. As shown, in one station both radio units with different RF channels are connected to a single antenna. The other station includes two antennas arranged in vertically spaced configuration each connected to only one radio unit. Assuming I_{SD} and I_{FD} as the space and frequency diversities related to the given link, then the improvement factor of hybrid diversity can be expressed by:

$$I_{\mathrm{HD}} \approx I_{\mathrm{SD}} + I_{\mathrm{FD}} \approx I_{\mathrm{SD}} \tag{9.25}$$

In this case, the radio path clearance criteria should be provided for both routes 1 and 2. The hybrid diversity is a good method, particularly in relay stations where antenna wind load is a dominant factor.

To minimize overall cost, a single antenna is used in repeater station with high-speed winds, while two antennas are installed in terminal station. This approach will decrease total expenses for high self-supporting towers and related civil works for their foundations.

9.4.9 Angle Diversity

Angle diversity, denoted by AD, is initially used to improve radiowaves reception in the troposcatter systems. It was employed for digital microwaves systems from mid-1980s. Concerns for the performance of digital radios, for which signal distortion

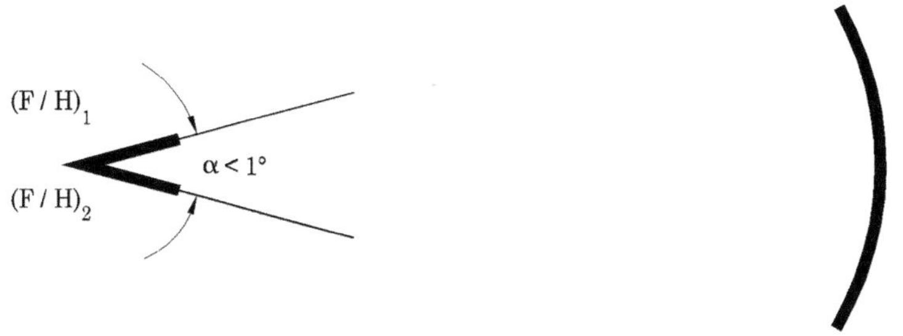

Fig. 9.21 Simple concept of angle diversity in LOS microwave link

is the dominant propagation impairments, have led to the introduction of diversity methods that rely on the nonuniform structure of the incident electromagnetic field near the main receiving antenna instead of large spatial separations to decorrelate the signal impairments.

In this method also sometimes called pattern diversity, the diversity signal is derived from a second antenna or beam that has a different directivity pattern or angular beamwidth in the vertical plane or a different boresight elevation angle. At present time, application of this method has been limited noting the existence of other improvement techniques. The angle diversity in the LOS communications is generally used when it is not possible to utilize SD and FD techniques due to space and/or frequency spectrum limitations.

This diversity method, which can be implemented with antennas that area the same or nearly the same, or with multiple feeds within a single antenna, allows a diversity reception to be added to an existing hop without requiring tower height extension to obtain path clearance for it. As depicted in Fig. 9.21 to fulfill this method by one parabolic antenna with dish reflector along with 2 feed horns are required.

The axis of feeder systems makes an angular spacing with a magnitude of around 1° or less on the vertical plane. This approach can be more effective when the fading relative to dispersive fading is dominant component and play a significant role with respect to the flat fading. Fading caused by dispersion generally occurs in coastal areas of warm seas having severe impacts on high-capacity systems in summer.

The range of improvement factor of angle diversity may reach more than 20. This value depends on terrain geometry of radio path, area climatic condition, RF channel frequency, and size of used antenna.

To achieve an optimum improvement factor, the most suitable and practical approach is the adjustment of antenna based on local statistics and experimental test results.

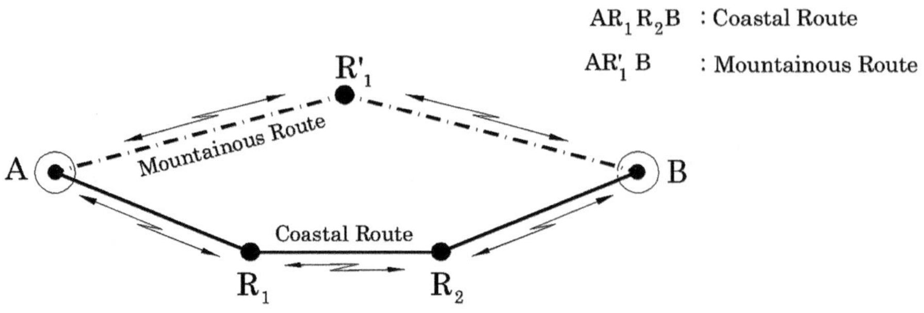

Fig. 9.22 Route diversity concept in a radio network

9.4.10 Route Diversity

Another improvement technique is to exchange of the radiowaves between the two terminals A and B, as depicted in Fig. 9.22, based on route diversity. The midway repeaters between A and B terminals are selected in a manner that the radiowaves propagate through two distinct and separate routes for which coastal and mountainous routes with great difference in geographic and environmental conditions is a good example. The former consists of low-altitude and smooth terrain with small K-factor, while the latter consists of highlands with standard K-factor.

In this approach, different routes are so selected to ensure reliable communications between terminals A and B. Indeed, when route diversity is used, fading mechanisms are completely different and independent ($\rho \approx 0$) for two routes. The route diversity, in addition to improving the adverse effects of radiowaves propagation, can increase the network traffic capacity including better service quality.

9.4.11 Adaptive Equalizer

Various experimental tests reveal that flat fading improvement techniques are not sufficient in the high-capacity digital radio networks to enhance the system performance. The effects of multipath fading in these networks depend on many factors such as the type of modulation, bandwidth, capacity, path length, and atmospheric conditions. Especially under these conditions, the selective fading has major impact on longer routes and for digital radio networks with capacities higher than 10 Mb/s, for which using adaptive equalizers is a suitable countermeasure.

In principle, fluctuation of 5 dB in the signal amplitude, even under conditions that a good SNR exists, will impair the digital system performance. For this reason, in such cases the value of net fade margin (NFM) is used in place of flat fade

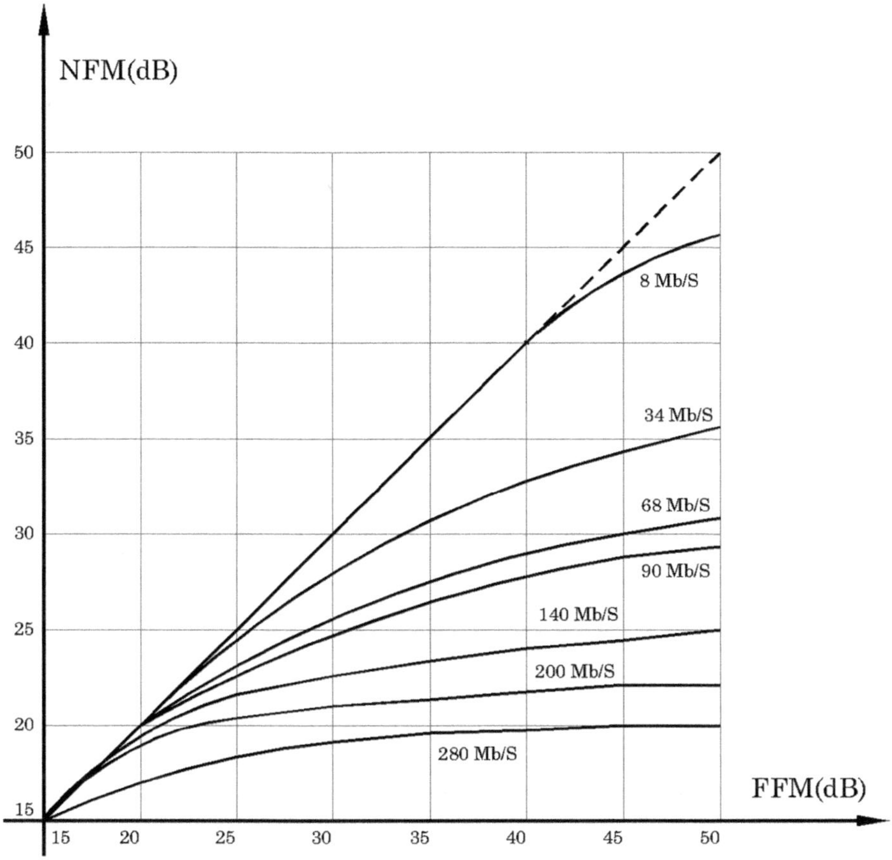

Fig. 9.23 NFM versus FFM for typical RF channel capacity

margin (FFM). Figure 9.23 typically demonstrates the relation between NFM and FFM for various capacities of a system with quadrature phase-shift keying or QPSK modulation for BER $= 10^{-5}$ at 11-GHz band for a path length of 50 km and without using any kind of improvement techniques. Adaptive equalizers, through sampling of different portions of the received signal spectrum and proper amplification of them, produce fairly a uniform input for demodulation stage, resulting in a great reduction of the system outage time.

Combining adaptive equalizers with space diversity produce (SD + AE) diversity system, along with a proper switching device, will make an efficient tool with synergetic effects to decrease the propagation outage time of the radio digital systems for very long routes and high capacities.

9.4.12 ITU-R Method for Digital Radio Systems

Radio section of ITU through recommendation P.530 has presented specific methods to calculate outage time percentage of digital radio link using improvement techniques of SD, FD, AD, and (SD + FD).

9.4.12.1 Space Diversity for Low-Capacity Systems

To calculate the improvement factor related to the vertical spacing between antennas in systems with low or medium capacities, the following relation is recommended:

$$I = [1 - \exp(-0.04S^{0.87} \cdot f^{-0.12} \cdot d^{0.48} \cdot P_0^{-1.04})] \times 10^{(FM-V)/10} \tag{9.26}$$

$$V = |G_1 - G_2| \tag{9.27}$$

The components and pertaining units in the above relations are:

FM : Fade margin without SD in dB
P_0 : Occurrence percentage of multipath fading
S : Distance between antennas axes in m
f : Radio channel frequency in GHz
d : Path length in km
G_1, G_2: Antennas gain in dB_i

Relation (9.26) is valid for the ranges specified below:

- 43 km $\leq d \leq$ 240 km
- 2 GHz $\leq f \leq$ 11 GHz
- 3 m $\leq S \leq$ 23 m

9.4.12.2 High-Capacity Systems

In high-capacity digital systems, selective fading is considerable and cannot be disregarded. For the following cases:

- Antenna space diversity, SD
- Frequency diversity of radio channels, FD
- Angle diversity arrival radiowaves in antennas, AD
- Combined space and frequency diversity (with 2 RX units)
- Combined space and frequency diversity (with 4 RX units)

ITU-R noting the correlation factor between different reception channels, has recommended complicated and specific methods in P-series (P.530) which may be referred to for more details.

9.5 Improvement Techniques for Reception of Satellite Radiowaves

9.5.1 *Fading and Its Cause*

Fading occurs in satellite communications due to unstable and temporary phenomena among which the following are outstanding within dedicated frequency bands:

- Rain
- Abrupt atmospheric changes
- Multipath routes

The fading level, in addition to the frequency of the radiowaves, depends on intensity of its constituent components. In contrast to the terrestrial LOS radio systems, utilization of improvement techniques in the satellite communications does not have wide applications. Considering the nature of propagation, adverse effects in the satellite communications, and their frequency dependence, it may be suitable to use one the following improvement techniques as a countermeasure:

- Cross band diversity, CBD
- Site diversity

Normally in satellite systems, application of site diversity has a major role in improving the quality of satellite communications for which certain techniques contributed by ITU-R for the calculation.

9.5.2 *Site Diversity of Ground Stations*

9.5.2.1 Introduction

One of the main reasons for fading of satellite radiowaves is severe raining, particularly on frequencies higher than 10 GHz. The geographic span of raining is normally limited, and it does not exceed from a radius of 50 km. As shown in and noting Fig. 9.24, by selection of two ground stations in different locations with distance d, it may be possible to make a great improvement to prevent reduction of signal level or quality of the ongoing communications.

Severe raining or the other acting elements can create excessive losses up to 20 dB or more which, in case of safety margin being low, causes full outage of the satellite radio link. In such a situation through handling of signals from the second station, not subject to a severe rain, it may be possible to establish the necessary communications with acceptable reliability and quality.

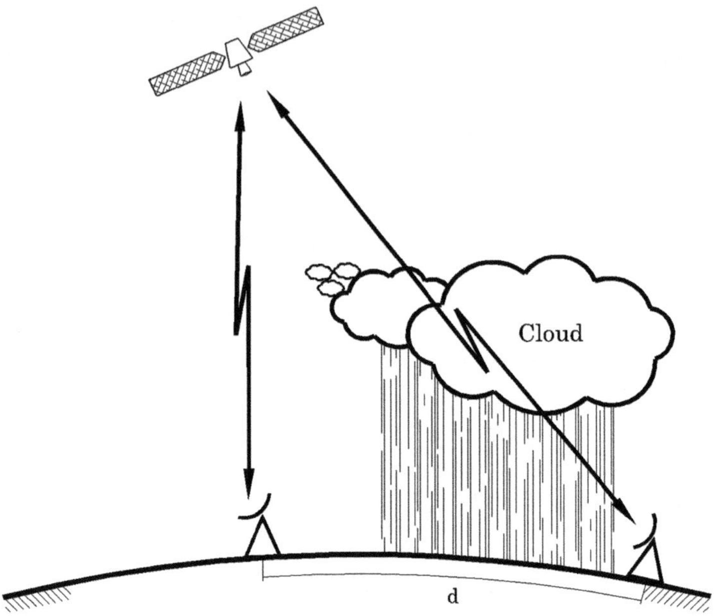

Fig. 9.24 Site diversity in satellite ground station

To express the effects of the diversity, the improvement factor, denoted by I, is used for raining with occurrence probability of less than 0.1%. Normally for time percentages of more than 0.1%, the rain intensity is low, and its losses may be ignored.

9.5.2.2 Improvement Factor

The improvement factor resulting from site diversity can be expressed as follows:

$$I = \frac{P_1}{P_2} \tag{9.28}$$

In the above relation, all the parameters are given below:

I : Link improvement factor
P_1: Time percentage of signal attenuation at a single ground station (relative to the reference level)
P_2: Time percentage of signal attenuation occurring at two ground stations simultaneously (relative to the reference level)

Value of I depends on the following factors:

- Distance of ground stations
- Working frequency

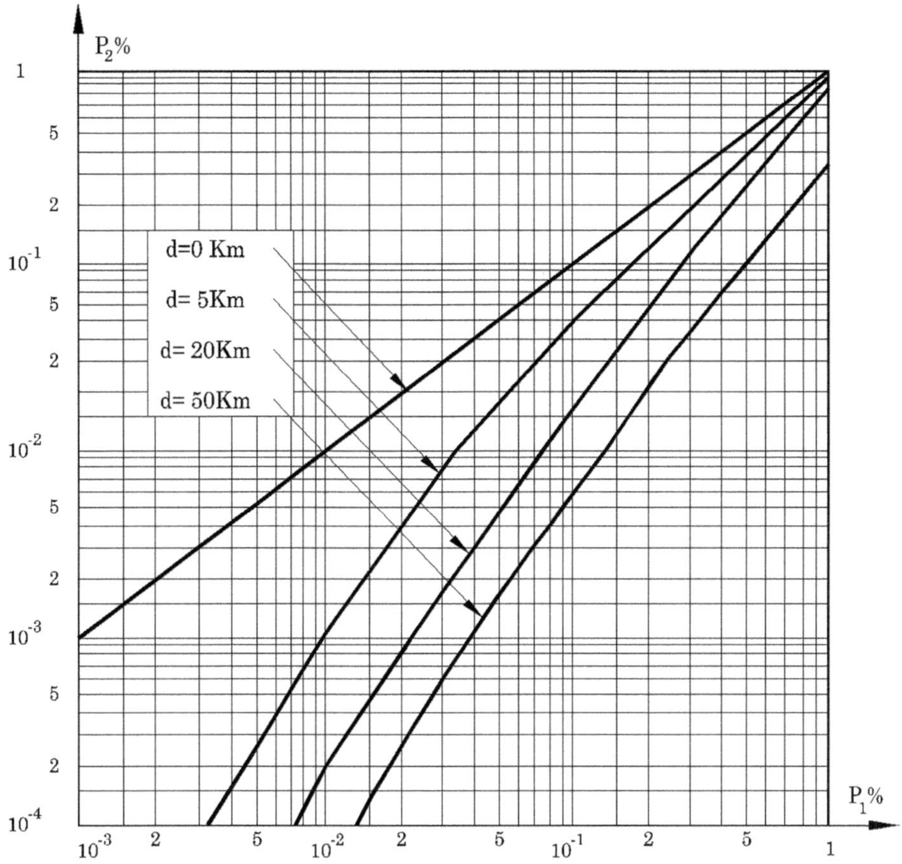

Fig. 9.25 Typical site diversity improvement for satellite ground station

- Elevation angle of the radiowaves
- Site locations

Figure 9.25 indicates the relation between P_1 and P_2 based on numerous test results from which it is possible to obtain the value of I for distance d. For the improvement factor indicated in the recommendation ITU-R, 618, the following relation is recommended:

$$I = \frac{P_1}{P_2} = \frac{1}{1+\beta^2}\left(1+\frac{100\beta^2}{P_1}\right) \approx 1+\frac{100\beta^2}{P_1} \tag{9.29}$$

In the above relation, P_1 and P_2 are the respective defined time percentages, and β is also a parameter which is related to the characteristics of satellite orbit. The right-hand side approximation of the above relation is also acceptable noting that the values of β^2 are small values less than one. Based on various measurements conducted on frequency span of 10–20 GHz especially in the satellite Ku band (11–13.6 GHz), it was specified that the value of β^2 depends on the distance of two ground satellite terminals as given below, and elevation angle does not have any serious effect on it:

$$\beta^2 = 10^{-4} \cdot d^{1.33} \quad (d \text{ in km}) \tag{9.30}$$

Example 9.14. The probability of the link outage in a ground station is less than 0.02%. Another station at a distance of 50 km is also used as a site diversity to enhance the above-mentioned outage time percentage, find:

1. Improvement factor using Fig. 9.25.
2. The same factor using ITU-R recommended expression.

Solution. 1.

$$(d = 50 \text{ km}, P_1 = 2 \times 10^{-2}\%) \ , \ P_2 = 2.5 \times 10^{-4}\%$$

This means that the outage probability has decreased to 0.00025% using site diversity in the satellite ground station, hence,

$$I = \frac{P_1}{P_2} = \frac{0.02}{0.00025} = 80$$

2.

$$P_1 = 0.02 \ , \ d = 50 \text{ km} \implies \beta^2 = 1.82 \times 10^{-2}$$

$$I' \approx 1 + \frac{100 \times 1.82 \times 10^{-2}}{0.02} = 92$$

■

9.5.2.3 Site Diversity Gain

Another method used to express the radiowaves reception improvement in satellite communications is entitled as site diversity gain designated by G in decibels. ITU-R represents in its recommendation P.618 different gains for main affecting parameters such as antenna spacing, frequency spacing, azimuth, and elevation angles. The following notations are used in the recommended equations:

- d: Distance of two ground stations in km
- A: Rain loss of radiowaves on the route between satellite and main ground station in dB
- f: Frequency of the satellite link in GHz
- θ: Elevation angle of the radiowaves in degree
- Ψ: Azimuth angle of the radiowaves route relative to the interconnecting line of the two terminals in degree.

Noting the explanations given above, the step-by-step procedure for calculating G is presented below:

Step 1: Calculate the effect of the site location distance of the two terminals by the following relation:

$$G_d = a(1 - \mathrm{e}^{-b.d}) \tag{9.31}$$

$$a = 0.78\,A - 1.94(1 - \mathrm{e}^{-0.11A}) \tag{9.32}$$

$$b = 0.59(1 - \mathrm{e}^{-0.1A}) \tag{9.33}$$

Step 2: Calculate the effect of frequency by the following relation:

$$G_f = \mathrm{e}^{-0.025\,f} \tag{9.34}$$

Step 3: Calculate the effect of elevation angle of the radiowaves route from the following relation:

$$G_\theta = 1 + 0.006\,\theta \tag{9.35}$$

Step 4: Calculate relative effect of the terminals position and azimuth of the radiowaves route:

$$G_\Psi = 1 + 0.002\,\Psi \tag{9.36}$$

Step 5: Finally the gain of site diversity denoted as G is obtained from the following relation:

$$G[\mathrm{dB}] = G_d \cdot G_f \cdot G_\theta \cdot G_\Psi \tag{9.37}$$

Comparing the results obtained from the above method with the tested values shows a difference of 0.14 dB for the average value and 0.96 dB for standard deviation which may cause an overall effective r.m.s error of less than 0.97 dB.

9.5.2.4 Reliability of Satellite Link

Reliability of a satellite link normally is expressed in the percentage of time and is an indication of desirable communications during a period of time. For example, a link of 99.8% reliability has a capability to provide acceptable service with quality not less than a reference criterion in 99.8% of operating time.

Desirable communications in digital systems are specified with the BER level and in analog systems with the SNR level. In order to achieve the required reliability, it is essential to consider some amount of safety margin for the link. The safety margin value will depend on the required reliability and also working

Table 9.1 Safety margin versus satellite link reliability for typical RF channel frequencies

Reliability (%)	Annual outage (in hours)	Required safety margin 11 GHz	20 GHz	30 GHz
99.5	44	1	3	6
99.9	8.8	3	10	20
99.95	4.4	5	20	30
99.99	0.88	15	30	–

The above values are valid for elevation angles of ground station antenna in the range of 30° ~ 50° based on 30 experimental tests results in temperate region 5.

frequency. Table 9.1 indicates the amount of essential safety margin to meet reliability requirements in terms of different frequencies.

Example 9.15. 1. Calculate the rain loss for an antenna elevation angle of 40° and a rain intensity of 25 mm/h for the frequencies of Ku = 12 GHz and K =20 GHz.
2. If the occurrence probability of the specified raining is less than 0.01% of time, what are the values of FM and related maximum amount of monthly link outage?

Solution. 1. Using Table 3.2 of the Chap. 3, the following result is obtained:

$$f_1 = 12\text{ GHz} \implies a = 0.0215\ ,\ b = 1.136$$

$$(A_{\text{rs}})_1 = aR^b = 0.833\text{ dB/km}$$

$$Le/Ku \xrightarrow{Graph(3.25)} 4\text{ km} \implies L_R = Le \times (A_{\text{rs}})_1 = 3.33\text{ dB}$$

$$f_2 = 20\text{ GHz} \implies a = 0.072\ ,\ b = 1.097$$

$$(A_{\text{rs}})_2 = aR^b = 2.6\text{ dB/km}$$

$$Le/Ku \xrightarrow{Graph(3.25)} 4\text{ km} \implies L_R = Le \times (A_{\text{rs}})_2 = 10.4\text{ dB}$$

The above results do not comply with the figures of Table 9.1 due to different conditions such as rain intensity.

2.

$$f_1 = 12\text{ GHz} \implies FM_1 = 3.33\text{ dB}$$

$$t_1 = 0.01 \times 10^{-2} \times 30 \times 24 \times 60 = 4.32\text{ min.}$$

$$f_2 = 20\text{ GHz} \implies FM_2 = 10.4\text{ dB}$$

$$t_2 = 0.01 \times 10^{-2} \times 30 \times 24 \times 60 = 4.32\text{ min.}$$

■

9.6 Improvement Techniques for Reception of Troposcatter Radiowaves

Troposcatter system performance is impaired severely during deep fading of the related radiowaves because of atmospheric adverse effects. For this mode of propagation, the common improvement techniques are:

- Antenna space diversity
- RF channel frequency diversity
- Antenna feed angle diversity

In the trans-horizon systems, using of antenna space, and angle and frequency diversity techniques either individually or in combination, to obtain sufficiently decorrelated replicas of the transmitted signal is a common practice. Based on the experimental test results, ITU-R has suggested some applied formulas for calculation of improvement factors in recommendation P.617.

In addition to the said techniques for reception of troposcatter radiowaves, a proper sitting of transmission links requires some care to avoid nearby obstructions. Antennas should be adjusted slightly over the horizon. The optimum elevation angle is a function of the radio path and atmospheric conditions which normally lies within about 0.2–0.6 antenna beamwidths above the horizon.

9.6.1 *Antenna Space Diversity*

Figures 9.26 and 9.27 illustrate this method using two antennas in each terminal. Since troposcatter antennas are generally large parabolic ones in addition to their vertical spacel diversity with different heights on the tower, they can be erected on the ground with horizontal distance and having short structures. The main

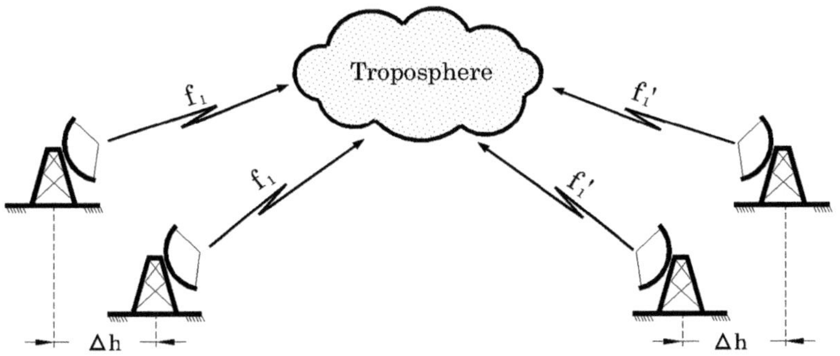

Fig. 9.26 Horizontal antenna space diversity concept in troposcattering radio link

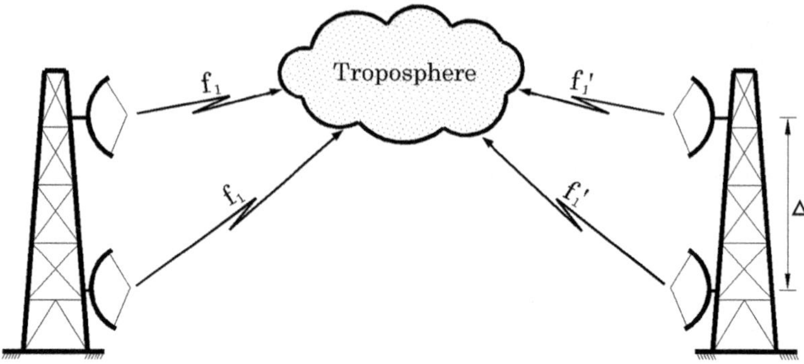

Fig. 9.27 Vertical antenna space diversity concept in troposcattering radio link

reason for using horizontal spacing is the complexity and very high cost of structures supporting large antennas in highlands and particularly during severe winds and gusts.

Selection of either horizontal or vertical spacing depends on the dimensions of the antennas, economic issues, and land availability. For frequencies above 1 GHz, the antenna spacing can be calculated using the following relations:

$$\Delta h = 0.36\sqrt{D^2 + 4I_h^2} \tag{9.38}$$

$$\Delta V = 0.36\sqrt{D^2 + 4I_V^2} \tag{9.39}$$

In the above relation, all of the components and their related units are described below:

Δh: Horizontal distance between antennas in m
ΔV: Vertical distance between antennas in m
D : Diameter of the antennas in m (if different antennas are used, larger antenna diameter should be used)
I_h : Empirical horizontal scale length, $I_h = 20\,\text{m}$
I_V : Empirical vertical scale length, $I_V = 15\,\text{m}$

9.6.2 *Frequency Diversity*

Figure 9.28 illustrates this method arranged with two TX/RX systems using one common antenna at each terminal. Each RF channel utilizes an independent carrier frequency at each terminal and with a proper frequency separation between them.

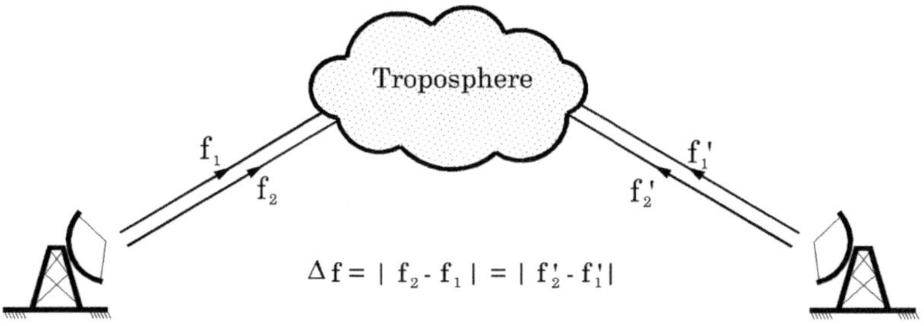

Fig. 9.28 Frequency diversity concept in troposcattering radio link

For troposcatter radio links where it is desired to employ frequency diversity, an adequate separation Δf is required for carrier frequencies greater than 1,000 MHz:

$$\Delta f = |f_2 - f_1| = |f_2' - f_1'| \tag{9.40}$$

$$\Delta f = (1.44f/\theta d)\sqrt{D^2} \tag{9.41}$$

All of the components and their pertaining units indicated in the above relations are expressed below:

Δf: Frequency separation of RF channels in MHz
f : Reference frequency of the selected band in MHz
θ : Scattering angle in mrad
d : Diameter of the antennas in m
I_V : Empirical vertical scale length, $I_V = 15$ m

Example 9.16. In a troposcatter system at 2-GHz frequency band, the (SD + FD) improvement technique is used. The antenna's diameter is 10 m, $d = 200$ km, and $\theta = 30$ mrad.

1. Calculate the minimum horizontal or vertical distance of antennas and indicate which technique is preferable.
2. Determine suitable radio channel separation for a frequency diversity reception.

Solution. 1.

$$\Delta h = 0.36\sqrt{10^2 + 4 \times 20^2} = 14.84 \text{ m}$$

$$\Delta V = 0.36\sqrt{10^2 + 15^2} = 6.49 \text{ m}$$

Since the diameter of the antennas is 10 m, for vertical spacing and using one tower, it will not be possible to mount them with a distance of 6.49 m and requires to select more than 10 m between antenna center lines. Besides, installing two antennas with 10 m diameter each on such heights requires very strong supporting structure and very high cost of installation. Thus, if adequate area is available in the site, horizontal space diversity is preferable. Otherwise, it is better to use angle diversity in place of SD.

2. The least RF channels frequency separation is given as:

$$\Delta f = \left(1.44 \times \frac{2000}{30 \times 200}\right)\sqrt{10^2 + 15^2} = 8.65\ \text{MHz}$$

Using of more frequency separation between main and protection channels will provide more improvement factor. ■

9.6.3 *Antenna Feed Angle Diversity*

Selecting two antenna feed systems with an elevation angle difference of $\Delta\theta_r$ creates an improvement in the reception of troposcatter radiowaves. The elevation angle difference of antenna feed systems will cause a variation in the troposcattering common volume with a vertical distance in the troposphere layer.

9.7 Exercises

Questions

1. Explain diversity reception of radiowaves and determine its common types for point to point and point to multipoint, satellite, and mobile radio systems.
2. Explain combination techniques of the received signals from various channels and specify which one provides higher gain.
3. Define and discuss the improvement factor in the reception of radiowaves using suitable graphs.
4. Name the types of improvement techniques in mobile radio systems.
5. Discuss application of site diversity in fixed stations of mobile radio network suitable to overcome long-term fading. What is the selection criterion of this technique and whether the concerning waves are correlated in this technique or not?
6. In space diversity, evaluate the condition for its effectiveness and state why it is not used in handheld mobile terminals.
7. Define frequency and polarization diversities in mobile radiowaves reception and state the amount of their improvements.
8. Explain how the radiowaves are exchanged between mobile terminal with linear polarization and base station with circular polarization. Determine whether in this type of reception any improvement is produced if the base station antenna is capable of receiving right-hand and left-hand circular waves simultaneously.
9. Define field components and time diversity techniques.

10. Noting Fig. 9.9 of the book, specify and discuss the radio link availability and quality of the received signals in each one of the indicated zones of the road.
11. List the main improvement techniques in terrestrial microwave systems giving a simple sketch to explain their concepts.
12. Evaluate characteristics of different improvement techniques in terrestrial microwave systems in terms of technical, operational, and economical aspects. Prepare a report stating the results and giving their relative improvement factors.
13. Explain why the occurrence probability of fading due to multipath is very low at the time of wind or rain. What is the application of the considered fade margin and whether its amount is always sufficient to overcome short-term fadings or not.
14. Describe the application of cross band diversity reception in the line-of-sight radio links.
15. Using Fig. 9.23, compare the relation of fading on digital systems having different capacities with flat fading and explain your conclusions.
16. Specify conditions for which space diversity is considered a suitable and cost-effective solution.
17. Define angle diversity technique in radiowaves reception giving a simple sketch to illustrate this concept. What kind of radio systems employs this type of improvement technique?
18. What are the main reasons for fading in satellite communications and what are the practical improvement techniques?
19. Discuss about improvement factor of site diversity reception using Fig. 9.25.
20. List diversity techniques used in trans-horizon radio links. Discuss about the advantages and applications of horizontal and vertical antenna space diversity.

Problems

1. Calculate the following values using the graphs of Fig. 9.3.

    (a) Improvement factor for a maximum occurrence probability of $P = 0.01\%$ of time and for $n = 2$ and $n = 3$
    (b) Improvement gain for a reception of 30 dB less than the median value for $n = 2$ and $n = 3$

2. The received power of a radio link for more than 0.1% of time is $P_r = 20\,\text{pW}$ without any protection. By utilizing improvement technique, it will increase to 2 nW. Calculate the improvement factor of the applied technique. Repeat the case if $P_r' = 8\,\text{pW}$ and $P_d' = 1\,\text{nW}$, respectively, for 0.01% of time.

3. Using graphs of Figs. 9.4 and 9.5, find:

   (a) The amount of improvement resulting from combining 2 channels by using of the maximal combiner for time percentages less than 0.1%.
   (b) The value of improvement factor if the channels are correlated with $\rho = 0.8$.

4. Calculate the minimum distance of antennas to achieve a proper correlation coefficient in a manner that it can operate as a space diversity improvement at frequencies of $f_1 = 300\,\text{MHz}$ and $f_2 = 1{,}500\,\text{MHz}$.
5. In order to achieve $\rho = 0.8$ with antennas at a height of 24 m and radio frequency of 850 MHz:

   (a) Calculate the least distance of antennas.
   (b) Obtain the least distance of antennas for 170 MHz frequency.

6. Assuming that the amount of standard deviation of time delays in a mobile radio link using frequency diversity is equal to 0.25:

   (a) Obtain proper frequency separation for radio channels with $\rho_f = 0.4$.
   (b) In case of achieving $I = 4$ which requires a correlation coefficient of 0.1, calculate the least channel separation.

7. Noting graphs of Fig. 9.14, compare and tabulate the results obtained for the main diversity techniques in terms of the number of TX(s), RX(s), antennas, and radio channels.
8. In a digital radio link at 4-GHZ band, similar antennas are used as space diversity. If $S = 12\,\text{m}$, FM $= 36\,\text{dB}$, $d = 40\,\text{km}$, and $\eta = 85\%$, then calculate the improvement factor.
9. Repeat Example 9.6 with the following assumptions.

$$f = 8\text{ GHz}\ ,\ \ S = 200\,\lambda\ ,\ \ FM = 37\text{ dB}$$

$$\eta = 80\%\ ,\ \ AD = 2.5\text{ dB}$$

10. For a link of 40 km length operating at 7.5-GHz band, the frequency separation of the main and diversity channels is 56 MHz. In case of 40 dB fade margin, calculate the value of frequency diversity improvement factor.
11. In a radio link with (2 + 1) configuration, 11-GHz frequency band is used for frequency planning with the following arrangement criteria:

$$f_n = f_0 - 500 + 40n\ \ ,\quad n = 1,\ldots,12$$

$$f_n' = f_0 + 20 + 40n\ \ ,\quad f_0 = 11{,}000\text{ MHz}$$

Assume path length and fade margin are 35 km and 38 dB, respectively. If channels 1 and 11 are used as the main channels and channel 6 is use as protection channel, find:

(a): Frequencies of TX and RX channels
(b): Improvement factor for the main channels (giving priority to the channel No. 1)

(a) To increase reliability of the radio link specified in the previous problem by using SD diversity, similar antennas are used. Then calculate the value of improvement factor. ($S = 200\,\lambda$).
(b) Plot a simple diagram to illustrate the related radio link according to the above assumption and specify the number of RX units.
(c) Noting Fig. 9.19 of the book, determine the number of TX and RX units, antennas, and radio channels.

12. Using graphs of Fig. 9.23 and for 3 values of 30, 35, and 40 dB for flat fading, specify the corresponding net fade margin (NFM) values for capacities of 140 Mb/s and 8 Mb/s and discuss about the results.
13. Noting Fig. 9.25 and 0.1% outage probability for a satellite ground station working at BER = 10E(−4), find:

(a) The maximum simultaneous outage probability when second ground station is employed at a distance of 30 km from the main station.
(b) The minimum distance of the second ground station to reduce the maximum outage probability to 5% of the initial value.

14. In order to obtain propagation reliability of 99.95% in a satellite link at 20 GHz, calculate:

(a) The fade margin of the radio link
(b) Maximum annual outage in terms of minutes

15. Using (SD + FD) diversity is recommended to increase propagation reliability in a troposcatter link. The applicable frequency arrangement is determined by the following relations:

$$f_n = f_0 - 37 + 5n \ , \ \ n = 1,2,\ldots,6$$

$$f'_n = f_0 + 2 + 5n \ , \ \ f_0 = 2,800 \text{ MHz}$$

Assuming that $D = 12$ m, $d = 250$ km, and $\theta = 30$ mrad, calculate:

(a) The least horizontal or vertical distance of antennas in SD situation and evaluate the feasibility conditions of each one.
(b) The least separation frequency of radio channels and specify whether the suggested arrangement can fulfill this condition or not.

Appendix A
Logarithmic System of Units

A.1 Introduction

This appendix deals with the concepts and fundamentals of the logarithmic units and their applications in the engineering of radiowaves propagation. Use of these units, expressing the formulas, and illustrating figures/graphs in the logarithmic systems are common and frequently referred in the book.

A.2 Definition

According to the definition, logarithm of the ratio of two similar quantities in decimal base is called "bel," and ten times of it is called "decibel."

$$\text{bel} = \log_{10}\left(\frac{p}{p_{\mathrm{r}}}\right)$$

$$\text{decibel} = [\text{dB}] = 10\ \log_{10}\left(\frac{p}{p_{\mathrm{r}}}\right)$$

As the above formula suggests, it is a logarithmic scale, and due to the persistent usage of decimal base in the logarithmic units, it is normally omitted from the related symbol for simplicity.

Since values in the logarithmic scale are the logarithm of the ratio of two similar quantities, hence they are dimensionless and measured in the same unit. In other words, the unit of denominator, p_{r}, is the base of comparison and includes the required unit.

To indicate the unit of the base in the logarithmic system of units, one or few characters are used according to the following examples:

dB_w: Decibel unit compared to 1 W power

A. Ghasemi et al., *Propagation Engineering in Radio Links Design*,
DOI 10.1007/978-1-4614-5314-7,

dB_m: Decibel unit compared to 1 mW power
dB_kw: Decibel unit compared to 1 kW power

Example A.1. Output power of a transmitter is 2 W, state its power in dB_m, dB_w, and dB_{kw}.

Solution.

$$P_t[dB_m] = 10 \log \frac{2 \times 10^3}{1\,\text{mw}} = 33\ dB_m$$

$$P_t[dB_w] = 10 \log \frac{2}{1\,\text{w}} = 3\ dB_w$$

$$P_t[dB_kw] = 10 \log \frac{2 \times 10^{-3}}{1\,\text{kw}} = -27\ dB_kw$$

■

A.3 Basic Formulas

Applying the decibel units necessitates the frequent usage of logarithm; thus, it is essential to review the following basic mathematical formulas:

1. $\log_{10}^{a} = P \iff a = 10^P \iff \text{Antilog}_{10} P = a$
2. $\log(a_1 \times a_2 \times \cdots \times a_n) = \log^{a_1} + \log^{a_2} + \cdots + \log^{a_n}$
3. $\log\ a^{(\frac{p}{q})} = \left(\frac{p}{q}\right) \log\ a$
4. $\log(\frac{a}{b}) = \log\ a - \log\ b$
5. $\log\ \sqrt[m]{a} = \frac{1}{m} \log\ a$

A.4 Common Logarithmic Quantities

Most of quantities used in the radio communications are defined in terms of logarithmic units. Main logarithmic units used in this book are indicated in Table A.1.

In addition to the aforementioned cases, all of the quantities mentioned in the sequel are normally defined in dB:

- Polarization loss, PL
- Cross-polarization discrimination Loss, XPD
- Antenna front to back ratio, F/B
- C/N, S/N, E/N
- Gain or loss factors such as FSL, passive reflector gain, feeder loss, and rain loss

Table A.1 Main logarithmic units

No.	Quantity	Base	Sign	No.	Quantity	Base	Sign
1	Gain/loss	Ratio	dB	8	Noise temperature	°K(kelvin)	dB/K
2	Power	Watt	dB_w	9	Bandwidth	Hertz	dB_{Hz}
3	Power	Milliwatt	dB_m	10	Bit rate	b/s	$dB_{b/s}$
4	Power	Kilowatt	dB_{kw}	11	Power flux	Watt/m^2	dB_{w/m^2}
5	Antenna gain	Isotropic	dB_i	12	Noise power	Watt	dB_{KTB}
6	Antenna gain	Dipole	dB_d	13	Field intensity (strength)	Microvolt per meter	$dB_{\mu V/m}$
7	Voltage	Microvolt	$dB_{\mu V}$	14	Boltzmann coefficient	Kelvin/Joule	$dB_{J/K}$

A.4 Principles of Logarithmic System of Units

Rule 1: The logarithmic system of units are dedicated only for scalar quantities. If this unit is used for a vector quantity, it implies the magnitude of the vector quantity.

Rule 2: In the logarithmic system of units, mathematical operations are simplified compared to the ordinary system of units (IS).

Rule 3: The logarithmic scale when compared with the ordinary system of units presents large quantities related to the reference unit in a compact form, while its role for small quantities is vice versa as follows:

$$\begin{array}{lll} 10 \Longleftrightarrow 10\text{ dB} & & 0.1 \Longleftrightarrow -10\text{ dB} \\ 100 \Longleftrightarrow 20\text{ dB} & & 0.01 \Longleftrightarrow -20\text{ dB} \\ 1{,}000 \Longleftrightarrow 30\text{ dB} & & 0.001 \Longleftrightarrow -30\text{ dB} \\ \vdots & & \vdots \\ 10^n \Longleftrightarrow 10n\text{ dB} & & 10^{-n} \Longleftrightarrow -10n\text{ dB} \end{array}$$

In radio communications, most of curves, graphs, and plots are normally illustrated on the logarithmic scaled axis for better presentation and simpler interpretation.

A.5 Advantages of Logarithmic System of Units

Applications of the logarithmic system of units have dominant advantages as specified below:

- Simplicity of the mathematical operations
- Much better presentation
- Better understanding and more simple interpretation

As an example, the following formula indicates the free-space loss in the ordinary (metric) system of units:

$$\mathrm{FSL} = \frac{(4\pi d)^2 \cdot f^2}{C^2}$$

The same formula in the logarithmic system of units is given as below:

$$\mathrm{FSL}\ [\mathrm{dB}] = 92.4 + 20\ \log\ f\ \cdot\ d$$

which is more popular in the technical calculations related to the radiowave propagation. It should be noted that every quantity presented in the logarithmic system of units can be easily transformed into the ordinary (metric) system of units or vice versa.

Example A.2. Equivalent isotropic radiated power of a transmitter is $43\,\mathrm{dB}_w$ (i.e., $\mathrm{EIRP} = 43\,\mathrm{dB}_w$), find

1. Equivalent radiated power in terms of kW.
2. Assuming $37\,\mathrm{dB}_i$ antenna gain, determine antenna directivity in the metric system of units.
3. Transmitter power in watts.

Solution.

1.

$$P_e = \mathrm{Antilog}\left(\frac{43}{10}\right) = 20{,}000\ \mathrm{W}\ \Rightarrow\ P_e = 20\ \mathrm{kW}$$

2.

$$G_\mathrm{t}[\mathrm{dB}] = 37\ \mathrm{dB}_i\ \Rightarrow\ G_\mathrm{t} = \mathrm{Antilog}\left(\frac{37}{10}\right) = 5{,}000$$

3.

$$\mathrm{EIRP}[\mathrm{dB}_w] = P_\mathrm{t}[\mathrm{dB}_w] + G_\mathrm{t}[\mathrm{dB}_i]$$

$$P_\mathrm{t}[\mathrm{dB}_w] = 43 - 37 = 6\ \mathrm{dB}_w\ \Rightarrow\ P_\mathrm{t} = \mathrm{Antilog}\left(\frac{6}{10}\right) = 4\ \mathrm{W}$$ ■

Due to the nature of mathematical operations in the ordinary and logarithmic systems, it is important to know the method to be used for every expression or formula.

Different methods may be employed in technical books and papers. The method which is applied in this book uses brackets [ ] for the logarithmic units as seen in the following examples:

$$P_\mathrm{r}[\mathrm{dB}_w] = P_\mathrm{t}[\mathrm{dB}_w] + G_\mathrm{t}[\mathrm{dB}_i] + G_\mathrm{r}[\mathrm{dB}_i] - \mathrm{FSL}[\mathrm{dB}] - L_\mathrm{m}[\mathrm{dB}]$$

or

$$\text{EIRP}[\text{dB}_m] = P_\text{t}[\text{dB}_m] + G_\text{t}[\text{dB}_i] - L_t[\text{dB}]$$

or

$$L_\text{bf}[\text{dB}] = P_\text{t}[\text{dB}_w] - E[\text{dB}_{\mu\text{V/m}}] + 20\ \log\ f(\text{GHz}) + 167.2$$

Appendix B
ITU-R Recommendations P-Series

Radiowave propagation	
P.310	Definitions of terms relating to propagation in non-ionized media
P.311	Acquisition, presentation, and analysis of data in studies of tropospheric propagation
P.313	Exchange of information for short-term forecasts and transmission of ionospheric disturbance warnings
P.341	The concept of transmission loss for radio links
P.368	Ground-wave propagation curves for frequencies between 10 KHz and 30 MHz
P.369	Reference atmosphere for refraction Note-Suppressed on 24/10/97 (RA-97)-This Recommendation has been replaced by Rec.ITU-R P.453-6
P.370	VHF and UHF propagation curves for the frequency range from 30–1,000 MHz. Broadcasting services Note-Suppressed on 22/10/01 (CACE/233)
P.371	Choice of indices for long-term ionospheric predictions
P.372	Radio noise
P.373	Definitions of maximum and minimum transmission frequencies
P.434	ITU-R reference ionospheric characteristics and methods of basic MUF, operational MUF and ray-path prediction Note-Suppressed on 24/10/97 (RA-97). This Recommendation has been replaced by Rec.ITU-R P.1239 and ITU-R P.1240
P.435	Sky-wave field-strength prediction method for the broadcasting service in the frequency range 150–1,600 kHz Note-Suppressed on 20/10/95 (RA-95)
P.452	Prediction procedure for the evaluation of microwave interference between stations on the surface of the Earth at frequencies above about 0.7 GHz
P.453	The radio refractive index: its formula and refractivity data Note-This Recommendation replaces Rec. ITU-R P.369-6
P.525	Calculation of free-space attenuation
P.526	Propagation by diffraction draft revision P.526-9(08/05)
P.527	Electrical characteristics of the surface of the Earth
P.528	Propagation curves for aeronautical mobile and radionavigation services using the VHF, UHF, and SHF bands
P.529	Prediction methods for the terrestrial land mobile service in the VHF and UHF bands Note-Suppressed on 22/10/01 (CACE/233)

(continued)

A. Ghasemi et al., *Propagation Engineering in Radio Links Design*,
DOI 10.1007/978-1-4614-5314-7,

(continued)

Radiowave propagation	
P.530	Propagation data and prediction methods required for the design of terrestrial line-of-sight systems
P.531	Ionospheric propagation data and prediction methods required for the design of satellite services and systems
P.532	Ionospheric effects and operational considerations associated with artificial modification of the ionosphere and the radiowave channel
P.533	HF propagation prediction method
P.534	Method for calculating sporadic-E field strength
P.581	The concept of "worst month"
P.616	Propagation data for terrestrial maritime mobile services operating at frequencies above 30 MHz Note-Suppressed on 22/10/01 (CACE/233)
P.617	Propagation prediction techniques and data required for the design of trans-horizon radio-relay systems
P.618	Propagation data and prediction methods required for the design of Earth-space telecommunication systems
P.619	Propagation data required for the evaluation of interference between stations in space and those on the surface of the Earth
P.620	Propagation data required for the evaluation of coordination distances in the frequency range 100 MHz to 105 GHz
P.676	Attenuation by atmospheric gases
P.678	Characterization of the natural variability of propagation phenomena
P.679	Propagation data required for the design of broadcasting-satellite systems
P.680	Propagation data required for the design of Earth-space maritime mobile telecommunication systems
P.681	Propagation data required for the design of Earth-space-land mobile telecommunication systems
P.682	Propagation data required for the design of Earth-space aeronautical mobile telecommunication systems
P.683	Sky-wave field-strength prediction method for propagation to aircraft at about 500KHz.
	Note-Suppressed on 20/10/95 (RA-95)
P.684	Prediction of field strength at frequencies below about 150 KHz
P.832	World Atlas of Ground Conductivities
P.833	Attenuation in vegetation. Draft revision P.833-5(08/05)
P.834	Effects of tropospheric refraction on radiowave propagation
P.835	Reference Standard Atmospheres
P.836	Water vapor: surface density and total columnar content
P.837	Characteristics of precipitation for propagation modeling
P.838	Specific attenuation model for rain for use in prediction methods
P.839	Rain height model for prediction methods
P.840	Attenuation due to clouds and fog
P.841	Conversion of annual statistics to worst-month statistics
P.842	Computation of reliability and compatibility of HF radio systems
P.843	Communication by meteor-burst propagation
P.844	Ionospheric factors affecting frequency sharing in the VHF and UHF bands (30 MHz–3 GHz)

(continued)

(continued)

Radiowave propagation	
P.845	HF field-strength measurement
P.846	Measurements of ionospheric and related characteristics
P.1057	Probability distributions relevant to radiowave propagation modeling
P.1058	Digital topographic databases for propagation studies
P.1059	Method for predicting sky-wave field strengths in the frequency range 1,605–1,705 KHz.
	Note-Suppressed on 20/10/95 (RA-95)
P.1060	Propagation factors affecting frequency sharing in HF terrestrial systems
P.1144	Guide to the application of the propagation methods of Radiocommunication Study Group3
P.1145	Propagation data for the terrestrial land mobile service in the VHF and UHF bands Note-Suppressed
P.1146	The prediction of field strength for land mobile and terrestrial broadcasting services in the frequency range from 1 to 3 GHz.
	Note-Suppressed on 22/10/01 (CACE/233)
P.1147	Prediction of sky-wave field strength at frequencies between about 150 and 1,700 KHz
P.1148	Standardized procedure for comparing predicted and observed HF sky-wave signal intensities and the presentation of such comparisons
P.1238	Propagation data and prediction methods for the planning of indoor radiocommunication systems and radio local area networks in the frequency range 900 MHz to 100 GHz
P.1239	ITU-R reference ionospheric characteristics
P.1240	ITU-R methods of basic MUF, operational MUF and ray-path prediction
P.1321	Propagation factors affecting systems using digital modulation techniques at LF and MF
P.1322	Radiometric estimation of atmospheric attenuation
P.1406	Propagation effects relating to terrestrial land mobile and broadcasting services in the VHF and UHF bands
P.1407	Draft revision of recommendation ITU-R P.1407-1.
	Multipath propagation and parameterization of its characteristics
P.1409	Propagation data and prediction methods required for the design of systems using high altitude platform stations at about 47 GHz
P.1410	Propagation data and prediction methods required for the design of terrestrial broadband millimetric radio access systems operating in a frequency range of about 20–50 GHz
P.1411	Propagation data and prediction methods for the planning of short-range outdoor radiocommunication systems and radio local area networks in the frequency range 300 MHz to 100 GHz
P.1412	Propagation data for the evaluation of coordination between Earth stations working in the bidirectionally allocated frequency bands
P.1510	Annual mean surface temperature
P.1511	Topography for Earth-to-space propagation modeling
P.1546	Method for point-to-area predictions for terrestrial services in the frequency range 30–3,000 MHz.
	Draft revision P.1546-2(08/05)

(continued)

(continued)

Radiowave propagation	
P.1621	Propagation data required for the design of Earth-space systems operating between 20 and 375 THz
P.1622	Prediction methods required for the design of Earth-space systems operating between 20 and 375 THz
P.1623	Prediction method of fade dynamics on Earth-space paths
P.1791	Propagation prediction methods for assessment of the impact of ultra-wideband devices
P.1812	A path-specific propagation prediction method for point-to-area terrestrial services in the VHF and UHF bands
P.1814	Prediction methods required for the design of terrestrial free-space optical links
P.1815	Differential rain attenuation
P.1816	The prediction of the time and the spatial profile for broadband land mobile services using UHF and SHF bands
P.1817	Propagation data required for the design of terrestrial free-space optical links
P.1853	Tropospheric attenuation time series synthesis
P.2001	A general purpose wide-range terrestrial propagation model in the frequency range 30 MHz–50 GHz

Appendix C
Acronyms

A Series

AD:	Angle diversity
AGL:	Above ground level
AM:	Amplitude modulation
AMSL:	Above mean sea level
AOR:	Atlantic ocean region
APD:	Amplitude probability distribution
ATPC:	Automatic transmitter power control
ATSC:	Advanced Television System Committee

B Series

BB:	Base band
BBS:	Base band signal
BER:	Bit error rate/block error rate
BSS:	Broadcasting-satellite service
BTS:	Base transceiver station

C Series

CATV:	Community antenna TV
CBD:	Cross band diversity
CCW:	Counterclockwise
CDMA:	Code division multiple access
CNR:	Carrier-to-noise ratio
CP:	Circular polarization
CT:	Cordless telephone
CW:	Clockwise/continuous waves

A. Ghasemi et al., *Propagation Engineering in Radio Links Design*,
DOI 10.1007/978-1-4614-5314-7,

D Series

DAB: Digital Audio Broadcasting
DBS: Direct broadcasting-satellite system
DCT: Digital cordless telephone
DECT: Digital enhanced cordless telephone
DTH: Direct to home
DTT: Digital terrestrial television
DWDM: Dense wavelength-division multiplexing

E Series

EHF: Extremely high frequency
EIRP: Equivalent isotropic radiated power
EM: Electromagnetic
ERP: Equivalent radiated power

F Series

F/B: Front to back (ratio)
FD: Fading depth/frequency diversity
FDMA: Frequency division multiple access
FEC: Forward error correction
FF: Far field
FFM: Flat fade margin
FM: Fade margin /frequency modulation
FNBW: First null beam width
FSL: Free-space loss

G Series

GEO: GEO-stationary Earth orbit/GEO-synchronous Earth orbit
GHz: Gigahertz
GO: Geometric optics
GPS: Global positioning system
GSO: Geostationary orbit

H Series

HBW: Half-power beamwidth
HD: Hybrid diversity
HDTV: High-definition television
HEO: Highly inclined Earth orbit
HF: High frequency
HP: High performance
HPA: High-power amplifier
HS: Hot standby

I Series

IBO: Input back-off
IDU: Indoor unit
IF: Improvement factor/intermediate frequency
IFRB: International Frequency Registration Board
IG: Improvement gain

INMARSAT:	International maritime satellite
INTELSAT:	International telecommunications satellite
IOR:	Indian Ocean region
IR:	Infrared
ISL:	Intersatellite link
ISO:	International Standard Organization
ITU:	International Telecommunications Union
ITU-R:	International telecommunications Union-Radio section

K Series

KHz:	Kilohertz

L Series

LAT:	Latitude
LEO:	Low Earth orbit
LF:	Low frequency
LHC:	Left-hand circular
LHCP:	Left-hand circular polarization
LHEP:	Left-hand elliptical polarization
LNA:	Low-noise amplifier
LNB:	Low-noise block
LON:	Longitude
LOS:	Line of sight
LP:	Linear polarization

M Series

MEO:	Medium Earth orbit
MF:	Medium frequency
MHz:	Megahertz
MPE:	Maximum permissible exposure
MSL:	Mean sea level
MSS:	Mobile satellite system

N Series

NAD:	Noise amplitude distribution
NDB:	Nondirectional beacon
NF:	Near field
NF:	Noise figure/noise factor
NFM:	Net fade margin
NLOS:	Non-line of sight
NRCS:	Normalized radar cross section

O Series

OBO:	Output back-off
ODU:	Outdoor unit

P Series

PD: Polarization diversity
PLF: Polarization loss factor
POR: Pacific Ocean region
PSDN: Public switched data network
PSTN: Public switched telephone network

R Series

RADAR: Radio detection and ranging
RCS: Radar cross section
RD: Research and development
RF: Radio frequency
RFI: Radio-frequency interference
RHC: Right-hand circular
RHCP: Right-hand circular polarization
RHEP: Right-hand elliptical polarization
RLAN: Radio local area network
RPE: Radiation pattern envelope/radiation peak envelope
RSL: Received signal level
RX: Receiver

S Series

S: Standard
SAR: Search and rescue
SCADA: Supervisory control and data acquisition
SCDMA: Synchronous code division multiple access
SD: Space diversity
SDH: Synchronous digital hierarchy
SDTV: Standard-definition television
SFD: Satellite flux density
SHF: Super high frequency
SHP: Super high performance
SNR: Signal-to-noise ratio
SWR: Standing wave ratio

T Series

TBW: Tenth power beamwidth
TDMA: Time division multiple access
TEC: Total electron content
TV: Television
TVRO: TV receive only
TX: Transmitter

U Series

UHF: Ultrahigh frequency
UHP: Ultrahigh performance
UTD: Uniform theory of diffraction
UV: Ultraviolet

V Series

VHF: Very high frequency
VLF: Very low frequency
VSB: Vestigial side band

W Series

WARC: World Administrative Radio Conference
WDM: Wavelength-division multiplexing
WLAN: Wireless local area network
WLL: Wireless local loop
WMO: World Meteorological Organization
WPBX: Wireless private branch exchange

X Series

XPD: Cross-polarization discrimination

References

1. Ghasemi A et al (2012) Propagation engineering in wireless communication. Springer, Berlin
2. Nathanson FE et al (2005) Radar design principles. Prentice-Hall of India
3. Skolnik ML (2005) Introduction to radar. McGraw-Hill, New York
4. Townsend AAR (1988) Digital line of sight radio links
5. Balanis CA (1989) Advanced engineering electromagnetics. Wiley, New York
6. Barclay L (2003) Propagation of radio waves, 2nd edn. The Institute of Electrical Engineers, Michael Faraday House, Stevenage, UK
7. Bernhard HW (1998) Mobile radio networks (Networking and protocols and traffic performance). Wiley, New York
8. Bhargara VK, Haccoum D, Matyas R, Nuspl PP (1981) Digital communications by satellite. Wiley, New York
9. Balanis CA (2005) Antenna theory, 3rd edn.
10. Cheng DK, Field and wave electromagnetics (1989) Addison-Wesley, Reading
11. Christensen G (2001) Wireless intelligent networking. Artech House, Boston
12. Collin RE (1985) Antenna and radio wave propagation. McGraw-Hill, New York
13. Collin RE (1992) Foundations for microwave engineering. McGraw-Hill, New York
14. Digital Microwave Systems Performance Calculations and Network Planning Siemens Telecommunicazioni, s.p.a. (1998)
15. Doble J (1996) Introduction to radio propagation for fixed and mobile communications. Artech House, Boston
16. Feher K (1983) Digital communications Satellite/Earth station engineering. Prentice-Hall, Englewood Cliffs
17. Ghassemi A, Ghassemi F (2004) Special topics in telecommunications, Persian version. Tarhe Ertebatat Consulting Engineers, Iran
18. Ghassemi, A. (2005) Principles of radio links design, Persian Version. Asrare-e-Danesh Publication Co., Iran
19. Harrington RF (1958) Introduction to electromagnetic engineering. McGraw-Hill, New York
20. International Telecommunications Union (2005) ITU Radio Regulations
21. International Telecommunications Union (2005) Radio Communication, ITU-R Recommendations
22. Irane RK (2003) Propagation handbook for wireless communication system design. CRC Press
23. Jerry W (2005) Standard handbook of broadcast engineering. McGraw-Hill, New York
24. Johnk CTA (1988) Engineering electromagnetic fields and waves. Wiley, New York
25. Lavegant J (2000) Radiowave propagation principles and techniques. Wiley, New York
26. Lee WCY (1982) Mobile communications engineering. McGraw-Hill, New York
27. Lee WCY (1986) Mobile communications design fundamentals. H.W. Sams, Indianapolis

A. Ghasemi et al., *Propagation Engineering in Radio Links Design*,
DOI 10.1007/978-1-4614-5314-7,

28. Lehpomer H (2004) Microwave transmission network. McGraw-Hill, New York
29. Leonov SA, Handbook of computer simulation in radio engineering, Communication and Radar. Artech-house
30. Levanon N (1998) Radar principles. Wiley, New York
31. Levanon N, Mozeson E (2004) Radar signals. Wiley, New York
32. Maral G (2003) VSAT Networks, J/W, 2nd edn.
33. Meeks ML (1982) Radar propagation at low altitudes. Artech-house
34. Microflect Co. Inc. (1989) Passive repeater engineering
35. Milion R (2006) RF propagation for wireless communications. McGraw-Hill, New York
36. Miya K (ed) (1981) Satellite communications technology. KDD Engineering and Consulting Inc., Tokyo
37. O'leary S (2000) Understanding digital terrestrial broadcasting. Artech-house
38. Parsons JD (2000) The mobile radio propagation channel, 2nd edn.
39. Parsons JD (John David) (2000) The mobile radio propagation channel. Wiley, New York
40. Peebles PZ (1998) Radar principles. Wiley, New York
41. Peyton ZP (1998) Radar principles. Wiley, New York
42. Plonsey R (1961) Principles and applications of electromagnetic fields. McGraw-Hill, New York
43. Product Specifications (1997) Andrew Catalog 37, System Planning
44. Riely MJ, Richardson IEG (1997) Digital video communications. Artech-house
45. Rihaczek AW, Hershkowitz SJ (2000) Theorian practice of radar target identification. Artech-house
46. Roddy D (2003) Satellite communications. McGraw-Hill, New York
47. Salema C (2005) Microwave radio links. Wiley, New York
48. Satellite Communication (2006) Rody
49. Seybold JS (2005) Introduction to RF propagation. Wiley, New York
50. Siwiak K, Radiowave propagation and antennas for personal communications. Artech House, London
51. Vaughand R (2003) Channels, propagation and antennas for mobile communications. IEE
52. Wiley (2002) Handbook on satellite communications. New York

Softwares

1. GIMS-Graphical Interference Management
2. SpaceCap-Space Data Capture on PC
3. SpaceCom-Space Comment Capture on PC
4. SpacePub-Space Publication
5. SpaceQry-Space Query and Extract System
6. SpaceVal-Space Filings Validation Software
7. SNS-Space Network Systems
8. Electronic Submission of Graphical Data
9. SRS-Database Conversion Utility
10. SRS -Fix Electronic Notification Database Utility
11. SRS-Space Radio Communications Stations on CD-ROM
12. SNL-Space Network List
13. Global Administration Data (GLAD)
14. Maritime Mobile Access and Retrieval System (MARS)
15. HFBC Planning Software
16. FXM Data Capture
17. TstTrs Attention to the File Size
18. Interference Calculations GE84 Attention to the File Size

19. Interference Calculations GE89 Attention to the File Size
20. SG 1-Programs on Spectrum Management
21. SG 3-Databanks and Computer Programs on Radiowave Propagation
22. SG 4-Databanks (Currently in Development)
23. ITU Patent Statement and Licensing Declaration Information
24. Ground-wave Propagation (GRWAVE)
25. UTD Formulation for Diffraction Loss due to Finitely Conducting Wedge
26. Point-to-area Prediction, 30–3000 MHz
27. Point-to-point (Interference) Propagation
28. Rain Scatter (SCAT)
29. Radio Refractivity
30. Atmospheric Gaseous Absorption
31. Atmospheric Profile Data
32. Water Vapour Data
33. Rainfall Rate Model
34. Specific Attenuation Model for Rain
35. Rain Height Model
36. Cloud Liquid Water Data
37. Annual Mean Surface Temperature
38. Topography (0.50 Resolution)
39. K9SE: VHF and UHF Propagation www.dxzone.
40. W6EL Prop: Predict Ionospheric (Sky-wave) Propagation (3–30 MHz) www.qsl.net/w6elprop
41. TAP: Terrain Analysis Package www.softwright.com
42. PA 3CQR Gray Line www.iri.tudelft.nl
43. HF Software www.elbert.its.bldrdoc.gov
44. Wincap Wizard 3 www.taborsoft.com
45. Win BASMS www.winpath.com
46. Win BASMS www.itu.int

Web Sites

1. INMARSAT www.inmarsat.com and www.Inmarsat.org
2. EUTELSAT www.eutelsat.org
3. INTELSAT www.intelsat.com
4. METEOSAT www.meteosat.com
5. IEC www.iec.chandwww.iec.org
6. ISO www.iso.ch
7. EIA www.eia.org
8. BTS www.bt.com
9. ITU www.itu.int
10. ETSI www.etsi.org
11. ASTM www.astm.org
12. WMO www.wmo.ch
13. IEEE www.ieee.org, www.ieeexplore.ieee.org
14. IEE www.iee.org
15. IMO www.imo.org
16. ICAO www.icao.int

Index

A. Ghasemi et al., *Propagation Engineering in Radio Links Design*,
DOI 10.1007/978-1-4614-5314-7,

GPSR Compliance

The European Union's (EU) General Product Safety Regulation (GPSR) is a set of rules that requires consumer products to be safe and our obligations to ensure this.

If you have any concerns about our products, you can contact us on ProductSafety@springernature.com

In case Publisher is established outside the EU, the EU authorized representative is:

Springer Nature Customer Service Center GmbH
Europaplatz 3
69115 Heidelberg, Germany

Batch number: 10620078

Printed by Printforce, the Netherlands